T0211475

**Functional Analysis for Probability and Stochastic Processes.**
**An Introduction**

This text is designed both for students of probability and stochastic processes and for students of functional analysis. For the reader not familiar with functional analysis a detailed introduction to necessary notions and facts is provided. However, this is not a straight textbook in functional analysis; rather, it presents some chosen parts of functional analysis that help understand ideas from probability and stochastic processes. The subjects range from basic Hilbert and Banach spaces, through weak topologies and Banach algebras, to the theory of semigroups of bounded linear operators. Numerous standard and non-standard examples and exercises make the book suitable for both a textbook for a course and for self-study.

ADAM BOBROWSKI is a Professor of Mathematics at Lublin University of Technology.

# Functional Analysis for Probability and Stochastic Processes

## An Introduction

A. BOBROWSKI

CAMBRIDGE
UNIVERSITY PRESS

# CAMBRIDGE
## UNIVERSITY PRESS

University Printing House, Cambridge CB2 8BS, United Kingdom

One Liberty Plaza, 20th Floor, New York, NY 10006, USA

477 Williamstown Road, Port Melbourne, VIC 3207, Australia

314-321, 3rd Floor, Plot 3, Splendor Forum, Jasola District Centre, New Delhi - 110025, India

79 Anson Road, #06-04/06, Singapore 079906

Cambridge University Press is part of the University of Cambridge.

It furthers the University's mission by disseminating knowledge in the pursuit of
education, learning and research at the highest international levels of excellence.

www.cambridge.org
Information on this title: www.cambridge.org/9780521539371

© Cambridge University Press 2005

First published 2005

*A catalogue record for this publication is available from the British Library*

ISBN 978-0-521-83166-6 Hardback
ISBN 978-0-521-53937-1 Paperback

*To the most enthusiastic writer ever – my son Radek.*

# Contents

# Preface

This book is an expanded version of lecture notes for the graduate course "An Introduction to Methods of Functional Analysis in Probability and Stochastic Processes" that I gave for students of the University of Houston, Rice University, and a few friends of mine in Fall, 2000 and Spring, 2001. It was quite an experience to teach this course, for its attendees consisted of, on the one hand, a group of students with a good background in functional analysis having limited knowledge of probability and, on the other hand, a group of statisticians without a functional analysis background. Therefore, in presenting the required notions from functional analysis, I had to be complete enough for the latter group while concise enough so that the former would not drop the course from boredom. Similarly, for the probability theory, I needed to start almost from scratch for the former group while presenting the material in a light that would be interesting for the latter group. This was fun. Incidentally, the students adjusted to this challenging situation much better than I.

In preparing these notes for publication, I made an effort to make the presentation self-contained and accessible to a wide circle of readers. I have added a number of exercises and disposed of some. I have also expanded some sections that I did not have time to cover in detail during the course. I believe the book in this form should serve first year graduate, or some advanced undergraduate students, well. It may be used for a two-semester course, or even a one-semester course if some background is taken for granted. It must be made clear, however, that this book is not a textbook in probability. Neither may it be viewed as a textbook in functional analysis. There are simply too many important subjects in these vast theories that are not mentioned here. Instead, the book is intended for those who would like to see some aspects of probability from the perspective of functional analysis. It may also serve as a (slightly long) introduction to such excellent and comprehensive expositions of probability and stochastic processes as Stroock's, Revuz's and Yor's, Kallenberg's or Feller's.

It should also be said that, despite its substantial probabilistic content, the book is not structured around typical probabilistic problems and methods. On the contrary, the structure is determined by notions that are functional analytic in origin. As it may be seen from the very chapters' titles, while the body is probabilistic, the skeleton is functional analytic.

Most of the material presented in this book is fairly standard, and the book is meant to be a textbook and not a research monograph. Therefore, I made little or no effort to trace the source from which I had learned a particular theorem or argument. I want to stress, however, that I have learned this material from other mathematicians, great and small, in particular by reading their books. The bibliography gives the list of these books, and I hope it is complete. See also the bibliographical notes to each chapter. Some examples, however, especially towards the end of the monograph, fit more into the category of "research".

A word concerning prerequisites: to follow the arguments presented in the book the reader should have a good knowledge of measure theory and some experience in solving ordinary differential equations. Some knowledge of abstract algebra and topology would not hurt either. I sketch the needed material in the introductory Chapter 1. I do not think, though, that the reader should start by reading through this chapter. The experience of going through prerequisites before diving into the book may prove to be like the one of paying a large bill for a meal before even tasting it. Rather, I would suggest browsing through Chapter 1 to become acquainted with basic notation and some important examples, then jumping directly to Chapter 2 and referring back to Chapter 1 when needed.

I would like to thank Dr. M. Papadakis, Dr. C. A. Shaw, A. Renwick and F. J. Foss (both PhDs soon) for their undivided attention during the course, efforts to understand Polish-English, patience in endless discussions about the twentieth century history of mathematics, and valuable impact on the course, including how-to-solve-it-easier ideas. Furthermore, I would like to express my gratitude to the Department of Mathematics at UH for allowing me to teach this course. The final chapters of this book were written while I held a special one-year position at the Institute of Mathematics of the Polish Academy of Sciences, Warsaw, Poland.

A final note: if the reader dislikes this book, he/she should blame F. J. Foss who nearly pushed me to teach this course. If the reader likes it, her/his warmest thanks should be sent to me at both addresses: bobrowscy@op.pl and a.bobrowski@pollub.pl. Seriously, I would like to thank Fritz Foss for his encouragement, for valuable feedback and for editing parts of this book. All the remaining errors are protected by my copyright.

# 1
# Preliminaries, notations and conventions

Finite measures and various classes of functions, including random variables, are examples of elements of natural Banach spaces and these spaces are central objects of functional analysis. Before studying Banach spaces in Chapter 2, we need to introduce/recall here the basic topological, measure-theoretic and probabilistic notions, and examples that will be used throughout the book. Seen from a different perspective, Chapter 1 is a big "tool-box" for the material to be covered later.

## 1.1 Elements of topology

1.1.1 *Basics of topology*    We assume that the reader is familiar with basic notions of topology. To set notation and refresh our memory, let us recall that a pair $(S, \mathcal{U})$ where $S$ is a set and $\mathcal{U}$ is a collection of subsets of $S$ is said to be a **topological space** if the empty set and $S$ belong to $\mathcal{U}$, and unions and finite intersections of elements of $\mathcal{U}$ belong to $\mathcal{U}$. The family $\mathcal{U}$ is then said to be the **topology** in $S$, and its members are called **open sets**. Their complements are said to be **closed**. Sometimes, when $\mathcal{U}$ is clear from the context, we say that the set $S$ itself is a topological space. Note that all statements concerning open sets may be translated into statements concerning closed sets. For example, we may equivalently define a topological space to be a pair $(S, \mathcal{C})$ where $\mathcal{C}$ is a collection of sets such that the empty set and $S$ belong to $\mathcal{C}$, and intersections and finite unions of elements of $\mathcal{C}$ belong to $\mathcal{C}$.

An open set containing a point $s \in S$ is said to be a **neighborhood** of $s$. A topological space $(S, \mathcal{U})$ is said to be **Hausdorff** if for all $p_1, p_2 \in S$, there exists $A_1, A_2 \in \mathcal{U}$ such that $p_i \in A_i, i = 1, 2$ and $A_1 \cap A_2 = \emptyset$. Unless otherwise stated, we assume that all topological spaces considered in this book are Hausdorff.

1

The **closure**, $cl(A)$, of a set $A \subset S$ is defined to be the smallest closed set that contains $A$. In other words, $cl(A)$ is the intersection of all closed sets that contain $A$. In particular, $A \subset cl(A)$. $A$ is said to be **dense** in $S$ iff $cl(A) = S$.

A family $\mathcal{V}$ is said to be a **base** of topology $\mathcal{U}$ if every element of $\mathcal{U}$ is a union of elements of $\mathcal{V}$. A family $\mathcal{V}$ is said to be a **subbase** of $\mathcal{U}$ if the family of finite intersections of elements of $\mathcal{V}$ is a base of $\mathcal{U}$.

If $(S, \mathcal{U})$ and $(S', \mathcal{U}')$ are two topological spaces, then a map $f : S \to S'$ is said to be **continuous** if for any open set $A'$ in $\mathcal{U}'$ its inverse image $f^{-1}(A')$ is open in $S$.

Let $S$ be a set and let $(S', \mathcal{U}')$ be a topological space, and let $\{f_t, t \in \mathbb{T}\}$ be a family of maps from $S$ to $S'$ (here $\mathbb{T}$ is an abstract indexing set). Note that we may introduce a topology in $S$ such that all maps $f_t$ are continuous, a trivial example being the topology consisting of all subsets of $S$. Moreover, an elementary argument shows that intersections of finite or infinite numbers of topologies in $S$ is a topology. Thus, there exists the smallest topology (in the sense of inclusion) under which the $f_t$ are continuous. This topology is said to be **generated** by the family $\{f_t, t \in \mathbb{T}\}$.

**1.1.2 Exercise**    Prove that the family $\mathcal{V}$ composed of sets of the form $f_t^{-1}(A'), t \in \mathbb{T}, A' \in \mathcal{U}'$ is a subbase of the topology generated by $f_t, t \in \mathbb{T}$.

**1.1.3** *Compact sets*    A subset $K$ of a topological space $(S, \mathcal{U})$ is said to be **compact** if every open cover of $K$ contains a finite subcover. This means that if $\mathcal{V}$ is a collection of open sets such that $K \subset \bigcup_{B \in \mathcal{V}} B$, then there exists a finite collection of sets $B_1, \dots, B_n \in \mathcal{V}$ such that $K \subset \bigcup_{1=1}^{n} B_i$. If $S$ is compact itself, we say that the space $(S, \mathcal{U})$ is compact (the reader may have noticed that this notion depends as much on $S$ as it does on $\mathcal{U}$). Equivalently, $S$ is compact if, for any family $C_t, t \in \mathbb{T}$ of closed subsets of $S$ such that $\bigcap_{t \in \mathbb{T}} C_t = \emptyset$, there exists a finite collection $C_{t_1}, \dots, C_{t_n}$ of its members such that $\bigcap_{i=1}^{n} C_{t_i} = \emptyset$. A set $K$ is said to be **relatively compact** iff its closure is compact. A topological space $(S, \mathcal{U})$ is said to be **locally compact** if for every point $p \in S$ there exist an open set $A$ and a compact set $K$, such that $s \in A \subset K$. The **Bolzano–Weierstrass Theorem** says that a subset of $\mathbb{R}^n$ is compact iff it is closed and bounded. In particular, $\mathbb{R}^n$ is locally compact.

1.1.4 *Metric spaces*    Let $\mathbb{X}$ be an abstract space. A map $d : \mathbb{X} \times \mathbb{X} \to \mathbb{R}^+$ is said to be a **metric** iff for all $x, y, z \in \mathbb{X}$

(a) $d(x, y) = d(y, x)$,
(b) $d(x, y) \le d(x, z) + d(z, y)$,
(c) $d(x, y) = 0$ iff $x = y$.

A sequence $x_n$ of elements of $\mathbb{X}$ is said to **converge** to $x \in \mathbb{X}$ if $\lim_{n \to \infty} d(x_n, x) = 0$. We call $x$ the **limit** of the sequence $(x_n)_{n \ge 1}$ and write $\lim_{n \to \infty} x_n = x$. A sequence is said to be **convergent** if it converges to some $x$. Otherwise it is said to be **divergent**.

An **open ball** $B(x, r)$ with radius $r$ and center $x$ is defined as the set of all $y \in \mathbb{X}$ such that $d(x, y) < r$. A closed ball with radius $r$ and center $x$ is defined similarly as the set of $y$ such $d(x, y) \le r$. A natural way to make a metric space into a topological space is to take all open balls as the base of the topology in $\mathbb{X}$. It turns out that under this definition a subset $A$ of a metric space is closed iff it contains the limits of sequences with elements in $A$. Moreover, $A$ is compact iff every sequence of its elements contains a converging subsequence and its limit belongs to the set $A$. (If $S$ is a topological space, this last condition is necessary but not sufficient for $A$ to be compact.)

A function $f : \mathbb{X} \to \mathbb{Y}$ that maps a metric space $\mathbb{X}$ into a normed space $\mathbb{Y}$ is continuous at $x \in \mathbb{X}$ if for any sequence $x_n$ converging to $x$, $\lim_{n \to \infty} f(x_n)$ exists and equals $f(x)$ ($x_n$ converges in $\mathbb{X}$, $f(x_n)$ converges in $\mathbb{Y}$). $f$ is called continuous if it is continuous at every $x \in \mathbb{X}$ (this definition agrees with the definition of continuity given in 1.1.1).

## 1.2 Measure theory

1.2.1 *Measure spaces and measurable functions*    Although we assume that the reader is familiar with the rudiments of measure theory as presented, for example, in [103], let us recall the basic notions. A family $\mathcal{F}$ of subsets of an abstract set $\Omega$ is said to be a $\sigma$-**algebra** if it contains $\Omega$ and complements and countable unions of its elements. The pair $(\Omega, \mathcal{F})$ is then said to be a **measurable space**. A family $\mathcal{F}$ is said to be an **algebra** or a **field** if it contains $\Omega$, complements and finite unions of its elements.

A function $\mu$ that maps a family $\mathcal{F}$ of subsets of $\Omega$ into $\mathbb{R}^+$ such that

$$\mu\left(\bigcup_{n \in \mathbb{N}} A_n\right) = \sum_{n=1}^{\infty} \mu(A_n) \tag{1.1}$$

for all pairwise-disjoint elements $A_n, n \in \mathbb{N}$ of $\mathcal{F}$ such that the union $\bigcup_{n \in \mathbb{N}} A_n$ belongs to $\mathcal{F}$ is called a **measure**. In most cases $\mathcal{F}$ is a $\sigma$-algebra but there are important situations where it is not, see e.g. 1.2.8 below. If $\mathcal{F}$ is a $\sigma$-algebra, the triple $(\Omega, \mathcal{F}, \mu)$ is called a **measure space**.

Property (1.1) is termed **countable additivity**. If $\mathcal{F}$ is an algebra and $\mu(S) < \infty$, (1.1) is equivalent to

$$\lim_{n \to \infty} \mu(A_n) = 0 \quad \text{whenever } A_n \in \mathcal{F}, A_n \supset A_{n+1}, \bigcap_{n=1}^{\infty} A_n = \emptyset. \quad (1.2)$$

The reader should prove it.

The smallest $\sigma$-algebra containing a given class $\mathcal{F}$ of subsets of a set is denoted $\sigma(\mathcal{F})$. If $\Omega$ is a topological space, then $\mathcal{B}(\Omega)$ denotes the smallest $\sigma$-algebra containing open sets, called the **Borel $\sigma$-algebra**. A measure $\mu$ on a measurable space $(\Omega, \mathcal{F})$ is said to be **finite** (or **bounded**) if $\mu(\Omega) < \infty$. It is said to be $\sigma$-**finite** if there exist measurable subsets $\Omega_n$, $n \in \mathbb{N}$, of $\Omega$ such that $\mu(\Omega_n) < \infty$ and $\Omega = \bigcup_{n \in \mathbb{N}} \Omega_n$.

A measure space $(\Omega, \mathcal{F}, \mu)$ is said to be **complete** if for any set $A \subset \Omega$ and any measurable $B$ conditions $A \subset B$ and $\mu(B) = 0$ imply that $A$ is measurable (and $\mu(A) = 0$, too). When $\Omega$ and $\mathcal{F}$ are clear from the context, we often say that the measure $\mu$ itself is complete. In Exercise 1.2.10 we provide a procedure that may be used to construct a complete measure from an arbitrary measure. Exercises 1.2.4 and 1.2.5 prove that properties of complete measure spaces are different from those of measure spaces that are not complete.

A map $f$ from a measurable space $(\Omega, \mathcal{F})$ to a measurable space $(\Omega', \mathcal{F}')$ is said to be $\mathcal{F}$ **measurable**, or just **measurable** iff for any set $A \in \mathcal{F}'$ the inverse image $f^{-1}(A)$ belongs to $\mathcal{F}$. If, additionally, all inverse images of measurable sets belong to a sub-$\sigma$-algebra $\mathcal{G}$ of $\mathcal{F}$, then we say that $f$ is $\mathcal{G}$ **measurable**, or more precisely $\mathcal{G}/\mathcal{F}'$ **measurable**. If $f$ is a measurable function from $(\Omega, \mathcal{F})$ to $(\Omega', \mathcal{F}')$ then

$$\sigma_f = \{A \in \mathcal{F} | A = f^{-1}(B) \text{ where } B \in \mathcal{F}'\}$$

is a sub-$\sigma$-algebra of $\mathcal{F}$. $\sigma_f$ is called the $\sigma$-algebra **generated by** $f$. Of course, $f$ is $\mathcal{G}$ measurable if $\sigma_f \subset \mathcal{G}$.

The $\sigma$-algebra of Lebesgue measurable subsets of a measurable subset $A \subset \mathbb{R}^n$ is denoted $\mathcal{M}_n(A)$ or $\mathcal{M}(A)$ if $n$ is clear from the context, and the Lebesgue measure in this space is denoted $leb_n$, or simply $leb$. A standard result says that $\mathcal{M} := \mathcal{M}(\mathbb{R}^n)$ is the smallest complete $\sigma$-algebra containing $\mathcal{B}(\mathbb{R}^n)$. In considering the measures on $\mathbb{R}^n$ we will always assume that they are defined on the $\sigma$-algebra of Lebesgue measurable sets, or Borel sets. The interval $[0, 1)$ with the family of its Lebesgue subsets and the Lebesgue measure restricted to these subsets is often referred to as **the standard probability space**. An $n$-dimensional **random vector** (or simply $n$-vector) is a measurable map from a probability space $(\Omega, \mathcal{F}, \mathbb{P})$ to the measurable space $(\mathbb{R}^n, \mathcal{B}(\mathbb{R}^n))$. A **complex-valued random variable** is simply a two dimensional random vector; we tend to use the former name if we want to consider complex products of two-dimensional random vectors. Recall that any random $n$-vector $\underline{X}$ is of the form $\underline{X} = (X_1, ..., X_n)$ where $X_i$ are random variables $X_i : \Omega \to \mathbb{R}$.

**1.2.2 Exercise**    Let $A$ be an open set in $\mathbb{R}^n$. Show that $A$ is union of all balls contained in $A$ with rational radii and centers in points with rational coordinates. Conclude that $\mathcal{B}(\mathbb{R})$ is the $\sigma$-algebra generated by open (resp. closed) intervals. The same result is true for intervals of the form $(a, b]$ and $[a, b)$. Formulate and prove an analog in $\mathbb{R}^n$.

**1.2.3 Exercise**    Suppose that $\Omega$ and $\Omega'$ are topological spaces. If a map $f : \Omega \to \Omega'$ is continuous, then $f$ is measurable with respect to Borel $\sigma$-fields in $\Omega$ and $\Omega'$. More generally, suppose that $f$ maps a measurable space $(\Omega, \mathcal{F})$ into a measurable space $(\Omega, \mathcal{F}')$, and that $\mathcal{G}'$ is a class of measurable subsets of $\Omega'$ such $\sigma(\mathcal{G}') = \mathcal{F}'$. If inverse images of elements of $\mathcal{G}'$ are measurable, then $f$ is measurable.

**1.2.4 Exercise**    Suppose that $(\Omega, \mathcal{F}, \mu)$ is a measure space, and $f$ maps $\Omega$ into $\mathbb{R}$. Equip $\mathbb{R}$ with the $\sigma$-algebra of Borel sets and prove that $f$ is measurable iff sets of the form $\{\omega | f(\omega) \leq t\}$, $t \in \mathbb{R}$ belong to $\mathcal{F}$. (Equivalently: sets of the form $\{\omega | f(\omega) < t\}$, $t \in \mathbb{R}$ belong to $\mathcal{F}$.) Prove by example that a similar statement is not necessarily true if Borel sets are replaced by Lebesgue measurable sets.

**1.2.5 Exercise**    Let $(\Omega, \mathcal{F}, \mu)$ be a *complete* measure space, and $f$ be a map $f : \Omega \to \mathbb{R}$. Equip $\mathbb{R}$ with the algebra of Lebesgue measurable sets and prove that $f$ is measurable iff sets of the form $\{\omega | f(\omega) \leq t\}$, $t \in \mathbb{R}$ belong to $\mathcal{F}$. (Equivalently: sets of the form $\{\omega | f(\omega) < t\}$, $t \in \mathbb{R}$ belong to $\mathcal{F}$.)

**1.2.6 Exercise**    Let $(S, \mathcal{U})$ be a topological space and let $S'$ be its subset. We can introduce a natural topology in $S'$, termed **induced**

**topology**, to be the family of sets $U' = U \cap S'$ where $U$ is open in $S$. Show that

$$\mathcal{B}(S') = \{B \subset S' | B = A \cap S', A \in \mathcal{B}(S)\}. \tag{1.3}$$

**1.2.7** *Monotone class theorem* A class $\mathcal{G}$ of subsets of a set $\Omega$ is termed a $\pi$-**system** if the intersection of any two of its elements belongs to the class. It is termed a $\lambda$-**system** if (a) $\Omega$ belongs to the class, (b) $A, B \in \mathcal{G}$ and $A \subset B$ implies $B \backslash A \in \mathcal{G}$ and (c) if $A_1, A_2, \ldots \in \mathcal{G}$, and $A_1 \subset A_2 \subset \ldots$ then $\bigcup_{n \in \mathbb{N}} A_n \in \mathcal{G}$. The reader may prove that a $\lambda$-system that is at the same time a $\pi$-system is also a $\sigma$-algebra. In 1.4.3 we exhibit a natural example of a $\lambda$-system that is not a $\sigma$-algebra. The **Monotone Class Theorem** or $\pi-\lambda$ **theorem**, due to W. Sierpiński, says that if $\mathcal{G}$ is a $\pi$-system and $\mathcal{F}$ is a $\lambda$-system and $\mathcal{G} \subset \mathcal{F}$, then $\sigma(\mathcal{G}) \subset \mathcal{F}$. As a corollary we obtain the uniqueness of extension of a measure defined on a $\pi$-system. To be more specific, if $(\Omega, \mathcal{F})$ is a measure space, and $\mathcal{G}$ is a $\pi$-system such that $\sigma(\mathcal{G}) = \mathcal{F}$, and if $\mu$ and $\mu'$ are two finite measures on $(\Omega, \mathcal{F})$ such that $\mu(A) = \mu'(A)$ for all $A \in \mathcal{G}$, then the same relation holds for $A \in \mathcal{F}$. See [5].

**1.2.8** *Existence of an extension of a measure* A standard construction involving the so-called outer measure shows the existence of an extension of a measure defined on a field. To be more specific, if $\mu$ is a finite measure on a field $\mathcal{F}$, then there exists a measure $\tilde{\mu}$ on $\sigma(\mathcal{F})$ such that $\tilde{\mu}(A) = \mu(A)$ for $A \in \mathcal{F}$, see [5]. It is customary and convenient to omit the "$\tilde{\phantom{x}}$" and denote both the original measure and its extension by $\mu$. This method allows us in particular to prove existence of the Lebesgue measure [5, 106].

**1.2.9** *Two important properties of the Lebesgue measure* An important property of the Lebesgue measure is that it is **regular**, which means that for any Lebesgue measurable set $A$ and $\epsilon > 0$ there exists an open set $G \supset A$ and a compact set $K \subset A$ such that $leb(G \setminus K) < \epsilon$. Also, the Lebesgue measure is **translation invariant**, i.e. $leb\, A = leb\, A_t$ for any Lebesgue measurable set $A$ and $t \in \mathbb{R}$, where

$$A_t = \{s \in \mathbb{R}; s - t \in A\}. \tag{1.4}$$

**1.2.10** **Exercise** Let $(\Omega, \mathcal{F})$ be a measure space and $\mu$ be a measure, not necessarily complete. Let $\mathcal{F}_0$ be the class of subsets $B$ of $\Omega$ such that there exists a $C \in \mathcal{F}$ such that $\mu(C) = 0$ and $B \subset C$. Let $\mathcal{F}_\mu = \sigma(\mathcal{F} \cup \mathcal{F}_0)$. Show that there exists a unique extension of $\mu$ to $\mathcal{F}_\mu$, and $(\Omega, \mathcal{F}_\mu, \mu)$ is a complete measure space. Give an example of two Borel measures $\mu$ and $\nu$ such that $\mathcal{F}_\mu \neq \mathcal{F}_\nu$.

**1.2.11** *Integral*   Let $(\Omega, \mathcal{F}, \mu)$ be a measure space. The integral $\int f \, d\mu$ of a **simple measurable function** $f$, i.e. of a function of the form $f = \sum_{i=1}^{n} c_i 1_{A_i}$ where $n$ is an integer, $c_i$ are real constants, $A_i$ belong to $\mathcal{F}$, and $\mu(A_i) < \infty$, is defined as $\int f \, d\mu = \sum_{i=1}^{n} c_i \mu(A_i)$. We check that this definition of the integral does not depend on the choice of representation of a simple function. The integral of a non-negative measurable function $f$ is defined as the supremum over integrals of non-negative simple measurable functions $f_s$ such that $f_s \leq f$ ($\mu$ a.e.). This last statement means that $f_s(\omega) \leq f(\omega)$ for all $\omega \in \Omega$ outside of a measurable set of $\mu$-measure zero. If this integral is finite, we say that $f$ is **integrable**.

Note that in our definition we may include functions $f$ such that $f(\omega) = \infty$ on a measurable set of $\omega$s. We say that such functions have their values in an extended non-negative half-line. An obvious necessary requirement for such a function to be integrable is that the set where it equals infinity has measure zero (we agree as it is customary in measure theory that $0 \cdot \infty = 0$).

If a measurable function $f$ has the property that both $f^+ = \max(f, 0)$ and $f^- = \max(-f, 0)$ are integrable then we say that $f$ is **absolutely integrable** and put $\int f \, d\mu = \int f^+ \, d\mu - \int f^- \, d\mu$. The reader may check that for a simple function this definition of the integral agrees with the one given initially. The integral of a complex-valued map $f$ is defined as the integral of its real part plus $\imath$ (the imaginary unit) times the integral of its imaginary part, whenever these integrals exist. For any integrable function $f$ and measurable set $A$ the integral $\int_A f \, d\mu$ is defined as $\int 1_A f \, d\mu$.

This definition implies the following elementary estimate which proves useful in practice:

$$\left| \int_A f \, d\mu \right| \leq \int_A |f| \, d\mu. \qquad (1.5)$$

Moreover, for any integrable functions $f$ and $g$ and any $\alpha$ and $\beta$ in $\mathbb{R}$, we have

$$\int (\alpha f + \beta g) \, d\mu = \alpha \int f \, d\mu + \beta \int g \, d\mu.$$

In integrating functions defined on $(\mathbb{R}^n, \mathcal{M}_n(\mathbb{R}^n), leb_n)$ it is customary to write $ds_1 ... ds_n$ instead of $d\,leb_n(\underline{s})$ where $\underline{s} = (s_1, ..., s_n)$. In one dimension, we write $ds$ instead of $d\,leb(s)$.

There are two important results concerning limits of integrals defined this way that we will use often. The first one is called **Fatou's Lemma** and the second **Lebesgue Dominated Convergence Theorem**. The former says that for a sequence of measurable functions $f_n$ with values in the extended non-negative half-line $\limsup_{n\to\infty} \int f_n \, d\mu \geq \int \limsup_{n\to\infty} f_n \, d\mu$, and the latter says that if $f_n$ is a sequence of measurable functions and there exists an integrable function $f$ such that $|f_n| \leq f$ ($\mu$ a.e.), then $\lim_{n\to\infty} \int f_n \, d\mu = \int g \, d\mu$, provided $f_n$ tends to $g$ pointwise, except perhaps on a set of measure zero. Observe that condition $|f_n| \leq f$ implies that $f_n$ and $g$ are absolutely integrable; the other part of the Lebesgue Dominated Convergence Theorem says that $\int |f_n - g| \, d\mu$ tends to zero, as $n \to \infty$. The reader may remember that both above results may be derived from the **Monotone Convergence Theorem**, which says that if $f_n$ is a sequence of measurable functions with values in the extended non-negative half-line, and $f_{n+1}(\omega) \geq f_n(\omega)$ for all $\omega$ except maybe on a set of measure zero, then $\int_A f_n \, d\mu$ tends to $\int_A \lim_{n\to\infty} f_n(\omega) \, d\mu$ regardless of whether the last integral is finite or infinite. Here $A$ is the set where $\lim_{n\to\infty} f_n(\omega)$ exists, and by assumption it is a complement of a set of measure zero.

Note that these theorems are true also when, instead of a sequence of functions, we have a family of functions indexed, say, by real numbers and consider a limit at infinity or at some point of the real line.

**1.2.12 Exercise**  Let $(a, b)$ be an interval and let, for $\tau$ in this interval, $x(\tau, \omega)$ be a given integrable function on a measure space $(\Omega, \mathcal{F}, \mu)$. Suppose furthermore that for almost all $\omega \in \Omega$, $\tau \to x(\tau, \omega)$ is continuously differentiable and there exists an integrable function $y$ such that $\sup_{\tau\in(a,b)} |x'(\tau, \omega)| \leq y(\omega)$. Prove that $z(\tau) = \int_\Omega x(\tau, \omega) \, \mu(d\omega)$ is differentiable and that $z'(\tau) = \int_\Omega x'(\tau, \omega) \, \mu(d\omega)$.

**1.2.13 Product measures**  Let $(\Omega, \mathcal{F}, \mu)$ and $(\Omega', \mathcal{F}', \mu')$ be two $\sigma$-finite measure spaces. In the Cartesian product $\Omega \times \Omega'$ consider the **rectangles**, i.e. the sets of the form $A \times A'$ where $A \in \mathcal{F}$ and $A' \in \mathcal{F}'$, and the function $\mu \otimes \mu'(A \times A') = \mu(A)\mu'(A')$. Certainly, rectangles form a $\pi$-system, say $\mathcal{R}$, and it may be proved that $\mu \otimes \mu'$ is a measure on $\mathcal{R}$ and that there exists an extension of $\mu \otimes \mu'$ to a measure on $\sigma(\mathcal{R})$, which is necessarily unique. This extension is called the **product measure** of $\mu$ and $\mu'$. The assumption that $\mu$ and $\mu'$ are $\sigma$-finite is crucial for the existence of $\mu \otimes \mu'$. Moreover, $\mu \otimes \mu'$ is $\sigma$-finite, and it is finite if $\mu$ and $\mu'$ are. The **Tonelli Theorem** says that if a function $f : \Omega \times \Omega' \to \mathbb{R}$ is $\sigma(\mathcal{R})$ measurable, then for all $\omega \in \Omega$ the function $f_\omega : \Omega' \to \mathbb{R}, f_\omega(\omega') = f(\omega, \omega')$ is $\mathcal{F}'$ measurable and the function $f^{\omega'} : \Omega \to \mathbb{R}, f^{\omega'}(\omega) = f(\omega, \omega')$ is $\mathcal{F}$ measurable. Furthermore, the **Fubini Theorem** says that for a $\sigma(\mathcal{R})$ measurable function $f : \Omega \times \Omega' \to \mathbb{R}^+$,

$$
\begin{aligned}
\int_{\Omega \times \Omega'} f \, \mathrm{d}(\mu \otimes \mu') &= \int_\Omega [\int_{\Omega'} f_\omega(\omega') \, \mu(d\omega')] \, \mu(d\omega) \\
&= \int_{\Omega'} [\int_\Omega f^{\omega'}(\omega) \, \mu(d\omega)] \, \mu(d\omega'),
\end{aligned}
$$

finite or infinite; measurability of the integrands is a part of the theorem. Moreover, this relation holds whenever $f$ is absolutely integrable.

1.2.14 *Absolute continuity*   Let $\mu$ and $\nu$ be two measures on a measure space $(\Omega, \mathcal{F})$; we say that $\mu$ is **absolutely continuous** (with respect to $\nu$) if there exists a non-negative (not necessarily integrable) function $f$ such that $\mu(A) = \int_A f \, d\nu$ for all $A \in \mathcal{F}$. In such a case $f$ is called the **density** of $\mu$ (with respect to $\nu$). Observe that $f$ is integrable (with respect to $\nu$) iff $\mu$ is finite, i.e. iff $\mu(\Omega) < \infty$. When it exists, the density is unique up to a set of $\nu$-measure zero.

1.2.15 *Change of variables formula*   Suppose that $(\Omega, \mathcal{F}, \mathbb{P})$ is a measure space and $f$ is a measurable map from $(\Omega, \mathcal{F})$ to another measurable space $(\Omega', \mathcal{F}')$. Consider the set function $\mu_f$ on $\mathcal{F}'$ defined by $\mu_f(A) = \mu(f^{-1}(A)) = \mu(f \in A)$. We check that $\mu_f$ is a measure in $(\Omega', \mathcal{F}')$. It is called the **transport of the measure** $\mu$ via $f$ or a measure **induced** on $(\Omega', \mathcal{F}')$ by $\mu$ and $f$. In particular, if $\mu$ is a probability measure, and $\Omega' = (\mathbb{R}^n, \mathcal{M}_n(\mathbb{R}^n))$, $\mu_f$ is called the **distribution** of $f$.

Note that a measurable function $x$ defined on $\Omega'$ is integrable with respect to $\mu_f$ iff $x \circ f$ is integrable with respect to $\mu$ and

$$
\int_{\Omega'} x \, \mathrm{d}\mu_f = \int_\Omega x \circ f \, \mathrm{d}\mu. \tag{1.6}
$$

To prove this relation, termed the **change of variables formula**, we check it first for simple functions, and then use approximations to show the general case. A particular case is that where a measure, say $\nu$, is already defined on $(\Omega', \mathcal{F}')$, and $\mu_f$ is absolutely continuous with respect to $\nu$. If $\phi$ is the density of $\mu_f$ with respect to $\nu$, then the change of variables formula reads:

$$\int_\Omega x \circ f \, d\mu = \int_{\Omega'} x \, d\mu_f = \int_{\Omega'} x\phi \, d\nu.$$

Of particular interest is the case when $\Omega' = \mathbb{R}^n$ and $\nu = leb_n$.

If $\mu = \mathbb{P}$ is a probability measure on $(\Omega, \mathcal{F})$ and $\Omega' = \mathbb{R}$, we usually denote measurable maps by the capital letter $X$. We say that $X$ has a first moment iff $X$ is integrable, and then write $E X \equiv \int X \, d\mathbb{P}$. $E X$ is called the **first moment** or **expected value** of $X$. The Hölder inequality (see 1.5.8 below) shows that if $X^2$ has a first moment then $X$ also has a first moment (but the opposite statement is in general not true). $E X^2$ is called the (non-central) **second moment** of $X$. If $E X^2$ is finite, we also define the central second moment or **variance** of $X$ as $D^2 X = \sigma_X^2 = E (X - E X)^2$. The reader will check that $\sigma_X^2$ equals $E X^2 - (E X)^2$.

If the distribution of a random variable $X$ has a density $\phi$ with respect to Lebesgue measure, than $E X$ exists iff $f(\xi) = \xi\phi(\xi)$ is absolutely integrable and then $E X = \int_{-\infty}^{\infty} \xi\phi(\xi) \, d\xi$.

**1.2.16** *Convolution of two finite measures*    Let $\mu$ and $\nu$ be two finite measures on $\mathbb{R}$. Consider the product measure $\mu \otimes \nu$ on $\mathbb{R} \times \mathbb{R}$, and a measurable map $f : \mathbb{R} \times \mathbb{R} \to \mathbb{R}$, $f(\varsigma, \tau) = \varsigma + \tau$. The **convolution** $\mu * \nu$ of $\mu$ with $\nu$ is defined as the transport of $\mu \otimes \nu$ via $f$. Thus, $\mu * \nu$ is a bounded measure on $\mathbb{R}$ and, by the change of variables formula,

$$\int x \, d(\mu * \nu) = \int \int x(\varsigma + \tau) \, \mu(d\varsigma)\nu(d\tau). \tag{1.7}$$

We have $\mu * \nu(\mathbb{R}) = \mu \otimes \nu(\mathbb{R} \times \mathbb{R}) = \mu(\mathbb{R})\nu(\mathbb{R})$. In particular, the convolution of two probability measures on $\mathbb{R}$ is a probability measure. Observe also that $\mu * \nu = \nu * \mu$, and that $(\mu * \mu') * \mu'' = \mu * (\mu' * \mu'')$ for all bounded measures $\mu, \mu'$ and $\mu''$.

**1.2.17** *Convolution of two integrable functions*    For two Lebesgue integrable functions $\phi$ and $\psi$ on $\mathbb{R}$ their convolution $\phi * \psi$ is defined by $\varphi(\xi) = \int_{-\infty}^{\infty} \phi(\xi - \varsigma)\psi(\varsigma) \, d\varsigma$. The reader will use the Fubini–Tonelli Theorem to check that $\phi * \psi$ is well-defined for almost all $\xi \in \mathbb{R}$.

**1.2.18 Exercise**    Suppose that $\mu$ and $\nu$ are two finite measures on $\mathbb{R}$, absolutely continuous with respect to Lebesgue measure. Let $\phi$ and $\psi$ be the densities of $\mu$ and $\nu$, respectively. Show that $\mu * \nu$ is absolutely continuous with respect to Lebesgue measure and has a density $\varphi = \phi * \psi$.

In particular, if both $\phi$ and $\psi$ vanish for $\xi < 0$ then so does $\varphi$ and $\varphi(\xi) = \int_0^\xi \phi(\xi - \varsigma)\psi(\varsigma) \, d\varsigma$.

**1.2.19 Exercise** A counting measure on $\mathbb{R}$ with support $\mathbb{Z}$ is a measure that assigns value one to any set $\{k\}$ where $k$ is an integer. Suppose that $\mu$ and $\nu$ are two measures on $\mathbb{R}$, absolutely continuous with respect to the counting measure. The densities of $\mu$ and $\nu$ may be identified with infinite sequences, say $(a_n)_{n \geq 1}$ and $(b_n)_{n \geq 1}$. Prove that $\mu * \nu$ is absolutely continuous with respect to counting measure and that its density may be identified with a sequence $c_n = \sum_{k=-\infty}^{\infty} a_{n-k} b_k$. In particular, if $a_n = b_n = 0$ for $n < 0$, then $c_n = 0$ for $n < 0$ and $c_n = \sum_{k=0}^{n} a_{n-k} b_k$ for $n \geq 0$.

**1.2.20 Proposition** Let $\mu$ and $\nu$ be two finite Borel measures on $\mathbb{R}$ and assume that $\int x \, d\mu = \int x \, d\nu$ for every bounded continuous function $x$. Then $\mu = \nu$, i.e. $\mu(A) = \nu(A)$ for all Borel sets $A$.

*Proof* It suffices to show that $\mu(a, b] = \nu(a, b], a < b \in \mathbb{R}$. Consider $x_t = \frac{1}{t} 1_{[0,t)} * 1_{(a,b]}, t > 0$. Since

$$x_t(\tau) = \frac{1}{t} \int_{-\infty}^{\tau} 1_{[0,t)}(\tau - \varsigma) 1_{(a,b]}(\varsigma) \, d\varsigma, \tag{1.8}$$

then $|x_t(\tau)| \leq 1$ and $|x_t(\tau + \varsigma) - x_t(\tau)| \leq \frac{\varsigma}{t}$, so that $x_t$ is bounded and continuous. Hence, by assumption

$$\int x_t \, d\mu = \int x_t \, d\nu. \tag{1.9}$$

If $\tau \leq a$, (1.8) implies $x_t(\tau) = 0$. If $\tau > b$, we write

$$x_t(\tau) = \frac{1}{t} \int_a^b 1_{[0,t)}(\tau - \varsigma) \, d\varsigma = \frac{1}{t} \int_{\tau-b}^{\tau-a} 1_{[0,t)}(\varsigma) \, d\varsigma \tag{1.10}$$

to see that $x_t(\tau) = 0$, if $\tau - b > t$. Finally, if $a < \tau \leq b$, $x_t(\tau) = \frac{1}{t} \int_0^t d\varsigma = 1$, for $t < \tau - a$. Consequently, $\lim_{t \to 0} x_t(\tau) = 1_{(a,b]}(\tau), \tau \in \mathbb{R}$. By the Lebesgue Dominated Convergence Theorem we may let $t \to 0$ in (1.9) to obtain $\mu(a, b] = \int 1_{(a,b]} \, d\mu = \int 1_{(a,b]} \, d\nu = \nu(a, b]$. $\qquad \square$

The reader should note how in the proof we have used the "smoothing property" of convolution and the family $x_t$ (which should be thought as approximating the Dirac measure at 0, see below). Also, a careful examination of the proof shows that our result is true as well when we replace the phrase "bounded continuous functions" by "bounded uniformly continuous functions" or even "continuous functions with compact support". This result will become even clearer in Chapter 5, see the celebrated Riesz Theorem 5.2.9.

**1.2.21 Exercise**   Let $(\Omega, \mathcal{F})$ be a measurable space and $\omega$ belong to $\Omega$. The measure $\delta_\omega$ given by $\delta_\omega(A) = 1$ if $\omega \in A$ and $\delta_\omega(A) = 0$ if $\omega \notin A$ is termed **Dirac (delta) measure** at the point $\omega$. Consider the Dirac measure $\delta_0$ in the measurable space $(\mathbb{R}, \mathcal{B}(\mathbb{R}))$ and show that for any Borel measure $\mu$ on $\mathbb{R}$, $\mu * \delta_0 = \mu$. How would you describe $\mu * \delta_t, t \in \mathbb{R}$? See (1.4).

**1.2.22** *Convolution on a topological semigroup*   The notion of convolution may be generalized if we introduce the notion of a topological semigroup. By definition, a **topological semigroup** $\mathbb{G}$ is a topological space and a semigroup at the same time, such that the multiplication in $\mathbb{G}$, $\cdot : \mathbb{G} \times \mathbb{G} \to \mathbb{G}$ is continuous ($\mathbb{G} \times \mathbb{G}$ is equipped with the product topology – see 5.7.4). This map, being continuous, is measurable with respect to appropriate Borel $\sigma$-fields. Therefore, for two bounded measures $\mu$ and $\nu$ on $\mathbb{G}$, we may define $\mu * \nu$ as the transport of $\mu \otimes \nu$ via this map. By the change of variables formula for any measurable $f : \mathbb{G} \to \mathbb{R}$,

$$\int f \, \mathrm{d}(\mu * \nu) = \int f(\xi \cdot \eta) \, \mu(\,\mathrm{d}\xi)\nu(\,\mathrm{d}\eta). \tag{1.11}$$

Convolution so defined is associative, multiplication in $\mathbb{G}$ being associative, but in general is not commutative, for neither must $\mathbb{G}$ be commutative.

If $\mathbb{G}$ is a group such that $(\xi, \eta) \mapsto \xi\eta$ and $\xi \mapsto \xi^{-1}$ are continuous, then $\mathbb{G}$ is called a **topological group**; $\xi^{-1}$ is, of course, the inverse of $\xi$ in $\mathbb{G}$.

**1.2.23 Example**   *Convolution on the Klein group.* The **Klein four-group** [50] (the Klein group for short) is a commutative group $\mathbb{G}$ with four elements, $g_1, ..., g_4$, and the following multiplication table:

| $\circ$ | $g_1$ | $g_2$ | $g_3$ | $g_4$ |
|---|---|---|---|---|
| $g_1$ | $g_1$ | $g_2$ | $g_3$ | $g_4$ |
| $g_2$ | $g_2$ | $g_1$ | $g_4$ | $g_3$ |
| $g_3$ | $g_3$ | $g_4$ | $g_1$ | $g_2$ |
| $g_4$ | $g_4$ | $g_3$ | $g_2$ | $g_1$ |

Table 1.1

$\mathbb{G}$ is a topological group when endowed with discrete topology (all sub-sets of $\mathbb{G}$ are open sets). Any finite measure $\mu$ on $\mathbb{G}$ may be identified with a vector $(a_i)_{i=1,\dots,4}$ with non-negative coordinates. For any $\nu = (b_i)_{i=1,\dots,4}$, the coordinates of convolution $\mu * \nu = (c_i)_{i=1,\dots,4}$, as the image of the measure $(a_i b_j)_{i,j=1,\dots,4}$ on $\mathbb{G} \times \mathbb{G}$, can be read off the multiplication table:

$$
\begin{aligned}
c_1 &= a_1 b_1 + a_2 b_2 + a_3 b_3 + a_4 b_4, \\
c_2 &= a_1 b_2 + a_2 b_1 + a_3 b_4 + a_4 b_3, \\
c_3 &= a_1 b_3 + a_2 b_4 + a_3 b_1 + a_4 b_2, \\
c_4 &= a_1 b_4 + a_2 b_3 + a_3 b_2 + a_4 b_1.
\end{aligned}
\tag{1.12}
$$

Note that convolution on the Klein group is commutative, for the group is commutative.

**1.2.24 Example**   *Convolution on the Kisyński group.* Consider $\mathbb{G} = \mathbb{R} \times \{-1, 1\}$, with multiplication rule

$$
(\tau, k) \circ (\varsigma, l) = (\tau l + \varsigma, kl).
\tag{1.13}
$$

We will leave it to the reader to check that $\mathbb{G}$ is a (non-abelian) group, and note the identity in $\mathbb{G}$ is $(0, 1)$ and the inverse of $(\tau, k)$ is $(-k\tau, k)$. $\mathbb{G}$ is also a topological space, even a metric space, when considered as a subspace of $\mathbb{R}^2$. Clearly, if $(\tau_n, k_n)$ converges to $(\tau, k)$ and $(\varsigma_n, l_n)$ converges to $(\varsigma, l)$ then $(\tau_n, k_n) \circ (\varsigma_n, l_n)$ converges to $(\tau, k) \circ (\varsigma, l)$, and $(-k_n \tau_n, k_n)$ converges to $(-k\tau, k)$, proving that $\mathbb{G}$ is a topological group.

If $\mu$ is a measure on $\mathbb{G}$ then we may define two measures on $\mathbb{R}$ by $\mu_i(A) = \mu(A \times \{i\}), i = 1, -1, A \in \mathcal{M}$. Conversely, if $\mu_i, i = 1, -1$ are two measures on $\mathbb{R}$, then for a measurable subset $B$ of $\mathbb{G}$ we may put $\mu(B) = \mu_1(B \cap (\mathbb{R} \times \{1\})) + \mu_{-1}(B \cap (\mathbb{R} \times \{-1\}))$, where $B \cap (\mathbb{R} \times \{1\})$ is identified with an appropriate subset of $\mathbb{R}$. This establishes a one-to-one correspondence between measures on $\mathbb{G}$ and pairs of measures on $\mathbb{R}$; to denote this correspondence we shall write $\mu = (\mu_1, \mu_{-1})$. We have

$$
\int_{\mathbb{G}} f \, d\mu = \int_{\mathbb{R}} f(\xi, 1) \, \mu_1(d\xi) + \int_{\mathbb{R}} f(\xi, -1) \, \mu_{-1}(d\xi),
\tag{1.14}
$$

for all bounded measurable functions $f : \mathbb{G} \to \mathbb{R}$.

Let $*_{\mathbb{G}}$ denote the convolution in $\mathbb{G}$ and let $*$ denote the convolution in $\mathbb{R}$. For $f$ as above,

$$
\int_{\mathbb{G}} f \, d(\mu *_{\mathbb{G}} \nu) = \int_{\mathbb{G}} \int_{\mathbb{G}} f\left((\tau, k) \circ (\varsigma, l)\right) \mu\left(d(\tau, k)\right) \nu\left(d(\varsigma, l)\right).
$$

Using (1.14) this may be rewritten as

$$\int_G \left[ \int_{\mathbb{R}} f(\tau l + \varsigma, l) \mu_1 \, (\mathrm{d}\tau) + \int_{\mathbb{R}} f(\tau l + \varsigma, -l) \mu_{-1} \, (\mathrm{d}\tau) \right] \nu \, (\mathrm{d}(\varsigma, l))$$

or, consequently, as

$$\int_{\mathbb{R}} \int_{\mathbb{R}} f(\tau + \varsigma, 1) \mu_1 \, (\mathrm{d}\tau) \nu_1 \, (\mathrm{d}\varsigma) + \int_{\mathbb{R}} \int_{\mathbb{R}} f(-\tau + \varsigma, -1) \mu_1 \, (\mathrm{d}\tau) \nu_{-1} \, (\mathrm{d}\varsigma)$$

$$+ \int_{\mathbb{R}} \int_{\mathbb{R}} f(\tau + \varsigma, -1) \mu_{-1} \, (\mathrm{d}\tau) \nu_1 \, (\mathrm{d}\varsigma)$$

$$+ \int_{\mathbb{R}} \int_{\mathbb{R}} f(-\tau + \varsigma, 1) \mu_{-1} \, (\mathrm{d}\tau) \nu_{-1} \, (\mathrm{d}\varsigma).$$

Hence, $(\mu *_G \nu)_1$ is a sum of two measures: one of them is $\mu_1 * \nu_1$ and the other one is $\tilde{\mu}_{-1} * \nu_{-1}$, where $\tilde{\mu}_{-1}(A) = \mu_{-1}(-A)$, and $-A \equiv \{\varsigma \in \mathbb{R}, -\varsigma \in A\}$. Similarly, we obtain a formula for $(\mu *_G \nu)_{-1}$. We may summarize this analysis by writing:

$$(\mu_1, \mu_{-1}) *_G (\nu_1, \nu_{-1}) = (\mu_1 * \nu_1 + \tilde{\mu}_{-1} * \nu_{-1}, \mu_{-1} * \nu_1 + \tilde{\mu}_1 * \nu_{-1}), \quad (1.15)$$

or in matrix notation,

$$(\mu_1, \mu_{-1}) *_G (\nu_1, \nu_{-1}) = \begin{pmatrix} \mu_1 & \tilde{\mu}_{-1} \\ \mu_{-1} & \tilde{\mu}_1 \end{pmatrix} \begin{pmatrix} \nu_1 \\ \nu_{-1} \end{pmatrix}. \quad (1.16)$$

**1.2.25 Exercise**    Let $\mu(t) = \frac{1}{2}(1 + e^{-2at}, 1 - e^{-2at}, 0, 0)$ where $a > 0$ is a constant, be a probability measure on the Klein group. Use (1.12) to check that $\mu(t) * \mu(s) = \mu(t + s), s, t \geq 0$.

**1.2.26 Exercise**    Let $\mu(t) = \frac{1}{2}((1 + e^{-2at}) \delta_t, (1 - e^{-2at}) \delta_t)$, where $a$ is a positive constant and $\delta_t$ is the Dirac measure at $t$, be a measure on the Kisyński group. Use (1.16) to check that $\mu(t) * \mu(s) = \mu(t + s), s, t \geq 0$.

**1.2.27 Exercise**    Find the formula for the convolution of two measures on a group $\mathbb{G}$ of congruences modulo $p$ where $p$ is a prime number. Recall that this group is composed of numbers $0, 1, ..., p - 1$ and the product of two elements $a$ and $b$ of this group is defined to be $ab$ (mod $p$).

**1.2.28 Exercise**    Use convolutions in $\mathbb{R}^n$ to extend the argument used in 1.2.20 and show that the same theorem holds for Borel measures in $\mathbb{R}^n$.

1.2.29 **Exercise** Let $\mathbf{C} = \{z \in \mathbb{C}; |z| = 1\}$ be the unit circle with usual multiplication. Check that a convolution of two integrable functions on $\mathbf{C}$ is given by $x * y(e^{i\alpha}) = \int_{-\pi}^{\pi} x(e^{i(\alpha-\theta)})y(e^{i\theta})\,d\theta$. Let

$$p_r(e^{i\alpha}) = \frac{1}{2\pi}\frac{1-r^2}{1-2r\cos\alpha+r^2} = \frac{1}{2\pi}\sum_{n=-\infty}^{\infty} r^{|n|}e^{in\alpha}, \qquad 0 \le r < 1$$

be the **Poisson kernel**. Check that $p_r \ge 0$, $\int_{-\pi}^{\pi} p_r(e^{i\theta})\,d\theta = 1$, and $p_r * p_s = p_{rs}$, $0 \le r, s < 1$. The equivalence of the two definitions of $p_r$ is a consequence of the formula for the sum of a geometric series and some simple (but intelligent) algebra – see [103, 111] or other textbooks.

1.2.30 **Definition** Even though the probability space is a basic notion of the theory of probability and stochastic processes, it is often the case that we are not able to say anything about the underlying "original" probability space upon which the random variable/process is defined. Neither do we need or intend to. Quite often all we need is the information on distributions. The following definitions are exactly in this spirit.

- A random variable $X$ is called **Gaussian/normal with parameters** $m \in \mathbb{R}, \sigma^2 > 0$ if its distribution is absolutely continuous with respect to *leb* and has density $\frac{1}{\sqrt{2\pi\sigma^2}}\exp\{-\frac{(s-m)^2}{2\sigma^2}\}$. We also write $X \sim N(m,\sigma^2)$. Sometimes it is convenient to allow $\sigma^2 = 0$ in this definition, and say that $X = m$ a.s. is a (degenerate) normal variable with parameters $m$ and 0.
- A random variable is called **exponential with parameter** $\lambda > 0$ if its distribution is absolutely continuous with respect to *leb* and has the density $\lambda 1_{\mathbb{R}_+}e^{-\lambda s}$. If its density equals $\frac{\lambda}{2}e^{-\lambda|s|}, s \in \mathbb{R}$, the random variable is called **bilateral exponential**.
- A random variable is called **uniform** on the interval $[a, b]$ if its distribution is absolutely continuous with respect to *leb* and has a density $\frac{1}{b-a}1_{[a,b]}$.
- A random variable is called **gamma with parameters** $a > 0$ **and** $b > 0$ if its distribution is absolutely continuous with respect to *leb* and has density $\frac{b^a}{\Gamma(a)}s^{a-1}e^{-bs}1_{\mathbb{R}_+}(s)$, where $\Gamma(a) = \int_0^{\infty} s^{a-1}e^{-s}\,ds$.
- A random variable is called **binomial with parameters** $n \in \mathbb{N}$ **and** $p \in [0, 1]$ if $P(X = k) = \binom{n}{k}p^k(1-p)^{n-k}, 0 \le k \le n$.
- A random variable $X$ is called **Poisson with parameter** $\lambda > 0$ if $\mathbb{P}(X = k) = e^{-\lambda}\frac{\lambda^k}{k!}, k = 0, 1, \ldots$

- A random variable $X$ is called **geometric with parameter** $p \in [0, 1]$ if $\mathbb{P}(X = k) = pq^k, q = 1 - p, k = 0, 1, \ldots$

The reader is encouraged to check that the first four functions listed above are indeed densities of probability measures on $\mathbb{R}$ (see Exercise 1.2.31 below), while the last three are probability mass functions of probability measures on $\mathbb{N}$. Notice also that a gamma variable with parameters 1 and $\lambda$ is an exponential variable. In the following sections we shall prove, not only that there exist random variables with any given distribution, but also that for any distribution there exist infinitely many independent random variables with this distribution. Some readers might find it surprising that all such variables may be defined on the standard probability space.

**1.2.31 Exercise**    Show that $\int_{\mathbb{R}} e^{-\frac{s^2}{2}} \, ds = \sqrt{2\pi}$.

**1.2.32 Exercise**    Prove that if $X$ is a normal variable with parameters 0 and $\sigma$, then $X^2$ is a gamma variable with parameters $\frac{1}{2}$ and $\frac{1}{2\sigma^2}$.

**1.2.33 Exercise**    Let $\mu$ be the distribution of a gamma variable with parameters $a$ and $b$, and let $\nu$ be the distribution of a gamma variable with parameters $a'$ and $b$. Show that $\mu * \nu$ has the same density as a gamma variable with parameters $a + a'$ and $b$.

**1.2.34 Exercise**    (*Poisson approximation to binomial*) Show that if $X_n$ is a sequence of random variables with binomial distributions having parameters $n$ and $p_n$ respectively, and if $\lim_{n \to \infty} np_n = \lambda > 0$, then $\lim \mathbb{P}[X_n = k] = \frac{\lambda^k}{k!} e^{-\lambda}, k \geq 0$.

**1.2.35 Exercise**    Show that if $X \sim N(\mu, \sigma^2)$ then $E\,X = \mu, \sigma_X^2 = \sigma^2$. Moreover, if $X \sim \Gamma(a, b)$ then $E\,X = \frac{a}{b}$, and $\sigma_X^2 = \frac{a}{b^2}$.

**1.2.36 Exercise**    Let $X$ be a non-negative random variable with finite expected value. Prove that $\mathbb{P}\{X \geq \epsilon\} \leq \frac{E\,X}{\epsilon}$ (Markov inequality). Also, deduce that for any random variable with a finite second moment, $\mathbb{P}\{|X - E\,X| \geq \epsilon\} \leq \frac{\sigma_X^2}{\epsilon^2}$ (Chebyshev's inequality).

**1.2.37 Exercise**    Use the Fubini Theorem to show that for any non-negative random variables $X$ and $Y$ and any numbers $\alpha, \beta > 0$ we have

$$E\,X^\alpha Y^\beta = \int_0^\infty \int_0^\infty \alpha s^{\alpha-1} \beta t^{\beta-1} \mathbb{P}\{X > s, Y > t\} \, ds \, dt. \qquad (1.17)$$

Take $X = 1$, to deduce

$$EY^\beta = \int_0^\infty \beta t^{\beta-1} \mathbb{P}\{Y > t\} \, dt. \tag{1.18}$$

Apply this to obtain

$$EY^\beta 1_{\{X>s\}} = \int_0^\infty \beta t^{\beta-1} \mathbb{P}\{X > s, Y > t\} \, dt, \qquad s > 0, \tag{1.19}$$

and, consequently,

$$E X^\alpha Y^\beta = \int_0^\infty \alpha s^{\alpha-1} E Y^\beta 1_{\{X>s\}} \, ds. \tag{1.20}$$

## 1.3 Functions of bounded variation. Riemann–Stieltjes integral

1.3.1 *Functions of bounded variation*    A function $y$ defined on a closed interval $[a, b]$ is said to be of **bounded variation** if there exists a number $K$ such that for every natural $n$ and every partition $a = t_1 \le t_2 \le \cdots \le t_n = b$,

$$\sum_{i=2}^n |y(t_i) - y(t_{i-1})| \le K.$$

The infimum over all such $K$ is then denoted $var[y, a, b]$. We do not exclude the case where $a = -\infty$ or $b = \infty$. In such a case we understand that $y$ is of bounded variation on finite subintervals of $[a, b]$ and that $var[y, -\infty, b] = \lim_{c \to -\infty} var[y, c, b]$ is finite and/or that

$$var[y, a, \infty] = \lim_{c \to \infty} var[y, a, c]$$

is finite. It is clear that $var[y, a, b] \ge 0$, and that it equals $|y(b) - y(a)|$ if $y$ is monotone. If $y$ is of bounded variation on $[a, b]$ and $a \le c \le b$, then $y$ is of bounded variation on $[a, c]$ and $[c, b]$, and

$$var[y, a, b] = var[y, a, c] + var[y, c, b]. \tag{1.21}$$

Indeed, if $a = t_1 \le t_2 \le \cdots \le t_n = c$ and $c = s_1 \le s_2 \le \cdots \le s_m = b$, then $u_i = t_i, i = 1, \ldots, n - 1$, $u_n = t_n = s_1$ and $u_{n+i} = s_{i+1}, i = 1, \ldots, m - 1$, is a partition of $[a, b]$, and

$$\sum_{i=2}^{m+n-1} |y(u_i) - y(u_{i-1})| = \sum_{i=2}^n |y(t_i) - y(t_{i-1})| + \sum_{i=2}^m |y(s_i) - y(s_{i-1})|.$$

This proves that the right-hand side of (1.21) is no greater than the left-hand side. Moreover, if $a = t_1 \leq t_2 \leq \cdots \leq t_n = b$ is a partition of $[a, b]$ then either $c = t_j$ for some $1 \leq j \leq n$, or there exists a $j$ such that $t_j < c < t_{j+1}$. In the first case we define the partition $a = t_1 \leq t_2 \leq \cdots \leq t_j = c$ of $[a, c]$ and the partition $c = s_1 \leq s_2 \leq \cdots \leq s_{n+1-j} = b$ of $[c, b]$ where $s_i = t_{j+i-1}$, to see that the right-hand side of (1.21) is no less than the left-hand side. In the other case we consider the partition $a = t_1' \leq t_2' \leq \cdots \leq t_{n+1}' = b$ of $[a, b]$, where $t_i' = t_i$ for $i \leq j$, $t_{j+1}' = c$, $t_{j+i}' = t_{j+i-1}$, $i = 1, \ldots, n+1-j$, and reduce the problem to the previous case by noting that

$$\sum_{i=2}^{n} |y(t_i) - y(t_{i-1})| \leq \sum_{i=2}^{n+1} |y(t_i') - y(t_{i-1}')|.$$

Equation (1.21) proves in particular that the function $v_+(t) = var[y, a, t]$ where $t \in [a, b]$ is non-decreasing. Define $v_-(t) = v_+(t) - y(t)$. For $s \leq t$, the expression $v_-(t) - v_-(s) = v_+(t) - v_+(s) - [y(t) - y(s)] = var[y, s, t] - [y(t) - y(s)]$ is non-negative. We have thus proved that any function of bounded variation is a difference of two non-decreasing functions $y(t) = v_+(t) - v_-(t)$. In particular, functions of bounded variation have right-hand and left-hand limits. The left-hand limit of $y$ at $t$ is denoted $y(t-)$ and the right-hand limit of $y$ at $t$ is denoted $y(t+)$. Note that the representation of a function of bounded variation as a difference of two non-decreasing functions is not unique. See 1.3.6 below.

**1.3.2 Lemma**    If $y(t)$ is a function of bounded variation on $[a, b]$ then there exists at most countable number of points of discontinuity of $y$, i.e. points $t \in [a, b]$ where $y(t-) \neq y(t)$, or $y(t+) \neq y(t)$.

*Proof*  Fix $m \in \mathbb{N}$. Note that there may be only a finite number of points such that $|y(t-) - y(t)| \geq \frac{1}{m}$. This shows that there exists at most countable number of points $t \in [a, b]$ where $y(t-) \neq y(t)$. The same argument proves that there exists at most countable number of points $t \in [a, b]$ where $y(t+) \neq y(t)$, and these two facts together imply our claim.                                                                      $\square$

**1.3.3 Exercise**    Let $y(t)$ be a function of bounded variation, and let $Dis(y)$ be the set of points of discontinuity of $y$. Let $y_r$ be defined as $y_r(t) = y(t+)$, $t \in [a, b]$ (note that $y$ and $y_r$ differ only on $Dis(y)$). Prove that $y_r$ is right-continuous and of bounded variation. Moreover, $var[y_r, a, b] \leq var[y, a, b]$.

The function $y_r$ is called the **regularization** of $y$. A function $y$ of bounded variation is said to be **regular** if $y = y_r$.

**1.3.4 Exercise**     Prove that if $y$ is of bounded variation and right-continuous, then so is $v_+(t) = var[y, a, t]$.

**1.3.5** *Monotone functions and finite measures*     In this subsection we will show that there is a one-to-one correspondence between regular non-decreasing functions satisfying (1.22) below, and finite measures on $\mathbb{R}$.

Let $\mu$ be a finite measure on $\mathbb{R}$. Define $y(t) = \mu(-\infty, t]$. It is clear that $y(t)$ is non-decreasing, has left limits and is right-continuous. As for the last two statements it suffices to note that if $t_n < t$ and $\lim_{n \to \infty} t_n = t$, then $\bigcup (-\infty, t_n] = (-\infty, t)$ and by continuity of measure, the limit of $y(t_n)$ exists and equals $\mu(-\infty, t)$; analogously, if $s_n > t$ and $\lim_{n \to \infty} s_n = t$, then $\bigcap (-\infty, t_n] = (-\infty, t]$ and the limit of $y(s_n)$ exists and equals $y(t) = \mu(-\infty, t]$. Also, note that

$$y(-\infty) = \lim_{t \to -\infty} y(t) = 0, \quad \text{and } y(\infty) = \lim_{t \to \infty} y(t) < \infty. \quad (1.22)$$

The last limit equals $\mu(\mathbb{R})$ and in particular, if $\mu$ is a probability measure, it equals 1.

Now, suppose that $y$ is a right-continuous non-decreasing function such that (1.22) holds. We will show that there exists a unique finite Borel measure $\mu$ such that $\mu(-\infty, t] = y(t)$.

If $\mu$ is such a measure, and $a_0 < a_1 \leq b_1 < a_2 \leq b_2 < ... < a_n \leq b_n < b_{n+1}$ are real numbers, then we must have

$$\mu \left( (-\infty, a_0] \cup \bigcup_{i=1}^{n} (a_i, b_i] \cup (b_{n+1}, \infty) \right)$$

$$= [y(a_0) - y(-\infty)] + \sum_{i=1}^{n} [y(b_i) - y(a_i)] + [y(\infty) - y(b_{n+1})]$$

with obvious modification if $(-\infty, a_0]$ and/or $(b_{n+1}, \infty)$ is not included in the union. Such finite unions form a field $\mathcal{F}$, and we see that the above formula defines an additive function on $\mathcal{F}$.

To show that $\mu$ thus defined is countably additive, assume that $A_n \supset A_{n+1}$ are members of $\mathcal{F}$ and that $\bigcap_{n=1}^{\infty} A_n = \emptyset$. We need to show that $\lim_{n \to \infty} \mu(A_n) = 0$.

Suppose that this is not the case, and that there exists an $\epsilon > 0$ such that $\mu(A_n) > \epsilon$ for all $n \geq 1$. By (1.22) and right-continuity of $y$, for every $A_n$ there exists a $B_n \in \mathcal{F}$ such that $cl(B_n) \subset A_n$ and

$\mu(A_n \setminus B_n) < \frac{\epsilon}{2^{n+1}}$. This is just saying that for a finite interval $(a, b] \in \mathcal{F}$ the right-continuity of $y$ implies that we may find $a' > a$ such that $y(b) - y(a')$ is arbitrarily close to $y(b) - y(a)$, and that (1.22) allows doing the same with infinite intervals. It will be convenient to treat $cl(B_n)$ as subsets of $\overline{\mathbb{R}} = \{-\infty\} \cup \mathbb{R} \cup \{+\infty\}$. $\overline{\mathbb{R}}$ equipped with the topology inherited from $\mathbb{R}$ plus neighborhoods of $\{-\infty\}$ and $\{+\infty\}$ of the form $[-\infty, s)$, and $(s, \infty]$, $s \in \mathbb{R}$ respectively, is a compact topological space. Since $\bigcap cl(B_n) = \emptyset$, there exists an $n \in \mathbb{N}$ such that $cl(B_1) \cap cl(B_2) \cap \ldots \cap cl(B_n) = \emptyset$. Now,

$$\mu(A_n) = \mu(A_n \setminus \bigcap_{i=1}^n B_i) + \mu(\bigcap_{i=1}^n B_i) = \mu(\bigcup_{i=1}^n (A_n \setminus B_i))$$

$$\leq \mu(\bigcup_{i=1}^n (A_i \setminus B_i)) \leq \sum_{i=1}^n \mu(A_i \setminus B_i) \leq \sum_{i=1}^n \frac{\epsilon}{2^{i+1}} < \frac{\epsilon}{2},$$

a contradiction. Hence, by 1.2.8 there exists an extension of $\mu$ to $\sigma(\mathcal{F})$ which is clearly equal to the Borel $\sigma$-algebra, and 1.2.7 shows that this extension is unique. Finally,

$$y(t) = \lim_{n \to \infty} [y(t) - y(-n)] = \lim_{n \to \infty} \mu(-n, t] = \mu(-\infty, t].$$

**1.3.6** *Functions of bounded variation and signed measures*    In this subsection we introduce the notion of a charge and discuss some properties of charges. In particular, we prove that there is a one-to-one correspondence between signed measures and regular functions of bounded variation satisfying (1.22).

A set function $\mu$ on a measurable space $(\Omega, \mathcal{F})$ is said to be a **charge** or a **signed measure** if there exist finite measures $\mu^+$ and $\mu^-$ such that $\mu = \mu^+ - \mu^-$. Of course such a representation is not unique; for any positive finite measure $\nu$ we have $\mu = (\mu^+ + \nu) - (\mu^- + \nu)$. Later on, we will see that there is representation of $\mu$ that is in a sense "minimal".

Given a Borel charge $\mu$ on $\mathbb{R}$, i.e. a signed measure which is the difference of two finite Borel measures on $\mathbb{R}$, we may define $y(t) = \mu(-\infty, t]$. Then $y$ is a regular function of bounded variation and satisfies (1.22), being the difference of $y^+(t) = \mu^+(-\infty, t]$ and $y^-(t) = \mu^-(-\infty, t]$. Conversely, if $y$ is a regular function of bounded variation satisfying (1.22), then there exists a unique Borel charge such that $\mu(-\infty, t] = y(t)$. To prove this, consider $x(t) = var[y, -\infty, t]$, $y_0^+ = \frac{x+y}{2}$ and $y_0^- = \frac{x-y}{2}$. For any $a \leq b$, $2[y_0^+(b) - y_0^+(a)] = x(b) - x(a) + y(b) - y(a) \geq 0$, proving that $y_0^+$ is non-decreasing. In a similar fashion we show that $y_0^-$ is non-decreasing also. Both $y_0^+$ and $y_0^-$ are regular and satisfy (1.22) since $x$

and $y$ do. Therefore, there exist two positive finite measures $\mu_0^+$ and $\mu_0^-$ such that $y_0^+(t) = \mu_0^+(-\infty, t]$ and $y_0^-(t) = \mu_0^-(-\infty, t]$. Since

$$y = y_0^+ - y_0^-, \tag{1.23}$$

the charge $\mu = \mu_0^+ - \mu_0^-$ satisfies our requirements.

Moreover, the functions $y_0^+$ and $y_0^-$ satisfy $x = y_0^+ + y_0^-$. The representation (1.23) is minimal in the following sense: if $y^+$ and $y^-$ are two regular non-decreasing functions satisfying (1.22) such that $y = y^+ - y^-$, then $y^+ + y^- \geq x$. Indeed, in such a case, for any $a \leq b$ we have

$$y(b) - y(a) = [y^+(b) - y^+(a)] - [y^-(b) - y^-(a)]$$

and consequently

$$|y(b) - y(a)| \leq [y^+(b) - y^+(a)] - [y^-(b) - y^-(a)].$$

Using this inequality for subintervals of $[a, b]$ we obtain

$$var[y, a, b] \leq [y^+(b) - y^+(a)] - [y^-(b) - y^-(a)] \tag{1.24}$$

and the claim follows by taking the limit as $a \to -\infty$.

By 1.2.8, this proves that

$$|\mu| \leq \mu^+ + \mu^- \tag{1.25}$$

where $|\mu|$ is a measure related to $x$ and $\mu^+$ and $\mu^-$ are related to $y^+$ and $y^-$, respectively. To be more specific, there exists a positive measure $\nu$ such that $|\mu| + 2\nu = \mu^+ + \mu^-$. Thus, the minimality of the representation (1.23) may also be rephrased as follows: for any charge $\mu$ there exists two positive measures $\mu_0^+$ and $\mu_0^-$ such that $\mu = \mu_0^+ - \mu_0^-$ and $|\mu| = \mu_0^+ + \mu_0^-$, and for any other measures $\mu^+$ and $\mu^-$ such that $\mu = \mu^+ - \mu^-$ there exists a positive measure $\nu$ such that $\mu^+ = \mu_0^+ + \nu$ and $\mu^- = \mu_0^- + \nu$.

Given a charge $\mu$ in the minimal representation $\mu = \mu_0^+ - \mu_0^-$ and a function $f$ that is absolutely integrable with respect to $\mu_0^+$ and $\mu_0^-$, we define $\int f \, d\mu$ as $\int f \, d\mu_0^+ - \int f \, d\mu_0^-$. It may be checked that if $\mu = \mu^+ - \mu^-$ for some other measures $\mu+$ and $\mu^-$ such that $f$ is absolutely integrable with respect to them, then $\int f \, d\mu = \int f \, d\mu^+ - \int f \, d\mu^-$. Obviously,

$$\left| \int f \, d\mu \right| \leq \int |f| \, d\mu^+ + \int |f| \, d\mu^-$$

and in particular

$$\left| \int f \, d\mu \right| \leq \int |f| \, d|\mu|.$$

1.3.7 *The Riemann–Stieltjes integral*     Let $[a, b]$ be an interval $(a < b)$ and let $y$ be a function of bounded variation on $[a, b]$. Suppose that $t \mapsto x(t) \in \mathbb{R}$ is another function on $[a, b]$. Consider two sequences, $\mathcal{T} = (t_i)_{i=0,\ldots,k}$ and $\Xi = (\xi_i)_{i=0,\ldots,k}$ of points of $[a, b]$, where $k$ is an integer, such that

$$a = t_0 < t_1 < \cdots < t_k = b, \qquad t_0 \leq \xi_0 \leq t_1 \leq \cdots \leq t_{k-1} \leq \xi_{k-1} \leq t_k.$$

Define the related numbers $\Delta(\mathcal{T}) = \sup_{0 \leq i \leq k}\{t_i - t_{i-1}\}$ and

$$S(\mathcal{T}, \Xi, x, y) = \sum_{i=0}^{k-1} x(\xi_i)[y(t_{i+1}) - y(t_i)]. \qquad (1.26)$$

If the limit $\lim_{n \to \infty} S(\mathcal{T}_n, \Xi_n, x, y)$ exists for any sequence of pairs $(\mathcal{T}_n, \Xi_n)$ such that $\lim_{n \to \infty} \Delta(\mathcal{T}_n) = 0$, and does not depend on the choice of the sequence of $(\mathcal{T}_n, \Xi_n)$, function $x$ is said to be **Riemann–Stieltjes integrable** with respect to $y$. The above limit is denoted

$$\int_a^b x_t \, dy(t)$$

and called the (**Riemann–Stieltjes**) **integral** of $x$ (with respect to $y$).

This definition has to be modified when either $a$ or $b$ is infinite. Assume for instance that $a$ is finite and $b = \infty$. It is clear then that the definition of $\Delta(\mathcal{T})$ has to be changed since $\Delta(\mathcal{T})$ as defined now is always infinite. We put therefore $\Delta(\mathcal{T}) = \sup_{0 \leq i \leq k-1}(t_i - t_{i-1})$, and then require that the limit $\lim_{n \to \infty} S(\mathcal{T}_n, \Xi_n, x, y)$ exists for any sequence of pairs $(\mathcal{T}_n, \Xi_n)$ such that $\lim_{n \to \infty} \Delta(\mathcal{T}_n) = 0$, and $\lim_{n \to \infty} t_{n,k_n - 1} = \infty$. Here $t_{n,k_n - 1}$ is the second to last element of partition $\mathcal{T}_n$; the last one is always $\infty$. Again the limit is not to depend on the choice of the sequence of $(\mathcal{T}_n, \Xi_n)$.

With this definition, it turns out that continuous functions are Riemann–Stieltjes integrable. First of all, Exercise 1.3.8 below shows that continuous functions are integrable with respect to a function of bounded variation iff they are integrable with respect to its regularization. Thus, it suffices to show that continuous functions are integrable with respect to regular functions of bounded variation. To this end, let $x$ be continuous on $[a, b]$. To focus our attention, we assume that $a = -\infty$ and $b$ is finite. We understand that the limit $\lim_{t \to -\infty} x(t)$ exists. Function $x$, being continuous, is measurable. Extend $y$ to the whole line by setting $y(t) = y(b)$ for $t \geq b$. Let $\mu$ be the unique finite (signed) measure corresponding to the regularization $y_r$ of such an extended function. Since $x$ is bounded, it is integrable and we may define $l = \int f \, d\mu$. Now, consider the sequence of partitions $\mathcal{T}_n$ with sets of midpoints $\Xi_n$. Fix $\epsilon > 0$. Since $x$ is uniformly continuous, we may choose an $R > 0$ and a $\delta > 0$ such that $|x(t) - x(s)| < \epsilon$ provided $|t - s| < \epsilon$, for $s, t \in (-\infty, b]$, or $s, t < -R$. Choose $n$ large enough so that $\Delta(\mathcal{T}_n) < \epsilon$ and $t_{n,2} < -R$. Again, $t_{n,2}$ is the second element of partition $\mathcal{T}_n$. Now

$$|l - S(\mathcal{T}_n, \Xi_n, x, y)|$$

$$= \left| \sum_{i=0}^{k_n} \int_{(t_{n,i}, t_{n,i+1}]} x\, d\mu - \sum_{i=0}^{k_n} x(\xi_{n,i})[y(t_{n,i+1}) - y(t_{n,i})] \right|$$

$$\leq \left| \sum_{i=0}^{k_n} \int_{(t_{n,i}, t_{n,i+1}]} |x(s) - x(\xi_{n,i})|\, \mu(ds) \right| \leq \epsilon \mu(\mathbb{R}).$$

This proves that the limit of $S(\mathcal{T}_n, \Xi_n, x, y)$ is $l$ and concludes the proof of our claim.

**1.3.8 Exercise**    Let $y$ be a function of bounded variation on an interval $[a, b]$, let $y_r$ be the regularization of $y$ and $x$ be a continuous function on $[a, b]$. Assume to fix attention that $a$ and $b$ are finite. Consider the sequence of partitions $\mathcal{T}_n$ with sets of midpoints $\Xi_n$, such that the corresponding $\Delta_n$ tends to zero. Prove that

$$\lim_{n \to \infty} S(\mathcal{T}_n, \Xi_n, x, y) = \lim_{n \to \infty} S(\mathcal{T}_n, \Xi_n, x, y_r) + x(a)[y(a+) - y(a)].$$

## 1.4 Sequences of independent random variables

**1.4.1 Definition**    Let $(\Omega, \mathcal{F}, \mathbb{P})$ be a probability space. Let $\mathcal{F}_t, t \in \mathbb{T}$ be a family of classes of measurable subsets ($\mathbb{T}$ is an abstract set of indexes). The classes are termed **mutually independent** (to be more precise: **mutually $\mathbb{P}$-independent**) if for all $n \in N$, all $t_1, ..., t_n \in \mathbb{T}$ and all $A_i \in \mathcal{F}_{t_i}, i = 1, ..., n$

$$\mathbb{P}(\bigcap_{i=1}^{n} A_i) = \prod_{i=1}^{n} \mathbb{P}(A_i). \tag{1.27}$$

The classes are termed **pairwisely independent** (to be more precise: **pairwisely $\mathbb{P}$-independent**) if for all $n \in N$, all $t_1, t_2 \in \mathbb{T}$ and all $A_i \in \mathcal{F}_{t_i}, i = 1, 2,$

$$\mathbb{P}(A_1 \cap A_2) = \mathbb{P}(A_1)\mathbb{P}(A_2).$$

It is clear that mutually independent classes are pairwisely independent. Examples proving that pairwise independence does not imply joint independence may be found in many monographs devoted to probability theory. The reader is encouraged to find one.

Random variables $X_t, t \in \mathbb{T}$ are said to be mutually (pairwisely) independent if the $\sigma$-algebras $\mathcal{F}_t = \sigma(X_t)$ generated by $X_t$ are mutually (pairwisely) independent.

From now on, the phrase "classes (random variables) are independent" should be understood as "classes (random variables) are mutually independent".

**1.4.2 Exercise**    Suppose that two events, $A$ and $B$, are independent, i.e. $\mathbb{P}(A \cap B) = \mathbb{P}(A)\mathbb{P}(B)$. Show that the $\sigma$-algebras

$$\{A, A^{\mathrm{C}}, \Omega, \emptyset\}, \quad \{B, B^{\mathrm{C}}, \Omega, \emptyset\}$$

are independent.

**1.4.3 Exercise**    Let $(\Omega, \mathcal{F}, \mathbb{P})$ be a probability space and $\mathcal{G} \in \mathcal{F}$ be a $\sigma$-algebra. Define $\mathcal{G}^{\perp}$ as the class of all events $A$ such that $A$ is independent of $B$ for all $B \in \mathcal{G}$. Show that $\mathcal{G}^{\perp}$ is a $\lambda$-system. To see that $\mathcal{G}^{\perp}$ is in general not a $\sigma$-algebra consider $\Omega = \{a, b, c, d\}$ with all simple events equally likely, and $\mathcal{G}$ a $\sigma$-algebra generated by the event $\{c, d\}$. Note that $A = \{a, c\}$ and $B = \{b, c\}$ are independent of $\mathcal{G}$ but that neither $A \cap B$ nor $A \cup B$ are independent of $\mathcal{G}$.

**1.4.4 Exercise**    Suppose that random variables $X$ and $Y$ are independent and that $f$ and $g$ are two Lebesgue measurable functions. Prove that $f(X)$ and $g(Y)$ are independent.

**1.4.5 Exercise**    Show that random variables $X$ and $Y$ are independent iff the distribution of the random vector $(X, Y)$ is $\mathbb{P}_X \otimes \mathbb{P}_Y$. Consequently, the distribution of the sum of two independent random variables is the convolution of their distributions:

$$\mathbb{P}_{X+Y} = \mathbb{P}_X * \mathbb{P}_Y.$$

**1.4.6 Exercise**    Suppose that $X_n, n \geq 1$ is a sequence of independent random variables with exponential distribution with parameter $\lambda > 0$. Show that $S_n = \sum_{k=1}^{n} X_k$ is a gamma variable with parameters $\lambda$ and $n$.

**1.4.7 Exercise**    Let $X \sim N(0, \sigma_1^2)$ and $Y \sim N(0, \sigma_2^2)$ be two independent random variables. Show that $X + Y \sim N(0, \sigma_1^2 + \sigma_2^2)$.

**1.4.8 Exercise**    Suppose that random variables $X$ and $Y$ are independent and have expected values. Show that $E\,XY = E\,X\,E\,Y$; the existence of $E\,XY$ is a part of the claim.

1.4.9 **Example**     It is easy to find an example showing that the converse statement to the one from the previous exercise is not true. Suppose however that $X$ and $Y$ are such that $E\,X f(Y) = E\,X E\,f(Y)$ for any Borel function $f$. Does that imply that $X$ and $Y$ are independent? The answer is still in the negative. As an example consider random variables $X$ that has only three possible values $0, 2$ and $4$ each with probability $\frac{1}{3}$, and $Y$ that attains values $a$ and $b$ ($a \neq b$) with probability $\frac{1}{2}$ each. Assume also that their joint probability mass function is given by:

| $Y \setminus X$ | 0 | 2 | 4 |
|:---:|:---:|:---:|:---:|
| $a$ | $\frac{1}{12}$ | $\frac{2}{6}$ | $\frac{1}{12}$ |
| $b$ | $\frac{3}{12}$ | $0$ | $\frac{3}{12}$ |

Table 1.2

Then the joint probability mass function of $X$ and $f(Y)$ for any $f$ is the same, except that $a$ is replaced with some real $\alpha$ and $b$ is replaced with a $\beta$ ($\alpha$ may happen to be equal $\beta$). Certainly $X$ and $Y$ are not independent, and yet $E\,X = 2, E\,f(Y) = \frac{\alpha+\beta}{2}$ and $E\,X f(Y) = \frac{8\alpha}{12} + \frac{4\alpha}{12} + \frac{12\beta}{12} = \alpha + \beta$, and so $E f(Y) X = E\,X E\,f(Y)$. The reader should be able to prove that if $E\,f(Y) g(X) = E\,f(Y) E\,g(X)$ for all Borel functions $f$ and $g$ then $X$ and $Y$ are independent.

1.4.10 **Exercise**     If random variables $X_i$ are exponential with parameter $\lambda_i, i = 1, 2$, and independent, then $Y = \min(X_1, X_2)$ is exponential with parameter $\lambda_1 + \lambda_2$.

1.4.11 **Exercise**     Show that if $X$ and $Y$ are independent exponential random variables with parameters $\lambda$ and $\mu$, respectively, then $\mathbb{P}[X \leq Y] = \frac{\lambda}{\lambda+\mu}$.

1.4.12 **Theorem**     Suppose that $\mathcal{F}_t, t \in \mathbb{T}$ are independent $\pi$-systems of measurable sets, and that $\mathbb{T}_u, u \in \mathcal{U}$ are disjoint subsets of $\mathbb{T}$. The $\sigma$-algebras $\mathcal{G}_u$ generated by $\mathcal{F}_t, t \in \mathbb{T}_u$ are independent.

*Proof*  Fix $n \in \mathbb{N}$, and choose indexes $u_1, ..., u_n \in \mathcal{U}$. (If the number of elements of $\mathcal{U}$ is finite, $n$ must be chosen no greater than the number of elements in $\mathcal{U}$.) We need to prove that (1.27) holds for all $A_i \in \mathcal{G}_{u_i}, i = 1, ..., n$. By assumption, (1.27) holds if all $A_i$ belong to the class $\mathcal{A}_i$ of events of the form

$$A_i = \bigcap_{j=1}^{k(i)} B_{i,j}, \qquad (1.28)$$

where $B_{i,j}$ belong to $\mathcal{F}_{t_{i,j}}$ for some $t_{i,j} \in \mathbb{T}_{u_i}$. Now, fix $A_i, i = 2, ..., n$, of this form and consider the class $\mathcal{B}_1$ of events $A_1$ such that (1.27) holds. As we have pointed out $\mathcal{B}_1$ contains $\mathcal{A}_1$, and it is easy to check that $\mathcal{B}_1$ is a $\lambda$-system (for example, to check that $A, B \in \mathcal{B}_1$ and $A \subset B$ implies $B \backslash A \in \mathcal{B}_1$ we use only the fact that for such sets $\mathbb{P}(B \backslash A) = \mathbb{P}(B) - \mathbb{P}(A))$. Since $\mathcal{A}_1$ is a $\pi$-system, by the Monotone Class Theorem $\mathcal{B}_1$ contains the $\sigma(\mathcal{A}_1)$, and, consequently $\mathcal{G}_{u_i}$. Thus, (1.27) holds for $A_1 \in \mathcal{G}_{u_1}$ and $A_i, i = 2, ..., n$, of the form (1.28).

Now, we fix $A_1 \in \mathcal{G}_{u_1}$ and $A_i, i = 3, ..., n$, of the form (1.28), and consider the class $\mathcal{B}_2$ of events $A_2$ such that (1.27) holds. Repeating the argument presented above we conclude the this class contains $\mathcal{G}_{u_2}$, which means that (1.27) holds if $A_1$ belongs to $\mathcal{G}_{u_1}$, $A_2$ belongs to $\mathcal{G}_{u_2}$ and the remaining $A_i$ are of the form (1.28). Continuing in this way we obtain our claim. $\qquad \square$

**1.4.13** *A sequence of independent random variables with two values*
Let $\eta(t) = \sum_{i=0}^{\infty} 1_{[\frac{1}{2},1)}(t - i)$ and $X_k, k \geq 1$, be random variables on the standard probability space, given by $X_k(t) = \eta(2^k t), t \in [0, 1)$. For any $k$, $X_k$ attains only two values, 0 and 1, both with the same probability $\frac{1}{2}$. To be more precise, $X_k = 1$ if $t \in [\frac{2i+1}{2^{k+1}}, \frac{2i+2}{2^{k+1}})$ for some $0 \leq i < 2^k$ and 0 otherwise. $X_k$ are also independent. To show this note first that for any $n \geq 0$, the set of $t$ such that $X_1 = \delta_1, ..., X_n = \delta_n$ (where $\delta_j$ are either 0 or 1), equals $[\sum_{i=0}^{n} \frac{\delta_i}{2^{i+1}}, \sum_{i=0}^{n} \frac{\delta_i}{2^{i+1}} + \frac{1}{2^n})$. Therefore,

$$leb\{X_1 = \delta_1, ..., X_n = \delta_n\} = \frac{1}{2^n} = \prod_{i=1}^{n} leb\{X_i = \delta_i\}. \qquad (1.29)$$

Moreover,

$$\begin{aligned} leb\{X_2 = \delta_2, ..., X_n = \delta_n\} &= leb\{X_1 = 1, X_2 = \delta_2, ..., X_n = \delta_n\} \\ &\quad + leb\{X_1 = 0, X_2 = \delta_2, ..., X_n = \delta_n\} \\ &= \frac{1}{2^n} + \frac{1}{2^n} = \frac{1}{2^{n-1}} = \prod_{i=2}^{n} leb\{X_i = \delta_i\}. \end{aligned}$$

In a similar fashion, we may remove any number of variables $X_j$ from formula (1.29) to get

$$leb\{X_{i_1} = \delta_1, ..., X_{i_k} = \delta_k\} = \frac{1}{2^n} = \prod_{j=1}^{n} leb\{X_{i_j} = \delta_j\},$$

which proves our claim.

**1.4.14 A sequence of independent uniformly-distributed random variables** The notion of independence does not involve the order in the index set $\mathbb{T}$. Reordering the sequence $X_k$ from 1.4.13 we obtain thus an infinite matrix $X_{n,m}, n, m \geq 1$, of independent random variables attaining two values 0 and 1 with the same probability $\frac{1}{2}$. Define $Y_n = \sum_{m=1}^{\infty} \frac{1}{2^m} X_{n,m}$. This series converges absolutely at any $t \in [0, 1)$, and the sum belongs to $[0, 1]$. Since for any $n \in \mathbb{N}$, $\sigma(Y_n)$ is included in the $\sigma$-algebra $\mathcal{F}_n$ generated by random variables $X_{n,m}$ $m \geq 1$, and $\sigma$-algebras $\mathcal{F}_n$ are independent by 1.4.12, the random variables $Y_n$ are independent. We claim that they are uniformly distributed on $[0, 1)$.

Let $\mu_n$ be the distribution of $Y_n$, and $\mu_{n,k}$ be the distribution of $\sum_{m=1}^{k} \frac{1}{2^m} X_{n,m}$. By the Lebesgue Dominated Convergence Theorem, for any continuous function $f$,

$$\int f \, d\mu_n = \int f(Y_n) \, dleb = \int_0^1 f(Y_n(s)) \, ds \qquad (1.30)$$

$$= \lim_{k \to \infty} \int_0^1 f\Big( \sum_{m=1}^{k} \frac{1}{2^m} X_{n,m}(s) \Big) \, ds =: \lim_{k \to \infty} \int f \, d\mu_{n,k}.$$

To determine the distribution $\mu_{n,k}$ and the above integral note that $\sum_{m=1}^{k} \frac{1}{2^m} X_{n,m}$ attains only $2^k$ values, from 0 to $1 - \frac{1}{2^k}$, each with probability $\frac{1}{2^k}$. Thus $\mu_{n,k}$ is the sum of point masses $\frac{1}{2^k}$ concentrated at the points $0, \frac{1}{2^k}, ..., 1 - \frac{1}{2^k}$. Hence, the last integral equals $\frac{1}{2^k} \sum_{i=0}^{2^k-1} f\big(\frac{i}{2^k}\big)$. This is, however, the approximating sum of the Riemann integral of the function $f$. Therefore, the limit in (1.30) equals $\int_0^1 f(s) \, ds$. On the other hand, this is the integral with respect to the distribution of a random variable that is uniformly distributed on $[0, 1)$. Our claim follows by 1.2.20.

**1.4.15 A sequence of independent normal random variables** From the random variables $Y_n$ we obtain easily a sequence of independent normal random variables. In fact, we consider $Z_n = \text{erf}(Y_n)$, where "erf" is the inverse of the increasing function $y(t) = \frac{1}{\sqrt{2\pi}} \int_{-\infty}^{t} e^{-\frac{s^2}{2}} \, ds$, that maps the real line into $(0,1)$.† $Z_n$ are independent since "erf" is continuous, so that $\sigma(Z_n) \subset \sigma(Y_n)$. Moreover, for any continuous function $f$ on $[0,1]$, changing the variables $s = y(t)$,

$$\int f\, d\mathbb{P}_{Z_n} = \int_{[0,1)} f(\mathrm{erf})\, d\mathbb{P} = \int_0^1 f(\mathrm{erf}(s))\, ds = \frac{1}{\sqrt{2\pi}} \int_{-\infty}^{\infty} f(t) e^{-\frac{t^2}{2}}\, dt,$$

and the claim follows.

**1.4.16 Exercise** Prove the existence of an infinite sequence of random variables with (a) exponential distribution (b) Poisson distribution.

**1.4.17 *A sequence of independent random variables*** If we are more careful, we may extend the argument from 1.4.15 to prove existence of a sequence of independent random variables with any given distribution. Indeed, it is just a question of choosing an appropriate function to play a role of "erf".

The distribution of a random variable $Y$ is uniquely determined by the non-decreasing, right-continuous function $y(t) = \mathbb{P}[Y \le t]$, satisfying (1.22) with $y(+\infty) = 1$, often called the **cumulative distribution function**. Therefore, it suffices, given a non-decreasing, right-continuous function $y$ satisfying (1.22) with $y(+\infty) = 1$, to construct a measurable function $x : [0,1] \to \mathbb{R}$ such that

$$leb\{s \in [0,1] : x(s) \le t\} = y(t), \quad t \in \mathbb{R}. \tag{1.31}$$

Indeed, if this can be done then for a sequence $Y_n, n \ge 1$ of independent random variables with uniform distribution in $[0,1)$ we may define $Z_n = x(Y_n)$, $n \ge 1$. Since $\sigma(Z_n) \subset \sigma(Y_n)$, $Z_n$ are independent and have cumulative distributions function equal to $y$, for we have

$$\mathbb{P}[Z_n \le t] = \mathbb{P}[x(Y_n) \le t] = \mathbb{P}_{Y_n}[s : x(s) \le t]$$
$$= leb\{s : x(s) \le t\} = y(t),$$

as desired.

Coming back to the question of existence of a function $y$ satisfying (1.31), note that if we require additionally that $x$ be non-decreasing and left-continuous, this relation holds iff

$$\{s \in [0,1]; x(s) \le t\} = [0, y(t)].$$

Thus, condition $s \le y(t)$ holds iff $x(s) \le t$, and for any $s$ we have

$$\{t : s \le y(t)\} = \{t; x(s) \le t\} = [x(s), \infty).$$

† This is somewhat non-standard but useful notation.

Therefore, we must have

$$x(s) = \inf\{t : s \le y(t)\}. \tag{1.32}$$

Note that for $x$ thus defined,

$$x(s) \le t \Leftrightarrow \inf\{u; y(u) \ge s\} \le t$$
$$\Leftrightarrow (t, \infty) \subset \{u; y(u) \ge s\}$$
$$\Leftrightarrow y(t) = y(t+) \ge s,$$

since $y$ is non-decreasing. This implies that $x$ is measurable and that (1.31) holds. Note that we do not need to know whether $x$ is or is not left-continuous; we have used the assumption of left-continuity to infer (1.32) from (1.31), but do not require it in proving that (1.32) implies (1.31).

**1.4.18 Exercise** We will say that a random variable $Y$ has a modified Bernoulli distribution with parameter $0 \le p \le 1$ iff $\frac{1}{2}(Y+1)$ is Bernoulli r.v. with the same parameter. In other words, $\mathbb{P}\{Y = 1\} = p, \mathbb{P}\{Y = -1\} = q = 1 - p$. Suppose that we have two sequences $X_n, n \ge 1$ and $Y_n, n \ge 1$ of mutually independent random variables such that all $X_n$ have the same distribution with $E X = m$ and $D^2 X_n = \sigma^2$, and that $Y_n$ all have the modified Bernoulli distribution with parameter $0 < p < 1$. Then $(X_n, Y_n)$ are random vectors with values in the Kisyński group. Let $Z_n, n \ge 1$ be defined by the formula: $(X_1, Y_1) \circ ... \circ (X_n, Y_n) = (Z_n, \prod_{i=1}^{n} Y_i)$. Show that (a) $\prod_{i=1}^{n} Y_i$ is a modified Bernoulli variable with parameter $p_n = \frac{1}{2}(p-q)^n + \frac{1}{2}$, (b) $E Z_n = \frac{m}{2q}(1 - (p-q)^n)$, and (c) $D^2 Z_n = n\sigma^2 + 4pq \sum_{i=1}^{n-1} (E Z_i)^2$, so that $\lim_{n \to \infty} \frac{D^2 Z_n}{n} = \sigma^2 + \frac{p}{q} m^2$.

## 1.5 Convex functions. Hölder and Minkowski inequalities

**1.5.1 Definition** Let $(a, b)$ be an interval (possibly unbounded: $a = -\infty$ and/or $b = \infty$). A function $\phi$ is termed **convex** if for all $u, v \in (a, b)$ and all $0 \le \alpha \le 1$,

$$\phi(\alpha u + (1 - \alpha)v) \le \alpha\phi(u) + (1 - \alpha)\phi(v). \tag{1.33}$$

**1.5.2 Exercise** Show that $\phi$ is convex in $(a, b)$ iff for all $a < u_1 \le u_2 \le u_3 < b$,

$$\phi(u_2) \le \frac{u_2 - u_1}{u_3 - u_1}\phi(u_3) + \frac{u_3 - u_2}{u_3 - u_1}\phi(u_2). \tag{1.34}$$

**1.5.3 Exercise** (a) Assume $\phi$ is convex in $(a, b)$. Define $\tilde{\phi}(u) = \phi(a + b - u)$. (If $a = -\infty, b = \infty$, put $a + b = 0$.) Show that $\tilde{\phi}$ is convex. (b) For convex $\phi$ on the real line and $t \in \mathbb{R}$, define $\overline{\phi}(u) = \phi(2t - u)$. Prove that $\overline{\phi}$ is convex.

**1.5.4 Lemma** Suppose $\phi$ is convex in $(a, b)$ and let $u \in (a, b)$. Define

$$f(s) = f_{\phi,u}(s) = \frac{\phi(u) - \phi(s)}{u - s}, \quad s \in (a, u),$$

$$g(t) = g_{\phi,u}(t) = \frac{\phi(t) - \phi(u)}{t - u}, \quad t \in (u, b). \tag{1.35}$$

Then (a) $f$ and $g$ are non-decreasing, and (b) $f(s) \le g(t)$ for any $s$ and $t$ from the domains of $f$ and $g$, respectively.

*Proof* To prove the statement for $f$, we take $a < s < s' < u$ and do some algebra using (1.34) with $u_1 = s_1, u_2 = s_2$ and $u_3 = u$. To prove the corresponding statement for $g$ we either proceed similarly, or note that $g_\phi(s) = -f_{\tilde{\phi}, a+b-u}(a + b - s)$. Indeed,

$$f_{\tilde{\phi}, a+b-u}(a + b - s) = \frac{\tilde{\phi}(a + b - u) - \tilde{\phi}(a + b - s)}{a + b - u - (a + b - s)} = \frac{\phi(u) - \phi(s)}{s - u}.$$

Finally, (b) follows from (1.34), with $u_1 = s, u_2 = u, u_3 = t$. $\qquad\square$

**1.5.5 Proposition** Convex functions are continuous.

*Proof* By 1.5.4, for any $u \in (a, b)$ there exist right-hand side and left-hand side derivatives of $\phi$ at $u$, which implies our claim. Note that the left-hand side derivative may be smaller than the right-hand side derivative: consider $\phi(u) = |u|$ at $u = 0$. $\qquad\square$

**1.5.6 Proposition** Let $\phi$ be a convex function on $(a, b)$, and let $S$ be the family of linear functions $\psi(t) = a + bt$ such that $\psi(t) \le \phi(t)$, $t \in (a, b)$. Furthermore, let $S_0 = \{\psi \in S | \psi(t) = at + b, a, b \in \mathbb{Q}\}$. Then (a) $\phi(t) = \sup_{\psi \in S} \psi(t)$, and (b) if $\phi$ is not linear itself, then $\phi(t) = \sup_{\psi \in S_0} \psi(t)$.

*Proof* (a) Obviously, $\phi(t) \ge \sup_{\psi \in S} \psi(t)$, so it is enough to show that for any $t \in (a, b)$ there exists a $\psi_t \in S$ such that $\psi_t(t) = \phi(t)$. We claim that we may take $\psi_t(s) = q(s - t) + \phi(t)$, where $q$ is any number bigger than the left-hand derivative $\phi'_-(t)$ of $\phi$ at $t$, and smaller than the right-hand derivative $\phi'_+(t)$ of $\phi$ at $t$. Indeed, $\psi_t(t) = \phi(t)$, and by 1.5.4, for $s \geq t$,

$$\frac{\psi_t(s) - \phi(t)}{s - t} = q \leq \phi'_+(t) \leq \frac{\phi(s) - \phi(t)}{s - t}$$

which implies $\psi_t(s) \leq \phi(s)$. For $s \leq t$, the argument is the same, or we could argue using the function $\tilde{\phi}$ defined in 1.5.3 to reduce the problem to the case $s \geq t$.

(b) Let $t \in (a, b)$ and let $\psi_t$ be the function defined in (a). Since $\phi$ is not linear, $\frac{\phi(s)-\phi(t)}{s-t}$ is not equal to $q$ for some $s \in (a, b)$. Without loss of generality, we may assume that $s > t$ (if $s < t$, consider the function $\tilde{\phi}(u) = \phi(2t - u)$, which is convex also, and note that it is enough to show (b) for $\tilde{\phi}$ instead of $\phi$). The claim will be proven if we show that for any $\epsilon > 0$, and sufficiently small $h > 0$, the function

$$\psi_{t,\epsilon,h}(u) = (q + h)(u - t) + \phi(t) - \epsilon = \psi_t(u) + h(u - t) - \epsilon \quad (1.36)$$

belongs to $S$. We take $h < \min(\frac{\epsilon}{s-t}, \frac{\phi(s)-\phi(t)}{s-t} - q)$. Note that $\frac{\phi(s)-\phi(t)}{s-t} > q$, by 1.5.4.

For $u \leq s$, $\psi_{t,\epsilon,h}(u) \leq \psi_t(u) \leq \phi(u)$ since $h(u-t)-\epsilon \leq h(s-t)-\epsilon \leq 0$. For $u > t$, by 1.5.4, $\frac{\phi(u)-\phi(t)}{u-t} \geq \frac{\phi(s)-\phi(t)}{s-t} > q + h$. Thus,

$$\psi_{t,\epsilon,h}(u) \leq \frac{\phi(u) - \phi(t)}{u - t}(u - t) + \phi(t) - \epsilon \leq \phi(u).$$

$\square$

**1.5.7 Proposition** If $\phi$ is continuously differentiable with $\phi'$ increasing, then $\phi$ is convex.

*Proof* Fix $u \in (a, b)$ and consider $f_{\phi,u}$ defined in (1.35). For $s < u$, there exists a $s < \theta < u$ such that $-\phi(\theta)(u - s) + \phi(u) - \phi(s) = 0$. Thus, $f'_{\phi,u}(s) = \frac{-\phi(s)(u-s)+\phi(u)-\phi(s)}{(u-s)^2} \geq 0$, proving that $f_{\phi,u}$ is non-decreasing in $(a, u)$. Reversing the argument from 1.5.4 we prove that this implies the thesis. $\square$

**1.5.8 *Hölder inequality*** Let $(\Omega, \mathcal{F}, \mu)$ be a measure space, and let $x, y$ be two measurable functions on $\Omega$ with values in $[0, \infty]$. Suppose that $\frac{1}{p} + \frac{1}{q} = 1, p > 1$. Then

$$\int_\Omega xy \, d\mu \leq \left(\int_\Omega x^p \, d\mu\right)^{\frac{1}{p}} \left(\int_\Omega y^q \, d\mu\right)^{\frac{1}{q}}.$$

*Proof* Let $K = \left(\int_\Omega x^p \, \mathrm{d}\mu\right)^{\frac{1}{p}}$ and $L = \left(\int_\Omega y^q \, \mathrm{d}\mu\right)^{\frac{1}{q}}$. Without loss of generality we may assume that both $K$ and $L$ are finite and non-zero since if $K > 0, L = \infty$ or $K = \infty, L > 0$, there is nothing to prove, or if $K = 0$ (respectively $L = 0$), then $x = 0 \; \mu$ a.e. ($y = 0 \; \mu$ a.e.) so that the left hand-side equals zero too.

Let, $X = x/K, Y = y/L$. Then $\int_\Omega X^p \, \mathrm{d}\mu = \int_\Omega Y^q \, \mathrm{d}\mu = 1$. Note also that $\int_\Omega XY \, \mathrm{d}\mu = \int_B XY \, \mathrm{d}\mu$, where $B = \{\omega | X(\omega)Y(\omega) > 0\}$. On $B$ we may define the functions $a(\omega)$ and $b(\omega)$ such that $X(\omega) = e^{a(\omega)/p}$ and $Y(\omega) = e^{b(\omega)/q}$. Since $\phi(s) = e^s$ is a convex function (see 1.5.7),

$$
\begin{aligned}
X(\omega)Y(\omega) &= e^{a(\omega)/p + b(\omega)/q} \leq \frac{1}{p} e^{a(\omega)} + \frac{1}{q} e^{b(\omega)} \\
&= \frac{1}{p} X^p(\omega) + \frac{1}{q} Y^q(\omega).
\end{aligned}
$$

Integrating over $B$ we get

$$
\begin{aligned}
\int_\Omega XY \, \mathrm{d}\mu &= \int_B XY \, \mathrm{d}\mu \leq \frac{1}{p} \int_B X^p \, \mathrm{d}\mu + \frac{1}{p} \int_B Y^q \, \mathrm{d}\mu \\
&= \frac{1}{p} \int_B X^p \, \mathrm{d}\mu + \frac{1}{p} \int_B Y^q \, \mathrm{d}\mu = \frac{1}{p} + \frac{1}{q} = 1,
\end{aligned}
$$

which gives the thesis. $\qquad\square$

1.5.9 *Minkowski inequality*   Under notations of 1.5.8,

$$
\left(\int_\Omega (x+y)^p \, \mathrm{d}\mu\right)^{\frac{1}{p}} \leq \left(\int_\Omega x^p \, \mathrm{d}\mu\right)^{\frac{1}{p}} + \left(\int_\Omega y^p \, \mathrm{d}\mu\right)^{\frac{1}{p}}.
$$

*Proof* By the Hölder inequality,

$$
\begin{aligned}
\int_\Omega (x+y)^p \, \mathrm{d}\mu &= \int_\Omega x(x+y)^{p-1} \, \mathrm{d}\mu + \int_\Omega y(x+y)^{p-1} \, \mathrm{d}\mu \\
&\leq \left(\int_\Omega x^p \, \mathrm{d}\mu\right)^{\frac{1}{p}} \left(\int_\Omega (x+y)^{q(p-1)} \, \mathrm{d}\mu\right)^{\frac{1}{q}} \\
&\quad + \left(\int_\Omega y^p \, \mathrm{d}\mu\right)^{\frac{1}{p}} \left(\int_\Omega (x+y)^{q(p-1)} \, \mathrm{d}\mu\right)^{\frac{1}{q}} \\
&= \left[\left(\int_\Omega x^p \, \mathrm{d}\mu\right)^{\frac{1}{p}} + \left(\int_\Omega y^p \, \mathrm{d}\mu\right)^{\frac{1}{p}}\right] \left(\int_\Omega (x+y)^p \, \mathrm{d}\mu\right)^{\frac{1}{q}}.
\end{aligned}
$$

The thesis follows by dividing both sides by the last term above (note that if this term is zero, there is nothing to prove.) $\qquad\square$

## 1.6 The Cauchy equation

The content of this section is not needed in what follows but it provides better insight into the results of Chapter 6 and Chapter 7. Plus, it contains a beautiful Theorem of Steinhaus. The casual reader may skip this section on the first reading (and on the second and on the third one as well, if he/she ever reads this book that many times). The main theorem of this section is 1.6.11.

**1.6.1 Exercise** Let $(\Omega, \mathcal{F}, \mu)$ be a measure space. Show that for all measurable sets $A, B$ and $C$

$$|\mu(A \cap B) - \mu(C \cap B)| \le \mu(A \div C).$$

Here $\div$ denotes the symmetric difference of two sets defined as $A \div B = (A \setminus B) \cup (B \setminus A)$.

**1.6.2 Lemma** If $A \subset \mathbb{R}$ is compact and $B \subset \mathbb{R}$ is Lebesgue measurable, than $x(t) = leb(A_t \cap B)$ is continuous, where $A_t$ is a translation of the set $A$ as defined in (1.4).

*Proof* By Exercise 1.6.1,

$$|leb(A_{t+h} \cap B) - leb(A_t \cap B)| \le leb(A_{t+h} \div A_t)$$
$$= leb(A_h \div A)_t = leb(A_h \div A), \quad t, h \in \mathbb{R},$$

since Lebesgue measure is translation invariant. Therefore it suffices to show that given $\epsilon > 0$ there exists a $\delta > 0$ such that $leb(A_h \div A) < \epsilon$ provided $|h| < \delta$. To this end let $G$ be an open set such that $leb(G \setminus A) < \frac{\epsilon}{2}$, and take $\delta = \min_{a \in A} \min_{b \in G^{\complement}} |a - b|$. This is a positive number since $A$ is compact, $G^{\complement}$ is closed, and $A$ and $G^{\complement}$ are disjoint (see Exercise 1.6.3 below). If $|h| < \delta$, then $A_h \subset G$. Hence,

$$leb(A_h \setminus A) < leb(G \setminus A) < \frac{\epsilon}{2},$$

and

$$leb(A \setminus A_h) = leb(A \setminus A_h)_{-h} = leb(A_{-h} \setminus A) < \frac{\epsilon}{2},$$

as desired. $\qquad\square$

**1.6.3 Exercise** Show that if $A$ and $B$ are disjoint subsets of a metric space $(\mathbb{X}, d)$, $A$ is compact, and $B$ is closed, then $\delta = \min_{a \in A} \min_{b \in B} |a - b|$ is positive. Show by example that the statement is not true if $A$ is closed but fails to be compact.

**1.6.4** *The Steinhaus Theorem*    If $A$ and $B$ are Lebesgue measurable subsets of $\mathbb{R}$ such that $leb(A) > 0$ and $leb(B) > 0$, then $A + B = \{c \in \mathbb{R} | c = a + b, a \in A, b \in B\}$ contains an open interval.

*Proof* Since Lebesgue measure is regular, there exists a compact subset of $A$ with positive measure. Therefore it suffices to show the theorem under additional assumption that $A$ is compact. Of course, $C = -A = \{u, -u \in A\}$ is then compact also. By 1.6.2, $x(t) = leb(C_t \cap B)$ is a continuous, in particular measurable, function. On the other hand, $x = 1_A * 1_B$, so that $\int_{-\infty}^{\infty} x(t)\, dt = leb(A)\, leb(B) > 0$.

This implies that there exists a point $t_0$ such that $x(t_0) > 0$. Since $x$ is continuous, there exists an interval $(t_0 - \delta, t_0 + \delta)$, $\delta > 0$ in which $x$ assumes only positive values. Hence $leb(C_t \cap B) > 0$ for $t$ in this interval, and in particular $C_t \cap B$ is non-empty. Thus, for any $t \in (t_0 - \delta, t_0 + \delta)$ there exists $b \in B$ and $a \in A$ such that $-a + t = b$. This shows that this interval is contained in $A + B$, as desired.                                    $\square$

**1.6.5** *The Cauchy equation*    A function $x : \mathbb{R}^+ \to \mathbb{R}$ is said to satisfy the **Cauchy equation** if

$$x(s + t) = x(s) + x(t), \qquad s, t > 0. \tag{1.37}$$

An example of such a function is $x(t) = at$, where $a \in \mathbb{R}$, and it turns out that there are no other simple examples (see 1.6.11 and 1.6.12 below). Functions that satisfy (1.37) and are not of this form are very strange (and thus very interesting for many mathematicians). In particular, it is easy to see that (1.37) implies that

$$x\left(\frac{k}{n}t\right) = \frac{k}{n}x(t), \quad t \in \mathbb{R}^+, k, n \in \mathbb{N}. \tag{1.38}$$

Therefore, if $x$ satisfies (1.37) and is continuous we may take $t = 1$ in (1.38) and approximate a given $s \in \mathbb{R}^+$ by rational numbers to obtain

$$x(s) = x(1)s. \tag{1.39}$$

We need, however, a stronger result. Specifically, we want to show that all *measurable* functions that satisfy (1.37) are of the form (1.39). To this end, we need the Steinhaus Theorem and the lemmas presented below. The reader should start by solving the next exercise.

**1.6.6 Exercise**    Prove that (1.37) implies (1.38).

**1.6.7 Lemma**    Suppose that $x$ satisfies (1.37) and is bounded from above in an interval $(t_0 - \delta, t_0 + \delta)$, where $\delta < t_0$, i.e. $x(t) \le M$ for some $M \in \mathbb{R}$ and all $t$ in this interval. Then $x$ is bounded from below in $(t_0 - \delta, t_0 + \delta)$, i.e. $x(t) \ge N$ for some $N \in \mathbb{R}$ and $t$ in this interval.

*Proof* For $t \in (t_0 - \delta, t_0 + \delta)$, let $t' = 2t_0 - t \in (t_0 - \delta, t_0 + \delta)$. We have $x(t_0) = \frac{1}{2}[x(t) + x(t')]$, so that $x(t) = 2x(t_0) - x(t') \ge 2x(t_0) - M$. In other words, we may choose $N = 2x(t_0) - M$.    □

**1.6.8 Exercise**    Show that if $x$ satisfies (1.37) and is bounded from below in an interval, then it is bounded from above in this interval.

**1.6.9 Lemma**    Suppose that $x$ satisfies (1.37) and is bounded in an interval $(t_0 - \delta, t_0 + \delta)$, i.e. $|x(t)| \le M$ for some $M \in \mathbb{R}$ and all $t$ in this interval. Then

$$|x(t) - x(t_0)| \le \frac{M}{\delta}|t - t_0|, \qquad t \in (t_0 - \delta, t_0 + \delta). \tag{1.40}$$

In particular, $x$ is continuous at $t_0$.

*Proof* Observe that if $t$ and $t'$ are as in the proof of Lemma 1.6.7, then $|x(t) - x(t_0)| = |x(t') - x(t_0)|$ and $|t - t_0| = |t' - t_0|$. Thus, we may restrict our attention to $t \in (t_0 - \delta, t_0)$. Let $t_n \in (t_0 - \delta, t_0), n \ge 1$, converge to $t_0 - \delta$. We may choose $t_n$ in such a way that $\alpha_n = \frac{t_0 - t}{t_0 - t_n}, n \ge 1$, are rational. Since $t = (1 - \alpha_n)t_0 + \alpha_n t_n$, $x(t) = (1 - \alpha_n)x(t_0) + \alpha_n x(t_n)$, and we obtain

$$x(t) - x(t_0) = \alpha_n[x(t_n) - x(t_0)] \le 2M\alpha_n = \frac{2M}{t_0 - t_n}(t_0 - t). \tag{1.41}$$

Letting $t_n \to t_0 - \delta$,

$$x(t) - x(t_0) \le \frac{2M}{\delta}(t_0 - t). \tag{1.42}$$

Analogously, if $t'_n \in (t_0, t_0 + \delta)$ tends to $t_0 + \delta$ in such a way that $\alpha_n = \frac{t_0 - t}{t'_n - t}$ is rational, then $t_0 = \alpha'_n t'_n + (1 - \alpha'_n)t$, and $x(t_0) - x(t) = \alpha'_n(x(t'_n) - x(t))$. Moreover,

$$x(t_0) - x(t) \le \lim_{n \to \infty} \frac{2M}{t'_n - t}(t_0 - t) = \frac{2M}{\delta}(t_0 - t). \tag{1.43}$$

□

**1.6.10 Lemma**    If $x$ satisfies (1.37) and is bounded from above in an interval $(t_0 - \delta, t_0 + \delta)$, where $t_0 > \delta > 0$, then it is also bounded from above in intervals $(t_1 - \delta_1, t_1 + \delta_1)$, where $t_1 \in \mathbb{R}$, and $\delta_1 = \delta_1(t_1) = \min(t_1, \delta)$.

*Proof* Let $t \in (t_1 - \delta, t_1 + \delta)$, where $t_1 > \delta$. Define $t' = t + t_0 - t_1 \in (t_0 - \delta, t_0 + \delta)$. We have

$$x(t) = x(t') + \text{sgn}(t_1 - t_0)x(|t_1 - t_0|) \tag{1.44}$$

$$\leq \sup_{t' \in (t_0 - \delta, t_0 + \delta)} x(t') + \text{sgn}(t_1 - t_0)x(|t_1 - t_0|), \tag{1.45}$$

i.e. $x$ is bounded from above in $t \in (t_1 - \delta, t_1 + \delta)$. Recall that

$$\text{sgn}(\tau) = \begin{cases} 1, & \tau > 0, \\ 0, & \tau = 0, \\ -1, & \tau < 0. \end{cases} \tag{1.46}$$

The case $t_1 \leq \delta$ is proved similarly.                          $\square$

**1.6.11 Theorem**    If a measurable function $x$ satisfies (1.37) then it is of the form (1.39).

*Proof* It suffices to show that $x$ is continuous. Since $\mathbb{R}^+ = \bigcup_{n=1}^{\infty}\{t; x(t) \leq n\}$, there exists an $n \in \mathbb{N}$ such that $leb\{t; x(t) \leq n\} > 0$. Let $A = \{t; x(t) \leq n\} + \{t; x(t) \leq n\}$. For $t \in A \subset \mathbb{R}^+$, $x(t) \leq 2n$. By the Steinhaus Theorem, $A$ contains an interval $(t_0 - \delta, t_0 + \delta)$, and so $x$ is bounded from above in this interval. By Lemma 1.6.10, for any $t_1 > 0$ there exists an interval $(t_1 - \delta_1, t_1 + \delta_1)$ where $\delta_1 > 0$ in which $x$ is bounded. By Lemma 1.6.7, this implies that $x$ is bounded in these intervals, and by Lemma 1.6.9, $x$ is continuous at every $t_1 > 0$.                          $\square$

**1.6.12 Corollary**    Our argument shows that a function $x$ that satisfies (1.37) and is not measurable must be unbounded in any open interval.

**1.6.13 Exercise**    Let $y : \mathbb{R}^+ \to \mathbb{R}$ be a measurable function such that $x(s) := \lim_{t \to \infty} \frac{y(t+s)}{y(s)}$ exists for all $s > 0$. Show that we must have $x(s) = e^{as}$ for some real $a$.

# 2

# Basic notions in functional analysis

A characteristic of functional analysis is that it does not see functions, sequences, or measures as isolated objects but as elements or points in a *space* of functions, a *space* of sequences, or a *space* of measures. In a sense, for a functional analyst, particular properties of a certain probability measure are not important; rather, properties of the whole space or of a certain *subspace* of such measures are important. To prove existence or a property of an object or a group of objects, we would like to do it by examining general properties of the whole space, not by examining these objects separately. There is both beauty and power in this approach. We hope that this crucial point of view will become evident to the reader while he/she progresses through this chapter and through the whole book.

## 2.1 Linear spaces

The central notion of functional analysis is that of a Banach space. There are two components of this notion: algebraic and topological. The algebraic component describes the fact that elements of a Banach space may be meaningfully added together and multiplied by scalars. For example, given two random variables, $X$ and $Y$, say, we may think of random variables $X + Y$ and $\alpha X$ (and $\alpha Y$) where $\alpha \in \mathbb{R}$. In a similar way, we may think of the sum of two measures and the product of a scalar and a measure. Abstract sets with such algebraic structure, introduced in more detail in this section, are known as linear spaces. The topological component of the notion of a Banach space will be discussed in Section 2.2.

37

**2.1.1 Definition**     Let $\mathbb{X}$ be a set; its elements will be denoted $x, y, z$, etc. A triple $(\mathbb{X}, +, \cdot)$, where $+$ is a map $+ : \mathbb{X} \times \mathbb{X} \to \mathbb{X}, (x, y) \mapsto x + y$ and $\cdot$ is a map $\cdot : \mathbb{R} \times \mathbb{X} \to \mathbb{X}, (\alpha, x) \mapsto \alpha x$, is called a **(real) linear space** if the following conditions are satisfied:

(a1)  $(x + y) + z = x + (y + z)$, for all $x, y, z \in \mathbb{X}$,

(a2)  there exists $\Theta \in \mathbb{X}$ such that $x + \Theta = x$, for all $x \in \mathbb{X}$,

(a3)  for all $x \in \mathbb{X}$ there exists an $x' \in \mathbb{X}$ such that $x + x' = \Theta$,

(a4)  $x + y = y + x$, for all $x, y \in \mathbb{X}$,

(m1)  $\alpha(\beta x) = (\alpha\beta)x$, for all $\alpha, \beta \in \mathbb{R}, x \in \mathbb{X}$,

(m2)  $1x = x$, for all $x \in \mathbb{X}$,

(d)  $\alpha(x + y) = \alpha x + \alpha y$, and $(\alpha + \beta)x = \alpha x + \beta x$ for all $\alpha, \beta \in \mathbb{R}$ and $x, y \in \mathbb{X}$.

Conditions (a1)–(a4) mean that $(\mathbb{X}, +)$ is a commutative group. Quite often, for the sake of simplicity, when no confusion ensues, we will say that $\mathbb{X}$ itself is a linear space.

**2.1.2 Exercise**     Conditions (a2) and (a4) imply that the element $\Theta$, called the zero vector, or the zero, is unique.

**2.1.3 Exercise**     Conditions (a1) and (a3)–(a4) imply that for any $x \in \mathbb{X}$, $x'$ is determined uniquely.

**2.1.4 Exercise**     Conditions (d), (a1) and (a3) imply that for any $x \in \mathbb{X}$, $0x = \Theta$.

**2.1.5 Exercise**     2.1.3, and 2.1.1 (d), (m2) imply that for any $x \in \mathbb{X}$, $x' = (-1)x$. Because of this fact, we will adopt the commonly used notation $x' = -x$.

**2.1.6 Example**     Let $S$ be a set. The set $\mathbb{X} = \mathbb{R}^S$ of real-valued functions defined on $S$ is a linear space, if addition and multiplication are defined as follows: $(x + y)(p) = x(p) + y(p)$, $(\alpha x)(p) = \alpha x(p)$, for all $x(\cdot), y(\cdot) \in \mathbb{R}^S$, $\alpha \in \mathbb{R}$, and $p \in S$. In particular, the zero vector $\Theta$ is a function $x(p) \equiv 0$, and $-x$ is defined by $(-x)(p) \equiv -x(p)$. This example includes a number of interesting subcases: (a) if $S = \mathbb{N}$, $\mathbb{R}^\mathbb{N}$ is the space of real-valued sequences, (b) if $S = \mathbb{R}$, $\mathbb{R}^\mathbb{R}$ is the space of real functions on $\mathbb{R}$, (c) if $S = \{1, ..., n\} \times \{1, 2, ..., k\}$, $\mathbb{R}^S$ is the space of real $n \times k$ matrices, etc.

2.1.7 *Linear maps* A map $L$ from a non-empty subset $\mathcal{D}(L)$ (the domain of $L$) of a linear space $\mathbb{X}$ to a linear space $\mathbb{Y}$ is **linear** if for all $\alpha, \beta \in \mathbb{R}$ and $x, y$ in $\mathcal{D}(L)$, $\alpha x + \beta y$ belongs to $\mathcal{D}(A)$, and $L(\alpha x + \beta y) = \alpha L(x) + \beta L(y)$. Note that the operations $+, \cdot$ on the left-hand side of this equation are performed in $\mathbb{X}$ while those on the right-hand side are operations in $\mathbb{Y}$. With linear operations it is customary to omit parentheses for the argument so that we write $Lx$ instead $L(x)$. Note that the definition implies that $\Theta$ belongs to $\mathcal{D}(L)$ and $L\Theta = \Theta$, where again the $\Theta$ on the left hand-side is the zero in $\mathbb{X}$, while that on the right-hand side is in $\mathbb{Y}$. In the sequel we shall write such equations without making these distinctions, and if the reader keeps these remarks in mind, there should be no confusion. A linear map $L : \mathbb{X} \to \mathbb{Y}$ is called an **algebraic isomorphism** of $\mathbb{X}$ and $\mathbb{Y}$, if it is one-to-one and onto. (In particular, $L^{-1}$ exists and is linear.) If such a map exists, $\mathbb{X}$ and $\mathbb{Y}$ are said to be **algebraically isomorphic.**

2.1.8 **Example** The collection $L(\mathbb{X}, \mathbb{Y})$ of linear maps from a linear space $\mathbb{X}$ (with domain equal to $\mathbb{X}$) to a linear space $\mathbb{Y}$ is a linear space itself, provided we define

$$(\alpha L)x = \alpha Lx \qquad \text{and} \qquad (L + M)x = Lx + Mx$$

for $L, M \in L(\mathbb{X}, \mathbb{Y})$ and $\alpha \in \mathbb{R}$. The reader is encouraged to check this.

2.1.9 *Algebraic subspace* A subset $\mathbb{Y}$ of a linear space $\mathbb{X}$ is called an **algebraic subspace** of $\mathbb{X}$ if for all $x, y \in \mathbb{Y}$, and $\alpha \in \mathbb{R}$, $x + y$ and $\alpha x$ belong to $\mathbb{Y}$. Observe that $\mathbb{Y}$ with addition and multiplication restricted to $\mathbb{Y}$ is itself a linear space.

2.1.10 **Example** Let $l^p$, $p > 0$ denote the space of sequences $x = (x_n)_{n \geq 1}$, such that $\sum_{n=1}^{\infty} |x_n|^p < \infty$. $l^p$ is a subspace of the space $\mathbb{R}^{\mathbb{N}}$. Indeed, denoting $f(x) = \sum_{n=1}^{\infty} |x_n|^p < \infty$, we have $f(\alpha x) = |\alpha|^p f(x)$ and $f(x+y) \leq 2^p(f(x)+f(y))$, where $x = (x_n)_{n \geq 1}, y = (y_n)_{n \geq 1}$. The last inequality follows directly from the estimate $|x + y|^p \leq 2^p(|x|^p + |y|^p)$, which can be proved by considering the cases $|y| \geq |x|$ and $|x| < |y|$ separately.

2.1.11 **Example** Recall that a function $x : S \to \mathbb{R}$ is said to be **bounded** if there exists an $M > 0$ such that $\sup_{p \in S} |x(p)| \leq M$. The space $B(S)$ of bounded functions is an algebraic subspace of $\mathbb{R}^S$ since if $x$ and $y$ are bounded by $M_x$ and $M_y$, respectively, then $\alpha x + \beta y$ is bounded by $|\alpha|M_x + |\beta|M_y$.

2.1.12 **Example**    Let $S$ be a topological space. The space $C(S)$ of real continuous functions is an algebraic subspace of $\mathbb{R}^S$. Similarly, if $S$ is a measurable space $(S, \mathcal{F})$, then the space $\mathcal{M}(S, \mathcal{F})$ (or $\mathcal{M}(S)$ if there is one obvious choice for $\mathcal{F}$) of real measurable functions is a subspace of $\mathbb{R}^S$. This just says that the sum of two continuous (measurable) functions is continuous (measurable), and that a continuous (measurable) function multiplied by a number is again continuous (measurable).

2.1.13 **Example**    Let $C^1(\mathbb{R})$ denote the set of differentiable functions. $C^1(\mathbb{R})$ is an algebraic subspace of $\mathbb{R}^{\mathbb{R}}$ and differentiation is a linear map from $C^1(\mathbb{R})$ to $\mathbb{R}^{\mathbb{R}}$.

2.1.14 **Exercise**    Let $L$ be a linear map from $\mathbb{X}$ to $\mathbb{Y}$. Show that (a) the domain $\mathcal{D}(L)$ is an algebraic subspace of $\mathbb{X}$, (b) the set $Ker\,L = \{x \in \mathbb{X} | Lx = 0\}$, called the **kernel** of $L$, is an algebraic subspace of $\mathbb{X}$, and (c) the set $Range\,L = \{y \in \mathbb{Y} | y = Lx, \text{ for some } x \in \mathbb{X}\}$, called the **range**, is an algebraic subspace of $\mathbb{Y}$.

2.1.15 **Definition**    Let $\mathbb{X}$ be a linear space and let $\mathbb{Y}$ be an algebraic subspace of $\mathbb{X}$. Consider the relation $\sim$ in $\mathbb{X}$, defined by

$$x \sim y \qquad \text{iff } x - y \in \mathbb{Y}.$$

Since $\mathbb{Y}$ is an algebraic subspace of $\mathbb{X}$, for any $x, y$ and $z \in \mathbb{X}$, we have (a) $x \sim y$ iff $y \sim x$, (b) $x \sim y$ and $y \sim z$ implies $x \sim z$, and (c) $x \sim x$. This means that $\sim$ is an **equivalence relation**. Let

$$[x] = \{y \in \mathbb{X} | x \sim y\}$$

be the **equivalence class** of $x$. (Note that for any $x$ and $y$ in $\mathbb{X}$, the classes $[x]$ and $[y]$ are either identical or disjoint, and that the union of all classes equals $\mathbb{X}$.) The set of equivalence classes is called the *quotient space* and denoted $\mathbb{X}/\mathbb{Y}$. We note that $\mathbb{X}/\mathbb{Y}$ is a linear space itself. Indeed, since $\mathbb{Y}$ is a subspace of $\mathbb{X}$, the classes $[x + y]$ and $[x' + y']$ coincide if $x \sim x'$ and $y \sim y'$, so that we may put $[x] + [y] = [x + y]$. Analogously, we note that we may put $\alpha[x] = [\alpha x]$ (in particular, that the definition does not depend on the choice of $x$ but only on $[x]$). It is easy to show that the conditions of Definition 2.1.1 are fulfilled; the zero of $\mathbb{X}/\mathbb{Y}$ is the space $\mathbb{Y}$. The map $x \mapsto [x]$ is called the **canonical map (canonical homomorphism)** of $\mathbb{X}$ onto $\mathbb{X}/\mathbb{Y}$. Notice that this map is linear.

**2.1.16 Definition**   If $A$ and $B$ are subsets of a linear subspace of $\mathbb{X}$, and $\alpha$ and $\beta$ are real numbers, then the set $\alpha A + \beta B$ is defined as $\{z \in \mathbb{X} | z = \alpha x + \beta y, \text{ for some } x \in A, y \in B\}$. In particular, if $A = \{x\}$, we write $x + \beta B$, instead of $\{x\} + \beta B$.

**2.1.17 Exercise**   Prove that any class in $\mathbb{X}/\mathbb{Y}$, different from $\mathbb{Y}$, may be represented as $x + \mathbb{Y}$ where $x \notin \mathbb{Y}$.

**2.1.18 Exercise**   Let $A$ and $B, A \subset B$ be subsets of a linear space $\mathbb{X}$, and let $x \in \mathbb{X}$. Show that $(B \setminus A) - x = (B - x) \setminus (A - x)$.

**2.1.19 Example**   Let $\mathbb{R}^{\mathbb{R}}$ be the space of real-valued functions defined on $\mathbb{R}$, and let $\mathbb{R}_{\mathrm{e}}^{\mathbb{R}}$ be its subset of even functions. We may check that $\mathbb{R}_{\mathrm{e}}^{\mathbb{R}}$ is an algebraic subspace of $\mathbb{R}^{\mathbb{R}}$. What is the quotient space $\mathbb{R}^{\mathbb{R}}/\mathbb{R}_{\mathrm{e}}^{\mathbb{R}}$? Note first that two functions are in the equivalence relation iff their difference is even. Secondly, in any equivalence class there exists at least one odd function. Indeed, any class contains at least one function $x$; and any function can be represented as $x = x_{\mathrm{e}} + x_{\mathrm{o}}$ where $x_{\mathrm{e}}(t) = [x(t) + x(-t)]/2$ is even and $x_{\mathrm{o}}(t) = [x(t) - x(-t)]/2$ is odd, so that $x$ is in relation with an odd function $x_{\mathrm{o}}$ (note that $x_{\mathrm{o}}$ may be zero, if $x$ itself is even). Moreover, there may not be more than one odd function in any class, for if there were two, their difference would have to be both even (by the definition of the equivalence relation) and odd (by the properties of odd functions), and hence zero. This suggests that $\mathbb{R}^{\mathbb{R}}/\mathbb{R}_{\mathrm{e}}^{\mathbb{R}}$ is algebraically isomorphic to the space $\mathbb{R}_{\mathrm{o}}^{\mathbb{R}}$ of odd functions on $\mathbb{R}$. The isomorphism maps a class to the unique odd function that belongs to the class. We have proved that this map is a one-to-one map and obviously it is onto. The reader is encouraged to check that it is linear.

**2.1.20 Exercise**   Let $S$ be a set and let $\mathbb{Y} \subset \mathbb{R}^{S}$ be the subspace of constant functions. Characterize $\mathbb{R}^{S}/\mathbb{Y}$.

**2.1.21 Exercise**   Let $L : \mathbb{X} \to \mathbb{Y}$ be a linear map. Show that $Range\, L$ is algebraically isomorphic to $\mathbb{X}/Ker\, L$.

**2.1.22 Example**   Suppose that $(\Omega, \mathcal{F})$ and $(\Omega', \mathcal{F}')$ are two measurable spaces, and let $f$ be a measurable map from $\Omega$ to $\Omega'$. Let $L : \mathcal{M}(\Omega') \to \mathcal{M}(\Omega)$ be given by $(Lx)(\omega) = x(f(\omega))$. $L$ is a linear map, and its range is the algebraic subspace $\mathcal{M}_f(\Omega)$ of $\mathcal{M}(\Omega)$ of functions $y(\omega)$ of the form $y(\omega) = x(f(\omega))$ where $x \in \mathcal{M}(\Omega')$. What is the kernel of $L$? It is the subspace of $\mathcal{M}(\Omega')$ of functions with the property that

$x(\omega') = 0$ for all $\omega' \in R_f$ (the range of $f$). The equivalence relation defined by means of $KerL$ identifies two functions that are equal on $R_f$. This suggests that $\mathcal{M}(\Omega')/KerL$ is algebraically isomorphic to the space of measurable functions on $R_f$. We have to be careful in stating this result, though, since $R_f$ may happen to be non-measurable in $\Omega'$. A natural way to make $R_f$ a measurable space is to equip it with the $\sigma$-algebra $\mathcal{F}_{R_f}$ of subsets of the form $R_f \cap B$ where $B \in \mathcal{F}'$. Using 2.1.21, we can then show that $\mathcal{M}_f(\Omega)$ is isomorphic to $\mathcal{M}(R_f, \mathcal{F}_{R_f})$.

**2.1.23 Exercise**   Take $\Omega = \Omega' = [0, 1]$, and let $\mathcal{F}$ be the Lebesgue measurable subsets, $\mathcal{F}' = \{\Omega, \emptyset\}$, and $f(\omega) = \frac{1}{2}$ and check to see that $f$ is measurable and that the range of $f$ is not measurable in $(\Omega', \mathcal{F}')$.

**2.1.24 *Doob–Dynkin Lemma***   A more fruitful and deeper result concerning $M_f(\Omega)$ is the following lemma due to Doob and Dynkin (see e.g. 3.2.5). With the notations of 2.1.22, $\mathcal{M}_f(\Omega)$ equals $\mathcal{M}(\Omega, \sigma(f))$.

*Proof* (Observe how the lattice structure of $\mathbb{R}$ is employed in the proof.) The relation $\mathcal{M}_f(\Omega) \subset \mathcal{M}(\Omega, \sigma(f))$ is obvious. To prove the opposite inclusion it is enough to show that any positive function from $\mathcal{M}(\Omega, \sigma(f))$ belongs to $\mathcal{M}_f(\Omega)$, since any function in this space is a difference of two positive functions. Now, the claim is true for simple functions $y = \sum_{i=1}^{n} a_i 1_{A_i}$ where the $a_i$ are constants and the $A_i$ belong to $\sigma(f)$. Indeed, the $A_i$ are of the form $f^{-1}(B_i)$ where the $B_i$ belong to $\mathcal{F}'$, so that we have $y(\omega) = x(f(\omega))$ where $x = \sum_{i=1}^{n} a_i 1_{B_i}$. Finally, if $y \in \mathcal{M}(\Omega, \sigma(f))$ is non-negative, then there exists a non-decreasing sequence of simple functions $y_n$ that converges pointwise to $y$. Let $x_n$ be the corresponding sequence of simple functions in $\mathcal{M}(\Omega')$ such that $x_n(f(\omega)) = y_n(\omega)$. Certainly, the idea is to prove that $x_n$ converges to some $x \in \mathcal{M}(\Omega')$ that satisfies $x \circ f = y$, so that $y = \mathcal{M}(\Omega)$. Note first that $x_n$ is non-decreasing on $R_f$. Indeed, for any $\omega'$ in this set there exists an $\omega \in \Omega$ such that $\omega' = f(\omega)$ and we have

$$x_n(\omega') = y_n(f(\omega)) \leq y_{n+1}(f(\omega)) = x_{n+1}(\omega).$$

However, it is hard to tell what happens outside of $R_f$; in particular we should not think that the sets $B_i$ defined above (for simple functions) are subsets of $R_f$; as a result $x_n$ may happen to be divergent outside of $R_f$. Certainly, the values of the limit function $x$ outside of $R_f$ do not matter as long as we make sure that $x$ is measurable. Were $R_f$ measurable we could bypass the difficulty by taking $x_n 1_{R_f}$ instead of

$x_n$. If $R_f$ is not measurable, we may consider the measurable set $C = \{\omega' | \lim_{n\to\infty} x_n(\omega') \text{ exists }\}$ and define $x(\omega) = \lim_{n\to\infty} x_n(\omega')$ for $\omega' \in C$, and zero otherwise. We already know that $R_f \subset C$, so that for $\omega \in \Omega$, $x(f(\omega)) = \lim_{n\to\infty} x_n(f(\omega)) = \lim_{n\to\infty} y_n(\omega) = y(\omega)$. $\qquad\square$

**2.1.25** *Linear combinations*    Let $x_i \in \mathbb{X}, i = 1, ..., n$ be given vectors, and let $\alpha_i \in \mathbb{R}$ be real numbers. A vector $y = \sum_{i=1}^{n} \alpha_i x_i$ is called a **linear combination** of $x_i \in \mathbb{X}, i = 1, ..., n$. If $\alpha_i$ are non-negative and satisfy $\sum_{i=1}^{n} \alpha_i = 1$, $y$ is called a **convex combination** of $x_i \in \mathbb{X}, i = 1, ..., n$. If $\mathbb{Y}$ is a subset of $\mathbb{X}$, then its **linear span**, or simply **span** $span\,\mathbb{Y}$ is defined to be the set of all linear combinations of vectors $x_i \in \mathbb{Y}, i = 1, ..., n$ (where $n \in \mathbb{N}$ and the vectors $x_i$ may vary from combination to combination). Certainly, $span\,\mathbb{Y}$ is an algebraic subspace of $\mathbb{X}$. The reader may check that it is actually the smallest algebraic subspace of $\mathbb{X}$ that contains $\mathbb{Y}$ in the sense that if $\mathbb{Z}$ is an algebraic subspace of $\mathbb{X}$ that contains $\mathbb{Y}$ then $span\,\mathbb{Y} \subset \mathbb{Z}$. Analogously, the **convex hull** of $\mathbb{Y}$, denoted $conv\,\mathbb{Y}$ is the set of all convex combinations of vectors $x_i \in \mathbb{Y}, i = 1, ..., n, n \in \mathbb{N}$ and is the smallest convex set that contains $\mathbb{Y}$. We say that $\mathbb{Y}$ is convex if $conv\,\mathbb{Y} = \mathbb{Y}$. Note that we always have $\mathbb{Y} \subset conv\,\mathbb{Y} \subset span\,\mathbb{Y}$, and $\mathbb{Y}$ is an algebraic subspace of $\mathbb{X}$ iff $span\,\mathbb{Y} = \mathbb{Y}$, in which case $\mathbb{Y} = conv\,\mathbb{Y} = span\,\mathbb{Y}$.

**2.1.26 Example**    Let $\mathbb{X} = \mathbb{R}^2$, and let $A$ and $B$ be two points in the plane $\mathbb{R}^2$. If 0 denotes the origin, then the interval $\overline{AB}$ is the convex hull of two vectors: $\vec{0A}$ and $\vec{0B}$. Indeed, $C \in \overline{AB}$ iff $\vec{0C} = \vec{0B} + \vec{BC} = \vec{0B} + \alpha\vec{BA}$ where $0 \leq \alpha \leq 1$. This means, however, that $\vec{0C} = (1-\alpha)\vec{0B} + \alpha\vec{0A}$ since $\vec{BA} = \vec{0A} - \vec{0B}$.

**2.1.27 Exercise**    Let $\mathbb{X} = \mathbb{R}^2$, and $\mathbb{Y}_1 = \{(x, y) | x^2 + (y - 2)^2 \leq 1\}$, $\mathbb{Y}_2 = \{(x, y) | x^2 + y^2 \leq 1\}$. Find the span and the convex hull of $\mathbb{Y}_i, i = 1, 2$.

**2.1.28 Exercise**    Let $\mathbb{X} = l^1$. Prove that $\mathbb{Y}_1 = \{(x_n)_{n\geq 1} \in l^1 | x_n \geq 0\}$ is convex. Define $\mathbb{Y}_2 = \{(x_n)_{n\geq 1} \in \mathbb{Y}_1 | \sum_{n=1}^{\infty} x_n = 1\}$. Find the convex hull and the span of $\mathbb{Y}_2$.

**2.1.29 Exercise**    Let $\mathbb{Y}_i, i = 1, ..., n$ be convex subsets of a linear space $\mathbb{X}$. Prove that the convex hull of $\bigcup_{i=1}^{n} \mathbb{Y}_i$ equals the set of $z \in \mathbb{X}$ of the form

$$z = \sum_{i=1}^{n} \alpha_i y_i, \qquad (2.1)$$

where $\alpha_i$ are non-negative, $\sum_{i=1}^n \alpha_i = 1$ and $y_i \in \mathbb{Y}_i$. Show by example that the claim is not true if $\mathbb{Y}_i$ are not convex.

**2.1.30 Exercise**    Let $f : \mathbb{R} \to \mathbb{R}$ be a function. Show that the subset of points in $\mathbb{R}^2$ lying above the graph of $f$ is convex iff $f$ is convex.

**2.1.31 Exercise**    Show that the functions of bounded variation on $\mathbb{R}$ form an algebraic subspace of $\mathbb{R}^{\mathbb{R}}$, and that the subset of non-decreasing functions is convex. Similarly, the set of signed measures on a measure space $(\omega, \mathcal{F})$ is a linear space, and probability measures form its convex subset.

**2.1.32 Exercise**    Let $\eta_{2^n+k}(s) = \eta(2^n s - k)$, $s \in \mathbb{R}^+, 0 \le k < 2^n$, where

$$\eta(s) = \left\{ \begin{array}{ll} 0, & s \in (-\infty, 0) \cup [1, \infty), \\ 1, & s \in [0, \frac{1}{2}), \\ -1, & s \in [\frac{1}{2}, 1). \end{array} \right.$$

Define vectors on $\mathbb{X} = \mathbb{R}^{[0,1)}$ by $z_m = (\eta_m)_{|[0,1)}, m \ge 1$ (restriction of $\eta_m$ to $[0, 1)$) and $z_0 = 1_{[0,1)}$. Also, let $y_{k,n} = 1_{[\frac{k}{2^n}, \frac{k+1}{2^n})}$. Finally, let $\mathbb{Z}_n = \{z_k | 0 \le k < 2^n\}$ and $\mathbb{Y}_n = \{y_{k,n} | 0 \le k < 2^n\}$. Prove that $span\,\mathbb{Z}_n = span\,\mathbb{Y}_n$.

## 2.2 Banach spaces

As we have mentioned already, the notion of a Banach space is crucial in functional analysis and in this book. Having covered the algebraic aspects of Banach space in the previous section, we now turn to discussing topological aspects. A natural way of introducing topology in a linear space is by defining a norm. Hence, we begin this section with the definition of a normed space (which is a linear space with a norm) and continue with discussion of Cauchy sequences that leads to the definition of a Banach space, as a normed space "without holes". Next, we give a number of examples of Banach spaces (mostly those that are important in probability theory) and introduce the notion of isomorphic Banach spaces. Then we show how to immerse a normed space in a Banach space and provide examples of dense algebraic subspaces of Banach spaces. We close by showing how the completeness of a Banach space may be used to prove existence of an element that satisfies some required property.

2.2.1 *Normed linear spaces*    Let $\mathbb{X}$ be a linear space. A function $\| \cdot \| :$ $\mathbb{X} \to \mathbb{R}, x \mapsto \|x\|$ is called a **norm**, if for all $x, y \in \mathbb{X}$ and $\alpha \in \mathbb{R}$,

(n1) $\|x\| \geq 0$,
(n2) $\|x\| = 0$, iff $x = \Theta$,
(n3) $\|\alpha x\| = |\alpha|\, \|x\|$,
(n4) $\|x + y\| \leq \|x\| + \|y\|$.

If (n2) does not necessarily hold, $\| \cdot \|$ is called a **semi-norm**. Note that if $\| \cdot \|$ is a semi-norm, then $\|\Theta\| = 0$ by (n3) and 2.1.4. A pair $(\mathbb{X}, \| \cdot \|)$, where $\mathbb{X}$ is a linear space and $\| \cdot \|$ is a norm in $\mathbb{X}$ called a **normed linear space**, and for simplicity we say that $\mathbb{X}$ itself is a normed linear space (or just **normed space**).

2.2.2 **Exercise**    (n3)–(n4) imply that for $x, y \in \mathbb{X}$,

$$\big|\|x\| - \|y\|\big| \leq \|x \pm y\|.$$

2.2.3 **Theorem**    Suppose $\mathbb{X}$ is a linear space, and $\| \cdot \|$ is a semi-norm. Then $\mathbb{Y}_0 = \{x \in \mathbb{X} : \|x\| = 0\}$ is an algebraic subspace of $\mathbb{X}$ and the pair $(\mathbb{X}/\mathbb{Y}_0, \||\cdot\||)$, where

$$\||[x]\|| = \inf_{y \in [x]} \|y\| = \|x\|, \tag{2.2}$$

is a normed linear space.

*Proof* That $\mathbb{Y}_0$ is an algebraic subspace of $\mathbb{X}$ follows directly from (n3)–(n4). By 2.2.2, if $x \sim y$ then $\big|\|x\| - \|y\|\big| \leq \|x - y\| = 0$, so that (2.2) holds. We need to show (n2)–(n4) of the definition for the function $\||\cdot\||$, (n1) being trivial. Condition (n2) follows directly from (2.2). Conditions (n3) and (n4) now follow from the fact that $\| \cdot \|$ is a semi-norm: indeed, for any $x' \in [x]$ and $y' \in [y]$,

$$\||[x] + [y]\|| = \||[x + y]\|| \leq \|x' + y'\| \leq \|x'\| + \|y'\| = \||[x]\|| + \||[y]\||.$$

and

$$\||\alpha[x]\|| = \||[\alpha x]\|| \leq \|\alpha x'\| = |\alpha|\, \|x'\| = |\alpha|\||[x]\||.$$

$\square$

2.2.4 **Exercise**  Suppose that $\mathbb{Y}$ is a subspace of a normed linear space $\mathbb{X}$. Extend the argument used in 2.2.3 to show that the quotient space $\mathbb{X}/\mathbb{Y}$ is a normed linear space if we introduce

$$\|[x]\|_* = \inf_{y \in [x]} \|y\|. \tag{2.3}$$

2.2.5 **Example**  Let $(\Omega, \mathcal{F}, \mu)$ be a measure space, and $p \geq 1$ be a real number. Let $L^p(\Omega, \mathcal{F}, \mu)$ be the set of measurable functions $x$ on $\Omega$ such that $\left(\int_\Omega |x|^p \, d\mu\right)^{1/p} < \infty$. An analogous argument to that presented for $l^p$ in 2.1.10 may be used to show that $L^p(\Omega, \mathcal{F}, \mu)$ is a linear space. We claim that

$$\|x\| = \left(\int_\Omega |x|^p \, d\mu\right)^{1/p}$$

is a semi-norm on this space. Indeed, (n1) and (n3) are trivial, and (n4) reduces to the Minkowski inequality (see 1.5.9) if $p > 1$. For $p = 1$, (n4) follows from the triangle inequality: $|x + y| \leq |x| + |y|$.

However, $\| \cdot \|$ is not a norm since $\|x\| = 0$ implies merely that $x = 0$ $\mu$ a.e. Thus, to obtain a normed linear space we need to proceed as in 2.2.3 and consider the quotient space $L^p(\Omega, \mathcal{F}, \mu)/\mathbb{Y}$, where $\mathbb{Y} = \{x | x = 0 \, \mu \text{ a.e.}\}$. In other words we do not distinguish two functions that differ on a set of measure zero.

It is customary, however, to write $L^p(\Omega, \mathcal{F}, \mu)$ for both $L^p(\Omega, \mathcal{F}, \mu)$ itself and for its quotient space defined above. Moreover, for simplicity, it is often said that a function $x$ belongs to $L^p(\Omega, \mathcal{F}, \mu)$ even though what is meant is that $x$ represents a class of functions in $L^p(\Omega, \mathcal{F}, \mu)/\mathbb{Y}$. This should not lead to confusion, although it requires using caution, at least initially. As a by-product of this notational (in)convenience we often encounter phrases like "Let $L^p(\Omega, \mathcal{F}, \mu)$ be the space of (equivalence classes of) functions integrable with the $p$th power".

2.2.6 *Normed spaces as metric spaces*  Note that if $\| \cdot \|$ is a norm, then $d(x, y) = \|x - y\|$ is a metric. This means that $(\mathbb{X}, d)$ is a metric space. We may thus introduce topological notions in $\mathbb{X}$; such as convergence of a sequence, open and closed sets etc. However, the structure of a normed linear space is richer than that of a metric space.

A subset $\mathbb{Y}$ of a normed linear space $\mathbb{X}$ is said to be **linearly dense** iff its linear span is dense in $\mathbb{X}$. $\mathbb{Y}$ is called a subspace of $\mathbb{X}$ if it is a closed algebraic subspace of $\mathbb{X}$. Note that a closure of an algebraic subspace is a subspace.

**2.2.7 Exercise** Let $x$ and $x_n, n \geq 1$ be elements of normed linear space. Suppose that $x = \lim_{n \to \infty} x_n$; show that

$$\|x\| = \lim_{n \to \infty} \|x_n\|.$$

**2.2.8 *Cauchy sequences*** A sequence $(x_n)_{n \geq 1}$ of elements of a normed linear space $\mathbb{X}$ is said to be **Cauchy** if for all $\epsilon > 0$ there exists an $n_0 = n_0(\epsilon)$ such that $d(x_n, x_m) = \|x_n - x_m\| < \epsilon$, for all $n, m \geq n_0$. We claim that every convergent sequence is Cauchy. For the proof, let $\epsilon > 0$ be given and let $x = \lim_{n \to \infty} x_n$. Choose $n_0$ large enough to have $\|x_n - x\| < \epsilon$ for all $n \geq n_0$. If $n, m \geq n_0$, then

$$\|x_n - x_m\| \leq \|x_n - x\| + \|x - x_m\| < 2\epsilon.$$

**2.2.9 Exercise** Show that every Cauchy sequence $(x_n)_{n \geq 1}$, $x_n \in \mathbb{X}$, is **bounded**, i.e. there exists an $M > 0$ such that $\|x_n\| \leq M$ for all $n \geq 1$. Moreover, the limit $\lim_{n \to \infty} \|x_n\|$ exists for all Cauchy sequences $(x_n)_{n \geq 1}$.

**2.2.10 *Not every Cauchy sequence is convergent*** Let $\mathbb{X} = C([0, 1])$ be the linear space of continuous real-valued functions on $[0, 1]$ equipped with the usual topology (see 2.1.12). Let $\|x(\cdot)\| = \int_0^1 |x(s)| \, ds$ (see 2.2.5) and define a sequence in $\mathbb{X}$ whose elements are given by

$$x_n(s) = \begin{cases} 0, & 0 \leq s \leq \frac{1}{2} - \frac{1}{n}, \\ \frac{1}{2} + \frac{n}{2}\left(s - \frac{1}{2}\right), & \frac{1}{2} - \frac{1}{n} \leq s \leq \frac{1}{2} + \frac{1}{n}, \\ 1, & \frac{1}{2} + \frac{1}{n} \leq s \leq 1. \end{cases}$$

For $m > n$, $\|x_m - x_n\| = \frac{1}{2}\left[\frac{1}{n} - \frac{1}{m}\right]$ (look at the graphs!), so that the sequence is Cauchy. However, it is not convergent. Indeed, if $\lim_{n \to \infty} \|x_n - x\| = 0$ for some $x \in C([0, 1])$, then for all $\epsilon > 0$, $\lim_{n \to \infty} \int_0^{\frac{1}{2} - \epsilon} |x_n(s) - x(s)| \, ds = 0$. Since for $n > \frac{1}{\epsilon}$, we have $x_n(s) = 0$ whenever $s \in [0, \frac{1}{2} - \epsilon]$, we have $\int_0^{\frac{1}{2} - \epsilon} |x(s)| = 0$, i.e. $x(s) = 0$ a.e. in $[0, \frac{1}{2} - \epsilon)$. By continuity, $x(s) = 0$ for all $s < \frac{1}{2}$. The same argument shows that $x(s) = 1$, for $s \in (\frac{1}{2}, 1]$. This contradicts continuity of $x$.

**2.2.11 Remark** There are at least two ways of explaining why, in the previous example, $(x_n)_{n \geq 1}$ failed to be convergent. Both are fruitful and lead to a better understanding of the phenomenon in question (actually, they are just different sides of the same coin). Note that *the notion of convergence (and of a Cauchy sequence) depends both on $\mathbb{X}$ and on the norm*. Thus, the first way of explaining 2.2.10 is to say that the norm

$\|x(\cdot)\| = \int_0^1 |x(s)| \, ds$ is not appropriate for $\mathbb{X} = C([0,1])$. This norm is too weak in the sense that it admits more Cauchy sequences than it should. If we define $\|x\|_{\sup} = \sup_{0 \le s \le 1} |x(s)|$, then $\|x_n - x_m\|_{\sup} = \frac{1}{2} - \frac{n}{2m}$ for $m \ge n$, so that the sequence is not Cauchy any more, and the problem with this sequence disappears. Moreover, we will not have problems with other sequences since *any* sequence that is Cauchy in this norm is convergent in this norm to an element of $C([0,1])$ (see 2.2.16 below). The second way of explaining 2.2.10 is to say that the space is not appropriate for the norm. Indeed, if we stick to this norm and, instead of $C([0,1])$ take the space $L^1[0,1]$ of (equivalence classes of) Lebesgue integrable functions, $(x_n)_{n \ge 1}$ will not only be Cauchy, but also convergent. Indeed, we have actually found the limit $x$ of our sequence: $x = 1_{(\frac{1}{2},1]}$.† The fact is that it does not belong to $C([0,1])$, but it does belong to $L^1[0,1]$. Moreover, we may prove that any Cauchy sequence in $L^1[0,1]$ is convergent (see below).

**2.2.12 Definition**    If every Cauchy sequence in a normed linear space $\mathbb{X}$ is convergent, $\mathbb{X}$ is called a Banach space. If we recall that a metric space is termed **complete** if every Cauchy sequence of its elements is convergent (see 2.2.6), we may say that a Banach space is a complete normed linear space. Note again that this notion involves both the space and the norm; and that this pair becomes a Banach space if both elements "fit" with each another.

Before we continue with examples of Banach spaces, the reader should solve the following two "warm-up" problems.

**2.2.13 Exercise**    Suppose that $\mathbb{Y}$ is a subspace of a Banach space. Show that $\mathbb{Y}$ is itself a Banach space, equipped with the norm inherited from $\mathbb{X}$.

**2.2.14 Exercise**    Let $\mathbb{X}$ be a normed linear space and $(x_n)_{n \ge 1}$ be a sequence of its elements. We say that a **series** $\sum_{n=1}^{\infty} x_n$ **converges**, if the sequence $y_n = \sum_{i=1}^{n} x_i$ converges. We say that this **series converges absolutely** if $\sum_{n=1}^{\infty} \|x_n\| < \infty$. Show that a normed linear space is a Banach space iff every absolutely convergent series converges.

**2.2.15 *The space of bounded functions***    Let $S$ be a set and let $B(S)$ be the linear space of bounded functions on $S$. Define the norm

$$\|x\| = \sup_{p \in S} |x(p)|$$

† We have not proven yet that $\lim x_n = x$, but this is a simple exercise.

(the supremum is finite by the definition of a bounded function; see 2.1.11). $B(S)$ is a Banach space.

*Proof* We check easily that conditions (n1)–(n4) of the definition of the norm are satisfied. The only non-trivial statement is that about completeness. Let, therefore, $x_n$ be the Cauchy sequence in $B(S)$. Let $p$ be a member of $S$. By the definition of the supremum norm the sequence $x_n(p)$ is a Cauchy sequence in $\mathbb{R}$. Let $x(p) = \lim_{n \to \infty} x_n(p)$. We claim that $x$ belongs to $B(S)$ and $x_n$ converges to $x$ in this space. Indeed, since $x_n$ is a Cauchy sequence in $B(S)$, given $\epsilon > 0$, we may choose an $n_0$ such that for all $p \in S$ and $n, m \geq n_0$, $|x_n(p) - x_m(p)| < \epsilon$. Taking the limit $m \to \infty$, we get $|x(p) - x_n(p)| \leq \epsilon$. In particular $\sup_{p \in S} |x(p)| \leq \sup_{p \in S} \{|x(p) - x_n(p)| + |x_n(p)|\} < \infty$, i.e. $x \in B(S)$, and $\|x_n - x\| \leq \epsilon$. This means that $x_n$ converges to $x$ in the supremum norm. $\qquad\square$

**2.2.16** *The space of continuous functions* Let $S$ be a compact Hausdorff topological space. The space $C(S)$ of continuous functions $x$ on $S$, equipped with the supremum norm:

$$\|x\| = \sup_{p \in S} |x(p)|$$

is a Banach space.

*Proof* It is enough to show that $C(S)$ is a subspace of $B(S)$. Note that continuous functions on a compact set are bounded, for the image of a compact set via a continuous function is compact and compact sets in $\mathbb{R}$ are necessarily bounded. It remains to show that the limit of a sequence $x_n \in C(S)$ does belong not only to $B(S)$ but also to $C(S)$. But this just means that the uniform limit of a sequence of continuous functions is continuous, which may be proven as follows. For $p \in S$ and $\epsilon > 0$, take $n$ such that $|x_n(p) - x(p)| \leq \frac{\epsilon}{3}$ for all $p \in S$. Moreover, let $U$ be the neighborhood of $p$ such that $q \in U$ implies $|x_n(p) - x_n(q)| \leq \frac{\epsilon}{3}$. The triangle inequality shows that

$$|x(p) - x(q)| \leq |x(p) - x_n(p)| + |x_n(p) - x_n(q)| + |x_n(q) - x(q)| \leq \epsilon,$$

as desired. $\qquad\square$

**2.2.17 Exercise** (a) Let $S$ be a Hausdorff topological space. The space $BC(S)$ of bounded continuous functions $x$ on $S$ equipped with the supremum norm is a Banach space. (b) Let $(\Omega, \mathcal{F})$ be a measurable space. The space $BM(\Omega)$ of bounded measurable functions on $\Omega$ equipped with the same norm is a Banach space.

**2.2.18** *The space* $L^\infty(\Omega, \mathcal{F}, \mu)$  Let $(\Omega, \mathcal{F}, \mu)$ be a measure space. As noted in Exercise 2.2.17 above, the space $BM(\Omega)$ of bounded measurable functions on $\Omega$, with the supremum norm is a Banach space. The fact that there exists a measure $\mu$ on $(\Omega, \mathcal{F})$ allows introducing a new Banach space that is closely related to $BM(\Omega)$. In many situations it is desirable to identify measurable functions that differ only on a set of $\mu$-measure zero. In other words, we introduce an equivalence relation: two measurable functions $x$ and $y$ are in this equivalence relation iff $x(\omega) = y(\omega)$ for all $\omega \in \Omega$ except maybe for a set of $\mu$-measure zero. Note that an unbounded function may belong to the equivalence class of a bounded function. Such functions are said to be **essentially bounded**, and we define the norm of the equivalence class $[x]$ of an essentially bounded function $x$ to be

$$\|[x]\| = \inf_{\substack{y \in [x] \\ y \in BM(\Omega)}} \|y\|_{BM(\Omega)}.$$

We shall prove later that the infimum in this definition is attained for a bounded $y \in [x]$. The space of (classes of) essentially bounded functions is denoted $L^\infty(\Omega, \mathcal{F}, \mu)$.

The reader may have noticed the similarity of the procedure that we are using here to the one in 2.2.3. Let us remark, however, that these procedures are not identical. Specifically, the linear space of essentially bounded measurable functions on $(\Omega, \mathcal{F})$, where the equivalence class was introduced above, is not a normed linear space in the sense of Definition 2.2.1, since $\|x\|_{BM(\Omega)}$ may be infinite for an $x$ in this space. Nevertheless, an argument almost identical to the one presented in 2.2.3 proves that $L^\infty(\Omega, \mathcal{F}, \mu)$ is a normed linear space.

Let us note that for any $x \in L^\infty(\Omega, \mathcal{F}, \mu)$ there exists a bounded $y \in [x]$ such that $\|[x]\|_{L^\infty(\Omega, \mathcal{F}, \mu)} = \|y\|_{BM(\Omega)}$. Indeed, let $y_n \in BM(\Omega)$ be such that $y_n = x$ for all $\omega \in \Omega \setminus A_n$ where $\mu(A_n) = 0$ and

$$\lim_{n \to \infty} \|y_n\|_{BM(\Omega)} = \|[x]\|_{L^\infty(\Omega, \mathcal{F}, \mu)}.$$

Define $y(\omega) = 0$ for $\omega \in A = \bigcup_{n \geq 1} A_n$ and $y(\omega) = x(\omega)$ for $\omega \notin A$. Certainly, $\mu(A) = 0$, and so $y \in [x]$. Moreover, $y$ is bounded because for any $n \in \mathbb{N}$ we have

$$\|y\|_{BM(\Omega)} = \sup_{\omega \in \Omega} |y(\omega)| = \sup_{\omega \in \Omega \setminus A_n} |y(\omega)|$$

$$\leq \sup_{\omega \in \Omega \setminus A_n} |y_n(\omega)| = \|y_n\|_{BM(\Omega)}.$$

Taking the limit as $n \to \infty$, this inequality also shows that

$$\|y\|_{BM(\Omega)} \leq \lim_{n \to \infty} \|y_n\|_{BM(\Omega)} = \|[x]\|_{L^\infty(\Omega, \mathcal{F}, \mu)}.$$

Since the reverse inequality is obvious, our claim is proved.

To show that $L^\infty(\Omega, \mathcal{F}, \mu)$ is complete, we use Exercise 2.2.14. Let $[x_n] \in L^\infty(\Omega, \mathcal{F}, \mu)$ be a sequence such that $\sum_{n=1}^{\infty} \|[x_n]\|_{L^\infty(\Omega, \mathcal{F}, \mu)} < \infty$. Let $y_n \in BM(\Omega)$ be such that $\|y_n\|_{BM(\Omega)} = \|[x_n]\|_{L^\infty(\Omega, \mathcal{F}, \mu)}$. Then the series $\sum_{n=1}^{\infty} y_n$ is absolutely convergent, and because $BM(\Omega)$ is a Banach space, $\sum_{n=1}^{\infty} y_n$ converges to a $y$ in this space. Since the class of $\sum_{i=1}^{n} y_i$ equals $\sum_{i=1}^{n} [x_i]$ we infer that $\sum_{i=1}^{n} [x_i]$ converges to the class of $y$, as desired.

**2.2.19** *The space $L^p(\Omega, \mathcal{F}, \mu)$, $p \geq 1$*   Let $(\Omega, \mathcal{F}, \mu)$ be a measure space. The space $L^p(\Omega, \mathcal{F}, \mu)$, $p \geq 1$, is a Banach space.

*Proof*  We shall use 2.2.14. Let $x_n \in L^p(\Omega, \mathcal{F}, \mu)$ be such that

$$\sum_{n=1}^{\infty} \|x_n\|_{L^p(\Omega, \mathcal{F}, \mu)} < \infty.$$

Consider the function $x_0 = \sum_{n=1}^{\infty} |x_n|$ which takes values in the extended positive half-line. By Fatou's Lemma and Minkowski's inequality

$$\int |x_0|^p \, d\mu \leq \lim_{n \to \infty} \int \left( \sum_{i=1}^{n} |x_i| \right)^p d\mu \leq \lim_{n \to \infty} \sum_{i=1}^{n} \|x_i\|^p < \infty.$$

In particular, the set of $\omega$ where $x_0(\omega) = \infty$ has measure zero. Therefore, the series $x(\omega) = \sum_{n=1}^{\infty} x_n(\omega)$ converges absolutely except maybe on a set of measure zero, where we put $x(\omega) = 0$. With such a definition, we have

$$\int \left| x - \sum_{i=1}^{n} x_i \right|^p d\mu \leq \lim_{k \to \infty} \int \sum_{i=n+1}^{k} |x_i|^p \, d\mu \leq \sum_{i=n+1}^{\infty} \|x_i\|_{L^p(\Omega, \mathcal{F}, \mu)}^p.$$

Hence, $x \in L^p(\Omega, \mathcal{F}, \mu)$ and $\lim_{n \to \infty} x_n = x$.  $\square$

**2.2.20 Remark**     In the proof of 2.2.19 we have used 2.2.14. This was a nice shortcut but because of its use we have lost valuable information (see the proof of 3.7.5 for instance). The result we are referring to says that any convergent sequence $x_n \in L^p(\Omega, \mathcal{F}, \mu), n \geq 1$ has a subsequence that converges almost surely; this result is a by-product of a direct proof of 2.2.19 (see e.g. [103]). To prove it, we note first that $x_n, n \geq 1$ is Cauchy. Next, we proceed as in the hint to 2.2.14 and find a subsequence $x_{n_k}, k \geq 1$ such $\|x_{n_{k+1}} - x_{n_k}\| < \frac{1}{2^k}$. Then the series $\sum_{k=1}^{\infty}(x_{n_{k+1}} - x_{n_k}) + x_{n_1}$ converges both a.s. and in the norm in $L^p(\Omega, \mathcal{F}, \mu)$. Since the partial sums of this series equal $x_{n_k}$, the sum of this series must be the limit of $x_n, n \geq 1$, which proves our claim.

**2.2.21 Corollary**     Let $(\Omega, \mathcal{F}, \mathbb{P})$ be a measure space, and let $\mathcal{G}$ be a sub-$\sigma$-algebra of $\mathcal{F}$. The space $L^p(\Omega, \mathcal{G}, \mathbb{P})$ is a subspace of $L^p(\Omega, \mathcal{F}, \mathbb{P})$, $p \geq 1$.

**2.2.22 Remark**     The proof of our Corollary is obvious, is it not? If a random variable is $\mathcal{G}$ measurable then it is also $\mathcal{F}$ measurable and since $L^p(\Omega, \mathcal{G}, \mathbb{P})$ is a Banach space itself, then it is a subspace of $L^p(\Omega, \mathcal{F}, \mathbb{P})$. Yes? No. We forgot that $L^p(\Omega, \mathcal{G}, \mathbb{P})$ is a space of *equivalence classes* and not functions. If $\mathcal{G}$ does not contain all of the sets $A \in \mathcal{F}$ with $\mathbb{P}(A) = 0$, then the equivalence classes in $L^p(\Omega, \mathcal{G}, \mathbb{P})$ are not equivalence classes of $L^p(\Omega, \mathcal{F}, \mathbb{P})$ and we may not even claim that $L^p(\Omega, \mathcal{G}, \mathbb{P})$ is a subset of $L^p(\Omega, \mathcal{F}, \mathbb{P})$! In other words, Corollary 2.2.21 is not true unless $\mathcal{G}$ contains all measurable sets of probability zero.

Without this assumption, the correct statement of Corollary 2.2.21 is that $L^p(\Omega, \mathcal{G}, \mathbb{P})$ is isometrically isomorphic to a subspace of $L^p(\Omega, \mathcal{F}, \mathbb{P})$ in the sense of 2.2.30, see 2.2.33 below.

**2.2.23 Corollary**     Suppose that $\mathbb{X}_0$ is a subspace of a Banach space $L^p(\Omega, \mathcal{F}, \mu)$, where $p \geq 1$ and $\mu$ is a finite measure such that $1_\Omega \in \mathbb{X}_0$. Then the collection $\mathcal{G}$ of events $A$ such that $1_A \in \mathbb{X}_0$ is a $\lambda$-system.

*Proof* $\Omega$ and $\emptyset$ belong to $\mathcal{G}$ by assumption. Moreover, if $A$ and $B$ belong to $\mathcal{G}$ and $A \subset B$ then $B \setminus A$ belongs to $\mathcal{G}$ since $1_{B\setminus A} = 1_B - 1_A \in \mathbb{X}_0$. Finally, if $A_n \in \mathcal{G}, n \geq 1$, is a non-decreasing sequence of events, then

$$\left\| 1_{\bigcup_{k=1}^{\infty} A_k} - 1_{A_n} \right\|_{L^p}^p = \mu\left( \bigcup_{k=1}^{\infty} A_k \setminus A_n \right)$$

converges to zero, as $n \to \infty$. Since $\mathbb{X}_0$ is closed (being a subspace) $1_{\bigcup_{k=1}^{\infty} A_k}$ belongs to $\mathbb{X}_0$, proving that $\bigcup_{k=1}^{\infty} A_k$ belongs to $\mathcal{G}$.     $\square$

**2.2.24** *The space of signed Borel measures*    For any Borel charge $\mu$ let $\mu = \mu_0^+ - \mu_0^-$ be its minimal representation, as described in 1.3.6. Define $\|\mu\| = |\mu|(\mathbb{R}) = \mu_0^+(\mathbb{R}) + \mu_0^-(\mathbb{R})$. Equivalently, let $y$ be a regular function of bounded variation such that (1.22) holds and define $\|y\|$ to be the variation of $y$ on $(-\infty, \infty)$. The space $\mathbb{BM}(\mathbb{R})$ of Borel charges (the space of regular functions of bounded variation such that (1.22) holds) with this norm is a Banach space.

*Proof*  Note that by definition $\mu = \mu^+ - \mu^-$, where $\mu^+$ and $\mu^-$ are positive measures, implies

$$\|\mu\| \le \mu^+(\mathbb{R}) + \mu^-(\mathbb{R}). \tag{2.4}$$

We need to start by showing that $\mathbb{BM}(\mathbb{R})$ is a normed space. If $\|\mu\| = |\mu|(\mathbb{R}) = 0$, then $\mu_0^+(\mathbb{R}) = 0$ and $\mu_0^-(\mathbb{R}) = 0$, so that $\mu_0^+(A) = 0$ and $\mu_0^-(A) = 0$, for all Borel subsets of $\mathbb{R}$, proving that $\mu = 0$. Of course $\mu = 0$ implies $\|\mu\| = 0$.

If $\alpha \in \mathbb{R}$ and $\mu \in \mathbb{BM}(\mathbb{R})$, then the variation of $y(t) = \alpha\mu(-\infty, t]$ on $(-\infty, \infty)$ equals $|\alpha|$ times the variation of $\mu(-\infty, t]$ in this interval. In other words $\|\alpha\mu\| = |\alpha|\|\mu\|$. Finally, if $\mu = \mu_0^+ - \mu_0^-$ and $\nu = \nu_0^+ - \nu_0^-$, then $\mu + \nu = (\mu_0^+ + \nu_0^+) - (\mu_0^- + \nu_0^-)$, so that by (2.4),

$$\|\mu + \nu\| \le \mu_0^+(\mathbb{R}) + \nu_0^+(\mathbb{R}) + \mu_0^-(\mathbb{R}) + \nu_0^-(\mathbb{R})$$
$$= \|\mu\| + \|\nu\|.$$

Turning to completeness of $\mathbb{BM}(\mathbb{R})$, let $\mu_n, n \ge 1$ be a sequence of charges such that $\sum_{n=1}^{\infty} \|\mu_n\| < \infty$. Let $\mu_n = \mu_{n,0}^+ - \mu_{n,0}^-$ be the minimal representation of $\mu_n$. By definition, for any Borel subset $A$ of $\mathbb{R}$,

$$\sum_{n=1}^{\infty} [\mu_{n,0}^+(A) + \mu_{n,0}^-(A)] \le \sum_{n=1}^{\infty} [\mu_{n,0}^+(\mathbb{R}) + \mu_{n,0}^-(\mathbb{R})] < \infty,$$

so that both series on the left converge absolutely. We may thus define

$$\mu^+(A) = \sum_{n=1}^{\infty} \mu_{n,0}^+(A), \qquad \mu^-(A) = \sum_{n=1}^{\infty} \mu_{n,0}^-(A).$$

Functions $\mu^+$ and $\mu^-$ are countably additive (this statement is a particular case of the Fubini Theorem). Thus we may introduce the charge

$\mu = \mu^+ - \mu^-$. By (2.4)

$$\|\mu - \sum_{i=1}^{n} \mu_i\| = \| \sum_{i=n+1}^{\infty} \mu_{i,0}^+ - \sum_{i=n+1}^{\infty} \mu_{i,0}^- \|$$
$$\leq \sum_{i=n+1}^{\infty} \mu_{i,0}^+(\mathbb{R}) + \sum_{i=n+1}^{\infty} \mu_{i,0}^-(\mathbb{R})$$

which tends to zero, as $n \to \infty$. This proves that $\mu$ is the sum of the series $\sum_{n=1}^{\infty} \mu_n$. □

**2.2.25 Exercise** Suppose that $\mathbb{X}$ is a linear normed space. Consider the set $b(\mathbb{X})$ of sequences $(x_n)_{n\geq1}$ with values in $\mathbb{X}$, that are bounded, i.e. $\|(x_n)_{n\geq1}\|_* = \sup_{n\geq1} \|x_n\| < \infty$. Prove that $b(\mathbb{X})$ is a linear, in fact normed space when equipped with $\|\cdot\|_*$. Moreover, it is a Banach space iff $\mathbb{X}$ is a Banach space. Finally, Cauchy sequences in $\mathbb{X}$ form a subspace, say $b_c(\mathbb{X})$, of $b(\mathbb{X})$.

**2.2.26 Exercise** Show directly that the following spaces of sequences are Banach spaces: (a) $c$ : the space of convergent sequences with the norm $\|(x_n)_{n\geq1}\| = \sup_{n\geq1} |x_n|$, (b) $l^p$ : the space of absolutely convergent sequences with the norm $\|(x_n)_{n\geq1}\| = \left( \sum_{n=1}^{\infty} |x_n|^p \right)^{\frac{1}{p}}$, $p \geq 1$. Show also that the space $c_0$ of sequences converging to zero is a subspace of $c$.

**2.2.27 Exercise** *Cartesian product* Prove that if $\mathbb{X}$ and $\mathbb{Y}$ are two Banach spaces then the space of ordered pairs $(x, y)$ where $x \in \mathbb{X}$ and $y \in \mathbb{Y}$ is a Banach space with the norm $\|(x, y)\| = \|x\| + \|y\|$, or $\|(x, y)\| = \sqrt{\|x\|^2 + \|y\|^2}$, or $\|(x, y)\| = \|x\| \vee \|y\|$, where $\|x\| \vee \|y\| = \max\{\|x\|, \|y\|\}$.

**2.2.28 Exercise** Let $S$ be a set and let $p \in S$. Show that the set of members $x$ of $B(S)$ such that $x(p) = 0$ is a subspace of $B(S)$.

**2.2.29 Exercise** Repeat 2.2.28 for a compact, Hausdorff topological space, with $B(S)$ replaced with $C(S)$. May we make a similar statement for $L^\infty(\mathbb{R})$ and some $p \in \mathbb{R}$?

**2.2.30 Definition** A linear map $I$ from a linear normed space $(\mathbb{X}, \|\cdot\|_{\mathbb{X}})$ onto a linear normed space $(\mathbb{X}, \|\cdot\|_{\mathbb{Y}})$ is an **isomorphism** if there exist two positive constants $m$ and $M$, such that $m\|x\| \leq \|Ix\|_{\mathbb{Y}} \leq M\|x\|_{\mathbb{X}}$. In particular, isomorphisms are bijections. In such a case, $\mathbb{X}$ and $\mathbb{Y}$ are said to be **isomorphic**. If $M = m = 1$, i.e. if $\|Ix\|_{\mathbb{Y}} = \|x\|_{\mathbb{X}}$,

$I$ is said to be an **isometric isomorphism** and $\mathbb{X}$ and $\mathbb{Y}$ are said to be isometrically isomorphic.

**2.2.31 Example** For any $a < b \in \mathbb{R}$, the space $C[a, b]$ is isometrically isomorphic to $C[0, 1]$. The isomorphism is given by $I : C[a, b] \to C[0, 1], y(\tau) = Ix(\tau) = x\left((1 - \tau)a + \tau b\right)$. (What is the inverse of $I$?) Analogously, $C[0, 1]$ is isometrically isomorphic to the space $C[-\infty, \infty]$ of continuous functions with limits at plus and minus infinity. The isomorphism is given by $y(\tau) = Ix(\tau) = x(\frac{1}{\pi}\arctan\tau + \frac{1}{2})$.

This result may be generalized as follows: if $S$ and $S'$ are two topological spaces such that there exists a homeomorphism $f : S \to S'$ and if $\alpha \in BC(S)$ is such that $|\alpha(p)| = 1$, then $Ix(p) = \alpha(p)x(f(p))$ is an isometric isomorphism of $BC(S)$ and $BC(S')$. The famous Banach–Stone Theorem says that if $S$ and $S'$ are compact, the inverse statement is true as well, i.e. all isometric isomorphisms have this form (see [22]).

**2.2.32 Exercise** Let $S$ and $S'$ be two sets. Suppose that $f : S \to S'$ is a bijection. Show that $B(S)$ is isometrically isomorphic to $B(S')$. In the case where $(S, \mathcal{F})$ and $(S, \mathcal{F}')$ are measurable spaces what additional requirement(s) on $f$ will guarantee that $BM(S)$ and $BM(S')$ are isometrically isomorphic?

**2.2.33 Example** Let $\mathcal{G}$ be a sub-$\sigma$-algebra of the $\sigma$-algebra $\mathcal{F}$ of events in a probability space $(\Omega, \mathcal{F}, \mathbb{P})$. In general $L^p(\Omega, \mathcal{G}, \mathbb{P})$ is not a subspace of $L^p(\Omega, \mathcal{F}, \mathbb{P})$, $p \geq 1$ (see 2.2.22). However, $L^p(\Omega, \mathcal{G}, \mathbb{P})$ is isometrically isomorphic to the subspace $L_0^p(\Omega, \mathcal{G}, \mathbb{P})$ of equivalence classes in $L^p(\Omega, \mathcal{F}, \mathbb{P})$ corresponding to integrable with $p$th power $\mathcal{G}$ measurable functions. To see this let us consider an equivalence class in $L^p(\Omega, \mathcal{G}, \mathbb{P})$. If $X$ is its representative, then $X$ is $\mathcal{G}$ measurable and all other elements of this class differ from $X$ on $\mathbb{P}$-null events that belong to $\mathcal{G}$. The equivalence class of $X$ in $L^p(\Omega, \mathcal{F}, \mathbb{P})$ is composed of all functions that differ from $X$ on $\mathbb{P}$-null events that are not necessarily in $\mathcal{G}$. Nevertheless, the norms of these classes of $X$ in $L^p(\Omega, \mathcal{G}, \mathbb{P})$ and $L^p(\Omega, \mathcal{F}, \mathbb{P})$ are equal. Let $I$ map the equivalence class of $X$ in $L^p(\Omega, \mathcal{G}, \mathbb{P})$ into the equivalence class of $X$ in $L^p(\Omega, \mathcal{F}, \mathbb{P})$. Since the range of $I$ is $L_0^p(\Omega, \mathcal{G}, \mathbb{P})$, our claim is proven.

Remark 2.2.11 suggests the following procedure for constructing Banach spaces from normed linear spaces: find the "limits" of Cauchy sequences and add them to the original space. Let us explain this idea in more detail. Since some Cauchy sequences in our normed linear space

$\mathbb{X}$ "cause trouble" by not being convergent, first we may immerse $\mathbb{X}$ in the space of Cauchy sequences $b_c$ by noting that an $x \in \mathbb{X}$ may be represented as a constant sequence in $b_c$. Next, we may note that we are really not interested in Cauchy sequences themselves but in their "limits", which may be thought of as equivalence classes of Cauchy sequences that become arbitrarily close to each other as $n \to \infty$. Some of these equivalence classes correspond to elements of $x$ and some do not. The point is however that they form "the smallest" Banach space that contains $\mathbb{X}$. We make this idea rigorous in the following theorem.

**2.2.34 "Filling holes" in a normed linear space**    Let $\mathbb{X}$ be a normed space. There exists a Banach space $\mathbb{Y}$ and a linear operator $L : \mathbb{X} \to \mathbb{Y}$ satisfying the following two conditions:

$$\|Lx\|_{\mathbb{Y}} = \|x\|_{\mathbb{X}}, \qquad cl(R(L)) = \mathbb{Y}. \tag{2.5}$$

*Proof* Consider the space of Cauchy sequences from Exercise 2.2.25, and its subspace $b_0(\mathbb{X})$ of sequences converging to $\Theta$. Let $\mathbb{Y} = b_c(\mathbb{X})/b_0(\mathbb{X})$ be the quotient space, and for any $x \in \mathbb{X}$, let $Lx$ be the equivalence class of a constant sequence $(x)_{n \geq 1}$. Two elements, $(x_n)_{n \geq 1}$ and $(x'_n)_{n \geq 1}$ of $b_c(\mathbb{X})$ are equivalent if $\lim_{n \to \infty} \|x_n - x'_n\| = 0$. This implies that $\lim_{n \to \infty} \|x_n\| = \lim_{n \to \infty} \|x'_n\|$, and this limit is the norm of the equivalence class to which they belong (see Exercise 2.2.4). In particular, the first condition in (2.5) holds. The map $L$ is linear, as a composition of two linear maps.

To complete the proof of (2.5), assume that $(x_n)_{n \geq 1} \in b_c(\mathbb{X})$. Let $y$ be the class of $(x_n)_{n \geq 1}$ in the quotient space $\mathbb{Y}$, and let $y_i = Lx_i$ be a sequence of elements of $\mathbb{Y}$. We have

$$\|y_i - y\|_{\mathbb{Y}} = \lim_{n \to \infty} \|x_i - x_n\|_{\mathbb{X}} \tag{2.6}$$

which implies that $\lim_{i \to \infty} \|y_i - y\|_{\mathbb{Y}} = 0$, as desired.

It remains to prove that $\mathbb{Y}$ is a Banach space. Let $y_n$ be a Cauchy sequence in $\mathbb{Y}$. There exists a sequence $x_n \in \mathbb{X}$ such that $\|Lx_n - y_n\| \leq \frac{1}{n}$. The sequence $(x_n)_{n \geq 1}$ is Cauchy in $\mathbb{X}$, for

$$\|x_n - x_m\|_{\mathbb{X}} = \|Lx_n - Lx_m\|_{\mathbb{Y}} \leq \frac{1}{n} + \|y_n - y_m\| + \frac{1}{m}.$$

Let $y$ be the class of $(x_n)_{n \geq 1}$ in $\mathbb{Y}$. Arguing as in (2.6) we see that $\lim_{n \to \infty} \|Lx_n - y\|_{\mathbb{Y}} = 0$ and hence $\lim_{n \to \infty} y_n = y$ as well.    $\square$

**2.2.35 Corollary** The space $\mathbb{Y}$ from the previous subsection is unique in the sense that if $\mathbb{Y}'$ is another space that satisfies the requirements of this theorem, then $\mathbb{Y}'$ is isometrically isomorphic to $\mathbb{Y}$. Therefore, we may meaningfully speak of *the completion of a normed space.*

*Proof* Let $L' : \mathbb{X} \to \mathbb{Y}'$ be a map such that $\|L'x\|_{\mathbb{Y}'} = \|x\|_{\mathbb{X}}, cl(R(L')) = \mathbb{Y}'$. For $y' \in \mathbb{Y}'$, there exists a sequence $x_n \in \mathbb{X}$ such that $\lim_{n \to \infty} \|Lx'_n - y'\|_{\mathbb{Y}'} = 0$. Since $\|L'x_n - L'x_m\|_{\mathbb{Y}'} = \|x_n - x_m\| = \|Lx_n - Lx_m\|_{\mathbb{Y}}$, $Lx_n$ is then a Cauchy sequence in $\mathbb{Y}$. Since $\mathbb{Y}$ is a Banach space, there exists a $y$ in $\mathbb{Y}$ such that $\lim_{n \to \infty} Lx_n = y$. This $y$ does not depend on the choice of the sequence $x_n$ but solely on $y'$, for if $x'_n$ is another sequence such that $L'x_n$ tends to $y'$ then

$$\|Lx_n - Lx'_n\|_{\mathbb{Y}} = \|x_n - x'_n\|_{\mathbb{X}} = \|L'x_n - L'x'_n\|_{\mathbb{Y}'}$$

tends to zero as $n \to \infty$.

Let us thus define $I : \mathbb{Y}' \to \mathbb{Y}$ by $Iy' = y$. Obviously, $I$ is linear. Moreover, it is onto for we could repeat the argument given above after changing the roles of $\mathbb{Y}'$ and $\mathbb{Y}$. Finally,

$$\|Iy'\|_{\mathbb{Y}} = \lim_{n \to \infty} \|Lx_n\|_{\mathbb{Y}} = \lim_{n \to \infty} \|x_n\|_{\mathbb{X}} = \lim_{n \to \infty} \|L'x_n\|_{\mathbb{Y}'} = \|y'\|_{\mathbb{Y}'}.$$

$\square$

**2.2.36 Example** If $\mathbb{X}$ is the space of sequences $x = (\xi_n)_{n \geq 1}$ that are eventually zero, with the norm $\|x\| = (\sum_{n=1}^{\infty} |\xi_n|^p)^{\frac{1}{p}}$, where the sum above is actually finite for each $x$, then $\mathbb{X}$ is a normed linear space but it is not complete. Its completion is $l^p$. Similarly, if in $\mathbb{X}$ we introduce the norm $\|x\| = \sup_{n \geq 1} |\xi_n|$, then the completion of the normed space $\mathbb{X}$ is the space $c_0$ of sequences converging to zero equipped with the supremum norm.

These two statements are equivalent to saying that $\mathbb{X}$, when equipped with the appropriate norm, is an algebraic subspace of $l^p$ and $c_0$ that is dense in these spaces.

**2.2.37 *The spaces $C_c(S)$ and $C_0(S)$*** Let $S$ be a locally compact Hausdorff space, and let $C_c(S)$ be the space of continuous functions $x$ on $S$ such that $x(p) \neq 0$ only on a compact subset $K$ of $S$. Note that $K = K(x)$ may be different for different $x$. The space $C_c(S)$ equipped with the supremum norm $\|x\| = \sup_{p \in S} |x(p)| = \sup_{p \in K(x)} |x(p)|$ is a normed linear space. In general, though, it is not complete. Its completion $C_0(S)$ and called the **space of functions vanishing at infinity**.

To explain this terminology, consider $S = \mathbb{N}$ equipped with discrete topology. The discrete topology is that one in which all subsets of $\mathbb{N}$ are open. Compact sets in $\mathbb{N}$ are sets with a finite number of elements. Therefore, $C_c(S)$ is a space of sequences $x = (\xi_n)_{n \geq 1}$ such that $\xi_n = 0$ for all but a finite number of $n$. In 2.2.36 we saw that $C_0(\mathbb{N})$ may be identified with the space $c_0$ of sequences converging to zero.

Similarly we may check that $C_0(\mathbb{R}^n)$ is the space of continuous functions $x$ such that $\lim_{|\underline{s}| \to \infty} x(\underline{s}) = 0$. Here $\underline{s} = (s_1, ...., s_n)$ and $|\underline{s}| = \sqrt{\sum_{i=1}^{n} s_i^2}$.

If $S$ is compact, then $C_0(S)$ coincides with $C(S)$. As an example one may take $S = \mathbb{N} \cup \{\infty\}$ with the topology defined by choosing its base to be the family of all singletons $\{n\}$ and neighborhoods of infinity of the form $\{n \in \mathbb{N} : n \geq k\} \cup \{\infty\}, k \geq 1$. $S$ is then compact, and continuous functions on $S$ may be identified with convergent sequences $(\xi_n)_{n \geq 1}$. The value of such a function at $\{\infty\}$ is the limit of the appropriate sequence. In topology, $S$ is called the **one-point compactification** of $\mathbb{N}$. ☐

**2.2.38 Exercise**  Show that $C_0(\mathbb{G})$ where $\mathbb{G}$ is the Kisyński group is isometrically isomorphic to the Cartesian product of two copies of $C_0(\mathbb{R})$ with the norm $\|(x_1, x_{-1})\| = \|x\|_{C_0(\mathbb{R})} \vee \|x\|_{C_0(\mathbb{R})}$.

We now continue with examples of dense algebraic subspaces of some Banach spaces.

**2.2.39 Proposition**  Let $(\Omega, \mathcal{F}, \mu)$ be a measure space. The simple functions that are non-zero only on a set of finite measure form a dense algebraic subspace of $L^1(\Omega, \mathcal{F}, \mu)$.

*Proof*  It suffices to show that for a non-negative $x \in L^1(\Omega, \mathcal{F}, \mu)$, there exists a sequence of simple functions approximating $x$ that are non-zero only on a set of finite measure. We know, however, that the integral of a non-negative function $x$ equals the supremum of the integrals of simple functions bounded above by $x$. In particular, for any $n > 0$ we may find a simple function $x_n$ such that $x_n \leq x$ and $\int_\Omega x_n \, d\mu > \int_\Omega x \, d\mu - \frac{1}{n}$. This implies that $\|x - x_n\|_{L^1(\Omega, \mathcal{F}, \mu)} = \int_\Omega (x - x_n) \, d\mu < \frac{1}{n}$ as desired. Furthermore, the set where $x_n$ is non-zero must be finite, for $\int x_n \, d\mu \leq \int x \, d\mu < \infty$.

☐

**2.2.40 Exercise**  Prove an analogous result for $L^p(\Omega, \mathcal{F}, \mu), \infty > p > 1$.

**2.2.41 Proposition**   Let $(\Omega, \mathcal{F}, \mu)$ be a measure space with a finite measure $\mu$, and let $1 < p < \infty$ be a given number. The space $L^p(\Omega, \mathcal{F}, \mu)$ is dense in $L^1(\Omega, \mathcal{F}, \mu)$.

*Proof*   We need to show first that $L^p(\Omega, \mathcal{F}, \mu)$ is a subset of the space $L^1(\Omega, \mathcal{F}, \mu)$; this follows from the Hölder inequality; if $x$ belongs to $L^p(\Omega, \mathcal{F}, \mu)$ then it belongs to $L^1(\Omega, \mathcal{F}, \mu)$ since

$$\int_\Omega |x| \, d\mu = \int_\Omega |x| \cdot 1_\Omega \, d\mu \leq \left[ \int_\Omega |x|^p \, d\mu \right]^{\frac{1}{p}} [\mu(\Omega)]^{\frac{1}{p}} .$$

is finite.

Since $\mu$ is finite, any indicator function of a set $A \in \mathcal{F}$ belongs to $L^p(\Omega, \mathcal{F}, \mu)$, and any simple function belongs to $L^p(\Omega, \mathcal{F}, \mu)$, as the linear combination of indicator functions. Thus, the claim follows from 2.2.39. $\qquad\square$

**2.2.42 Exercise**   Find a counterexample showing that the last proposition is not true if $\mu$ is not finite.

**2.2.43 Exercise**   Use the Hölder inequality to show that $L^r(\Omega, \mathcal{F}, \mu) \subset L^s(\Omega, \mathcal{F}, \mu)$ for all $1 \leq s \leq r \leq \infty$, provided $\mu$ is finite.

**2.2.44 Proposition**   Let $\Omega$ be a finite or an infinite interval in $\mathbb{R}$ (open or closed). Then $C_c(\Omega)$ is dense in $L^p(\Omega, \mathcal{M}(A), leb)$, $\infty > p \geq 1$.

*Proof*   By 2.2.39 and 2.2.40 it suffices to show that a function $1_A$, where $A$ is measurable with finite $leb(A)$, belongs to the closure of $C_c(\Omega)$. By 1.2.9, we may restrict our attention to compact sets $A$.

Let $A$ be a compact set and let $k$ be a number such that $A \subset [-k, k]$. Let $B = (-\infty, -(k+1)] \cup [k+1, \infty)$, and $x_n(\tau) = \frac{d(\tau, B)}{nd(\tau, A) + d(\tau, B)}$, where $d(\tau, B) = \min_{\sigma \in B} |\tau - \sigma|$ and $d(\tau, A) = \min_{\sigma \in A} |\tau - \sigma|$. Note that $d(\tau, A)$ and $d(\tau, B)$ may not be simultaneously zero, and that $x_n(\tau) = 0$ for $\tau \in B$, and $x_n(\tau) = 1$ for $\tau \in A$. Finally, $x_n$ are uniformly bounded by 1, supported in $[-(k+1), k+1]$ and tend to $1_A$ pointwise, for if $\tau \notin A$, then $d(\tau, A) \neq 0$. By the Lebesgue Dominated Convergence Theorem $\lim_{n \to \infty} \|x_n - 1_A\|_{L^p(\Omega, \mathcal{M}(A), leb)} = 0$. $\qquad\square$

**2.2.45 Corollary**   The completion of the space $C[0, 1]$ equipped with the norm $\|x\| = \int_0^1 |x| \, d\, leb$ is $L^1[0, 1]$.

**2.2.46 Corollary**  Another well-known consequence of the above theorem is that for any $x \in L^1(\mathbb{R}^+)$ the function $y(\tau) = \int_0^\infty x(\tau + \sigma) \, d\sigma$ is continuous in $\tau \in \mathbb{R}^+$. This result is obvious for continuous $x$ with compact support; if the support of $x$ is contained in the interval $[0, K], K > 0$, then

$$|y(\tau + \varsigma) - y(\tau)| \le K \sup_{0 \le \sigma \le K} |x(\tau + \varsigma + \sigma) - x(\tau + \sigma)| \underset{\varsigma \to 0}{\to} 0,$$

by uniform continuity of $x$. To prove the general case, let $x_n$ be a sequence of continuous functions with compact support approximating $x$ and let $y_n(\tau) = \int_0^\infty x_n(\tau + \sigma) \, d\sigma$. Then

$$\sup_{\tau \ge 0} |y(\tau) - y_n(\tau)| \le \int_0^\infty |x(\tau + \sigma) - x_n(\tau + \sigma)| \, d\sigma$$

$$\le \|x - x_n\|_{L^1(\mathbb{R}^+)},$$

and $y$ is continuous as a uniform limit of continuous functions.

We close by exhibiting some examples illustrating how the fact that $\mathbb{X}$ is a Banach space may be used to show existence of a particular element in $\mathbb{X}$. We shall use such arguments quite often later: see e.g. 2.3.13 and 7.1.2.

**2.2.47 Example**  Suppose that $\mathbb{R}^+ \ni t \mapsto x_t \in \mathbb{X}$ is a function taking values in a Banach space $\mathbb{X}$, and that for every $\epsilon > 0$ there exists a $v > 0$ such that $\|x_t - x_s\| < \epsilon$ provided $s, t > v$. Then there exists an $x \in \mathbb{X}$ such that for any $\epsilon > 0$ there exists a $\delta > 0$ such that $\|x_t - x\| < \epsilon$ for $t > v$. We then write, certainly, $x = \lim_{t \to \infty} x_t$. To prove this note first that if $u_n$ is a numerical sequence such that $\lim_{n \to \infty} u_n = \infty$, then $y_n = x_{u_n}$ is a Cauchy sequence. Let $x$ be its limit. For $\epsilon > 0$ choose $v$ in such a way that $t, u > v$ implies $\|x_t - x_u\| < \epsilon$. Since $\lim u_n = \infty$, almost all numbers $u_n$ belong to $(v, \infty)$ so that $\|x_t - x\| = \lim_{n \to \infty} \|x_t - x_{u_n}\| \le \epsilon$. This implies the claim. Finally, note that there may be no two distinct elements $x$ with the required property.

**2.2.48 *Riemann integral in a Banach space***  The completeness of Banach spaces allows us to extend the notion of the Riemann integral to the case of Banach space valued functions. Let $a < b$ be two real numbers. Suppose that $x. : [0, 1] \to \mathbb{X}, t \mapsto x_t$ is function on $[a, b]$ taking values in a normed linear space $\mathbb{X}$. Consider two sequences, $\mathcal{T} = (t_i)_{i=0,..,k}$ and

$\Xi = (\xi_i)_{i=0,\ldots,k-1}$ of points of $[a,b]$, where $k$ is an integer, such that

$$a = t_0 < t_1 < \cdots < t_k = b, \qquad t_0 \le \xi_0 \le t_1 \le \cdots \le t_{k-1} \le \xi_{k-1} \le t_k. \tag{2.7}$$

Define the related number $\Delta(\mathcal{T}) = \sup_{0 \le i \le k}(t_i - t_{i-1})$ and the element of $\mathbb{X}$ given by

$$S(\mathcal{T}, \Xi, x.) = \sum_{i=0}^{k} x_{\xi_i}(t_{i+1} - t_i).$$

If the limit $\lim_{n \to \infty} S(\mathcal{T}_n, \Xi_n, x.)$ exists for any sequence of pairs $(\mathcal{T}_n, \Xi_n)$ such that $\lim_{n \to \infty} \Delta(\mathcal{T}_n) = 0$, and does not depend on the choice of the sequence of $(\mathcal{T}_n, \Xi_n)$, function $x$ is said to be **Riemann integrable**. The above limit is denoted $\int_a^b x_t \, dt$ and called the **(Riemann) integral** of $x$. We shall prove that continuous functions taking values in a Banach space are Riemann integrable.

To this end, consider a continuous function $[a,b] \ni t \mapsto x_t \in \mathbb{X}$, and let $\epsilon > 0$ be given. Since $x$ is continuous on a compact interval, it is uniformly continuous and we may choose a $\delta > 0$ such that $|s - t| < \delta$ and $s,t \in [a,b]$ implies $\|x_s - x_t\| < \epsilon$. Let sequences $\mathcal{T} = (t_i)_{i=0,\ldots,k}$ and $\mathcal{T}' = (t'_i)_{i=0,\ldots,k'}$ be such that $\Delta(\mathcal{T}) < \delta$ and $\Delta(\mathcal{T}') < \delta$. Also, let $\mathcal{T}''$ be a sequence that contains all elements of $\mathcal{T}$ and $\mathcal{T}' : \mathcal{T}'' = (t''_i)_{i=1,\ldots,k''}$. We have $\Delta(\mathcal{T}'') < \delta$ and $k'' \le k + k' - 2$, for besides $t_0 = t'_0 = a$ and $t_k = t'_{k'} = b$ there may be some $t_i = t'_j$, $i = 1,\ldots,k-1, j = 1,\ldots,k'-1$. An interval $[t_i, t_{i+1}]$, $i = 0,\ldots,k$ either coincides with some $[t''_j, t''_{j+1}], j \in \{0,\ldots,k''-1\}$ or is a finite union of such intervals, say, $[t_i, t_{i+1}] = [t''_j, t''_{j+1}] \cup \ldots \cup [t''_{j+l}, t''_{j+l+1}]$ for some $l$. For any $\Xi = (\xi_i)_{i=0,\ldots,k}$ such that (2.7) holds,

$$\left\| x_{\xi_i}(t_{i+1} - t_i) - \sum_{m=0}^{l} x_{t''_{j+m}}(t''_{j+m+1} - t''_{j+m}) \right\|$$

$$= \left\| \sum_{m=0}^{l} [x_{\xi_i} - x_{t''_{j+m}}](t''_{j+m+1} - t''_{j+m}) \right\|$$

$$\le \epsilon \sum_{m=0}^{l} (t''_{j+m+1} - t''_{j+m}) = \epsilon(t_{i+1} - t_i),$$

since both $\xi_i$ and $t''_{j+m}$ belong to $[t_i, t_{i+1}]$, so that $|\xi_i - t''_{j+m}| < \delta$. Summing over $i$ we obtain

$$\|S(\mathcal{T}, \Xi, x) - S(\mathcal{T}'', \Xi'', x)\| \le \epsilon(b - a),$$

where $\Xi'' = (\xi_i'')_{i=0,\ldots,k''-1} \equiv (t_i'')_{i=0,\ldots,k''-1}$. This argument works for $\mathcal{T}'$ and $\mathcal{T}''$, as well. Hence

$$\|S(\mathcal{T},\Xi,x) - S(\mathcal{T}',\Xi',x)\| \leq 2\epsilon(b-a), \qquad (2.8)$$

for all sequences $\mathcal{T}$ and $\mathcal{T}'$ such that $\Delta(\mathcal{T}) < \delta$ and $\Delta(\mathcal{T}') < \delta$ and any sequences $\Xi$ and $\Xi'$ of appropriate midpoints. This proves that for any sequence $(\mathcal{T}_n, \Xi_n)$ such that $\lim_{n\to\infty} \Delta(\mathcal{T}_n) = 0$, $S(\mathcal{T}_n, \Xi_n, x)$ is a Cauchy sequence, and thus converges. Using (2.8) we prove that this limit does not depend on the choice of $(\mathcal{T}_n, \Xi_n)$.

More general, Lebesgue-type integrals may be introduced for functions with values in Banach spaces. We need to mention the Bochner and Pettis integrals here, see [54] for example. For our purposes, though, the Riemann integral suffices.

2.2.49 **Example**    Consider the elements $e_\lambda, \lambda > 0$, and $u(t), t \geq 0$ of $L^1(\mathbb{R}^+)$ defined by their representatives $e_\lambda(\tau) = e^{-\lambda\tau}$ and $u(t) = 1_{[0,t)}$, respectively. Let $\mathbb{R}^+ \ni t \mapsto u(t) = 1_{[0,t)}$ be the function with values in $L^1(\mathbb{R}^+)$. We will check that

$$\lambda \int_0^\infty e^{-\lambda t} u(t)\, dt = e_\lambda \quad \text{in } L^1(\mathbb{R}^+), \quad \text{for } \lambda > 0.$$

The above integral is an improper Riemann integral, i.e. we have

$$\int_0^\infty e^{-\lambda t} u(t)\, dt := \lim_{T\to\infty} \int_0^T e^{-\lambda t} u(t)\, dt.$$

We start by noting that $\|u(t) - u(s)\|_{L^1(\mathbb{R}^+)} = \|1_{[s,t)}\|_{L^1(\mathbb{R}^+)} = \int_s^t d\tau = (t-s), t \geq s$ so that $u$ is continuous, and so is the integrand above. Fix $T > 0$. We have:

$$\int_0^T e^{-\lambda t} u(t)\, dt = \lim_{n\to\infty} \frac{T}{n} \sum_{k=1}^n e^{-\lambda\frac{Tk}{n}} u\left(Tkn^{-1}\right) =: \lim_{n\to\infty} f_{T,n}$$

with the limit taken in $L^1(\mathbb{R}^+)$. Note that $f_{T,k}(\tau)$ equals

$$\frac{T}{n} \sum_{k=[\frac{\tau n}{T}]+1}^n e^{-\frac{\lambda Tk}{n}} 1_{[0,T)}(\tau) = \frac{\frac{T}{n}}{e^{\frac{\lambda T}{n}} - 1}\left[e^{-\frac{\lambda T}{n}[\frac{\tau n}{T}]} - e^{-\lambda T}\right] 1_{[0,T)}(\tau).$$

Certainly, the expression in brackets does not exceed $e^{-\lambda\tau}$ and tends to $e^{-\lambda\tau} - e^{-\lambda T}$ as $n \to \infty$, while the sequence before the brackets tends to $\frac{1}{\lambda}$. Hence, by the Dominated Convergence Theorem, $\lim_{n\to\infty} f_{T,n} =$

$\frac{1}{\lambda}(e_\lambda - e^{-\lambda T})u(T)$ in $L^1(\mathbb{R}^+)$. Thus,

$$\lambda \int_0^T e^{-\lambda t}u(t)dt = (e_\lambda - e^{-\lambda T})u(T).$$

Finally, $\|(e_\lambda - e^{-\lambda T})u(T) - e_\lambda\|_{L^1(\mathbb{R}^+)}$ does not exceed

$$\|e_\lambda u(T) - e_\lambda\|_{L^1(\mathbb{R}^+)} + e^{-\lambda T}\|u(T)\|_{L^1(\mathbb{R}^+)}$$
$$= \int_T^\infty e^{-\lambda \tau}\,d\tau + Te^{-\lambda T} = (1 + T)e^{-\lambda T}$$

which converges to 0 as $T \to \infty$.

**2.2.50 Exercise** Let $\mathbb{X}$ be a Banach space, and suppose that $t \mapsto x_t \in \mathbb{X}$ is continuous in an interval $[a, b]$. The scalar-valued function $t \to \|x_t\|$ is then continuous, and therefore integrable. Show that $\left\|\int_a^b x_t\,dt\right\| \leq \int_a^b \|x_t\|\,dt$.

## 2.3 The space of bounded linear operators

Throughout this section, $(\mathbb{X}, \|\cdot\|_\mathbb{X})$ and $(\mathbb{Y}, \|\cdot\|_\mathbb{Y})$ are two linear normed spaces. From now on, to simplify notation, we will denote the zero vector in both spaces by 0.

**2.3.1 Definition** A linear map $L : \mathbb{X} \to \mathbb{Y}$ is said to be **bounded** if $\|Lx\|_\mathbb{Y} \leq M\|x\|_\mathbb{X}$ for some $M \geq 0$. If $M$ can be chosen equal to 1, $L$ is called a **contraction**. In particular, isometric isomorphisms are contractions. Linear contractions, i.e. linear operators that are contractions are very important for the theory of stochastic processes, and appear often.

**2.3.2 Definition** As in 2.1.12, we show that the collection $\mathcal{L}(\mathbb{X}, \mathbb{Y})$ of continuous linear operators from $\mathbb{X}$ to $\mathbb{Y}$ is an algebraic subspace of $L(\mathbb{X}, \mathbb{Y})$. $\mathcal{L}(\mathbb{X}, \mathbb{Y})$ is called the space of **bounded (or continuous) linear operators** on $\mathbb{X}$ with values in $\mathbb{Y}$. The first of these names is justified by the fact that a linear operator is bounded iff it is continuous, as proved below. If $\mathbb{X} = \mathbb{Y}$ we write $\mathcal{L}(\mathbb{X})$ instead of $\mathcal{L}(\mathbb{X}, \mathbb{Y})$ and call this space the space of **bounded linear operators on** $\mathbb{X}$. If $\mathbb{Y} = \mathbb{R}$, we write $\mathbb{X}^*$ instead of $\mathcal{L}(\mathbb{X}, \mathbb{Y})$ and call it the space of **bounded linear functionals** on $\mathbb{X}$.

**2.3.3 Theorem**        Let $L$ belong to $L(\mathbb{X}, \mathbb{Y})$ (see 2.1.7). The following conditions are equivalent:

(a) $L$ is continuous ($L \in \mathcal{L}(\mathbb{X}, \mathbb{Y})$),
(b) $L$ is continuous at some $x \in \mathbb{X}$,
(c) $L$ is continuous at zero,
(d) $\sup_{\|x\|_{\mathbb{X}}=1} \|Lx\|_{\mathbb{Y}}$ is finite,
(e) $L$ is bounded.

Moreover, $\sup_{\|x\|_{\mathbb{X}}=1} \|Lx\|_{\mathbb{Y}} = \min\{M \in \mathcal{M}\}$ where $\mathcal{M}$ is the set of constants such that $\|Lx\|_{\mathbb{Y}} \leq M\|x\|_{\mathbb{X}}$ holds for all $x \in \mathbb{X}$.

*Proof* The implication $(a) \Rightarrow (b)$ is trivial. If a sequence $x_n$ converges to zero, then $x_n + x$ converges to $x$. Thus, if (b) holds, then $L(x_n + x)$, which equals $Lx_n + Lx$, converges to $Lx$, i.e. $Lx_n$ converges to 0, showing (c). To prove that (c) implies (d), assume that (d) does not hold, i.e. there exists a sequence $x_n$ of elements of $\mathbb{X}$ such that $\|x_n\|_{\mathbb{X}} = 1$ and $\|Lx_n\| > n$. Then the sequence $y_n = \frac{1}{\sqrt{n}} x_n$ converges to zero, but $\|Ly_n\|_{\mathbb{Y}} > \sqrt{n}$ must not converge to zero, so that (c) does not hold. That (d) implies (e) is seen by putting $M = \sup_{\|x\|_{\mathbb{X}}} \|Lx\|_{\mathbb{Y}}$; indeed, the inequality in the definition 2.3.1 is trivial for $x = 0$, and for a non-zero vector $x$, the norm of $\frac{1}{\|x\|} x$ equals one, so that $\|L\frac{1}{\|x\|} x\|_{\mathbb{Y}} \leq M$, from which (e) follows by multiplying both sides by $\|x\|$. Finally, (a) follows from (e), since $\|Lx_n - Lx\| \leq \|L(x_n - x)\| \leq M\|x_n - x\|$.

To prove the second part of the theorem, note that in the proof of the implication (d)$\Rightarrow$(e) we showed that $M_1 = \sup_{\|x\|_{\mathbb{X}}=1} \|Lx\|_{\mathbb{Y}}$ belongs to $\mathcal{M}$. On the other hand, if $\|Lx\|_{\mathbb{Y}} \leq M\|x\|_{\mathbb{X}}$ holds for all $x \in X$, then considering only $x$ with $\|x\|_{\mathbb{X}} = 1$ we see that $M_1 \leq M$ so that $M_1$ is the minimum of $\mathcal{M}$. $\qquad\square$

**2.3.4 Exercise**        Suppose that $a < b$ are two real numbers and that $[a, b] \ni t \mapsto x_t$ is a Riemann integrable function taking values in a Banach space $\mathbb{X}$. Let $A$ be a bounded linear operator mapping $\mathbb{X}$ into a Banach space $\mathbb{Y}$. Prove that $[a, b] \ni t \mapsto Ax_t \in \mathbb{Y}$ is Riemann integrable, and $A\int_a^b x_t \, dt = \int_a^b Ax_t \, dt$.

**2.3.5 Example**        Let $(\Omega, \mathcal{F}, \mu)$ and $(\Omega', \mathcal{F}', \nu)$ be two measure spaces. Suppose that $k(\omega, \omega')$ is bounded (say, by $M$) and measurable with respect to the product $\sigma$-algebra $\mathcal{F} \otimes \mathcal{F}'$ in $\Omega \otimes \Omega'$. Consider the linear operator $K : L^1(\Omega', \mathcal{F}', \nu) \to L^\infty(\Omega, \mathcal{F}, \mu)$ given by $(Kx)(\omega) = $

$\int_{\Omega'} k(\omega, \omega') x(\omega') \, d\nu(\omega')$. The estimate

$$|(Kx)(\omega)| \leq \int_{\Omega'} |k(\omega, \omega') x(\omega')| \, d\nu(\omega') \leq M \|x\|_{L^1(\Omega', \mathcal{F}', \nu)}$$

shows that $K$ is a bounded linear operator from the space $L^1(\Omega', \mathcal{F}', \nu)$ to $L^\infty(\Omega, \mathcal{F}, \mu)$.

**2.3.6 Example** For $\lambda > 0$ define $A_\lambda : BM(\mathbb{R}^+) \to BM(\mathbb{R}^+)$ by $(A_\lambda x)(\tau) = e^{-\lambda \tau} \sum_{n=0}^{\infty} \frac{(\lambda \tau)^n}{n!} x\left(\frac{n}{\lambda}\right)$. This series converges uniformly on all compact subintervals of $\mathbb{R}^+$ and its sum does not exceed $\|x\|$. Thus, $A_\lambda x$ belongs to $BM(\mathbb{R}^+)$ for all $\lambda > 0$; in fact it belongs to $BC(\mathbb{R}^+)$. Moreover, $A_\lambda$ maps the space $C_0(\mathbb{R}^+)$ of continuous functions on $\mathbb{R}^+$ that vanish at infinity into itself. To prove this, note that for any $x \in C_0(\mathbb{R}^+)$ and $\epsilon > 0$, we may find $T > 0$ and a function $x_T \in C_0(\mathbb{R}^+)$ such that $x_T(\tau) = 0$ whenever $\tau \geq T$ and $\|x - x_T\| \leq \epsilon$. Moreover, $A_\lambda x_T(\tau) = \sum_{n=0}^{[\lambda T]} x_T\left(\frac{n}{\lambda}\right) e^{-\lambda \tau} \frac{(\lambda \tau)^n}{n!}$ is a finite sum of members of $C_0(\mathbb{R}^+)$ and we have $\|A_\lambda x - A_\lambda x_T\| \leq \epsilon$. This proves that $A_\lambda$ belongs to the closure of $C_0(\mathbb{R}^+)$, which equals $C_0(\mathbb{R}^+)$.

Finally, $A_\lambda$ maps the space $C(\overline{\mathbb{R}^+})$ of continuous functions with limit at infinity into itself. To prove this consider an $x \in C(\overline{\mathbb{R}^+})$ and let $\kappa = \lim_{\tau \to \infty} x(\tau)$. Then $x - \kappa 1_{\mathbb{R}^+} \in C_0(\mathbb{R}^+)$ and

$$A_\lambda x = A_\lambda(x - \kappa 1_{\mathbb{R}^+}) + \kappa A_\lambda 1_{\mathbb{R}^+} = A_\lambda(x - \kappa 1_{\mathbb{R}^+}) + \kappa 1_{\mathbb{R}^+}$$

belongs to $C(\overline{\mathbb{R}^+})$, as desired.

Thus $A_\lambda$ is a linear contraction in $BM(\mathbb{R}^+), C_0(\mathbb{R}^+)$ and $C(\overline{\mathbb{R}^+})$.

**2.3.7 Definition** Let $L \in \mathcal{L}(\mathbb{X}, \mathbb{Y})$ be a bounded linear operator. The number $\|L\| = \sup_{\|x\|_{\mathbb{X}} = 1} \|Lx\|_{\mathbb{Y}}$, often denoted $\|L\|_{\mathcal{L}(\mathbb{X}, \mathbb{Y})}$ or simply $\|L\|$, is called the **norm of the operator**.

**2.3.8 Example** In 2.2.46 we showed that for any $x \in L^1(\mathbb{R}^+)$ the function $Tx(\tau) = \int_0^\infty x(\tau + \sigma) \, d\sigma$ is continuous. Obviously,

$$\sup_{\tau \geq 0} |Tx(\tau)| \leq \|x\|_{L^1(\mathbb{R}^+)}.$$

Hence $T$ maps $L^1(\mathbb{R}^+)$ into $BC(\mathbb{R}^+)$ and $\|T\| \leq 1$. Moreover, $Tx = x$ since $x(\tau) = e^{-\tau}$ and $\|x\|_{BC(\mathbb{R}^+)} = \|x\|_{L^1(\mathbb{R}^+)} = 1$, proving that $\|T\| = 1$.

**2.3.9 Exercise** Let $l^1$ be the space of absolutely summable sequences $x = (\xi_n)_{n \geq 1}$ with the norm $\|x\| = \sum_{n=1}^{\infty} |\xi_n|$. Let

$$L(\xi_n)_{n \geq 1} = (\xi_{n+1})_{n \geq 1}, \quad R(\xi_n)_{n \geq 1} = (\xi_{n-1})_{n \geq 1}$$

($\xi_0 = 0$, $L$ and $R$ stand for "left" and "right", respectively) be translation operators, and let $I$ be the identity in $l^1$. Show that $\|aL + bR + cI\| = |a| + |b| + |c|$, where $a, b$ and $c$ are real numbers.

**2.3.10 Proposition** The space $\mathcal{L}(\mathbb{X}, \mathbb{Y})$ of bounded linear operators equipped with the norm $\| \cdot \|_{\mathcal{L}(\mathbb{X}, \mathbb{Y})}$ is a normed linear space. Moreover, if $\mathbb{Y}$ is a Banach space, then so is $\mathcal{L}(\mathbb{X}, \mathbb{Y})$.

*Proof* We need to check conditions (n1)–(n4) of the definition of a norm. (n1)–(n3) are immediate. To prove (n4) we calculate:

$$\begin{aligned}
\|L + M\| &= \sup_{\|x\|_{\mathbb{X}}=1} \|Lx + Mx\|_{\mathbb{Y}} \leq \sup_{\|x\|_{\mathbb{X}}=1} \{\|Lx\|_{\mathbb{Y}} + \|Mx\|_{\mathbb{Y}}\} \\
&\leq \sup_{\|x\|_{\mathbb{X}}=1} \|Lx\|_{\mathbb{Y}} + \sup_{\|x\|_{\mathbb{X}}=1} \|Mx\|_{\mathbb{Y}} = \|L\| + \|M\|.
\end{aligned}$$

Let $L_n$ be a Cauchy sequence in $\mathcal{L}(\mathbb{X}, \mathbb{Y})$. For any $x \in \mathbb{X}$, the sequence $L_n x$ is Cauchy in $\mathbb{Y}$, since $\|L_n x - L_m x\|_{\mathbb{Y}} \leq \|L_n - L_m\|_{\mathcal{L}(\mathbb{X}, \mathbb{Y})} \|x\|_{\mathbb{X}}$. Let $Lx = \lim_{n \to \infty} L_n x$. It may be checked directly that $L$ is a linear operator. We repeat the argument from 2.2.15 to show that $L$ is the limit of $L_n$ in $\mathcal{L}(\mathbb{X}, \mathbb{Y})$. For arbitrary $\epsilon > 0$ there exists an $n_0$ such that $\|L_n x - L_m x\|_{\mathbb{Y}} \leq \epsilon \|x\|_{\mathbb{X}}$, for $n, m \geq n_0$. Taking the limit, as $m \to \infty$, we obtain $\|L_n x - Lx\|_{\mathbb{Y}} \leq \epsilon \|x\|_{\mathbb{X}}$, for $x \in \mathbb{X}, n \geq n_0$. Thus, $\sup_{\|x\|_{\mathbb{X}}=1} \|Lx\|_{\mathbb{Y}} \leq \epsilon + \sup_{\|x\|_{\mathbb{X}}=1} \|L_n x\|_{\mathbb{Y}} < \infty$, so that $L \in \mathcal{L}(\mathbb{X}, \mathbb{Y})$. Also $\|L_n x - Lx\|_{\mathbb{Y}} \leq \epsilon \|x\|_{\mathbb{X}}$ for $x \in \mathbb{X}$ is equivalent to $\|L - L_n\|_{\mathcal{L}(\mathbb{X}, \mathbb{Y})} \leq \epsilon$, which completes the proof. $\square$

**2.3.11 Exercise** Assume that $\mathbb{X}, \mathbb{Y}$ and $\mathbb{Z}$ are normed linear spaces, and let $L \in \mathcal{L}(\mathbb{X}, \mathbb{Y})$ and $K \in \mathcal{L}(\mathbb{Y}, \mathbb{Z})$. Then the composition $K \circ L$ of $K$ and $L$ (in the sequel denoted simply $KL$) is a bounded linear operator from $\mathbb{X}$ to $\mathbb{Z}$ and

$$\|KL\|_{\mathcal{L}(\mathbb{X}, \mathbb{Z})} \leq \|K\|_{\mathcal{L}(\mathbb{X}, \mathbb{Y})} \|L\|_{\mathcal{L}(\mathbb{Y}, \mathbb{Z})}.$$

**2.3.12 Exercise** Let $A_i, B_i, i = 1, ..., n$ be linear operators in a Banach space and let $M = \max_{i=1,...,n} \{\|A_i\|, \|B_i\|\}$. Then

$$\|A_n A_{n-1} ... A_1 - B_n B_{n-1} ... B_1\| \leq M^{n-1} \sum_{i=1}^{n} \|A_i - B_i\|. \tag{2.9}$$

In particular, for any $A$ and $B$ and $M = \max\{\|A\|, \|B\|\}$,

$$\|A^n - B^n\| \leq M^{n-1} n \|A - B\|. \tag{2.10}$$

**2.3.13 Exercise** Let $A$ be a bounded linear operator in a Banach space $\mathbb{X}$, and let $A^n = A \circ A^{n-1}, n \geq 2$ be its $n$th power. Prove that the series $\sum_{n=0}^{\infty} \frac{t^n A^n}{n!}$ converges for all $t \in \mathbb{R}$ in the operator topology, i.e. with respect to norm $\|\cdot\|_{\mathcal{L}(\mathbb{X})}$. Let $e^{tA}$ denote its sum. Show that for any real numbers $s$ and $t$, $e^{(t+s)A} = e^{tA} e^{sA}$. In other words, $\{e^{tA}, t \in \mathbb{R}\}$ is a *group* of operators (see Chapter 7). We often write $\exp(tA)$ instead of $e^{tA}$.

**2.3.14 Exercise** Let $A$ and $B$ be two bounded linear operators in a Banach space such that $AB = BA$. Show that $e^{t(A+B)} = e^{tA} e^{tB} = e^{tB} e^{tA}$.

**2.3.15 Exercise** Suppose that $A \in \mathcal{L}(\mathbb{X})$ is an operator such that $\|I - A\| < 1$. Then, we may define $\log A = -\sum_{n=1}^{\infty} \frac{1}{n}(I - A)^n$. Prove that $\exp(\log A) = A$.

**2.3.16 Exercise** Under notations of 2.3.9 show that

$$\|e^{aL + bR + cI}\| = e^{a+b+c}$$

where $a, b \in \mathbb{R}^+$ and $c \in \mathbb{R}$.

**2.3.17** *Measures as operators* In what follows $BM(\mathbb{R})$ will denote the space of bounded Borel measurable functions on $\mathbb{R}$, equipped with the supremum norm, and $BC(\mathbb{R})$ its subspace composed of bounded continuous functions. $BUC(\mathbb{R})$ will denote the subspace of bounded uniformly continuous functions on $\mathbb{R}$, and $C_0(\mathbb{R})$ the space of continuous functions that vanish at infinity.

Given a finite measure $\mu$ on $(\mathbb{R}, \mathcal{B}(\mathbb{R}))$ we may define an operator $T_\mu$ acting in $BM(\mathbb{R})$ by the formula

$$(T_\mu x)(\tau) = \int_{\mathbb{R}} x(\tau + \varsigma) \, d\mu(\varsigma). \tag{2.11}$$

Let us first check that $T_\mu$ indeed maps $BM(\mathbb{R})$ into itself. If $x = 1_{(a,b]}$ for some real numbers $a < b$, then $T_\mu x(\tau) = \mu(-\infty, b - \tau] - \mu(-\infty, a - \tau]$ is of bounded variation and hence measurable. The class $\mathcal{G}$ of measurable sets such that $T_\mu 1_A$ is measurable may be shown to be a $\lambda$-system. This class contains a $\pi$-system of intervals $(a, b]$ (plus the empty set). By the Sierpiński $\pi$–$\lambda$ theorem, the $\sigma$-algebra generated by such intervals is a subset of $\mathcal{G}$, and on the other hand, by 1.2.17 it is equal to the Borel $\sigma$-algebra. Hence, $T_\mu x$ is measurable for any $x = 1_A$, where $A \in \mathcal{B}(\mathbb{R})$. By linearity of the integral this is also true for simple functions $x$. Since pointwise limits of measurable functions are measurable, we extend this result to all $x \in BM(\mathbb{R})$. That $T_\mu x$ is bounded follows directly from the estimate given below. We have also noted that the map is linear (integration being linear). As for its boundedness, we have

$$\|T_\mu x\| \le \sup_{\tau \in \mathbb{R}} |(T_\mu x)(\tau)| \le \sup_{\tau \in \mathbb{R}} \sup_{\varsigma \in \mathbb{R}} |x(\tau + \varsigma)| \mu(\mathbb{R}) = \|x\| \mu(\mathbb{R}), \quad (2.12)$$

with equality for $x = 1_\mathbb{R}$. Thus, $\|T_\mu\| = \mu(\mathbb{R})$ and in particular $\|T_\mu\| = 1$ for a probability measure $\mu$. We note also that $T_\mu$ leaves the subspaces $BC(\mathbb{R})$ and $BUC(\mathbb{R})$ **invariant**, meaning that $T_\mu$ maps these spaces into themselves. The former assertion follows from the Lebesgue Dominated Convergence Theorem, and the latter from the estimate

$$|(T_\mu x)(\tau) - (T_\mu x)(\varsigma)| \le \sup_{\upsilon \in \mathbb{R}} |x(\tau + \upsilon) - x(\varsigma + \upsilon)| \, \mu(\mathbb{R}), \quad \tau, \varsigma \in \mathbb{R}.$$

Analogously, we may prove that $T_\mu$ maps $C_0(\mathbb{R})$ into itself.

An important property of $T_\mu$ (as an operator in $C_0(\mathbb{R})$) is that it determines $\mu$, meaning that if $T_\mu = T_\nu$ for two measures $\mu$ and $\nu$, then $\mu = \nu$. Indeed, $T_\mu = T_\nu$ implies in particular that for any $x \in C_0(\mathbb{R})$, $(T_\mu x)(0) = (T_\nu x)(0)$, i.e. $\int_\mathbb{R} x \, d\mu = \int_\mathbb{R} x \, d\nu$, which implies $\mu = \nu$ by 1.2.20. In other words, the map $\mu \mapsto T_\mu$ is a linear invertible map from $\mathbb{BM}(\mathbb{R})$ into $\mathcal{L}(BM(\mathbb{R}))$ (right now this map is defined only on a subset of $\mathbb{BM}(\mathbb{R})$, see 2.3.20 below, however). The same is true for $T_\mu$ as an operator in $BUC(\mathbb{R}), BC(\mathbb{R})$ and $BM(\mathbb{R})$.

Another important property of the map $\mu \mapsto T_\mu$ is related to the notion of the convolution of two finite measures on $\mathbb{R}$. Note that we have:

$$\begin{aligned} (T_\mu T_\nu x)(\tau) &= \int (T_\nu x)(\tau + \varsigma) \, \mu(d\varsigma) = \int \int x(\tau + \varsigma + \upsilon) \, \nu(d\upsilon) \mu(d\varsigma) \\ &= \int x(\tau + \rho)(\mu * \nu)(d\rho) = T_{\mu * \nu}(\tau). \end{aligned} \quad (2.13)$$

In words, $\mu \mapsto T_\mu$ changes convolution into operator composition. Functional analysts say that this map is a **homomorphism** of two Banach algebras (see Exercise 2.3.20 and Chapter 6).

**2.3.18 Exercise**    Find (a) the operator $T_\mu$ related to the normal distribution with parameters $m$ and $\sigma$, (b) the operator related to a uniform distribution, and (c) the operator related to the exponential distribution with parameter $\lambda > 0$.

**2.3.19 Exercise**   Let $X$ be a non-negative integer-valued random variable. Prove that $X$ is a Poisson variable with parameter $a > 0$ iff for any bounded function $g$ on the non-negative integers, $Xg(X)$ is integrable and $E\,Xg(X) = aE\,g(X+1)$. Analogously, show that $X$ has the geometric distribution with parameter $p$ iff for all bounded functions $g$ on the non-negative integers with $g(0) = 0$, $E\,g(X) = qE\,g(X+1)$, where $q = 1 - p$.

**2.3.20 Exercise**   Introduce the bounded linear operator related to a Borel charge $\mu$ on $\mathbb{R}$ and prove that for any two such Borel charges $\mu$ and $\nu$, $T_\mu T_\nu = T_{\mu * \nu}$ where

$$\mu * \nu = \mu^+ * \nu^+ + \mu^- * \nu^- - \mu^- * \nu^+ - \mu^+ * \nu^- \qquad (2.14)$$

with obvious notation. Of course, relation (2.14) is a result of viewing a signed measure as a difference of two positive measures and extending the operation of convolution by linearity. Note that for all $\mu, \nu \in \mathbb{BM}(\mathbb{R})$ the operators $T_\mu$ and $T_\nu$ commute:

$$T_\mu T_\nu = T_\nu T_\mu.$$

In particular, taking $\nu$ to be the Dirac measure at some point $t \in \mathbb{R}$, we see that all $T_\mu$ commute with translations. See 5.2.13 in Chapter 5 for a converse of this statement.

**2.3.21** *Operators related to random variables*   If $X$ is a random variable, we may assign to it the operator $T_X$ defined by $T_X = T_{\mathbb{P}_X}$ where $\mathbb{P}_X$ is the distribution of $X$. We thus have

$$T_X x(\tau) = \int_{\mathbb{R}} x(\tau + \varsigma)\,\mathbb{P}_X(\mathrm{d}\varsigma) = \int_{\Omega} x(\tau + X)\,\mathrm{d}\mathbb{P} = E\,x(\tau + X).$$

Note that if random variables $X$ and $Y$ are independent, then $\mathbb{P}_{X+Y} = \mathbb{P}_X * \mathbb{P}_Y$. Thus, $T_{X+Y} = T_X T_Y$.

However, while the map $\mu \to T_\mu$ preserves all information about $\mu$, $T_X$ does not determine $X$. In particular, we are not able to recover any information about the original probability space where $X$ was defined. In fact, as suggested by the very definition, all we can recover from $T_X$ is the distribution of $X$.

As an example, observe that the operator $A_\lambda x(\tau)$ described in 2.3.6 is related to the random variable $\frac{1}{\lambda} X_{\lambda\tau}$, where $X_{\lambda\tau}$ has the Poisson distribution with parameter $\lambda\tau$.

**2.3.22 Exercise** Find operators related to the random variables $X_n$ and $S_n$ described in 1.4.6. Prove that $T_{S_n} = (T_{X_n})^n$.

**2.3.23 *Measure as an operator on* $\mathbb{BM}(\mathbb{R})$** With a Borel charge $\mu$ on $\mathbb{R}$ we may also associate the operator $S_\mu$ on $\mathbb{BM}(\mathbb{R})$ defined by the formula $S_\mu \nu = \mu * \nu$. Using (2.14) we show that $\|\mu * \nu\| \leq \|\mu\| \, \|\nu\|$ so that $S_\mu$ is bounded and $\|S_\mu\| \leq \|\mu\|$. Moreover, $S_\mu \delta_0 = \mu$ where $\delta_0$ is the Dirac measure at 0, proving that $\|S_\mu\| = \|\mu\|$. Certainly, $S_\mu S_\nu = S_{\mu * \nu}$.

**2.3.24 Exercise** Find (a) the operator $S_\mu$ related to the normal distribution with parameters $m$ and $\sigma$, (b) the operator related to a uniform distribution, and (c) the operator related to the exponential distribution with parameter $\lambda > 0$.

**2.3.25 *Borel measures as operators on a locally compact group*** The results of the foregoing subsections may be generalized as follows. For a finite, possibly signed, measure $\mu$ on a locally compact group we define an operator $S_\mu$ on $\mathbb{BM}(\mathbb{G})$ as $S_\mu \nu = \mu * \nu$. Arguing as in 2.3.23, it can be shown that $S_\mu$ is a bounded linear operator with $\|S_\mu\| = \|\mu\|$ and $S_\mu S_\nu = S_{\mu * \nu}$.

We may also define the operators $T_\mu$ and $\tilde{T}_\mu$ by the formulae

$$T_\mu x(g) = \int x(hg) \, \mu(\mathrm{d}h)$$

and

$$\tilde{T}_\mu x(g) = \int x(gh) \, \mu(\mathrm{d}h).$$

Note that these formulae define two different operators unless $\mathbb{G}$ is commutative. In 5.3.1 we shall see that $S_\mu$ is related to $T_\mu$ as the operator $\tilde{S}_\mu \nu = \nu * \mu$ is related to $\tilde{T}_\mu$.

For now, we need only determine where the operator $T_\mu$ is defined. It may be shown that it maps $BM(\mathbb{G})$ into itself. The operator also maps the space $BUC(\mathbb{G})$ into itself. The space $BUC(\mathbb{G})$ is the space of bounded functions $x$ on $\mathbb{G}$ such that for every $\epsilon > 0$ there exists a neighborhood $\mathcal{U}$ of $e$ (the neutral element of $\mathbb{G}$) such that †

$$|x(g_1) - x(g_2)| < \epsilon, \qquad \text{whenever } g_1 g_2^{-1} \in \mathcal{U}. \tag{2.15}$$

---

† The space $BUC(\mathbb{G})$ thus defined is actually the space of bounded functions that are uniformly continuous with respect to the right uniform structure on $\mathbb{G}$; the space of functions that are uniformly continuous with respect to the left uniform structure is defined by replacing $g_1 g_2^{-1}$ in (2.15) with $g_2^{-1} g_1$ - see [51]. We will not distinguish between the two notions of uniform continuity because at all groups considered in this book these two notions are equivalent.

$BUC(\mathbb{G})$ is a Banach space, and to prove that $T_\mu$ maps $BUC(\mathbb{G})$ into itself it suffices to note that

$$|T_\mu x(g_1) - T_\mu x(g_2)| \leq \|\mu\| \sup_{h \in \mathbb{G}} |x(g_1 h) - x(g_2 h)|$$

and that $g_1 h(g_2 h)^{-1} = g_1 g_2^{-1}$, so that the right-hand side above does not exceed $\|\mu\|\epsilon$, if $g_1$ and $g_2$ are chosen as in (2.15).

It turns out that $C_0(\mathbb{G}) \subset BUC(\mathbb{G})$ and that $T_\mu$ maps $C_0$ into itself. Since the proof, although simple, requires more knowledge of topology of locally compact spaces we shall omit it here. Besides, the techniques used in it are not crucial here and the statement in question is obvious in the examples we consider below.

Note finally that if $\mu$ and $\nu$ are two finite measures on $\mathbb{G}$, then as in (2.13), we may calculate

$$
\begin{aligned}
(T_{\mu*\nu} x)(g) &= \int_{\mathbb{G}} x(hg)\,(\mu * \nu)(\mathrm{d}h) = \int_{\mathbb{G}} \int_{\mathbb{G}} x(h_1 h_2 g)\,\mu(\mathrm{d}h_1)\nu(\mathrm{d}h_2) \\
&= \int_{\mathbb{G}} T_\mu x(h_2 g)\nu(\mathrm{d}h_2) = T_\nu T_\mu x(g), \quad x \in BM(\mathbb{G}). \quad (2.16)
\end{aligned}
$$

The reader will check similarly that $\tilde{T}_{\mu*\nu} = \tilde{T}_\mu \tilde{T}_\nu$.

**2.3.26 Example**     Consider the space $\mathbb{BM}(\mathbb{G})$ of signed measures $\mu$ defined on the Klein group $\mathbb{G}$. Each measure $\mu$ on $\mathbb{G}$ may be identified with four real numbers $a_i = \mu(\{g_i\})$, $i = 1, 2, 3, 4$. The norm in this space is

$$
\|\mu\| = \left\| \begin{pmatrix} a_1 \\ a_2 \\ a_3 \\ a_4 \end{pmatrix} \right\| = \sum_{i=1}^{4} |a_i|, \qquad \mu = \begin{pmatrix} a_1 \\ a_2 \\ a_3 \\ a_4 \end{pmatrix}.
$$

Treating elements of this space as differences of two positive measures on $\mathbb{G}$, we define the convolution of two charges $\mu * \nu$ as in (2.14) and prove that (1.12) still holds. The operator $S_\mu(\nu) = \mu * \nu$ is a bounded linear operator given by

$$
S_\mu \begin{pmatrix} b_1 \\ b_2 \\ b_3 \\ b_4 \end{pmatrix} = \begin{pmatrix} a_1 & a_2 & a_3 & a_4 \\ a_2 & a_1 & a_4 & a_3 \\ a_3 & a_4 & a_1 & a_2 \\ a_4 & a_3 & a_2 & a_1 \end{pmatrix} \begin{pmatrix} b_1 \\ b_2 \\ b_3 \\ b_4 \end{pmatrix}, \qquad \text{where } \mu = \begin{pmatrix} a_1 \\ a_2 \\ a_3 \\ a_4 \end{pmatrix}. \quad (2.17)
$$

What is the form of $T_\mu$? A member $x$ of $C_0(\mathbb{G})$ may also be identified with four numbers $\xi_i = x(g_i), i = 1, 2, 3, 4$, but the norm in this space is $\|x\| = \max_{i=1,2,3,4} |\xi_i|$. Now

$$\eta_i := (T_\mu x)(g_i) = \int x(g_i h)\, \mu(\,\mathrm{d}h) = \sum_{j=1}^{4} x(g_i g_j) a_j.$$

Using the multiplication table from 1.2.23 we see that

$$T_\mu x = \begin{pmatrix} \eta_1 \\ \eta_2 \\ \eta_3 \\ \eta_4 \end{pmatrix} = \begin{pmatrix} a_1 & a_2 & a_3 & a_4 \\ a_2 & a_1 & a_4 & a_3 \\ a_3 & a_4 & a_1 & a_2 \\ a_4 & a_3 & a_2 & a_1 \end{pmatrix} \begin{pmatrix} \xi_1 \\ \xi_2 \\ \xi_3 \\ \xi_4 \end{pmatrix}.$$

Thus, except for acting in different spaces, $T_\mu$ and $S_\mu$ are represented by the same matrix. In general, if $\mathbb{G}$ is a finite but non-commutative group, the matrix that represents $S_\mu$ is the transpose of the matrix of $T_\mu$.

**2.3.27 Example**     In 1.2.24 we saw that any positive measure on the Kisyński group $\mathbb{G}$ may be identified with a pair of measures on $\mathbb{R}$. On the other hand, any signed measure on $\mathbb{G}$ is a difference of two positive measures. Hence, any charge on $\mathbb{G}$ may be identified with a pair of charges on $\mathbb{R}$. In other words, the space $\mathbb{BM}(\mathbb{G})$ is isometrically isomorphic to the Cartesian product of two copies of $\mathbb{BM}(\mathbb{R})$, with the norm $\|(\mu_1, \mu_{-1})\| = \|\mu_1\| + \|\mu_{-1}\|$. An easy argument shows that both (1.14) and (1.15) hold for charges as well. As a result of the latter equation

$$S_\mu \nu = \mu * \nu = (S_{\mu_1}\nu_1 + S_{\tilde{\mu}_{-1}}\nu_{-1}, S_{\mu_{-1}}\nu_1 + S_{\tilde{\mu}_1}\nu_{-1}),$$

where $S_{\mu_i}$ and $S_{\tilde{\mu}_i}$ are operators in $\mathbb{BM}(\mathbb{R})$ related to charges $\mu_i$ and $\tilde{\mu}_i$, $i = 1, -1$ on $\mathbb{R}$ ($\tilde{\mu}$ is defined in 1.2.24). This formula may be written in the matrix form as

$$S_\mu = \begin{pmatrix} S_{\mu_1} & S_{\tilde{\mu}_{-1}} \\ S_{\mu_{-1}} & S_{\tilde{\mu}_1} \end{pmatrix}. \tag{2.18}$$

As $\mathbb{BM}(\mathbb{G})$ is isometrically isomorphic to $\mathbb{BM}(\mathbb{R}) \times \mathbb{BM}(\mathbb{R})$, so $C_0(\mathbb{G})$ is isometrically isomorphic to $C_0(\mathbb{R}) \times C_0(\mathbb{R})$, with the norm $\|(x_1, x_{-1})\| = \|x_1\| \vee \|x_{-1}\|$ where $x_i \in C_0(\mathbb{R})$, $i = -1, 1$ (Exercise 2.2.38). Using (1.14) we see that

$$T_\mu x(\xi, l) = \int_{\mathbb{G}} x(\tau l + \xi, kl)\, \mu(\,\mathrm{d}(\tau, k))$$

equals

$$\int_{\mathbb{R}} x_1(\tau + \xi)\, \mu_1(\,\mathrm{d}\tau) + \int_{\mathbb{R}} x_{-1}(\tau + \xi)\, \mu_{-1}(\,\mathrm{d}\tau)$$

for $l = 1$ and

$$\int_{\mathbb{R}} x_1(-\tau + \xi) \, \mu_{-1}(\,d\tau) + \int_{\mathbb{R}} x_{-1}(-\tau + \xi) \, \mu_1(\,d\tau)$$

for $l = -1$. Hence, using matrix notation

$$T_\mu = \begin{pmatrix} T_{\mu_1} & T_{\mu_{-1}} \\ T_{\tilde{\mu}_{-1}} & T_{\tilde{\mu}_1} \end{pmatrix} \tag{2.19}$$

where $T_{\mu_i}$ and $T_{\tilde{\mu}_i}$ are operators on $C_0(\mathbb{R})$ related to measures $\mu_i$ and $\tilde{\mu}_i$ on $\mathbb{R}$. Notice that the matrix in (2.18) is a "conjugate" matrix to the matrix in (2.19), if we agree that the conjugate to $T_\nu$ is $S_\nu$ for any $\nu \in \mathbb{BM}(\mathbb{R})$.

**2.3.28 Exercise**     Find the form of the operators $\tilde{T}_\mu$ and $\tilde{S}_\mu$ on the Kisyński group.

**2.3.29** *Uniform topology versus strong topology. Weierstrass' Theorem*
Although it is nice to know that $\mathcal{L}(\mathbb{X}, \mathbb{Y})$ is a Banach space, in applications the mode of convergence related to the norm in $\mathcal{L}(\mathbb{X}, \mathbb{Y})$ is not very useful. The reason is that the requirement that operators $A_n$ converge to an operator $A$ in the norm of $\mathcal{L}(\mathbb{X}, \mathbb{Y})$ is very restrictive, and there are few interesting examples of such behavior. More often we encounter sequences such that $\|A_n x - Ax\|_{\mathbb{Y}} \to 0$ as $n \to \infty$ for all $x \in \mathbb{X}$. This mode of convergence is called **strong convergence**, in contrast to the one discussed above, called **convergence in the operator norm**, or **uniform convergence**. Indeed, convergence in the operator norm is a strong convergence that is uniform in any ball – see the definition of the norm in $\mathcal{L}(\mathbb{X}, \mathbb{Y})$. See also 5.4.18 in Chapter 5.

As an example consider the space $C[0, 1]$ of continuous functions on the unit interval, and a sequence of operators $A_n \in \mathcal{L}(C[0,1]), n \geq 1$, defined by

$$(A_n x)(s) = \sum_{j=1}^{n} x(j/n) \binom{n}{j} s^j (1 - s)^{n-j}. \tag{2.20}$$

Linearity of $A_n, n \geq 1$, is obvious, and

$$\|A_n x\| = \sup_{s \in [0,1]} |(A_n x)(s)| \leq \|x\| \sum_{j=1}^{n} \binom{n}{j} s^j (1 - s)^{n-j} = \|x\|, \tag{2.21}$$

so that $A_n$ are linear contractions in $\mathcal{L}(C[0,1])$. Taking $x(s) = 1$, we see that actually $\|A\| = 1$. We shall show that $A_n$ converges strongly but not uniformly to the identity operator $I$. For $n \geq 1$ let $x_n(s) =$

$0 \vee (1 - |2ns - 1|)$. We see that $\|x_n\| = 1$, but $x_n(j/n) = 0$, $j = 0, 1, ..., n$, so that $A_n x_n = 0$. Thus $\|A_n - I\| \geq \|A_n x_n - x_n\| = \|x_n\| = 1$ and we cannot have $\lim_{n \to \infty} \|A_n - I\| = 0$. On the other hand, from (2.21) we see that

$$(A_n x)(s) = E\, x(X_n/n)$$

where $X_n$ is a binomial variable $B(s, n)$. Recall that $E\, X_n/n = s$, and $D^2(X_n/n) = \frac{s(1-s)}{n}$. By Chebyshev's inequality

$$\mathbb{P}(|X_n/n - s| \geq \delta) \leq \frac{D^2(X_n/n)}{\delta^2} = \frac{s(1-s)}{n\delta^2} \leq \frac{1}{4n\delta^2}.$$

Moreover, $x$, as a continuous function on a compact interval, is uniformly continuous, i.e. for any $\epsilon > 0$ there exists a $\delta > 0$ such that $|x(s) - x(t)| < \epsilon/2$ provided $s, t \in [0, 1], |s - t| \leq \delta$. Therefore,

$$
\begin{aligned}
|(A_n x)(s) - x(s)| &\leq E\, |x(X_n/n) - x(s)| \\
&\leq E\, 1_{\{|X_n/n - s| \geq \delta\}} |x(X_n/n) - x(s)| \\
&\quad + E\, 1_{\{|X_n/n - s| < \delta\}} |x(X_n/n) - x(s)| \\
&\leq 2\|x\| \frac{1}{4n\delta^2} + \frac{\epsilon}{2}.
\end{aligned}
$$

Note that the $\delta$ on the right-hand side does not depend on $s$ but solely on $x$ and $\epsilon$. (Although the random variables $X_n$ and the events $\{|X_n/n - s| < \delta\}$ do!) Thus

$$\|A_n x - x\| \leq 2\|x\| \frac{1}{4n\delta^2} + \frac{\epsilon}{2}$$

and our claim follows: if we want to have $\|A_n x - x\|$ less than $\epsilon$, we take $n \geq \frac{\|x\|}{\epsilon \delta^2}$.

This proves also that polynomials form a dense set in $C[0, 1]$; indeed $A_n x$ is a polynomial regardless of what $x$ is, and $A_n x$ converges to $x$ in the supremum norm. This is the famous **Weierstrass Theorem**. The polynomials (2.20) were introduced by S. Bernstein, who also gave the proof of the Weierstrass Theorem reproduced above, and therefore are called **Bernstein polynomials**.

**2.3.30 Corollary**    If $x \in C[0, 1]$, and $\int_0^1 \tau^n x(\tau)\, d\tau = 0$, for all $n \geq 0$, then $x = 0$.

*Proof* By 2.3.29, for any $x \in \mathbb{X}$, and any $\epsilon > 0$ there exists a polynomial $x_\epsilon$ such that $\|x - x_\epsilon\| < \epsilon$. Our assumption implies that $\int_0^1 x_\epsilon(\tau) x(\tau)\, d\tau =$

0. Since

$$\left| \int_0^1 x^2(\tau)\, d\tau \right| = \left| \int_0^1 x^2(\tau)\, d\tau - \int_0^1 x_\epsilon(\tau)x(\tau)\, d\tau \right|$$

$$= \left| \int_0^1 [x(\tau) - x_\epsilon(\tau)]x(\tau)\, d\tau \right|$$

$$\leq \ \|x - x_\epsilon\|\,\|x\| \leq \epsilon\|x\|,$$

and since $\epsilon$ is arbitrary, $\int_0^1 x^2(\tau)\, d\tau = 0$. Therefore, $x(\tau) = 0$ almost surely, and by continuity, for all $\tau \in [0, 1]$. $\qquad\square$

**2.3.31 Exercise** Prove that $Ix(\tau) = x(-\ln\tau)$ with convention $\ln 0 = -\infty$ maps $C(\overline{\mathbb{R}^+})$ (the space of continuous functions with limit at infinity) isometrically isomorphically onto $C[0, 1]$. Conclude that the functions $e_\lambda(\tau) = e^{-\lambda\tau}$, $\lambda \geq 0$ form a linearly dense subset of $C(\overline{\mathbb{R}^+})$, and consequently that $\int_0^\infty e^{-\lambda\tau}x(\tau)\, d\tau = 0$, $\lambda > 0$, implies $x = 0$, for $x \in C(\overline{\mathbb{R}^+})$.

**2.3.32 *Two linear and continuous operators that coincide on a linearly dense subset are equal*** Suppose $\mathbb{X}_0$ is a subset of a normed linear space $\mathbb{X}$ and $\overline{span\,\mathbb{X}_0} = \mathbb{X}$. Let $L_i, i = 1, 2$, be two linear and continuous operators with values in a normed linear space $\mathbb{Y}$. If $L_1 x = L_2 x$ for all $x \in X_0$, then $L_1 = L_2$.

*Proof* If $x \in span\,\mathbb{X}_0$, then $x = \sum_{i=1}^n \alpha_i x_i$ for some scalars $\alpha_i \in \mathbb{R}$ and vectors $x_i \in \mathbb{X}_0$. Thus $L_1 x = L_2 x$ by linearity of $L_1$ and $L_2$. For $x \in \mathbb{X}$, we may find a sequence of vectors $x_n \in \mathbb{X}_0$ such that $\lim_{n\to\infty} x_n = x$. Thus, $L_1 x = L_2 x$ by continuity. $\qquad\square$

**2.3.33 *Existence of the extension of a linear operator defined on a linearly dense set*** Suppose $\mathbb{X}_0$ is a subset of a normed linear space $\mathbb{X}$, $span\,\mathbb{X}_0 = \mathbb{X}_0$, and $\overline{\mathbb{X}_0} = \mathbb{X}$. Let $L$ be a linear map from $\mathbb{X}_0$ into a Banach space $\mathbb{Y}$, and suppose that there exists a constant $C > 0$ such that for all $x \in \mathbb{X}_0$, $\|Lx\|_\mathbb{Y} \leq C\|x\|_\mathbb{X}$, where $\|\cdot\|_\mathbb{X}$ and $\|\cdot\|_\mathbb{Y}$ are norms in $\mathbb{X}$ and $\mathbb{Y}$, respectively. Then, there exists a unique linear and continuous operator $M : \mathbb{X} \to \mathbb{Y}$ such that $Lx = Mx$ for $x \in \mathbb{X}_0$ and $\|Mx\|_\mathbb{Y} \leq C\|x\|_\mathbb{X}, x \in \mathbb{X}$. (For obvious reasons, the operator M is usually denoted simply by $L$.)

*Proof* For any $x \in \mathbb{X}$ there exists a sequence of vectors $x_n \in \mathbb{X}_0$ such that $\lim_{n\to\infty} x_n = x$. The sequence $y_n = Lx_n$ is a Cauchy sequence in $\mathbb{X}$, since $x_n$ is a Cauchy sequence, because $\|y_n - y_m\|_\mathbb{Y} \leq C\|x_n - x_m\|_\mathbb{X}$.

Thus, $Lx_n$ converges to an element in $\mathbb{Y}$, since $\mathbb{Y}$ is complete. Moreover, the limit does not depend on the choice of the sequence $x_n$ but solely on $x$. Indeed, if for another sequence $x_n'$ we have $\lim_{n\to\infty} x_n' = x$, then by $\|Lx_n - Lx_n'\|_{\mathbb{Y}} \leq C\|x_n - x_n'\|_{\mathbb{X}}$, we have $\lim_{n\to\infty} Lx_n = \lim_{n\to\infty} Lx_n'$. We may thus put $Mx = \lim_{n\to\infty} Lx_n$. In particular, if $x \in \mathbb{X}_0$ we may take a constant sequence $x_n = x$ to see that $Mx = Lx$. The operator $M$ is linear: if $x, y \in \mathbb{X}$, then we pick sequences $x_n, y_n \in \mathbb{X}_0$ such that $\lim_{n\to\infty} x_n = x$, $\lim_{n\to\infty} y_n = y$; by linearity of $L$, we have

$$
\begin{aligned}
M(\alpha x + \beta y) &= \lim_{n\to\infty} L(\alpha x_n + \beta y_n) = \alpha \lim_{n\to\infty} Lx_n + \beta \lim_{n\to\infty} Ly_n \\
&= \alpha Mx + \beta My.
\end{aligned}
$$

Similarly,

$$
\|Mx\|_{\mathbb{Y}} = \lim_{n\to\infty} \|Lx_n\|_{\mathbb{Y}} \leq C \lim_{n\to\infty} \|x_n\|_{\mathbb{X}} = C\|x\|_{\mathbb{X}}.
$$

Uniqueness of $M$ follows from 2.3.32.                                    $\square$

**2.3.34 Exercise**    Prove that if $A_n, n \geq 0$ are bounded linear operators $A_n \in \mathcal{L}(\mathbb{X}, \mathbb{Y})$ such that $\|A_n\| \leq M, n \geq 0$ for some $M > 0$ and $A_n x$ converges to $A_0 x$, for all $x$ in a linearly dense set of $x \in \mathbb{X}$, then $A_n$ converges strongly to $A_0$.

**2.3.35 Definition**    A family $A_t, t \in \mathbb{T}$ of bounded linear operators where $\mathbb{T}$ is an index set is said to be a family of **equibounded operators** iff there exists a constant $M > 0$ such that $\|A_t\| \leq M$.

**2.3.36 Definition**    An operator $L \in \mathcal{L}(\mathbb{X})$ is said to **preserve a functional** $f \in \mathbb{X}^*$, if $f(Lx) = f(x)$ for all $x \in \mathbb{X}$. Note that, by 2.3.32, to check if $L$ preserves $f$ it is enough to prove that $f(Lx) = f(x)$ holds on a linearly dense set.

**2.3.37 *Markov operators***    Let $(\Omega, \mathcal{F}, \mu)$ be a measure space. Let $\mathbb{Y}$ be an algebraic subspace of $L^1(\Omega, \mathcal{F}, \mu)$ which is dense in $L^1(\Omega, \mathcal{F}, \mu)$, and such that $x^+ = \max(x, 0)$ belongs to $\mathbb{Y}$ for $x \in \mathbb{Y}$. Suppose that $P$ is a linear operator in $L(\mathbb{Y}, L^1(\Omega, \mathcal{F}, \mu))$ such that $Px \geq 0$ and

$$
\int_\Omega Px \, d\mu = \int_\Omega x \, d\mu, \tag{2.22}
$$

for all $x \in \mathbb{Y}$ such that $x \geq 0$. Then there exists a unique extension of $P$ to a contraction, denoted by the same letter, $P \in \mathcal{L}(L^1(\Omega, \mathcal{F}, \mu))$, such that $Px \geq 0$ if $x \geq 0$ and (2.22) holds for all $x \in L^1(\Omega, \mathcal{F}, \mu)$.

*Proof* Let us write $x = x^+ - x^-$, where $x^+ = \max(x, 0)$ and $x^- = \max(-x, 0)$. For $x \in \mathbb{Y}$, we have $P(x^+) - Px \geq 0$, since $x^+ - x \geq 0$. Thus,

$$(Px)^+ = \max(Px, 0) \leq P(x^+).$$

Since $x^- = (-x)^+$, we also have

$$(Px)^- = (-Px)^+ = [P(-x)]^+ \leq [P(-x)^+] = P(x^-).$$

Therefore,

$$
\begin{aligned}
\int_\Omega |Px| \, d\mu &= \int_\Omega [(Px)^+ + (Px)^-] \, d\mu \leq \int_\Omega [P(x^+) + P(x^-)] \, d\mu \\
&= \int_\Omega [x^+ + x^-] \, d\mu = \int_\Omega |x| \, d\mu.
\end{aligned}
$$

Thus, the existence of the extension to a contraction is secured by 2.3.33. Using linearity, we show that (2.22) holds for all $x \in \mathbb{Y}$, and not just for $x \geq 0$ in $\mathbb{Y}$, so that the bounded linear functional $F : x \to \int_\Omega x \, d\mu$ is preserved on $\mathbb{Y}$, and thus on $L^1(\Omega, \mathcal{F}, \mu)$.

It remains to show that $Px \geq 0$ if $x \geq 0$. If $x \geq 0$ and $x_n \in \mathbb{Y}$ converges to $x$, then $x_n^+$ belongs to $\mathbb{Y}$ and $\|x - x_n^+\|_{L^1(\Omega, \mathcal{F}, \mu)} \leq \|x - x_n\|_{L^1(\Omega, \mathcal{F}, \mu)}$ (since $\int_{\{x_n \leq 0\}} |x - x_n| \, d\mu = \int_{\{x_n \leq 0\}} (x + x_n) \, d\mu \geq \int_{\{x_n \leq 0\}} x \, d\mu$). Thus, $Px \geq 0$ as a limit of non-negative functions $x_n^+$. $\qquad\square$

**2.3.38 Remark** In 2.3.37, we may take $\mathbb{Y}$ to be the algebraic subspace of simple functions. If $\mu$ is finite, we may take $\mathbb{Y} = L^2(\Omega, \mathcal{F}, \mu)$.

**2.3.39 Definition** Suppose $L^1(\Omega, \mathcal{F}, \mu)$ is a space of absolutely integrable functions on a measure space $(\Omega, \mathcal{F}, \mu)$. A linear map $P : L^1(\Omega, \mathcal{F}, \mu) \to L^1(\Omega, \mathcal{F}, \mu)$ such that

(a) $Px \geq 0$ for $x \geq 0$,
(b) $\int_\Omega Px \, d\mu = \int_\Omega x \, d\mu$ for $x \geq 0$

is called a **Markov operator**. As we have seen, Markov operators are linear contractions. Condition (b) implies that Markov operators preserve the integral (which is a linear functional on $L^1(\Omega, \mathcal{F}, \mu)$).

**2.3.40 Exercise** Let $y$ be a non-negative element of $L^1(\mathbb{R}, \mathcal{M}, leb)$ such that $\int y \, dleb = 1$. Prove that $P_y$ defined on $L^1(\mathbb{R}, \mathcal{M}, leb)$ by $P_y x = y * x$ is a Markov operator.

**2.3.41 Example** The operator from 2.3.26 is a Markov operator in the space $L^1(\mathbb{G})$ of absolutely integrable functions with respect to the counting measure on the measure space $\Omega = \mathbb{G}$ composed of four elements, provided $\mu$ is non-negative and its coordinates add up to one. In other words, $S_\mu = \mu * \nu$ is a Markov operator provided $\mu$ is a probability measure.

Observe that the space $L^1(\mathbb{G})$ coincides with the space $\mathbb{BM}(\mathbb{G})$ of signed measures on $\mathbb{G}$, since all measures on $\mathbb{G}$ are absolutely continuous with respect to the counting measure.

**2.3.42 Exercise** Provide a similar example of a Markov operator on the space $L^1(\mathbb{G})$ of functions that are absolutely integrable with respect to Lebesgue measure on the Kisyński group $\mathbb{G}$; here we treat $\mathbb{G}$ as two parallel lines both equipped with one-dimensional Lebesgue measure.

**2.3.43 Exercise** Let $k(\tau, \sigma)$ ba a non-negative function on $\mathbb{R}^2$ such that $\int_{\mathbb{R}} k(\tau, \sigma) \, \mathrm{d}\sigma = 1$, for almost all $\tau$. Prove that the operator $K$ defined on $L^1(\mathbb{R}, \mathcal{M}, leb)$ by $Kx(\tau) = \int k(\tau, \sigma) \, \mathrm{d}\sigma$ is a Markov operator.

**2.3.44 Exercise** Let $(\Omega, \mathcal{F}, \mu)$ be a measure space and let $f$ be a measurable map into a measurable space $(\Omega', \mathcal{F}')$. Find a measure $\mu'$ on $(\Omega', \mathcal{F}')$ such that the operator $P : L^1(\Omega', \mathcal{F}', \mu') \to L^1(\Omega, \mathcal{F}, \mu)$ given by $(Px)(\omega) = x(f(\omega))$, $\omega \in \Omega$, is Markov. For another example of a Markov operator, see 3.4.5 below.

**2.3.45 *Campbell's Theorem*** We often encounter operators that map the space $L^1(\Omega, \mathcal{F}, \mu)$ into the space of integrable functions on another measure space $(\Omega', \mathcal{F}', \mu')$. If the operator $P$ maps (classes of) non-negative functions into (classes of) non-negative functions and the relation $\int_{\Omega'} Px \, \mathrm{d}\mu' = \int_\Omega x \, \mathrm{d}\mu$ holds we shall still call $P$ a Markov operator. Here is a famous example. Let $(\Omega, \mathcal{F}, \mathbb{P})$ be a probability space where a sequence $X_n$ of independent exponentially distributed random variables with parameter $a$ is defined. Let $S_n \equiv \sum_{j=1}^n X_j$. $S_n$ has the gamma distribution with parameters $n$ and $a$. For any absolutely integrable function $x$ on the right half-axis, let $(Px)(\omega) = \sum_{n=1}^\infty x(S_n(\omega))$. $P$ is a Markov operator mapping the space $L^1(\mathbb{R}^+, \mathcal{M}(\mathbb{R}^+), a \cdot leb)$ of functions that are absolutely integrable with respect to Lebesgue measure multiplied by $a$ (with the norm $\|x\| = a \int |x(s)| \, \mathrm{d}s$) into $L^1(\Omega, \mathcal{F}, \mathbb{P})$. Notice that for two functions, say $x$ and $y$, from the same equivalence class in $L^1(\mathbb{R}^+, \mathcal{M}(\mathbb{R}^+), a \cdot leb)$ the values of $Px$ and $Py$ evaluated at some $\omega$ may differ. Nevertheless, we may check that $P$ maps classes into classes.

We leave the proof as an exercise. It is clear that $P$ maps non-negative functions into non-negative functions. Moreover, for non-negative $x$,

$$\int_\Omega (Px)(\omega)\,d\omega = \int_\Omega \sum_{n=1}^\infty x\big(S_n(\omega)\big)\,d\omega$$

$$= \int_0^\infty \sum_{n=1}^\infty e^{-at}\frac{a^n t^{n-1}}{(n-1)!}x(t)\,dt = \int_0^\infty x(t)a\,dt.$$

We have proved a (small) part of **Campbell's Theorem** – see [41, 66]. We shall come back to this subject in 6.4.9.

**2.3.46 Exercise**    Consider the operator $D$ that maps the joint distribution of two random integer-valued variables into the distribution of their difference. Extend $D$ to the whole of $l^1(\mathbb{Z}\times\mathbb{Z})$, where $\mathbb{Z}\times\mathbb{Z}$ is equipped with the counting measure, find an explicit formula for $D$, and show that this operator is Markov.

# 3

# Conditional expectation

The space $L^2(\Omega, \mathcal{F}, \mathbb{P})$ of square integrable random variables on a probability space $(\Omega, \mathcal{F}, \mathbb{P})$ is a natural example of a Hilbert space. Moreover, for $X \in L^2(\Omega, \mathcal{F}, \mathbb{P})$ and a $\sigma$-algebra $\mathcal{G} \subset \mathcal{F}$, the conditional expectation $\mathbb{E}(X|\mathcal{G})$ of $X$ is the projection of $X$ onto the subspace of $\mathcal{G}$ measurable square integrable random variables. Hence, we start by studying Hilbert spaces and projections in Hilbert spaces in Section 3.1 to introduce conditional expectation in Section 3.2. Then we go on to properties and examples of conditional expectation and all-important martingales.

## 3.1 Projections in Hilbert spaces

**3.1.1 Definition**    A linear space with the binary operation $\mathbb{X} \times \mathbb{X} \to \mathbb{R}$, mapping any pair in $\mathbb{X} \times \mathbb{X}$ into a scalar denoted $(x, y)$, is called a **unitary space** or an **inner product space** iff for all $x, y, z \in \mathbb{X}$, and $\alpha, \beta \in \mathbb{R}$, the following conditions are satisfied:

(s1) $(x + y, z) = (x, z) + (y, z)$,
(s2) $(\alpha x, y) = \alpha(x, y)$,
(s3) $(x, x) \geq 0$,
(s4) $(x, x) = 0$ iff $x = 0$.
(s5) $(x, y) = (y, x)$.

The number $(x, y)$ is called the **scalar product** of $x$ and $y$. The vectors $x$ and $y$ in a unitary space are termed **orthogonal** iff their scalar product is 0.

**3.1.2 Example**    The space $l^2$ of square summable sequences with the scalar product $(x, y) = \sum_{n=1}^{\infty} \xi_n \eta_n$ is a unitary space; here $x = (\xi_n)_{n \geq 1}$, $y = (\eta_n)_{n \geq 1}$. The space $C_{[0,1]}$ of continuous functions on $[0, 1]$ with the scalar product $(x, y) = \int_0^1 x(s)y(s)\,\mathrm{d}s$ is a unitary space. Another important example is the space $L^2(\Omega, \mathcal{F}, \mu)$ where $(\Omega, \mathcal{F}, \mu)$ is a measure space, with $(x, y) = \int_\Omega xy\,\mathrm{d}\mu$. The reader is encouraged to check conditions (s1)–(s5) of the definition.

In particular, if $\mu$ is a probability space, we have $(X, Y) = E\,XY$. Note that defining, as customary, the covariance of two square integrable random variables $X$ and $Y$ as $cov(X, Y) = E\,(X - (E\,X)1_\Omega)(Y - (E\,Y)1_\Omega)$ we obtain $cov(X, Y) = (X, Y) - E\,X E\,Y$.

**3.1.3 *Cauchy–Schwartz–Bunyakovski inequality*** For any $x$ and $y$ in a unitary space,

$$(x, y)^2 \le (x, x)(y, y).$$

*Proof* Define the real function $f(t) = (x + ty, x + ty)$; by (s3) it admits non-negative values. Using (s1)–(s2) and (s5):

$$f(t) = (x, x) + 2t(x, y) + t^2(y, y); \tag{3.1}$$

so $f(t)$ is a second order polynomial in $t$. Thus, its discriminant must be non-positive, i.e. $4(x, y)^2 - 4(x, x)(y, y) \le 0$. $\qquad\square$

**3.1.4 Theorem** A unitary space becomes a normed space if we define $\|x\| = \sqrt{(x, x)}$. This norm is often called the **unitary norm**.

*Proof* (n1) follows from (s3), and (n2) follows from (s4). To show (n3) we calculate:

$$\|\alpha x\| = \sqrt{(\alpha x, \alpha x)} = \sqrt{\alpha^2(x, x)} = |\alpha|\sqrt{(x, x)} = |\alpha|\|x\|,$$

where we have used (s2) and (s5). Moreover, by (3.1) and 3.1.3,

$$\|x + y\|^2 = (x, x) + 2(x, y) + (y, y) \le \|x\|^2 + 2\|x\|\,\|y\| + \|y\|^2,$$

as desired. $\qquad\square$

**3.1.5 Example** In the case of $L^2(\Omega, \mathcal{F}, \mu)$ the norm introduced above is the usual norm in this space, $\|x\|^2 = \int_\Omega x^2\,\mathrm{d}\mu$.

**3.1.6 *Law of large numbers. First attempt*** Suppose that $X_n$ are identically distributed, uncorrelated, square integrable random variables in a probability space $(\Omega, \mathcal{F}, \mathbb{P})$, i.e. that $cov(X_i, X_j) = 0$ for $i \ne j$. Then $\frac{S_n}{n} = \frac{X_1 + \cdots + X_n}{n}$ converges in $L^2(\Omega, \mathcal{F}, \mathbb{P})$ to $(E\,X)1_\Omega$.

*Proof* It suffices to note that the squared distance between $\frac{S_n}{n}$ and $(E\,X)1_\Omega$ equals $\|\frac{1}{n}\sum_{i=1}^{n}(X_i - (E\,X)1_\Omega)\|^2$, which by a direct computation based on assumption of lack of correlation equals $\frac{1}{n^2}\sum_{i=1}^{n}\sigma_{X_i}^2 = \frac{1}{n}\sigma_{X_1}^2$. $\qquad\square$

**3.1.7 Remark**   Note that the Markov inequality implies that under the above assumptions, for any $\epsilon > 0$, $\mathbb{P}(|\frac{S_n}{n} - E\,X| > \epsilon) \le \frac{\sigma_{X_1}^2}{n\epsilon}$ tends to zero. This means by definition that $\frac{S_n}{n}$ converges to $E\,X$ *in probability* (see Chapter 5).

**3.1.8** *Parallelogram law*   In any unitary space $\mathbb{H}$,

$$\|x + y\|^2 + \|x - y\|^2 = 2\big[\|x\|^2 + \|y\|^2\big],$$

where $x, y \in \mathbb{H}$, and $\|\cdot\|$ is a unitary norm.

*Proof* Take $t = 1$ and $t = -1$ in (3.1) and add up both sides. $\qquad\square$

**3.1.9 Exercise** *(polarization formula)*   In any unitary space

$$(x, y) = \frac{\|x + y\|^2 - \|x - y\|^2}{4}.$$

**3.1.10 Definition**   Let $\mathbb{H}$ be a unitary space, and let $\|\cdot\|$ be the unitary norm. If $(\mathbb{H}, \|\cdot\|)$ is a Banach space, this pair is called a **Hilbert space**. Again, quite often we will say that $\mathbb{H}$ itself is a Hilbert space. A leading example of a Hilbert space is the space $L^2(\Omega, \mathcal{F}, \mu)$ where $(\Omega, \mathcal{F}, \mu)$ is a measure space.

**3.1.11** *Existence of the closest element from a closed convex set*   Let $C$ be a closed convex subset of a Hilbert space $\mathbb{H}$, and let $x \notin C$. There exists a unique element $y \in C$, such that

$$\|x - y\| = d := \inf_{z \in C}\|x - z\|.$$

*Proof* For any $z, z' \in C$, we have by 3.1.8,

$$\begin{aligned}
\|z - z'\|^2 &= \|(z - x) + (x - z')\|^2 \\
&= 2\{\|z - x\|^2 + \|z' - x\|^2\} - \|z + z' - 2x\|^2 \\
&= 2\{\|z - x\|^2 + \|z' - x\|^2\} - 4\left\|\frac{z + z'}{2} - x\right\|^2 \\
&\le 2\{\|z - x\|^2 + \|z' - x\|^2\} - 4d^2, \qquad\qquad (3.2)
\end{aligned}$$

since $\frac{z+z'}{2}$ belongs to $C$. By definition of $d$ there exists a sequence $z_n \in C$, such that $\lim_{n\to\infty} \|z_n - x\| = d$. We will show that $z_n$ is a Cauchy sequence. Choose $n_0$ such that $\|z_n - x\|^2 \leq d^2 + \frac{\epsilon}{4}$, for $n \geq n_0$. Then, using (3.2) with $z = z_n$ $z' = z_m$ we see that for $n, m \geq n_0$, $\|z_n - z_m\|^2 \leq \epsilon$, which proves our claim. Define $y = \lim_{n\to\infty} z_n$; we have $\|x - y\| = \lim_{n\to\infty} \|x - z_n\| = d$. Moreover, if $\|y' - x\| = d$, then (3.2) with $z = y$ and $z' = y'$ shows that $\|y - y'\| = 0$. $\qquad\square$

**3.1.12 *Existence of projection*** Let $\mathbb{H}_1$ be a subspace of a Hilbert space $\mathbb{H}$. For any $x$ in $\mathbb{H}$, there exists a unique vector $Px \in \mathbb{H}_1$ such that for any $z \in \mathbb{H}_1$, $(x - Px, z) = 0$. $Px$ is called the projection of $x$ in $\mathbb{H}_1$.

*Proof* If $x$ belongs to $\mathbb{H}_1$ we take $Px = x$, and it is trivial to check that this is the only element we can choose. Suppose thus that $x \notin \mathbb{H}_1$ and put $Px = y$ where $y$ is the element that minimizes the distance between $x$ and elements of $\mathbb{H}_1$. Let $z$ belong to $\mathbb{H}_1$. The function

$$f(t) = \|x - y + tz\|^2 = \|x - y\|^2 + 2t(z, x - y) + t^2\|z\|,$$

attains its minimum at $t_{\min} = 0$. On the other hand $t_{\min} = -\frac{2(z, x-y)}{\|x-y\|^2}$ so that $(x - y, z) = 0$. Suppose that $(x - y', z) = 0$ for some $y'$ in $\mathbb{H}_1$ and all $z \in \mathbb{H}_1$, and $y \neq y'$. Then

$$\|x - y\|^2 = \|x - y'\|^2 + 2(x - y', y' - y) + \|y' - y\|^2 = \|x - y'\|^2 + \|y' - y\|^2$$

since $y' - y$ belongs to $\mathbb{H}_1$. Thus $\|x - y\| > \|x - y'\|$, a contradiction. $\qquad\square$

**3.1.13 Corollary** Under assumptions and notations of 3.1.12, for all $x, y \in X$ and $\alpha, \beta \in \mathbb{R}$, $P(\alpha x + \beta y) = \alpha Px + \beta Py$, and $\|Px\| \leq \|x\|$. In other words, $P$ is a linear contraction.

*Proof* For the first part it suffices to show that $(\alpha x + \beta y - \alpha Px - \beta Py, z) = 0$ for $z \in \mathbb{H}_1$, but this follows directly from the definition of $Px$ and $Py$ and conditions (s1)–(s2) in 3.1.1. To complete the proof note that $\|x\|^2 = \|x - Px\|^2 + 2t(x - Px, Px) + \|Px\|^2 = \|x - Px\|^2 + \|Px\|^2$. $\qquad\square$

**3.1.14 Exercise** Show that for any projection $P$, $\|P\| = 1$.

**3.1.15 Corollary** Let $\mathbb{H}_1$ be a proper subspace of a Hilbert space $\mathbb{H}$. There exists a non-zero vector $y \in \mathbb{H}$ such that $(y, x) = 0$ for all $x \in \mathbb{H}_1$. We say that $y$ is perpendicular to $H_1$.

**3.1.16 Exercise**    Let $\mathbb{H}_1$ be the subspace of a Hilbert space $\mathbb{H}$. Define $\mathbb{H}_1^\perp$ as the set of all $y$ such that $(x, y) = 0$ for all $x \in \mathbb{H}_1$. Prove that $\mathbb{H}_1^\perp$ is a subspace of $\mathbb{H}$ and that any $z \in \mathbb{H}$ may be represented in a unique way in the form $z = x + y$ where $x \in \mathbb{H}_1$ and $y = \mathbb{H}_1^\perp$. Also, if $P$ is the projection on $\mathbb{H}_1$ then $I - P$ is the projection on $\mathbb{H}_1^\perp$.

**3.1.17 Example**    Let $\mathbb{H} = L^2(\Omega, \mathcal{F}, \mathbb{P})$ and $\mathbb{H}_1 = L^2(\Omega, \mathcal{G}, \mathbb{P})$ where $\mathcal{G}$ is a sub-$\sigma$-algebra of $\mathcal{F}$. If $X \in L^2(\Omega, \mathcal{F}, \mathbb{P})$ is independent of $\mathcal{G}$ then for any $Y \in L^2(\Omega, \mathcal{G}, \mathbb{P})$ we have $E\,XY = E\,X \cdot E\,Y$. Hence, $X - (E\,X)1_\Omega$ is perpendicular to $L^2(\Omega, \mathcal{G}, \mathbb{P})$. In particular, $(E\,X)1_\Omega$ is the projection of $X$ onto $L^2(\Omega, \mathcal{G}, \mathbb{P})$. Example 1.4.9 shows that it may happen that $X - (E\,X)1_\Omega$ belongs to $L^2(\Omega, \mathcal{G}, \mathbb{P})^\perp$ and yet $X$ is not independent of $\mathcal{G}$.

**3.1.18 *Properties of projections***    Let $\mathbb{H}_1$ and $\mathbb{H}_2$ be subspaces of a Hilbert space $\mathbb{H}$, and let $P_i, i = 1, 2$, denote corresponding projection operators. $\mathbb{H}_1 \subset \mathbb{H}_2$ iff $\|P_1 x\| \le \|P_2 x\|$ for all $x \in \mathbb{H}$. In such a case $P_1 P_2 = P_2 P_1 = P_1$.

*Proof*   For any $x \in \mathbb{H}$, $\|x\|^2 = \|x - P_1 x\|^2 + \|P_1 x\|^2$ since $x - P_1 x$ is perpendicular to $P_1 x$. Similarly, $\|x\|^2 = \|x - P_2 x\|^2 + \|P_2 x\|^2$. By definition, $P_2 x$ is the element of $\mathbb{H}_2$ that minimizes the distance between $x$ and an element of this subspace. Hence if $\mathbb{H}_1 \subset \mathbb{H}_2$ then $\|x - P_2 x\|^2 \le \|x - P_1 x\|^2$, and so $\|P_1 x\|^2 \le \|P_2 x\|^2$. Conversely, if the last inequality holds, then for $x \in \mathbb{H}_1$,

$$\|x - P_2 x\|^2 = \|x\| - \|P_2 x\|^2 \le \|x\| - \|P_1 x\|^2 = \|x - P_1 x\|^2 = 0,$$

proving that $x = P_2 x \in \mathbb{H}_2$.

If $\mathbb{H}_1 \subset \mathbb{H}_2$, then $P_2 P_1 = P_1$, since $P_1 x \in \mathbb{H}_1 \subset \mathbb{H}_2$ for any $x \in \mathbb{H}$. To calculate $P_1 P_2 x$ note that $P_1 x$ belongs to $\mathbb{H}_1$ and for any $z \in \mathbb{H}_1$, $(P_2 x - P_1 x, z) = (P_2 x - x, z) + (x - P_1 x, z)$. Now, $(P_2 x - x, z) = 0$ since $z \in \mathbb{H}_2$, and $(x - P_1 x, z) = 0$ since $z \in \mathbb{H}_1$. This implies that $P_1 x = P_1 P_2 x$.                                                                    □

**3.1.19 Definition**    A bounded linear operator $A$ in a Hilbert space $\mathbb{H}$ is said to be **self-adjoint** (see 5.3.1) if for any $x$ and $y$ in $\mathbb{H}$, $(Ax, y) = (x, Ay)$.

**3.1.20 Example**    Let $\mathbb{H}$ be the space $\mathbb{R}^n$ with the norm $\|(\xi_i)_{i=1,..n}\| = \sqrt{\sum_{i=1}^n \xi_i^2}$. A linear operator in this space may be identified with the

$n \times n$ matrix $(\alpha_{i,j})_{i,j=1,\ldots,n}$ with $\alpha_{i,j}$ being the $i$th coordinate in the vector $A(\delta_{i,j})_{i=1,\ldots,n}$. Such an operator is self-adjoint if the corresponding matrix is symmetric (see the proof of Theorem 4.1.3).

**3.1.21 Exercise**    Prove that for any projection $P$ and vectors $x$ and $y$ in $\mathbb{H}$ we have

$$(Px, y) = (Px, Py) = (x, Py); \qquad (3.3)$$

in particular, projections are self-adjoint.

**3.1.22** *The norm of a self-adjoint operator*    Let $A$ be a self-adjoint operator in a Hilbert space $\mathbb{H}$. Then $\|A\| = \sup_{\|x\|=1} |(Ax, x)|$.

*Proof*  Denote the supremum above by $\|A\|_0$. By the Cauchy inequality, $\|A\|_0 \leq \sup_{\|x\|=1} \|A\| \|x\| = \|A\|$. To prove the converse fix $x \in \mathbb{H}$ such that $Ax \neq 0$ and $\|x\| = 1$, and let $y = \frac{1}{\|Ax\|} Ax$. Then

$$\|Ax\| = (Ax, y) = \frac{1}{2}[(Ax, y) + (x, Ay)]$$

$$= \frac{1}{4}[(A(x+y), x+y) - (A(x-y), x-y)]$$

$$\leq \frac{1}{4}\|A\|_0[\|x+y\|^2 + \|x-y\|^2] = \frac{1}{2}\|A\|_0[\|x\|^2 + \|y\|^2].$$

Hence, $\|A\| = \sup_{\|x\|=1} \|Ax\| \leq \sup_{\|x\|=1} \frac{1}{2}\|A\|_0[\|x\|^2 + \|y\|^2] = \|A\|_0$, since $\|y\| = 1$. $\qquad \square$

**3.1.23** *A characterization of a projection operator*    We have seen that any projection $P$ is self-adjoint and that $P^2 = P$. The converse is also true: if $P$ is a self-adjoint operator in a Hilbert space $\mathbb{H}$ and if $P^2 = P$ then $\mathbb{H}_1 = Range\, P$ is a subspace of $\mathbb{H}$ and $P$ is the projection on $\mathbb{H}_1$.

*Proof*  Certainly, $\mathbb{H}_1$ is an algebraic subspace of $\mathbb{H}$. By definition, if $x_n$ belongs to $\mathbb{H}_1$ then there exists $y_n$ in $\mathbb{H}$ such that $Py_n = x_n$. Hence if $x_n$ converges to an $x$ then the calculation $Px = \lim_{n\to\infty} Px_n = \lim_{n\to\infty} P^2 y_n = \lim_{n\to\infty} Py_n = \lim_{n\to\infty} x_n = x$, proves that $x$ belongs to $\mathbb{H}_1$. Therefore, $\mathbb{H}_1$ is a subspace of $\mathbb{H}$.

Let $P_1$ be the projection on $\mathbb{H}_1$. By 3.1.16 it suffices to show that $P_1 x = Px$ for all $x \in \mathbb{H}_1$ and for all $x \in \mathbb{H}_1^\perp$. The first claim is true since for $x \in \mathbb{H}_1$ both $P_1 x$ and $Px$ are equal to $x$. Since $P_1 x = 0$ for all $\mathbb{H}_1^\perp$ we are left with proving that $Px = 0$ if $x$ belongs to $\mathbb{H}_1^\perp$. For such an $x$, however, we have $\|Px\|^2 = (Px, Px) = (x, P^2 x) = (x, Px) = 0$ since $Px$ belongs to $\mathbb{H}_0$. $\qquad \square$

**3.1.24 Exercise**   Suppose that $P_1$ and $P_2$ are two projections on the subspaces $\mathbb{H}_1$ and $\mathbb{H}_2$ of a Hilbert space $\mathbb{H}$, respectively. Prove that the following conditions are equivalent:

(a) $P_1 P_2 = P_2 P_1$,
(b) $P_3 = P_1 + P_2 - P_1 P_2$ is a projection operator,
(c) $P_4 = P_1 P_2$ is a projection operator.

If one of these conditions holds, then $\mathbb{H}_1 + \mathbb{H}_2$ is a subspace of $\mathbb{H}$ and $P_3$ is the projection on this subspace, and $P_4$ is a projection on $\mathbb{H}_1 \cap \mathbb{H}_2$.

**3.1.25 Example**   Suppose that $\mathcal{G}_1$ and $\mathcal{G}_2$ are independent $\sigma$-algebras of events in a probability space $(\Omega, \mathcal{F}, \mathbb{P})$. Let $P_1$ and $P_2$ denote the projections in $L^2(\Omega, \mathcal{F}, \mathbb{P})$ on the subspaces $L^2(\Omega, \mathcal{G}_1, \mathbb{P})$ and $L^2(\Omega, \mathcal{G}_2, \mathbb{P})$, respectively. For any $X$ in $L^2(\Omega, \mathcal{F}, \mathbb{P})$, $P_1 X$ is $\mathcal{G}_1$ measurable and so by Exercise 3.1.17, $P_2 P_1 X = (E\,X)\,1_\Omega$. Similarly, $P_1 P_2 = (E\,X)\,1_\Omega$. Thus $H_0 = L^2(\Omega, \mathcal{G}_1, \mathbb{P}) + L^2(\Omega, \mathcal{G}_2, \mathbb{P})$ is a subspace of $L^2(\Omega, \mathcal{F}, \mathbb{P})$, and in particular it is closed. The projection on this subspace is the operator $P_1 + P_2 - P_1 P_2$.

**3.1.26 *Direct sum*** Another important example is the case where $\mathbb{H}_1 \cap \mathbb{H}_2$ contains only the zero vector. In this case, $\mathbb{H}_1 + \mathbb{H}_2$ is termed the direct sum of $\mathbb{H}_1$ and $\mathbb{H}_2$. The representation of an $x \in \mathbb{H}_1 + \mathbb{H}_2$ as the sum of vectors $x_1 \in \mathbb{H}_1$ and $x_2 \in \mathbb{H}_2$ is then unique.

**3.1.27 Exercise**   Show that in the situation of 3.1.25, the $\lambda$-system of events $A$ such that $1_A \in L^2(\Omega, \mathcal{G}_1, \mathbb{P}) \cap L^2(\Omega, \mathcal{G}_2, \mathbb{P})$ is trivial, i.e. that $\mathbb{P}(A)$ equals either 0 or 1. In particular, if $\mathbb{P}$ is complete, $\mathcal{G}$ is a $\sigma$-algebra.

**3.1.28 *The form of a bounded linear functional on a Hilbert space***
Suppose that $\mathbb{H}$ is a Hilbert space and $f$ is a linear and continuous functional on $\mathbb{H}$. There exists a unique $y \in \mathbb{H}$ such that $f(x) = (x, y)$. In particular $\|x\| = \|f\|$.

*Proof*   Suppose that for $y, y' \in \mathbb{H}$ we have $(x, y) = (x, y')$ for all $x \in \mathbb{H}$. Put $x = y - y'$, to see that $\|y - y'\| = 0$, which implies the uniqueness assertion.

Next, consider $Ker\,f = \{x \in \mathbb{H}| f(x) = 0\}$. This is a linear subspace of $\mathbb{H}$, since $f$ is linear and continuous. If $Ker\,f = \mathbb{H}$, we put $y = 0$. In the other case, there exists a $z \notin Ker\,f$. The non-zero vector $y_0 = z - Pz$ is our candidate for $y$. (Just think: if we really have $f(x) = (x, y)$ then $y$ is orthogonal to $Ker\,f$.) If this is to work, we must have $(z - Pz, z - Pz) =$

$\|y_0\|^2 = f(y_0)$. If this formula does not hold, a little scaling will do the job: specifically, we take $y = ty_0$ where $t = \frac{f(y_0)}{\|y_0\|^2}$ so that $\|y\|^2 = f(y)$. It remains to prove that $f(x) = (x, y)$ for all $x \in \mathbb{H}$. Take $x \in \mathbb{H}$, and write $x = (x - \frac{f(x)}{\|y\|^2} y) + \frac{f(x)}{\|y\|^2} y$. The first term belongs to $Ker\, f$, which is orthogonal to $y$. Thus, $(x, y) = (\frac{f(y)}{\|y\|^2} y, y) = f(x)$. Finally, $|f(x)| \leq \|x\|\|y\|$ shows that $\|f\| \leq \|x\|$, and $\|x\|^2 = (x, x) = f(x) \leq \|f\|\|x\|$ gives $\|x\| \leq \|f\|$. $\qquad\square$

## 3.2 Definition and existence of conditional expectation

3.2.1 *Motivation*    Let $(\Omega, \mathcal{F}, \mathbb{P})$ be a probability space. If $B \in \mathcal{F}$ is such that $\mathbb{P}(B) > 0$ then for any $A \in \mathcal{F}$ we define **conditional probability** $\mathbb{P}(A|B)$ (probability of $A$ given $B$) as

$$\mathbb{P}(A|B) = \frac{\mathbb{P}(A \cap B)}{\mathbb{P}(B)}. \tag{3.4}$$

As all basic courses in probability explain, this quantity expresses the fact that a partial knowledge of a random experiment ("$B$ happened") influences probabilities we assign to events. To take a simple example, in tossing a die, the knowledge that an even number turned up excludes three events, so that we assign to them conditional probability zero, and makes the probabilities of getting $2, 4$ or $6$ twice as big. Or, if three balls are chosen at random from a box containing four red, four white and four blue balls, then the probability of the event $A$ that all three of them are of the same color is $3\binom{4}{3}/\binom{12}{3} = \frac{3}{55}$. However, if we know that at least one of the balls that were chosen is red, the probability of $A$ decreases and becomes $\binom{4}{3}[\binom{12}{3} - \binom{8}{3}]^{-1} = \frac{3}{130}$. By the way, if this result does not agree with the reader's intuition, it may be helpful to remark that the knowledge that there is no red ball among the chosen ones increases the probability of $A$, and that it is precisely the reason why the knowledge that at least one red ball was chosen decreases the probability of $A$.

An almost obvious property of $\mathbb{P}(A|B)$ is that, as a function of $A$, it constitutes a new probability measure on the measurable space $(\Omega, \mathcal{F})$. It enjoys also other, less obvious, and maybe even somewhat surprising properties. To see that, let $B_i, i = 1, ..., n, n \in \mathbb{N}$ be a collection of mutually disjoint measurable subsets of $\Omega$ such that $\bigcup_{i=1}^{n} B_i = \Omega$ and $\mathbb{P}(B_i) > 0$. Such collections, not necessarily finite, are often called **dissections**, or **decompositions**, of $\Omega$. Also, let $A \in \mathcal{F}$. Consider all functions $Y$ of the form

$$Y = \sum_{i=1}^{n} b_i 1_{B_i} \tag{3.5}$$

where $b_i$ are arbitrary constants. How should the constants $b_i, i = 1, ..., n$ be chosen for $Y$ to be the closest to $X = 1_A$? The answer depends, of course, on the way "closeness" is defined. We consider the distance

$$d(Y, X) = \sqrt{\int_{\Omega} (Y - X)^2 \, d\mathbb{P}} = \|Y - X\|_{L^2(\Omega, \mathcal{F}, \mathbb{P})}. \tag{3.6}$$

In other words, we are looking for constants $b_i$ such that the distance $\|Y - X\|_{L^2(\Omega, \mathcal{F}, \mathbb{P})}$ is minimal; in terms of 3.1.12 we want to find a projection of $X$ onto the linear span of $\{1_{B_i}, i = 1, ..., n\}$. Calculations are easy; the expression under the square-root sign in (3.6) is

$$\sum_{i=1}^{n} \int_{B_i} (Y - 1_A)^2 \, d\mathbb{P} = \sum_{i=1}^{n} \int_{B_i} (b_i - 1_A)^2 \, d\mathbb{P}$$

$$= \sum_{i=1}^{n} \left[ b_i^2 \mathbb{P}(B_i) - 2b_i \mathbb{P}(B_i \cap A) + \mathbb{P}(A) \right],$$

and its minimum is attained when $b_i$ are chosen to be the minima of the binomials $b_i^2 \mathbb{P}(B_i) - 2b_i \mathbb{P}(B_i \cap A) + \mathbb{P}(A)$, i.e. if

$$b_i = \frac{\mathbb{P}(A \cap B_i)}{\mathbb{P}(B_i)} = \mathbb{P}(A|B_i). \tag{3.7}$$

Now, this is very interesting! Our simple reasoning shows that in order to minimize the distance (3.6), we have to choose $b_i$ in (3.5) to be conditional probabilities of $A$ given $B_i$. Or: the conditional probabilities $\mathbb{P}(A|B_i)$ are the coefficients in the projection of $X$ onto the linear span of $\{1_{B_i}, i = 1, ..., n\}$. This is not obvious from the original definition at all.

This observation suggests both the way of generalizing the notion of conditional probability and the way of constructing it in much more complex situations. Why should we look for generalizations of the notion of conditional probability? First of all, the definition (3.4) is valid only under the condition that $\mathbb{P}(B) > 0$, which is very unpleasant in applications. Secondly, we want to have a way of constructing conditional probability of random variables more complex than $X = 1_A$ (in such cases we speak of **conditional expectation**). Lastly, we want to have a way of constructing conditional expectations with respect to $\sigma$-algebras.

To understand the modern concept of conditional expectation, the reader is advised to take notice that our analysis involves a fundamental change in approach. Specifically, instead of looking at $\mathbb{P}(A|B_i), i = 1, ..., n$ separately, we gather all information about them in one function $Y$. In a sense, $Y$ is more intrinsic to the problem, and, certainly, it conveys information in a more compact form. Thus, the modern theory focuses on $Y$ instead of $\mathbb{P}(A|B_i)$.

The function $Y$ defined by (3.5) and (3.7) is a prototype of such a conditional expectation; it is in fact the conditional expectation of $X = 1_A$ with respect to the $\sigma$-algebra $\mathcal{G}$ generated by the dissection $B_i, i = 1, 2, ..., n$. Let us, therefore, look closer at its properties. Notice that while $X$ is measurable with respect to the original $\sigma$-algebra $\mathcal{F}$, $Y$ is measurable with respect to a smaller $\sigma$-algebra $\mathcal{G}$. On the other hand, even though $Y$ is clearly different from $X$, on the $\sigma$-algebra $\mathcal{G}$ it mimics $X$ in the sense that the integrals of $X$ and $Y$ over any event $B \in \mathcal{G}$ are equal. Indeed, it suffices to check this claim for $B = B_i$, and we have

$$\int_{B_i} Y \, d\mathbb{P} = \int_{B_i} \mathbb{P}(A|B_i) \, d\mathbb{P} = \mathbb{P}(A \cap B_i) = \int_{B_i} 1_A \, d\mathbb{P}.$$

It suggests that the notion of conditional expectation should be considered in $L^1(\Omega, \mathcal{F}, \mathbb{P})$ rather than $L^2(\Omega, \mathcal{F}, \mathbb{P})$ and leads to Definition 3.2.5 below. Before it is presented, however, the reader should solve the following two exercises.

**3.2.2 Exercise**    Let $X$ be an exponential random variable. Check to see that $\mathbb{P}(X > t + s|X > s) = \mathbb{P}(X > t)$, where $s, t \geq 0$. This is often referred to as the **memoryless property** of the exponential distribution. Prove also the converse: if a non-negative random variable $T$ has the memoryless property, then it is exponential.

**3.2.3 Exercise**    Let $B_i, i \geq 0$ be a decomposition of $\Omega$, and let $A$ and $C$ with $\mathbb{P}(C) > 0$ be two events. Show that $\mathbb{P}(A) = \sum_{i \geq 1} \mathbb{P}(A|B_i)P(B_i)$ (the total probability formula) and $\mathbb{P}(A|C) = \sum_i \mathbb{P}(A|B_i \cap C)\mathbb{P}(B_i|C)$, where we sum over all $i$ such that $\mathbb{P}(B_i \cap C) > 0$.

**3.2.4 Exercise**    It is interesting to note that the reasoning presented at the beginning of 3.2.1 does not work in the context of the spaces $L^p(\Omega, \mathcal{F}, \mathbb{P})$ where $1 \leq p \leq \infty$ and $p \neq 2$. To be more exact, prove that (a) for $1 < p < \infty$, $p \neq 2$, the minimum of the distance $\|X - Y\|_{L^p(\Omega, \mathcal{F}, \mathbb{P})}$ is attained for $b_i = [\mathbb{P}(A \cap B_i)]^{\frac{1}{p-1}} / \left( [\mathbb{P}(A \cap B_i)]^{\frac{1}{p-1}} + [\mathbb{P}(A^{\complement} \cap B_i)]^{\frac{1}{p-1}} \right)$,

(b) for $p = 1$, $b_i$ must be chosen to be equal 1 or 0 according as $\mathbb{P}(A \cap B_i) > \mathbb{P}(A^C \cap B_i)$ or $\mathbb{P}(A \cap B_i) < \mathbb{P}(A^C \cap B_i)$; if $\mathbb{P}(A \cap B_i) = \mathbb{P}(A^C \cap B_i)$ the choice of $b_i$ does not matter as long as $0 \leq b_i \leq 1$, and (c) for $p = \infty$ the best approximation is $b_i = \frac{1}{2}$ for any $A$ and $B_i$ (unless the probability of the symmetric difference between $A$ and one of the events $B_i$ is zero).

**3.2.5 Definition**     Let $(\Omega, \mathcal{F}, \mathbb{P})$ be a probability space, $X$ belong to $L^1(\Omega, \mathcal{F}, \mathbb{P})$, and $\mathcal{G} \subset \mathcal{F}$ be a $\sigma$-algebra. A variable $Y \in L^1(\Omega, \mathcal{G}, \mathbb{P})$ such that for all $A \in \mathcal{G}$,

$$\int_A X \, \mathrm{d}\mathbb{P} = \int_A Y \, \mathrm{d}\mathbb{P} \tag{3.8}$$

is termed the **conditional expectation** (of $X$ with respect to $\mathcal{G}$) and denoted $\mathbb{E}(X|\mathcal{G})$. In words: the operator $P : X \mapsto \mathbb{E}(X|\mathcal{G})$ is a Markov operator in $L^1(\Omega, \mathcal{F}, \mathbb{P})$, with values in $L^1(\Omega, \mathcal{G}, \mathbb{P}) \subset L^1(\Omega, \mathcal{F}, \mathbb{P})$, that preserves all functionals $F_A : X \mapsto \int_A X \, \mathrm{d}\mathbb{P}$ where $A \in \mathcal{G}$. Note that $\mathbb{E}(X|\mathcal{G})$ depends on $\mathbb{P}$ as well (see e.g. 3.3.11 and 3.3.12, below) and if we want to stress that dependence we write $\mathbb{E}_{\mathbb{P}}(X|\mathcal{G})$. For $X \in L^1(\Omega, \mathcal{F}, \mathbb{P})$ and a random variable $Y$ we define $\mathbb{E}(X|Y)$ as $\mathbb{E}(X|\sigma(Y))$. By the Doob–Dynkin Lemma 2.1.24 $\mathbb{E}(X|Y) = f(Y)$ for a Lebesgue measurable function $f$. The conditional probability $\mathbb{P}(A|\mathcal{G})$ is defined as $\mathbb{E}(1_A|\mathcal{G})$. Note that $\mathbb{P}(A|\mathcal{G})$ is not a number, but a function (to be more specific: a class of functions).

**3.2.6 Theorem**     For any $X \in L^1(\Omega, \mathcal{F}, \mathbb{P})$, the conditional expectation exists. Moreover, the map $P : X \mapsto \mathbb{E}(X|\mathcal{G})$ is a Markov operator and, when restricted to $L^2(\Omega, \mathcal{F}, \mathbb{P})$, is a projection onto $L^2(\Omega, \mathcal{G}, \mathbb{P})$.

*Proof* Let $X \in L^2(\Omega, \mathcal{F}, \mathbb{P})$, and $Y = PX$ be the projection of $X$ on the subspace $L^2(\Omega, \mathcal{G}, \mathbb{P})$. For any $Z \in L^2(\Omega, \mathcal{G}, \mathbb{P})$, $(X - PX, Z) = 0$, i.e. $\int_\Omega (X - PX)Z \, \mathrm{d}\mathbb{P} = 0$. Taking $Z = 1_A$, $A \in \mathcal{G}$ we obtain (3.8). We have thus proved existence of conditional expectation for $X \in L^2(\Omega, \mathcal{F}, \mathbb{P})$. By 2.3.37–2.3.38, we will be able to extend the operator $P$ to a Markov operator on $L^1(\Omega, \mathcal{F}, \mathbb{P})$ if we prove that $P$ maps non-negative $X \in L^2(\Omega, \mathcal{F}, \mathbb{P})$ into a non-negative $PX$. Moreover, by 2.3.37, the extension of $P$ will preserve the integrals over $A \in \mathcal{G}$. Therefore, we will be done once we show the claim about images of non-negative $X \in L^2(\Omega, \mathcal{F}, \mathbb{P})$.

Let $X \in L^2(\Omega, \mathcal{F}, \mathbb{P})$ and $X \geq 0$. Assume the probability of the event

$A = \{\omega | PX(\omega) < 0\}$ is not zero. Since

$$A = \bigcup_{n=1}^{\infty} A_n := \bigcup_{n=1}^{\infty} \{\omega | PX(\omega) < -\frac{1}{n}\},$$

we have $\mathbb{P}(A_n) = \delta > 0$ for some $n \geq 1$. Thus, since $A_n \in \mathcal{G}$ we have by (3.8),

$$-\frac{\delta}{n} > \int_{A_n} PX \, d\mathbb{P} = \int_{A_n} X \, d\mathbb{P} \geq 0,$$

a contradiction. $\qquad\qquad\square$

## 3.3 Properties and examples

**3.3.1 Theorem** Let $\alpha, \beta \in \mathbb{R}$, $X, Y$ and $X_n$, $n \geq 1$, belong to the space $L^1(\Omega, \mathcal{F}, \mathbb{P})$, and let $\mathcal{H}, \mathcal{G}$, $\mathcal{H} \subset \mathcal{G}$ be sub-$\sigma$-algebras of $\mathcal{F}$. Then,

(a) $\mathbb{E}(X|\mathcal{G})$ is $\mathcal{G}$ measurable,
(b) $\int_A \mathbb{E}(X|\mathcal{G}) \, d\mathbb{P} = \int_A X \, d\mathbb{P}$, for $A \in \mathcal{G}$,
(c) $E[\mathbb{E}(X|\mathcal{G})] = EX$,
(d) $\mathbb{E}(\alpha X + \beta Y|\mathcal{G}) = \alpha \mathbb{E}(X|\mathcal{G}) + \beta \mathbb{E}(Y|\mathcal{G})$,
(e) $\mathbb{E}(X|\mathcal{G}) \geq 0$ if $X \geq 0$,
(f) $E|\mathbb{E}(X|\mathcal{G})| \leq E|X|$, or, which is the same:

$$\|\mathbb{E}(X|\mathcal{G})\|_{L^1(\Omega, \mathcal{F}, \mathbb{P})} \leq \|X\|_{L^1(\Omega, \mathcal{F}, \mathbb{P})},$$

(g) $\mathbb{E}(\mathbb{E}(X|\mathcal{G})|\mathcal{H}) = \mathbb{E}(\mathbb{E}(X|\mathcal{H})|\mathcal{G}) = \mathbb{E}(X|\mathcal{H})$,
(h) $\mathbb{E}(X|\mathcal{G}) = (E\,X)1_\Omega$ if $X$ is independent of $\mathcal{G}$,
(i) if $X$ is $\mathcal{G}$ measurable, then $\mathbb{E}(X|\mathcal{G}) = X$; in particular, $\mathbb{E}(1_\Omega|\mathcal{G}) = 1_\Omega$,
(j) if $\mathcal{G}$ and $\mathcal{H}$ are independent then $\mathbb{E}(X|\sigma(\mathcal{G}\cup\mathcal{H})) = \mathbb{E}(X|\mathcal{G}) + \mathbb{E}(X|\mathcal{H}) - (E\,X)1_\Omega$; if, additionally, $X$ is independent of $\mathcal{H}$ then $\mathbb{E}(X|\sigma(\mathcal{G}\cup\mathcal{H})) = \mathbb{E}(X|\mathcal{G})$,
(k) if $\lim_{n\to\infty} X_n = X$ in $L^1(\Omega, \mathcal{F}, \mathbb{P})$, then $\lim_{n\to\infty} \mathbb{E}(X_n|\mathcal{G}) = \mathbb{E}(X|\mathcal{G})$ in $L^1(\Omega, \mathcal{G}, \mathbb{P})$,
(l) $|\mathbb{E}(X|\mathcal{G})| \leq \mathbb{E}(|X| \,|\mathcal{G})$,
(m) if $X_n \geq 0$ and $X_n \nearrow X$ (a.s.), then $\mathbb{E}(X_n|\mathcal{G}) \nearrow \mathbb{E}(X|\mathcal{G})$,
(n) if $XY$ is integrable, and $X$ is $\mathcal{G}$ measurable, then

$$\mathbb{E}(XY|\mathcal{G}) = X\mathbb{E}(Y|\mathcal{G}),$$

(o) $X$ is independent of $\mathcal{G}$ iff for any Lebesgue measurable function $f$ such that $f(X) \in L^1(\Omega, \mathcal{F}, \mathbb{P})$, $\mathbb{E}(f(X)|\mathcal{G}) = (Ef(X))1_\Omega$.

(p) If $\phi$ is convex and $\phi(X)$ is integrable, then

$$\phi\big(\mathbb{E}(X|\mathcal{G})\big) \leq \mathbb{E}(\phi(X)|\mathcal{G}).$$

Conditions (d)–(e),(g)–(k), and (m)–(p) hold $\mathbb{P}$ a.s.

*Proof* (a)–(b) is the definition, repeated here for completeness of the list. It is worth noting here in particular that (a)–(b) imply that $\mathbb{E}(X|\mathcal{G}) \in L^1(\Omega, \mathcal{F}, \mathbb{P})$. (d)–(f) have already been proved. (c) follows from (b) on putting $A = \Omega$.

(g) If we introduce the notation $P_1 X = \mathbb{E}(X|\mathcal{G}), P_2 X = \mathbb{E}(X|\mathcal{H})$, (g) reads $P_1 P_2 X = P_2 P_1 X = P_2 X$. Our claim follows thus by 3.1.18, for $X \in L^2(\Omega, \mathcal{F}, \mathbb{P})$. To complete the proof we apply 2.3.33 and 2.2.41.

(h) By density argument it suffices to show that our formula holds for all square integrable $X$. This, however, has been proved in 3.1.17.

(i) The first part is obvious by definition. For the second part note that $1_\Omega$ is $\mathcal{G}$ measurable for any $\mathcal{G}$.

(j) Again, it is enough to consider square integrable $X$. Under this assumption the first part follows from Example 3.1.25 and the second from (h).

(k) By (f),

$$\|\mathbb{E}(X_n|\mathcal{G}) - \mathbb{E}(X|\mathcal{G})\|_{L^1(\Omega, \mathcal{G}, \mathbb{P})} \leq \|X_n - X\|_{L^1(\Omega, \mathcal{F}, \mathbb{P})}.$$

(l) Apply (e) to $|X| \pm X$ and use linearity.

(m) By (e), $\mathbb{E}(X_n|\mathcal{G}) \nearrow$ to some $\mathcal{G}$ measurable $Y$ (be careful, these inequalities hold only a.s.!) Moreover, for $A \in \mathcal{G}$,

$$\int_A Y \, d\mathbb{P} = \lim_{n \to \infty} \int_A \mathbb{E}(X_n|\mathcal{G}) \, d\mathbb{P} = \lim_{n \to \infty} \int_A X_n \, d\mathbb{P} = \int_A X \, d\mathbb{P}.$$

This implies that $Y \in L^1(\Omega, \mathcal{G}, \mathbb{P})$, and $Y = \mathbb{E}(X|\mathcal{G})$.

(n) This result may be proved by considering indicator functions first, and then applying linearity and continuity of conditional expectation. Let us, however, take another approach and prove that (n) is actually (i) in a different probability space.

Note that we may assume that $Y \geq 0$. Let $\mathbb{P}^\sharp$ be the probability measure in $\{\Omega, \mathcal{F}\}$ defined by $\mathbb{P}^\sharp(A) = \frac{1}{k} \int_A Y \, d\mathbb{P}$, where $k = \int_\Omega Y \, d\mathbb{P}$. A random variable $Z$ on $\Omega$ belongs to $L^1(\Omega, \mathcal{F}, \mathbb{P}^\sharp)$ iff $ZY$ belongs to $L^1(\Omega, \mathcal{F}, \mathbb{P})$, and we have

$$\int Z \, d\mathbb{P}^\sharp = \frac{1}{k} \int ZY \, d\mathbb{P}.$$

For $A \in \mathcal{G}$,

$$\mathbb{P}^{\sharp}(A) = \frac{1}{k} \int_A \mathbb{E}_{\mathbb{P}}(Y|\mathcal{G}) \, d\mathbb{P};$$

i.e. the restriction of $\mathbb{P}^{\sharp}$ to $\mathcal{G}$ has a density $\frac{1}{k}\mathbb{E}_{\mathbb{P}}(Y|\mathcal{G})$. As above, a $\mathcal{G}$ measurable random variable $Z$ belongs to $L^1(\Omega, \mathcal{G}, \mathbb{P}^{\sharp})$ iff $Z\mathbb{E}_{\mathbb{P}}(Y|\mathcal{G})$ belongs to $L^1(\Omega, \mathcal{G}, \mathbb{P})$, and we have

$$\int Z \, d\mathbb{P}^{\sharp} = \frac{1}{k} \int Z\mathbb{E}_{\mathbb{P}}(Y|\mathcal{G}) \, d\mathbb{P}.$$

By assumption, $X \in L^1(\Omega, \mathcal{F}, \mathbb{P}^{\sharp})$. Let $Z = \mathbb{E}_{\mathbb{P}^{\sharp}}(X|\mathcal{G})$. We have

$$\int_A Z \, d\mathbb{P}^{\sharp} = \int_A X \, d\mathbb{P}^{\sharp}, \qquad A \in \mathcal{G}.$$

The right-hand side equals $\frac{1}{k} \int_A XY \, d\mathbb{P}$, and since $Z$ is $\mathcal{G}$ measurable, the left-hand side equals $\frac{1}{k} \int Z\mathbb{E}_{\mathbb{P}}(Y|\mathcal{G}) \, d\mathbb{P}$. Moreover, $Z\mathbb{E}_{\mathbb{P}}(Y|\mathcal{G})$ is $\mathcal{G}$ measurable. Thus $\mathbb{E}_{\mathbb{P}}(XY|\mathcal{G})$ equals $Z\mathbb{E}_{\mathbb{P}}(Y|\mathcal{G})$. Furthermore, $Z = X$ ($\mathbb{P}^{\sharp}$ a.s.), $X$ being $\mathcal{G}$ measurable. Therefore, $\mathbb{E}_{\mathbb{P}}(XY|\mathcal{G}) = X\mathbb{E}_{\mathbb{P}}(Y|\mathcal{G})$ $\mathbb{P}$ a.s., since if $\mathbb{P}^{\sharp}(A) = 0$, for some $A \in \mathcal{G}$, then either $\mathbb{P}(A) = 0$, or $Y = 0$ on $A$ ($\mathbb{P}$ a.s.); and in this last case $\mathbb{E}_{\mathbb{P}}(Y|\mathcal{G}) = 0$ ($\mathbb{P}$ a.s.).

(o) The necessity follows from (h), $f(X)$ being independent of $\mathcal{G}$ if $X$ is. If $X$ is not independent of $\mathcal{G}$, then there exist sets $A \in \sigma(X), B \in \mathcal{G}$ such that $\mathbb{P}(A \cap B) \neq \mathbb{P}(A)\mathbb{P}(B)$. Let $f = 1_C$, where $X^{-1}(C) = A$. We have $f(X) = 1_A$, so that $Ef(X) = \mathbb{P}(A)1_{\Omega}$. Taking $B$ introduced above we have $\int_B \mathbb{P}(A)1_{\Omega} \, d\mathbb{P} = \mathbb{P}(A)\mathbb{P}(B)$ while $\int_B f(X) \, d\mathbb{P} = \int_B 1_A \, d\mathbb{P} = \mathbb{P}(A \cap B)$. Thus

$$\mathbb{E}(f(X)|\mathcal{G}) \neq (Ef(X))1_{\Omega}.$$

(p) If $\phi$ is linear, (p) reduces to (d). In the other case, for $\phi \in S_0$, (see 1.5.6 for notations), we have $\psi(X) \leq \phi(X)$, thus

$$\psi(\mathbb{E}(X|\mathcal{G})) = \mathbb{E}(\psi(X)|\mathcal{G}) \leq \mathbb{E}(\phi(X)|\mathcal{G}),$$

almost surely. Since $S_0$ is countable, the set of $\omega \in \Omega$ where the last inequality does not hold for some $\psi \in S_0$ also has probability zero. Taking the supremum over $\psi \in S_0$, we obtain the claim. $\qquad\square$

**3.3.2 Remark**   Point (h) above says that conditioning $X$ on a $\sigma$-algebra that is independent from $\sigma(X)$ is the same as conditioning on a trivial $\sigma$-algebra $\{\Omega, \emptyset\}$. This is related to the so-called 0–1 law; see 3.6.11. Condition (g), called the tower property, is quite important and useful in proving results pertaining to the conditional expectation. Note that (c) may be viewed as a particular case of (g). Condition (p) is called Jensen's inequality; note that (f) is a particular case of (p) for $\phi(t) = |t|$.

**3.3.3** *Conditional expectation in $L^p(\Omega, \mathcal{F}, \mathbb{P})$*    Let $X$ be in $L^p(\Omega, \mathcal{F}, \mathbb{P})$ where $p > 1$ and $\mathcal{G}$ be a sub-$\sigma$-algebra of $\mathcal{F}$. By the Hölder inequality, $X \in L^1(\Omega, \mathcal{F}, \mathbb{P})$, too, and so $\mathbb{E}(X|\mathcal{G})$ exists. By Jensen's inequality $|\mathbb{E}(X|\mathcal{G})|^p \leq \mathbb{E}(|X|^p|\mathcal{G})$. Therefore,

$$\int |\mathbb{E}(X|\mathcal{G})|^p \, d\mathbb{P} \leq \int \mathbb{E}(|X|^p|\mathcal{G}) \, d\mathbb{P} = \int |X|^p \, d\mathbb{P} = \|X\|_{L^p}^p.$$

This shows that $\mathbb{E}(X|\mathcal{G}) \in L^p(\Omega, \mathcal{F}, \mathbb{P})$ and that conditional expectation is a contraction operator from $L^p(\Omega, \mathcal{F}, \mathbb{P})$ to $L^p(\Omega, \mathcal{F}, \mathbb{P})$.

**3.3.4 Exercise**    In 3.2.1 we have seen that if $\mathcal{G}$ is generated by a finite dissection $(B_i)_{i=1,\ldots n}$, then $\mathbb{E}(X|\mathcal{G}) = \sum_{i=1}^n b_i 1_{B_i}$ where $X = 1_A, b_i = \frac{\mathbb{P}(A \cap B_i)}{\mathbb{P}(B_i)}$. Prove that for $X \in L^1(\Omega, \mathcal{F}, \mathbb{P})$ the formula is the same except that $b_i = \frac{1}{\mathbb{P}(B_i)} \int_{B_i} X \, d\mathbb{P}$.

**3.3.5 Example**    A die is tossed twice. Let $X_i, i = 1, 2$, be the number on the die in the $i$th toss. We will find $\mathbb{E}(X_1|X_1 + X_2)$. The space $\Omega$ is a set of ordered pairs $(i, j)$, $1 \leq i, j \leq 6$. $\sigma(X_1 + X_2)$ is generated by the dissection $(B_i)_{i=2,\ldots,12}$ where $B_i$ is composed of pairs with coordinates adding up to $i$. In other words $B_i$ are diagonals in the "square" $\Omega$. We have

$$\mathbb{E}(X_1|X_1 + X_2) = \sum_{i=2}^{12} b_i 1_{B_i}, \quad b_i = \frac{1}{\mathbb{P}(B_i)} \int_{B_i} X_1 \, d\mathbb{P}.$$

For example $b_2 = 1$; which means that if $X_1 + X_2 = 2$, then $X_1 = 1$. Similarly, $b_3 = \frac{3}{2}$, which can be interpreted by saying that if $X_1 + X_2 = 3$, then $X_1 = 1$ or $2$ with equal probability. Similarly, $b_8 = 4$, which means that the knowledge that $X_1 + X_2 = 8$ increases the expected result on the first die.

**3.3.6 Exercise**    Let $B_i, i = 1, 2, \ldots$, be an infinite dissection of $\Omega$. Consider $A \in \mathcal{F}$, and functions of the form

$$\phi = \sum_{i=1}^{\infty} b_i 1_{B_i}. \tag{3.9}$$

What are necessary and sufficient conditions for $\phi \in L^2(\Omega, \mathcal{F}, \mathbb{P})$ (in terms of $b_i$)? Choose $b_i$ in such a way that the distance in (3.6) is minimal. Check to see that $\phi$ with such coefficients belongs to $L^2(\Omega, \mathcal{F}, \mathbb{P})$ and satisfies $\int_{B_i} 1_A \, d\mathbb{P} = \int_{B_i} \phi \, d\mathbb{P}$.

**3.3.7 Exercise** Imitating the proof of the Cauchy–Schwartz– Bunya-kovski inequality (see 3.1.3) show that

$$\mathbb{E}(XY|\mathcal{G})^2 \leq \mathbb{E}(X^2|\mathcal{G})\mathbb{E}(Y^2|\mathcal{G}) \qquad \mathbb{P} \text{ a.s.,} \qquad (3.10)$$

for $X, Y \in L^2(\Omega, \mathcal{F}, \mathbb{P})$.

**3.3.8 Exercise** State and prove analogs of the Lebesgue Dominated Convergence Theorem and Fatou's Lemma for conditional expectation.

**3.3.9 Exercise** Prove the following Markov inequality (c.f. 1.2.36):

$$\mathbb{P}(X \geq a|\mathcal{G}) \leq \frac{1}{a}\mathbb{E}(X|\mathcal{G}), \quad a > 0, X \geq 0 \ \mathbb{P} \text{ a.s.}$$

**3.3.10 Exercise** Let $\mathbb{VAR}(X|\mathcal{G}) = \mathbb{E}(X^2|\mathcal{G}) - \mathbb{E}(X|\mathcal{G})^2$. Show that

$$D^2(X) = E\left[\mathbb{VAR}(X|\mathcal{G})\right] + D^2[\mathbb{E}(X|\mathcal{G})].$$

In calculating conditional expectation, quite often the difficult part is to guess the answer; checking that this is a correct one is usually a much easier matter. The fact that conditional expectation is a projection in a Hilbert space of square integrable functions may facilitate such guessing. We hope the following examples illustrate this idea. By the way, we do not *have to* check that our answer is correct since conditional expectation in $L^1(\Omega, \mathcal{F}, \mathbb{P})$ is uniquely determined by its values in $L^2(\Omega, \mathcal{F}, \mathbb{P})$.

**3.3.11 Example** Let $\Omega = [0, 1]$, $\mathcal{F}$ be the $\sigma$-algebra of Lebesgue measurable subsets of $[0, 1]$, and let $\mathbb{P}$ be the restriction of the Lebesgue measure to this interval. Let $\mathcal{G}$ be the collection of Lebesgue measurable subsets such that $1 - A = A$ ($\mathbb{P}$ a.s.). By definition, $1 - A = \{\omega | \omega = 1 - \omega', \omega' \in A\}$, and $A = B$ ($\mathbb{P}$ a.s.) iff the probability of the symmetric difference of $A$ and $B$ is 0. In other words $\mathcal{G}$ is the family of sets that are symmetric with respect to $\frac{1}{2}$. We claim that $\mathcal{G}$ is a $\sigma$-algebra.

(a) Obviously $1 - \Omega = \Omega$, so that $\Omega \in \mathcal{G}$.
(b) Note that $\omega \in 1 - A^{\complement} \Leftrightarrow 1 - \omega \in A^{\complement} \Leftrightarrow 1 - \omega \notin A \Leftrightarrow \omega \notin 1 - A$; i.e. $1 - A^{\complement} = (1 - A)^{\complement}$. Moreover, $A^{\complement} = B^{\complement}$ ($\mathbb{P}$ a.s.) whenever $A = B$ ($\mathbb{P}$ a.s.). Thus, $A \in \mathcal{G} \Rightarrow A^{\complement} \in \mathcal{G}$ for $1 - A^{\complement} = (1 - A)^{\complement} = A^{\complement}$ ($\mathbb{P}$ a.s.).
(c) As above we show that $1 - \bigcup_{n \geq 1} A_n = \bigcup_{n \geq 1}(1 - A_n)$. Moreover, if $A_n = B_n$ ($\mathbb{P}$ a.s.), then $\bigcup_{n \geq 1} A_n = \bigcup_{n \geq 1} B_n$ ($\mathbb{P}$ a.s.). Thus $A_n \in \mathcal{G} \Rightarrow \bigcup_{n \geq 1} A_n \in \mathcal{G}$.

What is $L^2(\Omega, \mathcal{G}, \mathbb{P})$? Because any $Y \in L^2(\Omega, \mathcal{G}, \mathbb{P})$ is a pointwise limit of a sequence of linear combinations of indicator functions of sets in $\mathcal{G}$, and $1_A, A \in \mathcal{G}$ satisfies $1_A(\omega) = 1_A(1 - \omega)$, ($\mathbb{P}$ a.s.), $Y$ belongs to $L^2(\Omega, \mathcal{G}, \mathbb{P})$ iff $Y(\omega) = Y(1-\omega)$, ($\mathbb{P}$ a.s). Let $X \in L^2(\Omega, \mathcal{F}, \mathbb{P})$. What is $Z = \mathbb{E}(X|\mathcal{G})$? $Z$ must belong to $L^2(\Omega, \mathcal{G}, \mathbb{P})$ and minimize the integral

$$\int_0^1 (X(\omega) - Z(\omega))^2 \, d\omega$$

$$= \int_0^{\frac{1}{2}} (X(\omega) - Z(\omega))^2 \, d\omega + \int_{\frac{1}{2}}^1 (X(\omega) - Z(\omega))^2 \, d\omega$$

$$= \int_0^{\frac{1}{2}} \left\{ (X(\omega) - Z(\omega))^2 + (X(1 - \omega) - Z(\omega))^2 \right\} \, d\omega;$$

the last equality resulting from the change of variables and $Z(\omega) = Z(1 - \omega)$ ($\mathbb{P}$ a.s.). Now, for two numbers $a, b \in \mathbb{R}$, $(a-\tau)^2 + (b-\tau)^2$ is minimal for $\tau = \frac{a+b}{2}$. Thus, we must have $\mathbb{E}(X|\mathcal{G})(\omega) = Z(\omega) = \frac{1}{2}[X(\omega) + X(1-\omega)]$, ($\mathbb{P}$ a.s.). Certainly, this formula holds also for $X \in L^1(\Omega, \mathcal{F}, \mathbb{P})$, which can be checked directly.

**3.3.12** *Conditional expectation depends on probability measure* Let the space $(\Omega, \mathcal{F})$ and the $\sigma$-algebra $\mathcal{G}$ be as in 3.3.11, and let $\mathbb{P}(A) = 2 \int_A \omega \, d\omega$. To calculate $Y = \mathbb{E}(X|\mathcal{G})$, $X \in L^2(\Omega, \mathcal{F}, \mathbb{P})$, observe that we must have $Y(1 - \omega) = Y(\omega)$, $\mathbb{P}$ a.s. Noting that $\mathbb{P}$ a.s. is the same as *leb* a.s., we consider the distance

$$d(X, Y) = \int_\Omega (X - Y)^2 \, d\mathbb{P}$$

and choose $Y$ in such a way that this distance is minimal. Calculating as in 3.3.11, we see that this distance equals

$$2 \int_0^{\frac{1}{2}} \left\{ [X(\omega) - Y(\omega)]^2 \omega + 2[X(1 - \omega) - Y(\omega)]^2 (1 - \omega) \right\} \, d\omega. \quad (3.11)$$

For fixed $\omega$, we treat $X(\omega)$ and $X(1 - \omega)$ as given and minimize the integrand. The minimum will be attained for

$$\mathbb{E}(X|\mathcal{G})(\omega) = Y(\omega) = \omega X(\omega) + (1 - \omega) X(1 - \omega).$$

To check that this formula is valid for $X \in L^1(\Omega, \mathcal{F}, \mathbb{P})$ we calculate:

$$\int_A Y(\omega) \, d\mathbb{P}(\omega) = 2 \int_A \left\{ \omega X(\omega) + (1 - \omega) X(1 - \omega) \right\} \omega \, d\omega$$

$$= 2 \int_A \omega^2 X(\omega) \, d\omega + 2 \int_{1-A} \omega X(1 - \omega)(1 - \omega) \, d\omega$$

where the last relation holds since $A \in \mathcal{G}$; changing variables, this equals:

$$2 \int_A \omega^2 X(\omega) \, d\omega + 2 \int_A (\omega - \omega^2) X(\omega) \, d\omega$$

$$= 2 \int_A \omega X(\omega) \, d\omega = \int_A X(\omega) \, d\mathbb{P}.$$

**3.3.13 Example**  Let $\Omega = \{(x, y) \mid |x| + |y| \leq 1\}$, $\mathcal{F}$ be the $\sigma$-algebra of its Lebesgue subsets, and let $\mathbb{P}$ be $\frac{1}{2}$ times the restriction of the Lebesgue measure to $\Omega$. Let $X(\omega) = x, Y(\omega) = y$ where $\omega = (x, y)$. The problem is to find $\mathbb{E}(X^2 | Y)$. To this end, we will find a function $F(y)$ such that $\mathbb{E}(X^2 | Y)(\omega) = F(Y(\omega))$. Note that $X \in L^2(\Omega, \mathcal{F}, \mathbb{P})$. We have

$$\int_\Omega (F(Y) - X)^2 \, d\mathbb{P} = \int_{-1}^1 \int_{-1+|x|}^{1-|x|} (F(y) - x^2)^2 \, dx \, dy$$

$$= \int_{-1}^1 \left\{ 2F(y)(1 - |y|) - \frac{4}{3} F(y)(1 - |y|)^3 + \frac{1}{5}(1 - |y|)^5 \right\} dy.$$

The minimum of the integrand is attained if $F(y)$ equals $\frac{1}{3}(1 - |y|)^2$ for $|x| \leq 1$. For $|x| > 1$ the value of the function $F$ can be defined arbitrarily. It remains to check that $F(Y(\omega))$ equals $\mathbb{E}(X^2 | \mathcal{G})$; calculations are straightforward.

**3.3.14 Example**  Suppose that random variables $X, Y$ have a joint density $f(x, y)$, which is a Lebesgue measurable function, and $g(x)$ is a Lebesgue measurable function such that $g(X) \in L^1(\Omega, \mathcal{F}, \mathbb{P})$. Again, the problem is to find $\mathbb{E}(g(X) | Y)$. To see what our guess should be, assume first that $g(X) \in L^2(\Omega, \mathcal{F}, \mathbb{P})$, and minimize the distance $d(Z, X)$ where $Z \in L^2(\Omega, \mathcal{G}, \mathbb{P})$. Recalling that $Z$ must be of the form $Z = z(Y)$, we minimize

$$
\begin{aligned}
d(Z, X) &= \iint_{\mathbb{R}^2} (z(y) - g(x))^2 f(x, y) \, dx \, dy \\
&= \int_{\mathbb{R}} \left[ z^2(y) \int_{\mathbb{R}} f(x, y) \, dx - 2z(y) \int_{\mathbb{R}} g(x) f(x, y) \, dx \right. \\
&\quad \left. + \int_{\mathbb{R}} g^2(x) f(x, y) \, dx \right] dy. \tag{3.12}
\end{aligned}
$$

The minimum of the integrand is attained for $z(y) = \frac{\int_{\mathbb{R}} g(x) f(x, y) \, dx}{\int_{\mathbb{R}} f(x, y) \, dx}$ provided $\int_{\mathbb{R}} f(x, y) \, dx \neq 0$. If this last integral is zero, then $f(x, y) = 0$ for almost all $x \in \mathbb{R}$. Thus, the integral (3.12), taken over the set of

$y \in \mathbb{R}$ such that $\int_{\mathbb{R}} f(x, y)\, \mathrm{d}x = 0$, equals zero regardless of what $z(y)$ is defined to be. For example, we may put

$$
z(y) = \begin{cases} \dfrac{\int_{\mathbb{R}} g(x) f(x,y)\, \mathrm{d}x}{\int_{\mathbb{R}} f(x,y)\, \mathrm{d}x}, & \int_{\mathbb{R}} f(x,y)\, \mathrm{d}x \neq 0, \\ 0, & \text{otherwise.} \end{cases} \tag{3.13}
$$

**3.3.15 Exercise**    State and prove an analogous proposition for discrete random variables.

**3.3.16 Exercise**    Let $Z_1 = X_1 + Y$ and $Z_2 = X_2 + Y$ where $E\, X_1 = E\, X_2$, and $X_1, X_2$ and $Y$ are independent absolutely integrable random variables. Show that $\mathbb{E}(Z_1 | Z_2)$ equals $Z_2$.

**3.3.17** *Conditional distribution*    So far, we have interpreted the formula (3.13) to mean that if $Y$ is known, then the "average" outcome of $g(X)$ is $z(Y)$. Notice that class of functions $g$ is quite large here, and therefore should determine more than just the expected value of $X$: it should determine "the distribution of $X$ given $Y = y$". Note that we may not speak of "the distribution of $X$ given $Y = y$" in terms of (3.4) since $\mathbb{P}\{Y = y\}$ is quite often zero. In particular, right now it is unclear where the variable $(X|Y = y)$ (in words: "$X$ given that $Y = y$") is defined. It is convenient and reasonable, though, to interpret the first formula in (3.13) to be $\int_{\mathbb{R}} g(x) h(x, y)\, \mathrm{d}x$ where

$$
h(x, y) = \frac{f(x, y)}{\int_{\mathbb{R}} f(z, y)\, \mathrm{d}z}, \tag{3.14}
$$

is a density of $(X|Y = y)$. Here is an example. Let $X_i$, $i = 1, 2$, be two exponential random variables with parameter $\lambda$. It is often said that if $Y = X_1 + X_2$ is known to be $y$ then $X = X_1$ is uniformly distributed on $[0, y]$. This is a simple case of (3.13). To see this note that the joint distribution of $X_1$ and $X_2$ is given by the density $f_0(x_1, x_2) = \lambda^2 e^{-\lambda(x_1 + x_2)}$, $x_1, x_2 \geq 0$, and that, by the change of variables formula, the density of $X$ and $Y$ is given by $f(x, y) = \lambda^2 e^{-\lambda y}$, $y > x > 0$. The numerator in (3.13), where $g(x) = x$, is therefore equal to $\lambda^2 e^{-\lambda y} y^2 / 2$ and the denominator equals $\lambda^2 e^{-\lambda y} y$. Thus, in the first interpretation we infer that if $Y = y$ then $X$ is on average expected to be $y/2$. In the second however, we calculate $h(x, y)$ given in (3.14) to be $1/y$ for $x < y$ and 0 for $x > y$. This is exactly what we have claimed.

Note, finally, that even though the probability space where $(X|Y = y)$ is defined may have nothing to do with the probability space where

$X_i$ were defined, the existence of such a probability space follows from the single fact that $h(x,y)$ is a density function for all $y$ such that $\int_{\mathbb{R}} f(x,y)\, dx \neq 0$.

**3.3.18** *Use of symmetries* Suppose that random variables $X_1, ..., X_n$ are integrable and that their joint distribution $\mathbb{P}_{X_1,...,X_n}$ does not change when the random variables are permuted, i.e. that they are **exchangeable**. In particular, for any Borel measurable set $C \subset \mathbb{R}^n$ we have

$$\int_C x_i\, d\mathbb{P}_{X_1,...,X_n} = \int_C x_j\, d\mathbb{P}_{X_1,...,X_n}, \qquad 1 \le i,j \le n. \tag{3.15}$$

Suppose now that in an experiment we are not able to observe the variables $X_i$ themselves but only their sum $S$. Since the random variables are exchangeable, our best bet on the value of $X_i$, given $S$, is of course $\frac{1}{n}S$. This is an intuitive guess that $\mathbb{E}(X_i|S) = \frac{1}{n}S$, $i = 1, ..., n$. To prove that our intuition is right we note first that

$$\mathbb{E}(X_i|S) = \mathbb{E}(X_j|S), \qquad 1 \le i,j \le n. \tag{3.16}$$

Indeed, it suffices to show that for any $A \in \sigma(S)$

$$\int_A X_i\, d\mathbb{P} = \int_A X_j\, d\mathbb{P}, \qquad 1 \le i,j \le n. \tag{3.17}$$

Now, if $A \in \sigma(S)$ then there exists a Borel measurable $B \subset \mathbb{R}$ such that $A = \{\omega \in \Omega; S(\omega) \in B\}$ and so there exists a Borel measurable $C \subset \mathbb{R}^n$ such that $A = \{\omega \in \Omega; (X_1(\omega), ..., X_n(\omega)) \in C\}$. By the change of variables formula $\int_A X_i\, d\mathbb{P} = \int_C x_i\, d\mathbb{P}_{X_1,...,X_n}$. Hence (3.15) forces (3.17) and (3.16).

Hence, as predicted:

$$\mathbb{E}(X_i|S) = \frac{1}{n}\mathbb{E}\left(\sum_{j=1}^n X_j \Big| S\right) = \frac{1}{n}\mathbb{E}(S|S) = \frac{1}{n}S.$$

This result is particularly useful when applied to an infinite sequence $X_n, n \ge 1$ of independent, identically distributed random variables. The assumption of independence forces $X_1, ..., X_n$ to be exchangeable for any $n$ and our result gives

$$\mathbb{E}(X_i|S_n) = \frac{S_n}{n}, \qquad 1 \le i \le n, \tag{3.18}$$

where $S_n = \sum_{i=1}^n X_i$. The same assumption implies also a stronger result:

$$\mathbb{E}(X_i|\sigma(S_n, S_{n+1}, ...)) = \frac{S_n}{n}, \qquad 1 \le i \le n. \tag{3.19}$$

Indeed, $\sigma(S_n, S_{n+1}, ...) = \sigma(S_n, X_{n+1}, X_{n+2}, ...)$ because $S_k, k \geq n$ may be expressed as linear combinations of $S_n$ and $X_k, k \geq n$, and vice versa. Moreover, $\sigma(S_n, X_{n+1}, X_{n+2}, ...) = \sigma(S_n, \sigma(X_{n+1}, X_{n+2}, ...))$ and the $\sigma$-algebra $\sigma(X_{n+1}, X_{n+2}, ...)$ is independent of both $S_n$ and $X_i, 1 \leq i \leq n$. Hence (3.19) follows from 3.3.1 (j).

What is nice about (3.19), and what is its advantage over (3.18) is that it shows that $\frac{S_n}{n}$ is a sequence of conditional expectations of $X_1$ with respect to a non-increasing sequence of $\sigma$-algebras, for there are well-known theorems concerning convergence of such sequences. Later on, in 3.7.5, we will be able to use (3.19) to prove the all-important Strong Law of Large Numbers.

In the remainder of this section we focus on the tower property.

**3.3.19 Exercise** Using (c), (h), and (n) in 3.3.1 show that if $X$ and $Y$ are independent and $X, Y$, and $XY$ are integrable then $E\,XY = E\,X \cdot E\,Y$.

**3.3.20 Example** Let $\mathcal{G}, \mathcal{H},$ and $\mathcal{I}$ be $\sigma$-algebras in $(\Omega, \mathcal{F}, \mathbb{P})$ such that $\mathcal{I} \subset \mathcal{H} \cap \mathcal{G}$. We say that given $\mathcal{I}, \mathcal{G}$ and $\mathcal{H}$ are independent iff:

$$\mathbb{P}(A \cap B | \mathcal{I}) = \mathbb{P}(A | \mathcal{I})\mathbb{P}(B | \mathcal{I}), \quad A \in \mathcal{G}, B \in \mathcal{H}. \qquad (3.20)$$

We will show that this condition is equivalent to each of the following two:

$$\mathbb{P}(B | \mathcal{G}) \;=\; \mathbb{P}(B | \mathcal{I}), \qquad B \in \mathcal{H}, \qquad (3.21)$$
$$\mathbb{P}(A | \mathcal{H}) \;=\; \mathbb{P}(A | \mathcal{I}), \qquad A \in \mathcal{G}. \qquad (3.22)$$

These conditions are often expressed by saying that $\mathcal{H}$ depends on $\mathcal{G}$ only through $\mathcal{I}$ and $\mathcal{G}$ depends on $\mathcal{H}$ only through $\mathcal{I}$, respectively.

By symmetry, it is enough to prove that (3.20) is equivalent to (3.21). To prove that (3.20) implies (3.21), note first that $\mathbb{P}(B | \mathcal{I})$ is a $\mathcal{G}$ measurable function, so it suffices to show that

$$\int_\Omega 1_A \mathbb{P}(B | \mathcal{G})\,d\mathbb{P} = \int_\Omega 1_A \mathbb{P}(B | \mathcal{I})\,d\mathbb{P}, \qquad A \in \mathcal{G}$$

(this is obvious for $A \in \mathcal{I} \subset \mathcal{G}$). Since $1_A$ is $\mathcal{G}$ measurable, the left-hand side equals

$$\int_\Omega \mathbb{E}(1_{A \cap B} | \mathcal{G})\,d\mathbb{P} = \int_\Omega 1_{A \cap B}\,d\mathbb{P} = \mathbb{P}(A \cap B),$$

and the right-hand side, by (3.20),

$$\int_\Omega 1_A \mathbb{E}(1_B|\mathcal{I})\, d\mathbb{P} = \int_\Omega \mathbb{E}\left[1_A \mathbb{E}(1_B|\mathcal{I})\,|\mathcal{I}\right] d\mathbb{P} = \int_\Omega \mathbb{E}(1_A|\mathcal{I})\mathbb{E}(1_B|\mathcal{I})\, d\mathbb{P}$$
$$= \int_\Omega \mathbb{E}(1_{A\cap B}|\mathcal{I})\, d\mathbb{P} = \mathbb{P}(A\cap B).$$

Conversely, if (3.21) holds, the left-hand side in (3.20) equals:

$$\begin{aligned}
\mathbb{E}(1_{A\cap B}|\mathcal{I}) &= \mathbb{E}\left\{[\mathbb{E}(1_{A\cap B}|\mathcal{G})]\,|\mathcal{I}\right\} = \mathbb{E}\left\{\mathbb{E}(1_A 1_B|\mathcal{G})|\mathcal{I}\right\} \\
&= \mathbb{E}(1_A \mathbb{E}(1_B|\mathcal{G})|\mathcal{I}) = \mathbb{E}(1_A \mathbb{E}(1_B|\mathcal{I})|\mathcal{I}) \\
&= \mathbb{E}(1_A|\mathcal{I})\mathbb{E}(1_B|\mathcal{I}).
\end{aligned}$$

**3.3.21 Exercise** The previous example is important in the theory of Markov processes. To consider the simplest case; a sequence $X_n, n \geq 0$ of random variables with values in $\mathbb{N}$ is said to be a *Markov chain* if it enjoys the *Markov property*: for all $n \geq 0$, and $i_k \in \mathbb{N}, 0 \leq k \leq n+1$,

$$\mathbb{P}[X_{n+1} = i_{n+1}|X_k = i_k, 0 \leq k \leq n] = \mathbb{P}[X_{n+1} = i_{n+1}|X_n = i_n]$$
$$=: p_{i_n, i_{n+1}}$$

provided $\mathbb{P}[X_k = i_k, 0 \leq k \leq n] > 0$. Here the $p_{i,j}, i, j \in \mathbb{N}$ are called the transition probabilities (or jump probabilities) of the chain and satisfy $\sum_{j\in\mathbb{N}} p_{i,j} = 1, i \in \mathbb{N}$ and $p_{i,j} \geq 0$. Existence of a Markov chain with a given matrix of transition probabilities is proved by a straightforward construction, e.g. in [5]. The reader will show that the Markov property may be defined in an equivalent way by any of the following three conditions:

$$\begin{aligned}
\mathbb{P}(A \cap B|X_n) &= \mathbb{P}(A|X_n)\mathbb{P}(B|X_n), & A \in \mathcal{G}_n, B \in \mathcal{H}_n, \\
\mathbb{P}(B|\mathcal{G}_n) &= \mathbb{P}(B|X_n), & B \in \mathcal{H}_n, \\
\mathbb{P}(A|\mathcal{H}_n) &= \mathbb{P}(A|X_n), & A \in \mathcal{G}_n,
\end{aligned}$$

where $\mathcal{G}_n = \sigma(X_n, X_{n+1}, ...)$ and $\mathcal{H}_n = \sigma(X_0, ..., X_n)$.

## 3.4 The Radon–Nikodym Theorem

**3.4.1** *Lebesgue decomposition* Let $\lambda, \mu$ be two finite measures on a measurable space $(\Omega, \mathcal{F})$. Then, there exists a non-negative function $f \in L^1(\Omega, \mathcal{F}, \mu)$ and a measure $\nu$ singular to $\mu$ (i.e. such that there exists a set $S \in \mathcal{F}$ with $\mu(S) = 0$ and $\nu(\Omega \setminus S) = 0$) such that

$$\lambda(A) = \int_A f\, d\mu + \nu(A), \quad \text{for all } A \in \mathcal{F}.$$

*Proof* Consider a linear functional $Fx = \int x \, d\lambda$, acting in the space $L^2(\Omega, \mathcal{F}, \lambda + \mu)$. The estimate

$$|Fx| \le \sqrt{\lambda(\Omega)} \sqrt{\int_\Omega |x|^2 \, d\lambda} \le \sqrt{\lambda(\Omega)} \|x\|_{L^2(\Omega, \mathcal{F}, \lambda + \mu)}$$

shows that $F$ is well-defined and bounded. Therefore, there exists a function $y \in L^2(\Omega, \mathcal{F}, \lambda + \mu)$ such that $Fx = \int_\Omega xy \, d(\lambda + \mu)$. Taking $x = 1_A, A \in \mathcal{F}$, we see that

$$\lambda(A) = \int_A y \, d\lambda + \int_A y \, d\mu. \tag{3.23}$$

This in turn proves that $y \ge 0$, $(\lambda + \mu)$ a.e., and $y \le 1$, $\lambda$ a.e. Let $S = \{\omega | y(\omega) = 1\} \in \mathcal{F}$. By (3.23), $\mu(S) = 0$. Rewriting (3.23) in the form $\int_\Omega (1-y) 1_A \, d\lambda = \int_\Omega y 1_A \, d\mu$, we see that for any non-negative measurable function $x$ on $\Omega$, $\int_\Omega (1-y) x \, d\lambda = \int_\Omega yx \, d\mu$. Define $f(\omega) = \frac{y(\omega)}{1-y(\omega)}$ on $S^{\complement}$, and zero on $S$. If $A \in \mathcal{F}$, and $A \subset S^{\complement}$, we may take $x = 1_A \frac{1}{1-y}$ to see that $\lambda(A) = \int_A f \, d\mu$. Also, let $\nu(A) = \lambda(S \cap A)$. Thus, $\nu(S^{\complement}) = 0$, i.e. $\mu$ and $\nu$ are singular. Moreover,

$$\lambda(A) = \lambda(A \cap S^{\complement}) + \lambda(A \cap S) = \int_{A \cap S^{\complement}} f \, d\mu + \nu(A) = \int_A f \, d\mu + \nu(A).$$

Finally, $f$ belongs to $L^1(\Omega, \mathcal{F}, \mu)$, since it is non-negative and $\int_\Omega f \, d\mu = \int_{S^{\complement}} f \, d\mu = \lambda(S^{\complement}) < \infty$. $\qquad\square$

3.4.2 *The Radon–Nikodym Theorem*    Under assumptions of 3.4.1, suppose additionally that $\mu(A) = 0$ for some $A \in \mathcal{F}$ implies that $\lambda(A) = 0$. Then $\nu = 0$; i.e. $\lambda$ is absolutely continuous with respect to $\mu$.

*Proof* We know that $\mu(S) = 0$, so that $\nu(S) = \lambda(S) = 0$. On the other hand, $\nu(S^{\complement}) = 0$ so that $\nu = 0$. $\qquad\square$

3.4.3 **Remark**    With the Radon–Nikodym Theorem at hand, we may prove existence of conditional expectation very easily. We note first that it is enough to restrict ourselves to non-negative variables. Consider the space $(\Omega, \mathcal{G}, \mathbb{P})$ and the measure $\mu(A) = \int_A X \, d\mathbb{P}$ where $X \ge 0, X \in L^1(\Omega, \mathcal{F}, \mathbb{P})$. The measure $\mu$ is absolutely continuous with respect to $\mathbb{P}$ and so there exists a non-negative $Y \in L^1(\Omega, \mathcal{G}, \mathbb{P})$ such that

$$\int_A X \, d\mathbb{P} = \int_A Y \, d\mathbb{P},$$

as desired. This method of approaching conditional expectations is found in most probabilistic monographs. (Following Kolmogorov, who was the first to investigate conditional expectation in this generality and used the Radon–Nikodym Theorem to prove its existence.) See e.g. [5, 46, 57]. Analysts, however, prefer the approach via projections [107]. We should mention that there are of course purely measure-theoretic proofs of the Lebesgue decomposition theorem and the Radon–Nikodym Theorem [49, 103].

**3.4.4 Exercise** Using the Radon–Nikodym Theorem prove that the conditional expectation $\mathbb{E}(X|Y)$ equals $f(Y)$ for some Lebesgue measurable function $f$. (We have proved it before using the Doob–Dynkin Lemma.)

**3.4.5** *Application: Frobenius–Perron operators* Let $(\Omega, \mathcal{F}, \mu)$ be a measure space. Suppose that a measurable map $f : \Omega \to \Omega$ is **non-singular**, i.e. that for any $A \in \mathcal{F}$ such that $\mu(A) = 0$, we also have $\mu(f^{-1}(A)) = 0$. In studying asymptotic behavior of iterates $f^{\circ n+1}(\omega) = f^{\circ n}(f(\omega))$, $n \geq 1$, $\omega \in \Omega$, we may use a linear operator in $L^1(\Omega, \mathcal{F}, \mu)$, called the **Frobenius–Perron operator**, related to $f$ [80]. To define it, assume first that (a representation of) an element of $x \in L^1(\Omega, \mathcal{F}, \mu)$ is non-negative, and define a set function on $\mathcal{F}$ by $\mu_x(A) = \int_{f^{-1}(A)} x \, d\mu$. It is easy to see that $\mu_x$ is a measure. Furthermore, since $f$ is non-singular, $\mu_x$ is absolutely continuous with respect to $\mu$. Hence, there exists a non-negative element $Px$ of $L^1(\Omega, \mathcal{F}, \mu)$ such that $\int_{f^{-1}(A)} x \, d\mu = \int_A Px \, d\mu$. Note that $Px$ is defined only as an equivalence class of functions and that $P$ maps the set of (equivalence classes of) non-negative functions into itself. For arbitrary $x \in L^1(\Omega, \mathcal{F}, \mu)$ we define $Px$ as the difference of $Px^+$ and $Px^-$. It is easy to check that $Px$ is linear. Moreover, for non-negative $x$ we have $\int_\Omega x \, d\mu = \int_{f^{-1}\Omega} x \, d\mu = \int_\Omega Px \, d\mu$, so that $P$ preserves the integral, and therefore is a Markov operator.

## 3.5 Examples of discrete martingales

3.5.1 **Definition** Let $(\Omega, \mathcal{F}, \mathbb{P})$ be a probability space and let $\mathcal{F}_n, n \geq 1$, be an increasing sequence of $\sigma$-algebras of measurable sets: $\mathcal{F}_n \subset \mathcal{F}_{n+1} \subset \mathcal{F}$; such a sequence is called a **filtration**. A sequence $X_n, n \geq 1$ of random variables $X_n \in L^1(\Omega, \mathcal{F}, \mathbb{P})$ is termed a **martingale** if $X_n$ is $\mathcal{F}_n$ measurable and $\mathbb{E}(X_n|\mathcal{F}_{n-1}) = X_{n-1}$ for all $n \geq 1$. To be more specific, we should say that $X_n$ is a martingale with respect to $\mathcal{F}_n$ and

$\mathbb{P}$. However, $\mathcal{F}_n$ and $\mathbb{P}$ are often clear from the context and for simplicity we omit the phrase "with respect to $\mathcal{F}_n$ and $\mathbb{P}$". Similarly, a sequence $X_n \in L^1(\Omega, \mathcal{F}, \mathbb{P}), n \geq 1$ is termed a **submartingale** (with respect to $\mathcal{F}_n$ and $\mathbb{P}$) if $X_n$ are $\mathcal{F}_n$ measurable and $\mathbb{E}(X_{n+1}|\mathcal{F}_n) \geq X_n, n \geq 1$. If $-X_n$ is a submartingale, $X_n$ is called a **supermartingale**. Filtrations, martingales, supermartingales and submartingales indexed by a finite ordered set are defined similarly.

**3.5.2 Exercise** Show that $X_n, n \geq 1$, is a submartingale iff there exists a martingale $M_n, n \geq 1$, and a previsible sequence $A_n, n \geq 1$ (i.e. $A_1 = 0$ and $A_{n+1}, n \geq 1$, is $\mathcal{F}_n$ measurable) such that $A_{n+1} \geq A_n$ (a.s.) and $X_n = M_n + A_n$. This decomposition, called the **Doob decomposition**, is unique in $L^1(\Omega, \mathcal{F}, \mathbb{P})$.

**3.5.3** *Sum of independent random variables* If $X_n, n \geq 1$ are (mutually) independent random variables, and $E X_n = 0$ for $n \geq 1$, then $S_n = \sum_{i=1}^n X_i$ is a martingale with respect to $\mathcal{F}_n = \sigma(X_1, ..., X_n)$. Indeed, by 3.3.1 (h)–(i), $\mathbb{E}(S_{n+1}|\mathcal{F}_n) = \mathbb{E}(X_{n+1}+S_n|\mathcal{F}_n) = \mathbb{E}(X_{n+1}|\mathcal{F}_n) + S_n = E X_{n+1} 1_\Omega + S_n = S_n$, since $X_{n+1}$ is independent of $\sigma(X_1, ..., X_n)$.

**3.5.4** *Polya's urn scheme* Suppose that in a box there are $w$ white balls and $b$ black balls. One ball is drawn at random and returned to the box together with $k$ balls of the same color. Let $X_n$ be the proportion of white balls in the box after the $n$th draw. We will show that $X_n$ is a martingale with respect to the filtration $\mathcal{F}_n = \sigma(X_1, ..., X_n)$. Note that we do not know $(\Omega, \mathcal{F}, \mathbb{P})$; all we know (by description) is a joint distribution of $(X_n, X_{n+1})$.

$X_n$ is a simple function (in particular: bounded and integrable) that admits $n + 1$ values $\frac{w+ik}{w+b+nk}$, $i = 0, 1, ..., n$, on sets $B_i$, and any set in $\mathcal{F}_n$ is a finite disjoint union of $B_i$s. If $X_n = \frac{w+ik}{w+b+nk}$, then $X_{n+1} = \frac{w+(i+1)k}{w+b+(n+1)k}$ with probability $\frac{w+ik}{w+b+nk}$ and $X_{n+1} = \frac{w+ik}{w+b+(n+1)k}$ with probability $\frac{b+(n-i)k}{w+b+nk}$. Therefore,

$$\int_{B_i} X_{n+1} \, d\mathbb{P} =$$

$$\left[ \frac{w+(i+1)k}{w+b+(n+1)k} \frac{w+ik}{w+b+nk} + \frac{w+ik}{w+b+(n+1)k} \frac{b+(n-i)k}{w+b+nk} \right] \mathbb{P}(B_i),$$

which by simple algebra equals

$$\frac{(w+ik)(w+b+(n+1)k)}{(w+b+(n+1)k)(w+b+nk)} \mathbb{P}(B_i) = \frac{w+ik}{w+b+nk} \mathbb{P}(B_i) = \int_{B_i} X_n \, d\mathbb{P}.$$

3.5.5 *Galton–Watson process* The famous **Galton–Watson process** describes the number of individuals in a population of descendants of a single ancestor in which individuals give birth, independently from other individuals of this population, to a random number of children; the distribution of the number of children is the same for all individuals.

Formally, let $X_k^n, n, k \geq 1$, be an infinite matrix of independent, identically distributed random variables with values in $\{0, 1, ...\}$. Assume that $E X_k^n = m < \infty$. The Galton–Watson process is defined inductively by $Z_0 = 1$ and $Z_{n+1} = \sum_{i=1}^{Z_n} X_i^{n+1}, n \geq 0$, if $Z_n \geq 1$, and $Z_{n+1} = 0$ if $Z_n = 0$.

Let $\mathcal{F}_n = \sigma(Z_1, ..., Z_n)$, $n \geq 1$. We will prove that $M_n = \frac{Z_n}{m^n}$ is a martingale. Indeed,

$$E Z_{n+1} = \sum_{k=0}^{\infty} E Z_{n+1} 1_{Z_n = k} = \sum_{k=1}^{\infty} E \sum_{i=1}^{k} X_i^{n+1} 1_{Z_n = k} \qquad (3.24)$$

$$= \sum_{k=1}^{\infty} E 1_{Z_n = k} E \sum_{i=1}^{k} X_i^{n+1} = m \sum_{k=1}^{\infty} k \, \mathbb{P}\{Z_n = k\} = m E \, Z_n.$$

Thus, $E \, Z_n$ is integrable and $E Z_n = m^n$ for all $n \geq 0$. Therefore $M_n$ is integrable and $E \, M_n = 1$.

Now, each member of the $\sigma$-algebra $\sigma(Z_1, ..., Z_n)$ is a disjoint union of sets of the form $\{Z_1 = k_1, ..., Z_n = k_n\}$. The random variable $M_n$ is $\sigma(Z_1, ..., Z_n)$ measurable and equals $\frac{k_n}{m^n}$ on such sets. Also, since $X_i^{n+1}$ are independent of $Z_1, ..., Z_n$,

$$\int_{\{Z_1 = k_1, ..., Z_n = k_n\}} M_{n+1} \, d\mathbb{P} = \frac{1}{m^{n+1}} E \sum_{i=1}^{k_n} X_i^{n+1} 1_{\{Z_1 = k_1, ..., Z_n = k_n\}}$$

$$= \frac{m k_n}{m^{n+1}} \mathbb{P}\{Z_1 = k_1, ..., Z_n = k_n\}$$

$$= \int_{\{Z_1 = k_1, ..., Z_n = k_n\}} M_n \, d\mathbb{P}$$

provided $k_n \geq 1$. In the other case both the first and the last integrals in this formula are zero. This proves our claim.

3.5.6 *Wright–Fisher model of population genetics* Imagine a population of $2N$, $N \in \mathbb{N}$, individuals; each individual belonging to either of two distinct classes, say $a$ and $A$. (Individuals are actually chromosomes – that's why we have $2N$ of them – and $a$ and $A$ denote two possible alleles at a particular locus.) The population evolves in discrete time: a next generation is formed by selecting its $2N$ members from the previous one independently and with replacement. The state of such a population in generation $n$ is conveniently described by a single random variable $X(n)$ equal to the number of individuals of type $a$, say. Note that if $X(n) = k$ then $X(n+1)$ is a binomial random variable with parameters $\frac{k}{2N}$ and $2N$:

$$\mathbb{P}\{X(n+1) = l | X(t) = k\} = \binom{2N}{l} p^l (1-p)^{2N-l}.$$

In particular $\mathbb{E}\left(X(n+1)|X(n)\right) = 2N\frac{X(n)}{2N}$. This proves that $X(n)$ is a non-negative martingale. See 3.7.6 for an application.

**3.5.7 Exercise**    Calculating as in (3.24) show that if $X_k^n, n, k \geq 1$ are square integrable then

$$D^2(Z_{n+1}) = m^2 D^2(Z_n) + \sigma^2 m^n \tag{3.25}$$

where $\sigma^2 = D^2(X_k^n)$ and $m = E X_k^n$. In particular $Z_n \in L^2(\Omega, \mathcal{F}, \mathbb{P})$, $n \geq 1$. Conclude that for $m \neq 1$,

$$D^2(Z_{n+1}) = \frac{\sigma^2}{m(m-1)} m^{2n} - \frac{\sigma^2}{m(m-1)} m^n = \frac{\sigma^2 m^{n-1}(m^n - 1)}{m - 1}, \tag{3.26}$$

and when $m = 1$,

$$D^2(Z_{n+1}) = n\sigma^2. \tag{3.27}$$

**3.5.8 Exercise**    Let $X \in L^1(\Omega, \mathcal{F}, \mathbb{P})$ and $\mathcal{F}_n$ be a filtration. Show that $X_n = E\left(X|\mathcal{F}_n\right)$ is a martingale with respect to $\mathcal{F}_n$.

**3.5.9 Exercise**    Suppose that $X_n$ is a martingale with respect to filtration $\mathcal{F}_n$. Let $\mathcal{F}_0$ be the trivial $\sigma$-algebra and assume that $Y_n, n \geq 0$ are $\mathcal{F}_n$ measurable, absolutely integrable random variables such that $Y_n(X_{n+1} - X_n)$ are absolutely integrable. Show that $Z_n, n \geq 1$, defined by $Z_n = \sum_{i=1}^{n} Y_{i-1}(X_i - X_{i-1}), n \geq 0$, where $X_0 = 0$, is a martingale.

## 3.6 Convergence of self-adjoint operators

3.6.1 *Motivation*    In the previous section we have already encountered examples of theorems concerning convergence of conditional expectations. In Theorem 3.3.1 point (m) and in Exercise 3.3.8 we saw that if the $\sigma$-algebra $\mathcal{G}$ is fixed, then the conditional expectation with respect to this $\sigma$-algebra behaves very much like an integral. In this section we devote ourselves to a short study of theorems that involve limit behavior of conditional expectation $E\left(X|\mathcal{F}_n\right)$ where $X$ is fixed and $\mathcal{F}_n$ is a family of $\sigma$-algebras. This will lead us in a natural way to convergence theorems for martingales presented in Section 3.7.

If $\mathcal{F}_n$ is a filtration in a probability space $(\Omega, \mathcal{F}, \mathbb{P})$, then $L^1(\Omega, \mathcal{F}_n, \mathbb{P})$ is a non-decreasing sequence of subspaces of $L^1(\Omega, \mathcal{F}, \mathbb{P})$, and $L^2(\Omega, \mathcal{F}_n, \mathbb{P})$ is a non-decreasing sequence of subspaces of $L^2(\Omega, \mathcal{F}, \mathbb{P})$. If $X$ is a square integrable random variable, then the sequence $X_n = E\left(X|\mathcal{F}_n\right)$ of conditional expectations of $X$ is simply the sequence of projections of $X$ onto this sequence of subspaces. Thus, it is worth taking a closer look at asymptotic behavior of a sequence $x_n = P_n x$, where $x$ is a member of an abstract Hilbert space $\mathbb{H}$ and $P_n$ are projections on a non-decreasing sequence of subspaces $\mathbb{H}_n$ of this space. In view of Theorem 3.1.18, the assumption that $\mathbb{H}_n$ is a non-decreasing sequence may be conveniently expressed as $(P_n x, x) \leq (P_{n+1} x, x) \leq (x, x)$.

As an aid in our study we will use the fact that projections are self-adjoint operators (see 3.1.19). Self-adjoint operators are especially important in quantum mechanics, and were extensively studied for decades. Below, we will prove a well-known theorem on convergence of self-adjoint operators and then use it to our case of projections. Before we do that, however, we need to introduce the notion of a **non-negative** operator and establish a lemma.

**3.6.2 Definition** A self-adjoint operator $A$ is said to be non-negative if $(Ax, x) \geq 0$ for all $x \in \mathbb{H}$; we write then $A \geq 0$. If $A$ and $B$ are two self-adjoint operators such that $A - B \geq 0$ we often write $A \geq B$ or $B \leq A$.

**3.6.3 Exercise** Prove that if $A$ is non-negative then so are all its powers. Moreover, all even powers of any self-adjoint operator are non-negative.

**3.6.4 Lemma** Let $A$ be a non-negative, self-adjoint operator in a Hilbert space $\mathbb{H}$. Then $(A^2 x, x) \leq \|A\|(Ax, x)$.

*Proof* If $\|A\| = 0$, there is nothing to prove. In the other case our relation is equivalent to $(B^2 x, x) \leq (Bx, x)$ where $B = \frac{1}{\|A\|}A$ is a self-adjoint contraction. Certainly, $(Bx, x) \geq 0$. Note that $I - B$, where $Ix = x$, is also self-adjoint as a difference of two self-adjoint operators, and $((I - B)y, y) = (y, y) - (By, y) \geq 0$, for any $y \in \mathbb{H}$. Since $B - B^2 =$

$B(I - B)B + (I - B)B(I - B)$, we have

$$((B - B^2)x, x) = ((I - B)Bx, Bx) + (B(I - B)x, (I - B)x) \geq 0$$

which proves our claim. □

3.6.5 *Convergence theorem for a monotone sequence of self-adjoint operators* If $A_n$ is a sequence of self-adjoint operators in a Hilbert space $\mathbb{H}$, such that $A_n \leq A_{n+1} \leq MI$, for all natural $n$, where $M$ is a constant and $Ix = x$, then there exists the strong limit $Ax = \lim_{n \to \infty} A_n x$ of $A_n$ and $A$ is self-adjoint.

*Proof* For any $x$, the numerical series $(A_n x, x)$ is non-decreasing and bounded by $M\|x\|^2$, and therefore converges to a real number, say $F(x)$. Hence, for all $x$ and $y$ in $\mathbb{H}$ there exists the limit

$$G(x, y) = \lim_{n \to \infty} (A_n x, y) \tag{3.28}$$

since $(A_n x, y)$ equals

$$\frac{1}{4} [(A_n x + A_n y, x + y) - (A_n x - A_n y, x - y)] \tag{3.29}$$

which tends to $\frac{1}{4} [F(x + y) - F(x - y)]$. Since for any $n \geq 1$

$$-\|A_1\| \, \|x\|^2 \leq (A_n x, x) \leq M\|x\|^2, \tag{3.30}$$

3.1.22 shows that $\|A_n\| \leq M'$, where $M' = M \vee \|A_1\|$. Thus, $|G(x, y)| \leq M'\|x\| \, \|y\|$. Fix $x$. As a function of $y$, $G(x, y)$ is a linear functional on $\mathbb{H}$. Moreover, by (3.28), this functional is bounded with norm less than $M'\|x\|$. Therefore, there exists an element $Ax$ of $\mathbb{H}$ such that $G(x, y) = (Ax, y)$. By (3.28) the map $x \mapsto Ax$ is linear. Since $\|Ax\| \leq M'\|x\|$, $A$ is bounded and $\|A\| \leq M'$. $A$ is also self-adjoint since $(Ax, y) = \lim_{n \to \infty}(A_n x, y) = \lim_{n \to \infty}(A_n y, x) = (Ay, x)$. Finally, $A - A_n$ is self-adjoint and by Lemma 3.6.4, for any $x \in \mathbb{H}$,

$$\|Ax - A_n x\|^2 = ((A - A_n)^2 x, x) \leq \|A - A_n\|((A - A_n)x, x)$$
$$\leq 2M'((A - A_n)x, x)$$

and the last sequence converges to zero, as $n \to \infty$. □

3.6.6 **Exercise** Show that if $A_n$ is a sequence of self-adjoint operators in a Hilbert space, such that $(A_n x, x) \geq (A_{n+1}x, x)$ and $(A_n x, x) \geq M\|x\|^2$ for all $x \in \mathbb{H}$ and natural $n$, then there exists the strong limit $Ax = \lim_{n \to \infty} A_n x$ of $A_n$ and $A$ is self-adjoint.

**3.6.7** *An application: the square root* If $A$ is a self-adjoint operator such that $(Ax, x) \geq 0$ for all $x \geq 0$ then there exists a self-adjoint operator $B$ that commutes with all operators that commute with $A$ and such that $B^2 = A$.†

*Proof* Without loss of generality we may assume that $0 \leq A \leq I$, for if this is not the case then either $A = 0$ (in which case the theorem is obvious) or the operator $A' = \frac{1}{\|A\|} A$ satisfies this condition and then the square root of $A$ may be found as $\sqrt{\|A\|}$ times the square root of $A'$. Let $C = I - A$; observe that $0 \leq C \leq I$. Consider the following sequence of operators defined inductively: $A_0 = 0, A_{n+1} = \frac{1}{2}(C + A_n^2)$. An induction argument shows that $A_n$ are self-adjoint and non-negative, and commute with $A$. Since

$$A_{n+1} - A_n = \frac{1}{2}(A_n^2 - A_{n-1}^2) = \frac{1}{2}(A_n - A_{n-1})(A_n + A_{n-1}),$$

$A_{n+1} - A_n$ is a linear combination of powers of the operator $C$ with positive coefficients and hence $A_{n+1} \geq A_n$. An induction argument shows also that $A_n \leq I$. Indeed, $A_0 \leq I$, and if $A_n \leq I$ then $\|A_n\| \leq 1$ and by 3.6.4 we have $A_n^2 \leq \|A_n\| A_n \leq A_n$ which in turn implies $A_{n+1} \leq \frac{1}{2}(C + A_n) \leq I$. Hence, there exists the strong limit $A_\infty$ of $A_n, n \geq 1$, and $A_\infty x = \lim_{n \to \infty} A_{n+1} x = \frac{1}{2} Cx + \frac{1}{2} \lim_{n \to \infty} A_n^2 x = \frac{1}{2} Cx + \frac{1}{2} A_\infty^2 x$. Let $B = I - A_\infty$. Then $B^2 x = x - 2A_\infty x + A_\infty^2 x = x - Cx = Ax$, as desired. Finally, $B$ commutes with all operators commuting with $A$ since $A_n, n \geq 1$ and $A_\infty$ do. $\square$

**3.6.8 Corollary** (a) Suppose that $\mathbb{H}_n$ is an increasing sequence of subspaces of a Hilbert space $\mathbb{H}$. Projections $P_n$ on $\mathbb{H}_n$ converge strongly to the projection $P_\infty$ on $\mathbb{H}_\infty = cl(\bigcup_{n \geq 1} \mathbb{H}_n)$. (b) If $\mathbb{H} = L^2(\Omega, \mathcal{F}, \mathbb{P})$ and $\mathbb{H}_n = L^2(\Omega, \mathcal{F}_n, \mathbb{P})$ where $\mathcal{F}_n$ is a filtration, then $\mathbb{H}_\infty = L^2(\Omega, \mathcal{F}_\infty, \mathbb{P})$ where $\mathcal{F}_\infty = \sigma(\bigcup_{n \geq 1} \mathcal{F}_n)$.

*Proof* (a) The strong convergence of $P_n$ to a self-adjoint operator $A_\infty$ has been proved in 3.6.5, so it suffices to identify $A_\infty$ as the projection $P_\infty$.

Note first that we have $A_\infty^2 x = \lim_{n \to \infty} P_n A_\infty x = \lim_{n \to \infty} P_n P_n x = \lim_{n \to \infty} P_n x = A_\infty x$ since $P_n^2 = P_n$ and $\|P_n P_n x - P_n A_\infty x\| \leq \|P_n x - A_\infty x\|$ tends to zero. By 3.1.23, the range $\tilde{\mathbb{H}}_\infty$ of $A_\infty$ is closed and $A_\infty$ is the projection on $\tilde{\mathbb{H}}_\infty$.

† See e.g. Theorem 4.1.3 for an application. This operator is uniquely determined [90], but we will not need this fact here and will not prove it either.

We need to prove that $\tilde{\mathbb{H}}_\infty = \mathbb{H}_\infty$. It is clear that all $\mathbb{H}_n$ are subsets of $\tilde{\mathbb{H}}_\infty$, for if $x \in \mathbb{H}_n$ then $P_m x = x$ for all $m \geq n$ and so $A_\infty x = \lim_{m\to\infty} P_m x = x$ forces $x \in \tilde{\mathbb{H}}_\infty$. Hence, $\mathbb{H}_\infty \subset \tilde{\mathbb{H}}_\infty$. On the other hand, if $x \in \tilde{\mathbb{H}}_\infty$, then $x = \lim_{n\to\infty} P_n x$ where $P_n$ belong to $\mathbb{H}_n$. Thus, $\mathbb{H}_\infty \supset \tilde{\mathbb{H}}_\infty$ and so $\mathbb{H}_\infty = \tilde{\mathbb{H}}_\infty$.

(b) The inclusion $\mathbb{H}_\infty \subset L^2(\Omega, \mathcal{F}_\infty, \mathbb{P})$ is easy to prove: the limit of a sequence $X_n$ of random variables that are $\mathcal{F}_n$ measurable, respectively, must be $\mathcal{F}_\infty$ measurable. To prove the other inclusion we note that $1_\Omega \in \mathbb{H}_\infty$ (for $1_\Omega \in \mathbb{H}_n, n \geq 1$) and recall that by 2.2.23 the collection $\mathcal{G}$ of events $A$ such that $1_A \in \mathbb{H}_\infty$ is a $\lambda$-system. On the other hand, $\mathbb{H}_n \subset \mathbb{H}_\infty$, i.e. $\mathcal{F}_n \subset \mathcal{G}$ and so $\bigcup_{n=1}^\infty \mathcal{F}_n \subset \mathcal{G}$. By the $\pi$–$\lambda$ theorem, $\mathcal{F}_\infty \subset \mathcal{G}$, and 2.2.39 forces $L^2(\Omega, \mathcal{F}_\infty, \mathbb{P}) \subset \mathbb{H}_\infty$. $\qquad\square$

**3.6.9 Corollary**   Let $\mathcal{F}_n, n \geq 1$, be a filtration in a probability space $(\Omega, \mathcal{F}, \mathbb{P})$. For any $X \in L^1(\Omega, \mathcal{F}, \mathbb{P})$ the sequence $X_n = \mathbb{E}(X|\mathcal{F}_n)$ converges in $L^1(\Omega, \mathcal{F}, \mathbb{P})$ to $X_\infty = \mathbb{E}(X|\mathcal{F}_\infty)$ where $\mathcal{F}_\infty = \sigma(\bigcup_{n\geq 1} \mathcal{F}_n)$.

*Proof* If $X$ is square integrable then $\|X_n - X_\infty\|_{L^1}$, which is less than $\|X_n - X_\infty\|_{L^2}$ by the Hölder inequality, converges to zero by 3.6.8. Since conditional expectation is a Markov operator in $L^1(\Omega, \mathcal{F}, \mathbb{P})$, and in particular has norm one, convergence of a sequence of conditional expectations on a dense set such as $L^2(\Omega, \mathcal{F}, \mathbb{P})$ implies convergence on the whole of $L^1(\Omega, \mathcal{F}, \mathbb{P})$. $\qquad\square$

**3.6.10 Exercise**   Let $(\Omega, \mathcal{F}, \mathbb{P})$ be a standard probability space. For $n \geq 1$, let $\mathcal{F}_n$ be the $\sigma$-algebra generated by the partition $A_0 = [\frac{1}{2}, 1)$, $A_i = [\frac{i-1}{2^n}, \frac{i}{2^n})$, $i = 1, ..., 2^{n-1}$. In other words, the space $L^1(\Omega, \mathcal{F}_n, \mathbb{P})$ consists of (equivalence classes of) bounded functions that are constant on each of $A_i$, $i = 0, ..., 2^{n-1}$. Show that $\mathbb{E}(X|\mathcal{F}_n)$ converges in $L^1(\omega, \mathcal{F}, \mathbb{P})$ to $X_\infty = X1_{[0,\frac{1}{2})} + \int_{[\frac{1}{2},1)} X \, d\mathbb{P}1_{[\frac{1}{2},1)}$.

**3.6.11 *The 0–1 law***   Corollary 3.6.9 has the following nice application known as the 0–1 law: if $\mathcal{G}_n$ is a sequence of independent $\sigma$-algebras, then the tail algebra $\mathcal{T} = \bigcap_{n=1}^\infty \sigma(\bigcup_{k=n}^\infty \mathcal{G}_k)$ contains only events $A$ such that $\mathbb{P}(A)$ is either zero or one. To prove this law it suffices to show that all $X \in L^1(\Omega, \mathcal{T}, \mathbb{P})$ are constants a.e., and this follows from the following calculation:

$$X = \mathbb{E}(X|\sigma(\bigcup_{n=1}^\infty \mathcal{G}_n)) = \lim_{n\to\infty} \mathbb{E}(X|\sigma(\bigcup_{i=1}^n \mathcal{G}_i)) = (E\,X)1_\Omega. \qquad (3.31)$$

The first equality above follows from the fact that $\mathcal{T} \subset \sigma(\bigcup_{n=1}^{\infty} \mathcal{G}_n)$, the second one from Corollary 3.6.9 and the fact that $\sigma(\bigcup_{n=1}^{\infty} \sigma(\bigcup_{i=1}^{n} \mathcal{G}_i)) = \sigma(\bigcup_{n=1}^{\infty} \mathcal{G}_n)$ and the third one from the fact that $X$ is independent of $\sigma(\bigcup_{i=1}^{n} \mathcal{G}_i))$ being $\sigma(\bigcup_{i=k+1}^{\infty} \mathcal{G}_i)$ measurable.

**3.6.12 Example** Let $X_n, n \geq 1$ be independent random variables. We claim that $S = \limsup_{n\to\infty} \frac{S_n}{n}$ where $S_n = \sum_{i=1}^{n} X_i$ is tail measurable. Indeed, first of all for any $a \in \mathbb{R}$, $\{S < a\} = \bigcup_{k\geq 1}\bigcap_{n\geq k}\{\frac{S_n}{n} < a\}$ is $\sigma(X_n, n \geq 1)$ measurable. Secondly, for any $k \geq 1$, $\limsup_{n\to\infty} \frac{S_n}{n}$ is equal to the upper limit of $\frac{n-k}{n} \frac{S_{n-k}^{(k)}}{n-k}$ where $S_{n-k}^{(k)} = X_{k+1} + \cdots + X_n$. Since $S_{n-k}^{(k)}$ is the $(n-k)$th sum of the sequence $X_{k+n}, n \geq 1$, the upper limit of $\frac{n-k}{n} \frac{S_{n-k}n^{(k)}}{n-k}$ is $\sigma(X_{n+k}, n \geq 1)$ measurable and so is $S$. Since $k$ is arbitrary, the claim is proved.

The random variable $\liminf_{n\to\infty} \frac{S_n}{n}$ is also tail measurable, being equal to $(-1)$ times the upper limit of the sequence $-X_n, n \geq 1$. Therefore, the set $\{\omega| \lim_{n\to\infty} \frac{S_n(\omega)}{n} \text{ exists}\}$ has probability either zero or one, for it belongs to the tail $\sigma$-algebra as the set where $\liminf$ and $\limsup$ are equal.

**3.6.13 Corollary** Suppose that $\mathcal{F}_n \supset \mathcal{F}_{n+1}, n \geq 1$ is a non-decreasing sequence of $\sigma$-algebras of events in a probability space $(\Omega, \mathcal{F}, \mathbb{P})$.

Exercise 3.6.6 and the argument used in 3.6.8 prove that for any $X \in L^2(\Omega, \mathcal{F}, \mathbb{P})$, $\mathbb{E}(X|\mathcal{F}_n)$ converges in $L^2(\Omega, \mathcal{F}, \mathbb{P})$ to $PX$ where $PX$ is the projection on $H_\infty = \bigcap_{n=1}^{\infty} \mathbb{H}_n = L^2(\Omega, \bigcap_{n=1}^{\infty} \mathcal{F}_n, \mathbb{P})$. Thus $PX = \mathbb{E}(X| \bigcap_{n=1}^{\infty} \mathcal{F}_n)$. Arguing as in 3.6.9, we obtain that for any integrable $X$, $\mathbb{E}(X|\mathcal{F}_n)$ converges to $\mathbb{E}(X| \bigcap_{n=1}^{\infty} \mathcal{F}_n)$ in $L^1(\Omega, \mathcal{F}, \mathbb{P})$.

**3.6.14 *Almost sure convergence*** Although it is proper to view 3.6.8 and 3.6.13 as particular cases of 3.6.5, such a broader view comes at the cost of losing important details of the picture. To see what we have overlooked, let us consider again a sequence of projections $P_n$ on a non-decreasing sequence of subspaces $\mathbb{H}_n$ of a Hilbert space $\mathbb{H}$. Given $x \in \mathbb{H}$, let $y_1 = P_1 x$ and $y_n = P_n x - P_{n-1}x, n \geq 2$. Then $P_n x = \sum_{i=1}^{n} y_i$ and $y_n \perp P_{n-1}x$ so that $\|P_n x\|^2 = \sum_{i=1}^{n} \|y_i\|^2$ and hence $\sum_{i=1}^{\infty} \|y_i\|^2 \leq \|x\|^2 < \infty$. Thus, if $\mathbb{H} = L^2(\Omega, \mathcal{F}, \mathbb{P})$ for some probability space $(\Omega, \mathcal{F}, \mathbb{P})$ then $\lim_{n\to\infty} P_n x = \sum_{i=1}^{\infty} y_i$ converges to $Px$ not only in the norm but also a.s. (see the proof of 2.2.19). In other words, convergence in 3.6.8 and the first part of 3.6.13 is not only in $L^2(\Omega, \mathcal{F}, \mathbb{P})$ but also a.s. It is worth stressing here that in general convergence in $L^p$ does not imply a.s. convergence, as it may be seen from Example 3.6.15 below, and that neither does a.s. convergence imply convergence in $L^p$. Hence the information that convergence in 3.6.8 is both in $L^2$ and a.s. is non-trivial.

Similarly, under assumptions of 3.6.13, $P_n' = I - P_n$ is a sequence of non-decreasing projections and thus converges a.s. and in $L^2$ to a projection $P'$. Therefore, $P_n = I - P_n$ converge to $P = I - P'$ not only in $L^2$ but also a.s.

**3.6.15 Example**      *Convergence in $L^1$ does not imply a.s. convergence*
Let $x_{k,n} = 1_{[\frac{k}{2^n}, \frac{k+1}{2^n})}$, $0 \le k \le 2^n - 1, n \in \mathbb{N}$, be a doubly indexed triangular array of vectors in $L^1(0,1)$. Note that $\|x_{k,n}\| = 2^{-n}$. Hence, if we reorder these vectors in a natural way (going row by row), we obtain a sequence converging to zero. On the other hand, for $\tau \in [0,1)$ the values of corresponding functions do not converge to zero, attaining the value 1 infinitely many times.

**3.6.16 Example**      By Exercise 3.5.7, if $Z_n$ is the Galton–Watson process with square integrable $X_k^n$ then $Z_n, n \ge 1$ are square integrable also. Moreover, if $n > 1$, (3.26) and (3.27) show that the martingale $M_n = \frac{Z_n}{m}^n$ is bounded in $L^2$, for $\lim_{n \to \infty} D^2(M_n) = \frac{\sigma^2}{m(m-1)}$. Therefore, $M_n$ converges pointwise and in $L^2$ to a random variable $M$.

**3.6.17 *The Law of Large Numbers. Second attempt***      As a consequence of 3.6.13 we obtain the following form of the Law of Large Numbers. Suppose that $X_n, n \ge 1$ are independent, square integrable, identically distributed random variables in a probability space $(\Omega, \mathcal{F}, \mathbb{P})$. Then $\frac{S_n}{n} = \frac{X_1 + \cdots + X_n}{n}$ converges to $E\,X_1 1_\Omega$ in $L^2$ and a.s.

*Proof* By (3.19), $\frac{S_n}{n} = \mathbb{E}(X_1 | \sigma(S_n, S_{n+1}, ...))$ and $\mathcal{F}_n = \sigma(S_n, S_{n+1}, ...)$ is a non-increasing sequence of $\sigma$-algebras. Thus $\frac{S_n}{n}$ converges a.s. and in $L^2$. Since $\limsup_{n \to \infty} \frac{S_n}{n} = \lim_{n \to \infty} \frac{S_n}{n}$ is tail $\mathcal{T}$ measurable, it must be constant, and this constant must be $E\,X$.          □

## 3.7    ... and of martingales

A natural way to generalize 3.6.17 is to abandon the assumption that $X_n$ are square integrable and suppose that they are absolutely integrable instead. Then Corollary 3.6.13 still applies to show that $\frac{S_n}{n}$ converges to $(EX) 1_\Omega$ in $L^1$. However, our argument does not prove that the convergence is also a.s. We could try to prove a.s. convergence using truncation/density arguments, the most elegant proof, however, and the one that leads to a more general result, namely the Martingale Convergence Theorem, leads through Doob's Upcrossing Inequality to be presented below. The situation is typical in that there is much more to probability theory than functional analysis (at least in the narrow sense of a mixture of linear algebra and topology) can explain in a straightforward way. In fact, probability theory is a beautiful and independent field of mathematics, and the rapid development of direct probabilistic methods that was witnessed in the past century allows the use of them in an essential way in other branches of mathematics to obtain new deep results and to give new surprisingly short proofs of old results or to make old reasonings more transparent. Therefore, even in this book, functional analytic argument should sometimes give way to a purely probabilistic one ...

**3.7.1** *Upcrossing Inequality*    Let $N$ be a natural number, $X_n, 1 \leq n \leq N$, be a martingale with respect to a filtration $\mathcal{F}_n, 1 \leq n \leq N$, and let $a < b$ be two real numbers. We will be interested in the number $U_n = U_n(a, b), 1 \leq n \leq N$, of times the value of $X_n$ crosses from below $a$ up above $b$. To be more specific $U_n = m$ if $m$ is the largest integer such that we may find $1 \leq l_1 < k_1 < l_2 < k_2 < ... < l_m < k_m = n$ such that $X_{l_i} < a$ and $X_{k_i} > b$ for $i = 1, ..., m$. The reader should convince himself that $U_n$, termed the **number of upcrossings**, is a random variable.

This variable is related to a betting strategy in the following hazard game. Suppose that at time $k$, $1 \leq k \leq N$, only numbers $X_n(\omega), 0 \leq n \leq k - 1$ are known to a player, who basing on this knowledge decides to take or not to take part in the game at this moment of time; by definition we put $X_0 \equiv E X_1$. If the player chooses to play, he places a bet and gains or loses $X_k - X_{k-1}$ dollars, depending on the sign of $X_k - X_{k-1}$. Consider the following betting strategy. We do not bet at time $k$ until $X_{k-1}$ is less than $a$, and from that time on we bet until $X_{k-1}$ is more than $b$, at which time we stop betting and wait until $X_{k-1}$ is less than $a$ to repeat the whole procedure. In other words, we bet at time 1 only if $X_0 \equiv E X_1 < a$ and then bet at time $k > 2$ if we did bet at time $k - 1$ and $X_{k-1}$ is less than $b$ or if we did not bet at $k - 1$ and $X_{k-1}$ is less than $a$. With this strategy, our total gain or loss at time $n$ is $Z_n = \sum_{i=1}^{n} Y_i(X_i - X_{i-1})$, where $Y_1 = 1$ if $E X < a$ and $Y_1 = 0$ if $E X_1 \geq a$, and $Y_n = 1_{\{Y_{n-1}=1\}} 1_{\{X_{n-1} \leq b\}} + 1_{\{Y_{n-1}=0\}} 1_{\{X_{n-1} < a\}}, 1 \leq n \leq N$. By 3.5.9, $Z_n, 1 \leq n \leq N$ is a martingale and we have $E Z_n = 0$.

Suppose now that we play only up to time $n$ and that up to that time there were $m$ upcrossings. Suppose also that the last upcrossing has ended at time $k_m$, so that we stopped betting at $k_m + 1$. Up to time $k_m$ we have gained at least $\$(b-a)m$, but what has happened after that? There are two possible scenarios. In the first scenario, all $X_k$, $k_m < k < n$ are bigger than or equal to $a$. In this case, we do not bet after time $k_m$ and hence our total gain is the same as it was at time $k_m$. As a consequence:

$$Z_n \geq (b-a)U_n. \tag{3.32}$$

In the second scenario one of the numbers $X_k$, $k_m < k < n$ is less than $a$, say that $i$ is the smallest number with this property. In this case, we start betting at time $i+1$ and bet for the rest of the game, since there can be no more upcrossings. In the last part of the game we gain or lose $X_n - X_i$ dollars. Therefore,

$$Z_n \geq (b-a)U_n + X_n - X_i \geq (b-a)U_n + X_n - a. \tag{3.33}$$

Consequently, in both scenarios we have:

$$Z_n \geq (b-a)U_n - (X_n - a)^-. \tag{3.34}$$

Taking expectations, we obtain the **upcrossing inequality**:

$$(b-a)E\,U_n \leq E\,(X_n - a)^- \leq E\,|X_n - a| \leq E\,|X_n| + a. \tag{3.35}$$

**3.7.2 Exercise**   Show that the upcrossing inequality is true also for supermartingales.

**3.7.3 *Martingale Convergence Theorem***   Suppose that $X_n, n \geq 1$ is a martingale such that $E\,|X_n| \leq M$ for some $M > 0$ and all $n \geq 1$. Then there exists an absolutely integrable $X_\infty$ such that $X_n$ converges to $X_\infty$ a.s.

*Proof*  By the upcrossing inequality, for any $a < b$ the number of upcrossings $U(a,b) = \lim_{n\to\infty} U_n(a,b)$ from below $a$ up above $b$ is an integrable random variable. In particular, $U(a,b)$ is finite a.s. so that the set $A(a,b)$ of all $\omega$ such that $U(a,b) = \infty$ has probability zero. On the other hand, if the limit $\lim_{n\to\infty} X_n(\omega)$ does not exist then the upper limit of the sequence $X_n(\omega), n \geq 1$ is strictly larger than its lower limit. As a consequence, for some rational numbers $a$ and $b$ there are infinitely many upcrossings of the sequence $X_n(\omega), n \geq 1$ from below $a$ up above $b$, i.e. $\omega$ belongs to $A(a,b)$. In other words, the set of all $\omega$ such that $X_n(\omega)$ does not converge is contained in the set $\bigcup_{a,b\in\mathbb{Q}} A(a,b)$ which has measure zero as a countable union of sets of measure zero. The assertion that $X_\infty$ is absolutely integrable follows by Fatou's Lemma.  □

**3.7.4 Remark** Note that we *do not* claim that $X_n$ converges to $X_\infty$ in $L^1(\Omega, \mathcal{F}, \mathbb{P})$.

**3.7.5** *The Strong Law of Large Numbers* Suppose that $X_n, n \geq 1$ are independent, integrable, identically distributed random variables in a probability space $(\Omega, \mathcal{F}, \mathbb{P})$. Then $\frac{S_n}{n} = \frac{X_1 + \cdots + X_n}{n}$ converges to $(E\, X_1)1_\Omega$ in $L^1$ and a.s.

*Proof* We proceed as in 3.6.17, and write $\frac{S_n}{n}$ as the conditional expectation of $X_1$ with respect to the non-increasing sequence of $\sigma$-algebras $\sigma(S_n, S_{n+1}, \ldots), n \geq 1$. Corollary 3.6.13 allows us then to prove convergence of $\frac{S_n}{n}$ to $(E\, X_1)1_\Omega$ in $L^1(\Omega, \mathcal{F}, \mathbb{P})$. We check that for any natural $N$, $\mathcal{F}_m = \sigma(S_{-m}, S_{-m+1}, \ldots), -N \leq m \leq -1$, is a filtration, and $Y_m, -N \leq m \leq -1$, where $Y_m = \frac{S_{-m}}{-m}$ is a martingale with respect to this filtration. Moreover, $\sup_{m \leq -1} E\, |Y_m| = \sup_{n \geq 1} E\, \left|\frac{S_n}{n}\right| \leq E\, |X_1|$. Hence, arguing as in 3.7.3, we see that, as $m \to -\infty$, $Y_{-m}$ converges almost surely to an absolutely integrable random variable $X_\infty$. This means that as $n \to \infty$, $\frac{S_n}{n}$ it converges almost surely to $X_\infty$. By 2.2.20, $\frac{S_n}{n}$ has a subsequence that converges a.s. to $(E\, X_1)1_\Omega$ and so $(E\, X_1)1_\Omega = X_\infty$ almost surely. $\qquad \square$

**3.7.6 Example** The Wright–Fisher model 3.5.6 describes one of the forces of population genetics, called **genetic drift**, which is, roughly speaking, a random loss of variation in a population. In our simple case genetic drift is expressed in the fact that the states $X = 0$ and $X = 2N$ are absorbing, i.e. genetic drift forces one of the alleles to be fixed in the population. The Martingale Convergence Theorem may be used to find the probability $p_k$ of fixation of allele $a$ given that $X(0) = k$. Indeed, its assumptions are satisfied, so that $X(\infty) = \lim_{n \to \infty} X(n)$ exists almost surely and we have $p_k \cdot 2N + (1 - p_k) \cdot 0 = \mathbb{E}\, X(\infty) = \mathbb{E}\, X(0) = k$ and, consequently, $p_k = \frac{k}{2N}$.

**3.7.7** *Martingale convergence in $L^1$* As we have seen, a martingale that is bounded in $L^1$ converges almost surely but not necessarily in $L^1$. To ensure convergence in $L^1$ one needs to assume additionally that the random variables involved are **uniformly integrable**. There are two equivalent definitions of uniform integrability of a sequence $X_n, n \geq 1$ of absolutely integrable functions on a measure space $L^1(\Omega, \mathcal{F}, \mu)$ with finite measure $\mu$. The first definition requires that the numerical sequence $\sup_{n \geq 1} E\, |X_n| 1_{|X_n| \geq k}$ converges to zero, as $k \to \infty$. The second requires that $X_n, n \geq 1$ be bounded in $L^1$ and that for every $\epsilon > 0$ there would exist a $\delta > 0$ such that $E|X_n|1_A < \epsilon$ for all $n \geq 1$ provided $\mu(A) < \delta$. In problem 3.7.9 the reader is asked to prove equivalence of these definitions.

It is easy to see that uniform integrability is a necessary condition for convergence of a sequence in $L^1$. Indeed, for any measurable set $A$ we have

$$E|X_n|1_A \leq E|X_n - X|1_A + E|X|1_A$$
$$\leq \|X_n - X\|_{L^1} + E|X|1_A.$$

Hence, if $X_n$ converges to $X$ in $L^1$, then given $\epsilon > 0$ we may choose $n_0$ such that $\|X_n - X\|_{L^1} \leq \frac{\epsilon}{2}$ for all $n \geq n_0$, and for such $n$ make $E|X_n|1_A$ less than $\epsilon$ by choosing $A$ of small measure. For the remaining finite number of variables $X_n, 1 \leq n < n_0$, $E|X_n|1_A$ may be made small uniformly in $n$ by choosing sets $A$ of possibly smaller measure (see Exercise 3.7.10).

In general, the sole uniform integrability of a sequence does not imply its convergence in $L^1$.† Nevertheless, uniform integrability implies convergence when coupled with **convergence in measure**. A sequence $X_n$ on $(\Omega, \mathcal{F}, \mu)$ is said to converge in measure to $X$ if for every $\epsilon > 0$, the measure of the set where $|X_n - X|$ is greater than $\epsilon$ converges to 0, as $n \to \infty$. As it was with uniform integrability, convergence in measure is necessary for convergence in $L^1$; this can be seen from the Markov inequality 1.2.36:

$$\mu\{|X_n - X| > \epsilon\} \leq \frac{E|X_n - X|}{\epsilon}. \tag{3.36}$$

Furthermore, we have the following criterion for convergence in $L^1$ :

*a sequence $X_n \in L^1, n \geq 1$ converges to an $X \in L^1$ iff it is uniformly integrable and converges to $X$ in measure.*

We have already proved the "only if part" of this theorem. Before giving the proof of the "if part", let us note that as a direct consequence we obtain that

*if a martingale $X_n, n \geq 1$ (with respect to a filtration $\mathcal{F}_n, n \geq 1$) is uniformly integrable then it converges to an integrable limit $X_\infty$ almost surely and in $L^1$; in such a case, $X_n = \mathbb{E}(X_\infty|\mathcal{F}_n)$.*

---

† Rather than with convergence, uniform integrability is related to compactness and weak compactness [32, 37]. See e.g. [5] for more information on uniform integrability. We will come back to this question and in particular explain the notion of weak compactness in Chapter 5 – see 5.7.1. In Subsection 5.8 of that same chapter we will also give more information on convergence in measure, to be mentioned in the next sentence.

Indeed, since $X_n, n \geq 1$, is bounded, it converges a.s. Hence, $\mathbb{P}\{|X_n - X_\infty| > \epsilon\} = E\,1_{|X_n - X_\infty| > \epsilon}$ converges to 0, as $n \to \infty$, by the inequality $1_{|X_n - X_\infty| > \epsilon} < 1_\Omega$ and Dominated Convergence Theorem. This proves convergence in measure (i.e. convergence in probability), and since uniform integrability is given, our criterion applies, proving convergence in $L^1$. Finally, since $X_n = \mathbb{E}(X_m | \mathcal{F}_n)$ for all $m \geq n$, Theorem 3.3.1 (k) shows that $X_n = \lim_{m \to \infty} \mathbb{E}(X_m | \mathcal{F}_n) = \mathbb{E}(X_\infty | \mathcal{F}_n)$, as desired.

We are left with proving that convergence in measure and uniform integrability of a sequence imply its convergence. Thus, let us assume that $X_n \in L^1, n \geq 1$ is uniformly integrable and converges in measure to an $X \in L^1$. For $k \geq 1$, define the function $\phi_k : \mathbb{R} \to \mathbb{R}$ by $\phi_k(\tau) = \tau$ for $|\tau| < k$ and $\phi_k(\tau) = k$ or $-k$ according as $\tau \geq k$ or $\tau \leq -k$. Random variables $X_{n,k} = \phi_k(X_n)$ are thus bounded by $k$. Moreover, $\lim_{k \to \infty} \sup_{n \geq 1} \|X_{n,k} - X_n\|_{L^1} \leq \sup_{n \geq 1} E\,|X_n|1_{|X_n| \geq k} = 0$, and it is easy to see that $\phi_k(X)$ converges in $L^1$ to $X$, as $k \to \infty$. Hence, of the three terms appearing on the right-hand side of the inequality

$$\|X_n - X\|_{L^1} \leq \|X_n - X_{n,k}\|_{L^1} + \|X_{n,k} - \phi_k(X)\|_{L^1} + \|\phi_k(X) - X\|,$$

the first and the last may be made arbitrarily small by choosing a large $k$ and this can be done uniformly with respect to $n$. In other words, given $\epsilon > 0$ we may choose a $k$ such that

$$\|X_n - X\|_{L^1} \leq \frac{\epsilon}{2} + \|X_{n,k} - \phi_k(X)\|_{L^1}.$$

Let us fix this $k$. Since $|\phi_k(\tau) - \phi_k(\sigma)| \leq |\tau - \sigma|$, the event $\{|X_{n,k} - \phi_k(X)| > \epsilon\}$ is a subset of $\{|X_n - X| > \epsilon\}$, which proves that $X_{n,k}$ converges to $\phi_k(X)$ in measure. Hence, we may choose an $n_0$ so that for all $n \geq n_0$, $\mu\{|X_{n,k} - \phi_k(X)| \geq \frac{\epsilon}{4\mu(\Omega)}\} < \frac{\epsilon}{4k}$. As a result,

$$\|X_{n,k} - \phi_k(X)\|_{L^1} = E\,|X_{n,k} - \phi_k(X)|1_{|X_{n,k} - \phi_k(X)| \geq \frac{\epsilon}{4\mu(\Omega)}}$$
$$+ E\,|X_{n,k} - \phi_k(X)|1_{|X_{n,k} - \phi_k(X)| < \frac{\epsilon}{4\mu(\Omega)}}$$
$$\leq 2k\mu\{|X_{n,k} - \phi_k(X)| \geq \frac{\epsilon}{4\mu(\Omega)}\} + \frac{\epsilon}{4\mu(\Omega)}\mu(\Omega) < \epsilon$$

and we are done.

**3.7.8 Corollary**  Let $X_n, n \geq 1$ be a martingale with respect to a filtration $\mathcal{F}_n, n \geq 1$. The following are equivalent:

(a)  there exists an $X \in L^1$ such that $X_n = \mathbb{E}(X | \mathcal{F}_n)$,
(b)  $X_n, n \geq 1$ converges a.s. and in $L^1$ to $X_\infty \in L^1$,
(c)  $X_n, n \geq 1$ is uniformly integrable.

*Proof* By 3.6.9, condition (a) implies that $X_n, n \geq 1$ converges to some $X_\infty$ in $L^1$. In particular, our martingale is bounded in $L^1$ and so 3.7.3 proves almost sure convergence (we use 2.2.20 to make sure that the limits spoken of in 3.6.9 and 3.7.3 are the same a.s.). Implications (b)$\Rightarrow$(c) and (c)$\Rightarrow$(a) have been proved in 3.7.7. $\qquad\square$

**3.7.9 Exercise** Prove equivalence of the two definitions of uniform integrability.

**3.7.10 Exercise** Prove that a constant sequence $X_n = X \in L^1, n \geq 1$, is uniformly integrable. More generally, a sequence that admits only a finite number of values is uniformly integrable. Deduce that if a sequence $X_n, n \geq 1$, is dominated by an integrable non-negative random variable $Y$, i.e. $|X_n| \leq Y$ a.s., then it is uniformly integrable.

**3.7.11** *Doob's Optional Sampling Theorem* Let $\mathcal{F}_n, n \geq 1$, be a filtration in a measurable space $(\Omega, \mathcal{F})$, and let $\mathcal{F}_\infty := \bigcup_{n \geq 1} \mathcal{F}_n$. An $\mathcal{F}_\infty$ measurable random variable $\tau$ with values in $\mathbb{N} \cup \{\infty\}$ is said to be a **Markov time** (or optional time, or stopping time) if $\{\tau = n\} \in \mathcal{F}_n$. Intuitively, if the filtration comes from a random process $X_n, n \geq 1$, then by observing the process up to time $n$, we are able to tell whether $\tau$ has happened yet, i.e. iff $\tau \leq n$ or not. A straightforward argument shows that the collection $\mathcal{F}_\tau$ of events $A \in \mathcal{F}_\infty$ such that $A \cap \{\tau = k\} \in \mathcal{F}_k$ for all $k$ is a $\sigma$-algebra. Intuitively, for $A \in \mathcal{F}_\tau$, by time $\tau$ we are able to tell if $A$ happened or not. Some properties of $\mathcal{F}_\tau$ are almost obvious. For example, if $\tau \equiv n$, then $\mathcal{F}_\tau = \mathcal{F}_n$. Moreover, $\tau$ is $\mathcal{F}_\tau$ measurable and if $\tau$ and $\sigma$ are two Markov times with $\tau \leq \sigma$ then $\mathcal{F}_\tau \subset \mathcal{F}_\sigma$. Indeed, if $A \in \mathcal{F}_\tau$ then

$$A \cap \{\sigma = n\} = \bigcup_{m=1}^{n} A \cap \{\tau = m\} \cap \{\sigma = n\}$$

with $A \cap \{\tau = m\} \in \mathcal{F}_m \subset \mathcal{F}_n$ and $\{\sigma = n\} \in \mathcal{F}_n$.

Let $X_n, \mathcal{F}_n, n \geq 1$ be a uniformly integrable martingale so that there exists an integrable $X_\infty$ such that $\mathbb{E}(X_\infty | \mathcal{F}_n) = X_n$. Moreover, let $\tau$ and $\sigma$ be two Markov times with $\tau \leq \sigma$. The random variable $X_\tau = \sum_{n \in \mathbb{N} \cup \{\infty\}} 1_{\tau=n} X_n$ is well-defined with $\|X_\tau\|_{L^1} \leq \|X_\infty\|_{L^1}$; $X_\sigma$ is defined similarly. The **Optional Sampling Theorem** says that in such circumstances $\mathbb{E}(X_\sigma | \mathcal{F}_\tau) = X_\tau$. Since $X_\tau$ is integrable and $\mathcal{F}_\tau$ measurable, to prove this theorem it suffices to check that $\int_A X_\tau \, d\mathbb{P} = \int_A X_\sigma \, d\mathbb{P}$ for all $A \in \mathcal{F}_\tau$. Fix $n \leq m$. By assumption $A \cap \{\tau = n\} \in \mathcal{F}_n$ and

$A \cap \{\tau = n\} \cap \{\sigma = m\} \in \mathcal{F}_m$. Hence, $\int_{A \cap \{\tau=n\}} X_n \, d\mathbb{P} = \int_{A \cap \{\tau=n\}} X_\infty \, d\mathbb{P}$ and $\int_{A \cap \{\tau=n\} \cap \{\sigma=m\}} X_m \, d\mathbb{P} = \int_{A \cap \{\tau=n\} \cap \{\sigma=m\}} X_\infty \, d\mathbb{P}$. Consequently,

$$
\begin{aligned}
\int_A X_\sigma \, d\mathbb{P} &= \sum_{n \in \mathbb{N} \cup \{\infty\}} \sum_{m \in \{n,\dots,\infty\}} \int_{A \cap \{\tau=n\} \cap \{\sigma=m\}} X_m \, d\mathbb{P} \\
&= \sum_{n \in \mathbb{N} \cup \{\infty\}} \sum_{m \in \{n,\dots,\infty\}} \int_{A \cap \{\tau=n\} \cap \{\sigma=m\}} X_\infty \, d\mathbb{P} \\
&= \sum_{n \in \mathbb{N} \cup \{\infty\}} \int_{A \cap \{\tau=n\}} X_\infty \, d\mathbb{P} \\
&= \sum_{n \in \mathbb{N} \cup \{\infty\}} \int_{A \cap \{\tau=n\}} X_n \, d\mathbb{P} = \int_A X_\tau \, d\mathbb{P},
\end{aligned}
$$

as desired.

**3.7.12 Exercise**    Let $X_n, n \geq 1$, be a sequence of integer-valued random variables and $\mathcal{F}_n, n \geq 1$ be the natural filtration. Fix a number $k \in \mathbb{N}$. Show that $\tau_k = \min\{n \in \mathbb{N} | X_n = k\}$ is a Markov time while $\sigma_k = \max\{n \in \mathbb{N} | X_n = k\}$, in general, is not. (We put $\min \emptyset = \infty$.)

**3.7.13 Exercise**    Let $\mathcal{F}_n, n \geq 1$ be a filtration in a measurable space $(\Omega, \mathcal{F})$. (a) Show that an $\mathcal{F}_\infty$ measurable random variable $\tau$ is a Markov time iff $\{\tau \leq n\} \in \mathcal{F}_n, n \geq 1$. (b) Show that if a sequence $X_n, n \geq 1$, of random variables is **adapted** in that $X_n$ is $\mathcal{F}_n$ measurable for $n \geq 1$, and $\tau$ is a finite Markov time, then $X_\tau = \sum_{n=1}^\infty 1_{\tau=n} X_n$ is an $\mathcal{F}_\tau$ measurable random variable. (c) Show that for any finite Markov time $\mathcal{F}_\tau$ is the $\sigma$-algebra generated by all $X_\tau$ constructed from adapted processes $X_n, n \geq 1$.

**3.7.14 *The maximal inequality***    Suppose that $X_n, n \geq 1$ is a positive submartingale with respect to a filtration $\mathcal{F}_n, n \geq 1$, and let $X_n^* = \sup_{1 \leq k \leq n} X_k, n \geq 1$. Then for any number $t > 0$,

$$
t \mathbb{P}\{X_n^* > t\} \leq E X_n 1_{\{X_n^* > t\}}.
$$

*Proof* The event $\{X_n^* > t\}$ is a disjoint union of sets $A_k, 1 \leq k \leq n$ where $\omega \in A_k$ iff $X_k(\omega)$ is the first variable with the property that $X_k(\omega) > t$. Since on $A_k$, $X_k$ is greater than $t$, we have $t \mathbb{P}\{X_n^* > t\} = \sum_{k=1}^n t \, \mathbb{P}(A_k) \leq \sum_{k=1}^n E X_k 1_{A_k} \leq \sum_{k=1}^n E X_n 1_{A_k} = E X_n 1_{\{X_n^* > t\}}$ where in the second-to-the-last step we have used the submartingale property and the fact that $A_k$ are $\mathcal{F}_k$ measurable. $\qquad \square$

3.7.15 *Martingale convergence in $L^p$, $p > 1$*    Suppose that there exists a $p > 1$ such that a martingale $X_n, n \geq 1$, is integrable with the $p$th power and $\sup_{n \geq 1} E\,|X_n|^p =: M < \infty$. Then, by the Hölder inequality, this martingale is also bounded in $L^1$. Moreover, it is uniformly integrable. Indeed, if $\tau \geq k$ then $1 \leq (\frac{\tau}{k})^{p-1}$, and $\tau \leq \frac{\tau^p}{k^{p-1}}$. Hence, $E\,|X_n|1_{|X_n|\geq k} \leq \frac{1}{k^{p-1}}E\,|X_n|^p \leq \frac{M}{k^{p-1}} \to 0$ as $k \to \infty$. Therefore, by 3.7.3 and 3.7.7 the martingale converges a.s. and in $L^1$ to an integrable variable $X_\infty$. By Fatou's Lemma, $X_\infty \in L^p$. Is it true that $X_n, n \geq 1$ converges to $X_\infty$ in $L^p$ as well? The answer in the affirmative follows from the following **Doob's $L^p$ inequality**:

$$\|\sup_{n \geq 1}|X_n|\|_{L^p} \leq qM, \tag{3.37}$$

which implies in particular that $X^* = \sup_{n \geq 1}|X_n|$ is finite a.s. and belongs to $L^p$; $q$, as always, is such that $\frac{1}{p} + \frac{1}{q} = 1$. If (3.37) is established, our claim follows by $|X_n - X_\infty|^p \leq 2^p(X^*)^p$ and the Dominated Convergence Theorem.

We have done all the toil needed to prove (3.37) in 3.7.14 and Exercise 1.2.37. To complete the proof note that by Jensen's inequality $Y_n = |X_n|$ is a submartingale. Hence, if we let $Y_n^* = \sup_{1 \leq k \leq n} Y_k$, then by 3.7.14 we have $t\mathbb{P}\{Y_n^* > t\} \leq E\,Y_n 1_{\{Y_n^* > t\}}$. By (1.18) and (1.20),

$$\|Y_n^*\|_{L^p}^p = E\,(Y_n^*)^p = \int_0^\infty pt^{p-1}\mathbb{P}\{Y_n^* > t\}\,dt$$

$$\leq \frac{p}{p-1}\int_0^\infty (p-1)t^{p-2}E\,Y_n 1_{\{Y_n^* > t\}}\,dt$$

$$= qE\,Y_n(Y_n^*)^{p-1} \leq q(E\,Y_n^p)^{\frac{1}{p}}\left(E\,(Y_n^*)^{(p-1)q}\right)^{\frac{1}{q}}$$

$$\leq qM(E\,(Y_n^*)^p)^{1-\frac{1}{p}}$$

where we have used the Hölder inequality and $(p-1)q = p$. Since $Y_n^*$ belongs to $L^p$ we may divide by $(E\,(Y_n^*)^p)^{1-\frac{1}{p}}$ to obtain

$$(E\,(Y_n^*)^p)^{\frac{1}{p}} \leq qM,$$

which implies (3.37) by $\lim_{n \to \infty} Y_n^* = X^*$ and the Monotone Convergence Theorem.

Since all convergent sequences are bounded, the main result of this subsection can be phrased as follows: *a martingale in $L^p$ converges in $L^p$ iff it is bounded in $L^p$.*

# 4

# Brownian motion and Hilbert spaces

The Wiener mathematical model of the phenomenon observed by an English botanist Robert Brown in 1828 has been and still is one of the most interesting stochastic processes. Kingman [66] writes that the deepest results in the theory of random processes are concerned with the interplay of the two most fundamental processes: Brownian motion and the Poisson process. Revuz and Yor [100] point out that the Wiener process "is a good topic to center a discussion around because Brownian motion is in the intersection of many fundamental classes of processes. It is a continuous martingale, a Gaussian process, a Markov process or more specifically a process with independent increments". Moreover, it belongs to the important class of diffusion processes [58]. It is actually quite hard to find a book on probability and stochastic processes that does not describe this process at least in a heuristic way. Not a serious book, anyway.

Historically, Brown noted that pollen grains suspended in water perform a continuous swarming motion. Years (almost a century) later Bachelier and Einstein derived the probability distribution of a position of a particle performing such a motion (the Gaussian distribution) and pointed out its Markovian nature – lack of memory, roughly speaking. But it took another giant, notably Wiener, to provide a rigorous mathematical construction of a process that would satisfy the postulates of Einstein and Bachelier.

It is hard to overestimate the importance of this process. Even outside of mathematics, as Karatzas and Shreve [64] point out "the range of application of Brownian motion (...) goes far beyond a study of microscopic particles in suspension and includes modelling of stock prices, of thermal noise in electric circuits (...) and of random perturbations in a variety of other physical, biological, economic and management systems". In

121

mathematics, Wiener's argument involved a construction of a measure in the infinite-dimensional space of continuous functions on $\mathbb{R}^+$, and this construction was given even before establishing firm foundations for the mathematical measure theory.

To mention just the most elementary and yet so amazing properties of this measure let us note that it is concentrated on functions that are not differentiable at any point. Hence, from its perspective, functions that are differentiable at a point form a negligible set in the space of continuous functions. This should be contrasted with quite involved proofs of existence of a single function that is nowhere differentiable and a once quite common belief (even among the greatest mathematicians) that nowhere differentiable functions are not an interesting object for a mathematician to study and, even worse, that all continuous functions should be differentiable somewhere. On the other hand, if the reader expects this process to have only strange and even peculiar properties, he will be surprised to learn that, to the contrary, on the macroscopic level it has strong smoothing properties. For example, if we take any, say bounded, function $x : \mathbb{R} \to \mathbb{R}$ and for a given $t > 0$ consider the function $x_t(\tau) = Ex(\tau + w(t)), \tau \in \mathbb{R}$ where $w(t)$ is the value of a Wiener process at time $t$, this new function turns out to be infinitely differentiable! Moreover, $x(t, \tau) = x_t(\tau)$ is the solution of a famous heat equation $\frac{\partial u}{\partial t} = \text{const.} \frac{\partial^2 u}{\partial \tau^2}$ with the initial condition $u(0, \tau) = x(\tau)$. And this fact is just a peak of a huge iceberg of connections between stochastic processes and partial differential equations of second order.

Finally, let us mention that the notion of Itô stochastic integral (to be discussed in brief in the final section of this chapter) would not emerge if there was no model of Brownian motion and its properties were not described. And without stochastic integral there would be no stochastic analysis and no way for a mathematician to approach important problems in physics, biology, economy etc.

For more information on Brownian motion see the excellent introductory chapter to Rogers and Williams's book [102].

We will define Brownian motion in the first section and then in Section 3 use Hilbert space theory to rewrite the original Wiener's proof of the existence of Brownian motion thus defined. The necessary background from the Hilbert space is provided in Section 2. The final Section 4 is devoted to a discussion of the Itô integral.

## 4.1 Gaussian families & the definition of Brownian motion

**4.1.1 Definition** A family $\{X_t, t \in \mathbb{T}\}$ of random variables defined on a probability space $(\Omega, \mathcal{F}, \mathbb{P})$ where $\mathbb{T}$ is an abstract set of indexes is called a **stochastic process**. The cases $\mathbb{T} = \mathbb{N}, \mathbb{R}, \mathbb{R}^+, [a, b], [a, b), (a, b], (a, b)$ are very important but do not exhaust all cases of importance. For example, in the theory of point processes, $\mathbb{T}$ is a family of measurable sets in a measurable space [24, 66]. If $\mathbb{T} = \mathbb{N}$ (or $\mathbb{Z}$), we say that our process is time-discrete (hence, a time-discrete process is a sequence of random variables). If $\mathbb{T} = \mathbb{R}, \mathbb{R}^+$ etc. we speak of time-continuous processes. For any $\omega \in \Omega$, the function $t \to X_t(\omega)$ is referred to as realization/sample path/trajectory/path of the process.

*4.1.2 Gaussian random variables* An $n$-dimensional random vector

$$\underline{X} = (X_1, ..., X_n)$$

is said to be **normal or Gaussian** iff for any $\underline{\alpha} = (\alpha_1, \ldots, \alpha_n) \in R^n$ the random variable $\sum_{j=1}^n \alpha_j X_j$ is normal. It is said to be **standard normal** if $X_i$ are independent and normal $N(0, 1)$. A straightforward calculation (see 1.4.7) shows that convolution of normal densities is normal, so that the standard normal vector is indeed normal. In general, however, for a vector to be normal it is not enough for its coordinates to be normal. For instance, let us consider a $0 < p < 1$ and a vector $(X_1, X_2)$ with the density

$$f(x_1, x_2) = p\frac{1}{\pi}e^{-\frac{x_1^2+x_2^2}{2}} + (1-p)\frac{1}{\pi}\frac{2\sqrt{3}}{3}e^{-\frac{4}{3}\frac{x_1^2+x_1x_2+x_2^2}{2}}.$$

Then $X_1$ and $X_2$ are normal but $(X_1, X_2)$ is not. To see that one checks for example that $X_1 - \frac{1}{2}X_2$ is not normal.

**4.1.3 Theorem** An $n$-dimensional vector $\underline{X}$ is normal iff there exists an $n \times n$ matrix $A$ and a vector $\underline{m}$ such that for any $n$-dimensional standard normal vector $\underline{Y}$, vectors $\underline{X}$ and $A\underline{Y} + \underline{m}$ have the same distribution. The vector $\underline{m}$ and the matrix $R = AA^{\mathrm{T}}$ are determined, by $\underline{m} = (\mu_i)_{1 \leq i \leq n} = (E\, X_i)_{1 \leq i \leq n}$ and $R = (\, cov(X_i, X_j)\,)_{1 \leq i, j \leq n}$.

*Proof* The space $\mathbb{R}^n$ is a Hilbert space when equipped with scalar product $(\underline{\alpha}, \underline{\beta}) = \sum_{i=1}^n \alpha_i \beta_i, \underline{\alpha} = (\alpha_i)_{1 \leq i \leq n}, \underline{\beta} = (\beta_i)_{1 \leq i \leq n}$. The space of linear operators on the Hilbert space $\mathbb{R}^n$ is isometrically isomorphic to the space of $n \times n$ matrices $(a_{i,j})_{1 \leq i, j \leq n}$ with the norm $\|(a_{i,j})_{1 \leq i, j \leq n}\| = \sum_{i=1}^n \sum_{j=1}^n a_{i,j}^2$. The isomorphism maps an operator $A$ into a matrix

$(a_{i,j})_{1\leq i,j\leq n}$, where $a_{i,j}$ is the $i$th coordinate of the vector $Ae_j$, $e_j = (\delta_{i,j})_{1\leq i\leq n}$, and we have $A\underline{\alpha} = (\sum_{j=1}^{n} a_{i,j}\alpha_j)_{1\leq i,j\leq n}$. Moreover, $A$ is self-adjoint iff the corresponding matrix is symmetric.

Consider the symmetric matrix $R = (cov(X_i, X_j))_{1\leq i,j\leq n}$, and let the same letter $R$ denote the corresponding operator in $\mathbb{R}^n$. For any $\underline{\alpha} \in \mathbb{R}^n$, by linearity of expectation,

$$(R\underline{\alpha}, \alpha) = \sum_{i=1}^{n}\sum_{j=1}^{n} E(X_i - \mu_i)(X_j - \mu_j)\alpha_i\alpha_j = E\left[\sum_{k=1}^{n}\alpha_k(X_k - \mu_k)\right]^2$$
$$= D^2(\underline{\alpha}, \underline{X}). \tag{4.1}$$

Hence, $(R\underline{\alpha}, \alpha)$ is non-negative and by 3.6.7 there exists a non-negative square root of $R$, i.e. a symmetric matrix $A$ such that $R = A^2$ and $(A\underline{\alpha}, \underline{\alpha}) \geq 0, \underline{\alpha} \in \mathbb{R}^n$.

Let $\underline{Y}$ be any standard normal vector and define $\underline{Z} = A\underline{Y} + \underline{m}$. Clearly, $\underline{Z}$ is Gaussian. We claim that $\underline{Z}$ has the same distribution as $\underline{X}$. By the result to be proved in Subsection 6.6.11, it suffices to show that the distribution of $(\underline{\alpha}, \underline{Z})$ is the same as that of $(\underline{\alpha}, \underline{X}), \underline{\alpha} \in \mathbb{R}^n$. Since both variables have normal distributions, it remains to prove that $E(\underline{\alpha}, \underline{Z}) = E(\underline{\alpha}, \underline{X})$, and $D^2(\underline{\alpha}, \underline{Z}) = D^2(\underline{\alpha}, \underline{X})$. The former relation is obvious, for both its sides equal $(\underline{\alpha}, \underline{m})$. To prove the latter, note first that by 1.4.7, $D^2(\beta_1 Y_1 + \beta_2 Y_2) = \beta_1^2 + \beta_2^2$, $\beta_1, \beta_2 \in \mathbb{R}$, and, more generally, $D^2(\underline{\beta}, \underline{Y}) = D^2(\sum_{i=1}^{n}\beta_i Y_i) = \|\underline{\beta}\|_{\mathbb{R}^n}^2$. Moreover, since $A$ is self-adjoint,

$$D^2(\underline{\alpha}, \underline{Z}) = E[(\underline{\alpha}, \underline{Z}) - (\underline{\alpha}, \underline{m})]^2 = E[(\underline{\alpha}, A\underline{Y})]^2 = E[(A\underline{\alpha}, \underline{Y})]^2$$
$$= D^2[(A\underline{\alpha}, \underline{Y})] = \|A\underline{\alpha}\|_{\mathbb{R}^n}^2 = (A\underline{\alpha}, A\underline{\alpha}) = (R\underline{\alpha}, \alpha),$$

completing the proof by (4.1).                                  □

*Note* As a result of our proof and 6.6.11, $R$ and $\underline{m}$ determine the distribution of $\underline{X}$; we write $\underline{X} \sim N(\underline{m}, R)$. In particular, normal variables are independent iff they are uncorrelated, i.e. iff the appropriate covariances are zero.

4.1.4 *An auxiliary result that is of its own importance*    Our next theorem, Theorem 4.1.5 below, concerns $L^2$ convergence of Gaussian vectors. In its proof we will use the following result that will turn out to be of its own importance in Chapter 5, where various modes of convergence of random variables will be discussed.

If $X$ and $X_n, n \geq 1$ are random variables defined on the same probability space $(\Omega, \mathcal{F}, \mathbb{P})$ and $\lim_{n\to\infty} \mathbb{P}(|X - X_n| > \epsilon) = 0$ for all $\epsilon > 0$, then

$\lim_{n\to\infty} E\,f(X_n) = E\,f(X)$ for all bounded uniformly continuous functions $f$ on $\mathbb{R}$. Indeed, for any $\delta > 0$ the absolute value of the difference between $E\,f(X_n)$ and $E\,f(X)$ is estimated by

$$\int_{|X_n-X|\geq\delta} |f(X_n) - f(X)|\,d\mathbb{P} + \int_{|X_n-X|<\delta} |f(X_n) - f(X)|\,d\mathbb{P}.$$

Moreover, given $\epsilon > 0$ one may choose a $\delta$ so that $|f(\tau) - f(\sigma)| < \epsilon$ provided $|\tau - \sigma| < \delta$. For such a $\delta$ the second integral above is less than $\epsilon\mathbb{P}\{|X_n-X| < \delta\} \leq \epsilon$ while the first is less than $2\|f\|_{\sup}\mathbb{P}\{|X_n-X| \geq \delta\}$, which may be made arbitrarily small by taking large $n$.

**4.1.5 Theorem**    If a sequence $X_n \sim N(\mu_n, \sigma_n^2)$ of normal variables converges in $L^2$ to a variable $X$, then there exist limits $\mu = \lim_{n\to\infty} \mu_n$ and $\sigma_n^2 = \lim_{n\to\infty} \sigma_n^2$, the variable $X$ is Gaussian and $X \sim N(\mu, \sigma^2)$.

*Proof* The sequence $\mu_n, n \geq 1$ must converge because $\mu_n = E\,X_n$ and by the Hölder inequality expectation is a bounded linear functional on $L^2$. Moreover, by continuity of the norm, the sequence $\|X_n\|_{L^2}$ converges as well. Since $\|X_n\|_{L^2}^2 = \mu_n^2 + \sigma_n^2$, there exists the limit of $\sigma_n^2$.

By 4.1.4 and Chebyshev's inequality (cf. 3.1.7), $\lim_{n\to\infty} E\,f(X_n) = E\,f(X)$, for any bounded uniformly continuous function $f$. On the other hand,

$$\lim_{n\to\infty} E\,f(X_n) = \frac{1}{\sqrt{2\pi\sigma_n^2}} \int_{-\infty}^{\infty} \exp\left\{-\frac{(\tau-\mu_n)^2}{2\sigma_n^2}\right\} f(\tau)\,d\tau$$

$$= \frac{1}{\sqrt{2\pi\sigma}} \int_{-\infty}^{\infty} \exp\left\{-\frac{(\tau-\mu)^2}{2\sigma^2}\right\} f(\tau)\,d\tau,$$

which identifies the distribution of $X$ by 1.2.20. To justify the passage to the limit under the integral above, change the variable to $\xi = (\tau - \mu_n)\sigma_n^{-1}$, use Lebesgue's Dominated Convergence Theorem, and come back to the $\tau$ variable. This calculation is valid only if $\sigma^2 \neq 0$. In the other case, $\lim_{n\to\infty} E\,f(X_n) = f(\mu)$ which identifies $X$ as a degenerate Gaussian random variable concentrated at $\mu$.    □

**4.1.6 Definition**    A random process $w_t, t \geq 0$ on a probability space $(\Omega, \mathcal{F}, \mathbb{P})$ is said to be a (one-dimensional) Wiener process (Brownian motion) on $\mathbb{R}^+$ starting at 0, iff

(a) the vector $\underline{w}(t_1, ..., t_k) = (w_{t_1}, ...., w_{t_k})$ is a $k$-dimensional Gaussian vector, for any $k \in \mathbb{N}$ and all $t_1, t_2, ..., t_k \geq 0$,

(b) $E\,w_t w_s = s \wedge t, E\,w_t = 0$ for all $s, t \geq 0$,

(c) the trajectories/sample paths of $w_t$ start at zero and are continuous, i.e. the map $t \to w_t$ is continuous, and $w_0(\omega) = 0$, for any $\omega \in \Omega$, except perhaps for $\omega$ in an event of probability zero.

If the process is defined only for $t$ in a subinterval $[0, a], a > 0$ of $\mathbb{R}^+$, and conditions (a)–(c) are satisfied for $t \in [0, a]$, $w_t$ is called a Brownian motion on $[0, a]$.

**4.1.7** *Equivalent conditions* **1** Condition $E\, w_t w_s = t \wedge s$ implies $E\,(w_t - w_s)^2 = |t - s|$. Conversely, if $w_0 = 0$ and $E\,(w_t - w_s)^2 = |t - s|$ then $E\, w_t w_s = t \wedge s$. This follows directly from $E\,(w_t - w_s)^2 = E\, w_t^2 - 2E\, w_t w_s + E\, w_s^2$.

**2** Condition (b) determines the vector of expectations and the covariance matrix of the Gaussian vector $\underline{w}(t_1, ..., t_k) = (w_{t_1}, ..., w_{t_k})$. If we assume $w_0 = 0$, conditions (a)–(b) are equivalent to

(d) for any $k \in N$ and $0 \le t_1 < t_2 ... < t_k$, the vector

$$\underline{v}(t_1, ..., t_k) = \left( \frac{w_{t_2} - w_{t_1}}{\sqrt{t_2 - t_1}}, ..., \frac{w_{t_k} - w_{t_{k-1}}}{\sqrt{t_k - t_{k-1}}} \right)$$

is a standard Gaussian vector, or, equivalently, random variables $w_{t_i} - w_{t_{i-1}}, i = 1, ..., k$, are independent with normal distribution $N(0, t_i - t_{i-1})$.

Indeed, observe that $\underline{v}(t_1, ..., t_k) = A[\underline{w}(t_1, ..., t_k)]^\mathsf{T}$ where $A$ is a matrix

$$\begin{pmatrix} \frac{1}{\sqrt{t_k - t_{k-1}}} & -\frac{1}{\sqrt{t_k - t_{k-1}}} & 0 & 0 & \cdots & 0 \\ 0 & \frac{1}{\sqrt{t_{k-1} - t_{k-2}}} & -\frac{1}{\sqrt{t_{k-1} - t_{k-2}}} & 0 & \cdots & 0 \\ \vdots & \vdots & \ddots & \ddots & \ddots & \vdots \\ 0 & \cdots & \cdots & \cdots & \frac{1}{\sqrt{t_2 - t_1}} & -\frac{1}{\sqrt{t_2 - t_1}} \end{pmatrix}$$

and the superscript $T$ denotes the transpose of a vector. Hence, if (a) holds, $\underline{v}(t_1, ..., t_n)$ is Gaussian. Moreover, by (b), $E\left( \frac{w_{t_i} - w_{t_{i-1}}}{t_i - t_{i-1}} \right) = 0$, for $1 \le i \le k$, and if $t_i > t_j$, then expanding the product under the expectation sign we obtain that $E\left( \frac{w_{t_i} - w_{t_{i-1}}}{\sqrt{t_i - t_{i-1}}} \frac{w_{t_j} - w_{t_{j-1}}}{\sqrt{t_j - t_{j-1}}} \right)$ equals

$\frac{1}{\sqrt{t_i - t_{i-1}}} \frac{1}{\sqrt{t_j - t_{j-1}}} (t_j - t_{j-1} - t_j - t_{j-1}) = 0$. Finally, $E\left( \frac{w_{t_i} - w_{t_{i-1}}}{t_i - t_{i-1}} \right)^2 = 1$.

This shows that the vector of expectations of $\underline{v}(t_1, ..., t_k)$ is zero and its covariance matrix is the identity matrix.

To prove the converse implication note first that it is enough to show that (d) implies that (a) holds for distinct $t_i \in \mathbb{R}^+$ and this follows from the fact that $\underline{w}(t_1, ..., t_k)$ equals

$$\begin{pmatrix} \sqrt{t_1} & 0 & 0 & \cdots & 0 \\ \sqrt{t_2 - t_1} & \sqrt{t_2 - t_1} & 0 & \cdots & 0 \\ \vdots & \vdots & \ddots & \ddots & \\ \sqrt{t_k - t_{k-1}} & \cdots & \cdots & \cdots & \sqrt{t_k - t_{k-1}} \end{pmatrix} [\underline{v}(0, t_1, ..., t_k)]^{\mathrm{T}}.$$

Moreover, if (d) holds and $s \leq t$ then $E\, w_t = E\,(w_t - w_0) = 0$ for $w_t - w_0 \sim \mathrm{N}(0, t)$, and $E\, w_t w_s = E\,(w_t - w_s)(w_s - w_0) + E\, w_s^2 = E\,(w_t - w_s)E\,(w_s - w_0) + E\, w_s^2 = s$, which means that (d) implies (b).

## 4.2 Complete orthonormal sequences in a Hilbert space

4.2.1 *Linear independence*    Vectors $x_1, ..., x_n$ in a linear space $\mathbb{X}$ are said to be **linearly independent** iff the relation

$$\alpha_1 x_1 + \cdots + \alpha_n x_n = 0 \tag{4.2}$$

where $\alpha_i \in \mathbb{R}$ implies $\alpha_1 = \alpha_2 = \cdots = \alpha_n = 0$. In other words, $x_1, ..., x_n$ are independent iff none of them belongs to the subspace spanned by the remaining vectors. In particular, none of them may be zero.

We say that elements of an infinite subset $\mathbb{Z}$ of a linear space are linearly independent if any finite subset of $\mathbb{Z}$ is composed of linearly independent vectors.

4.2.2 *Orthogonality and independence*    A subset $\mathbb{Y}$ of a Hilbert space is said to be composed of **orthogonal vectors**, or to be an **orthogonal set**, if for any distinct $x$ and $y$ from $\mathbb{Y}$, $(x, y) = 0$, and if $0 \notin \mathbb{Y}$. If, additionally $\|x\| = 1$ for any $x \in \mathbb{Y}$, the set is said to be composed of **orthonormal vectors**, or to be an **orthonormal set**. By a usual abuse of language, we will also say that a sequence is orthonormal (orthogonal) if its values form an orthonormal (orthogonal) set.

Orthogonal vectors are linearly independent, for if (4.2) holds, then $0 = (\sum_{i=1}^n \alpha x_i, \sum_{i=1}^n \alpha x_i) = \sum_{i=1}^n \alpha^2 \|x_i\|^2$, which implies $\alpha_i = 0$, for $i = 1, ..., n$. On the other hand, if $x_1, ..., x_n$ are linearly independent then one may find a sequence $y_1, ..., y_n$, of orthonormal vectors such that $span\,\{x_1, ..., x_n\} = span\,\{y_1, ..., y_n\}$. The proof may be carried by induction. If $n = 1$ there is nothing to prove; all we have to do is take $y_1 = \frac{x_1}{\|x_1\|}$ to make sure that $\|y_1\| = 1$. Suppose now that vectors $x_1, ..., x_{n+1}$ are linearly independent; certainly $x_1, ..., x_n$ are independent also. Let $y_1, ..., y_n$ be orthonormal vectors such that $\mathbb{Y} := span\,\{x_1, ..., x_n\} = span\,\{y_1, ..., y_n\}$. The vector $x_{n+1}$ does not belong to $\mathbb{Y}$ and so we may take $y_{n+1} = \frac{x_{n+1} - Px_{n+1}}{\|x_{n+1} - Px_{n+1}\|}$ where $P$ denotes projection on $\mathbb{Y}$. (The reader will check that $\mathbb{Y}$ is a subspace of $\mathbb{H}$; consult 5.1.5 if needed.) Certainly, $y_1, ..., y_{n+1}$ are orthonormal. Moreover, since $y_{n+1} \in span\,\{x_1, ..., x_{n+1}\}$ (since $Px_{n+1} \in span\{x_1, ..., x_n\}$), $span\{y_1, ..., y_{n+1}\} \subset span\{x_1, ..., x_{n+1}\}$; analogously we prove the converse inclusion. The above procedure of constructing orthonormal vectors from linearly independent vectors is called **the Gram–Schmidt orthonormalization procedure.** There are a number of examples of sequences of orthogonal polynomials that can be obtained via the Gram–Schmidt orthonormalization procedure, including (scalar multiples) of Legendre, Hermite and Laguerre polynomials that are of importance both in mathematics and in physics (see [83], [53], [75] Section 40).

**4.2.3 Exercise**     Let $\{x_1, ..., x_n\}$ be an orthonormal set in a Hilbert space $\mathbb{H}$, and let $x \in \mathbb{H}$. Show that the projection $Px$ of $x$ on the linear span of $\{x_1, ..., x_n\}$ is given by

$$Px = \sum_{k=1}^{n} (x, x_k) x_k.$$

If $\{x_1, ..., x_n\}$ is orthogonal, then

$$Px = \sum_{k=1}^{n} (x, x_k) \frac{x_k}{\|x_k\|^2}.$$

**4.2.4** *Least square regression*     Let $X$ and $Y$ be given square integrable random variables. The problem of finding constants $a$ and $b$ such that the square distance between $Y$ and $aX + b$ is the least is known as the problem of **least square regression.** An equivalent formulation of this problem is to find the projection $PY$ of $Y$ on the $span\{1_\Omega, X\}$. Since this span is the same as that of $\{1_\Omega, X - (E\,X)1_\Omega\}$ and $1_\Omega$ and $X - (E\,X)1_\Omega$ are orthogonal, we find that

$$
\begin{aligned}
PY &= (Y, X - (E\,X)1_\Omega)\frac{X - (E\,X)1_\Omega}{\|X - (E\,X)1_\Omega\|^2} + (Y, 1_\Omega)\frac{1_\Omega}{\|1_\Omega\|^2} \\
&= \frac{E\,XY - E\,X\,E\,Y}{\sigma_X^2}(X - (E\,X)1_\Omega) + (E\,Y)1_\Omega \\
&= \frac{cov(X, Y)}{\sigma_X^2}(X - (E\,X)1_\Omega) + (E\,Y)1_\Omega.
\end{aligned}
$$

The quantity $\rho = \rho_{X,Y} = \frac{cov(X,Y)}{\sigma_X \sigma_Y}$ is called the **correlation coefficient**. The above formula is often written as

$$PY - (EY)1_{\Omega} = \rho \frac{\sigma_X}{\sigma_Y}(X - (EX)1_{\Omega}).$$

**4.2.5 Exercise**    Let $x_k, k \geq 1$ be an orthonormal sequence in a Hilbert space $\mathbb{H}$. Prove the following **Bessel's inequality**: for any $x \in \mathbb{H}$,

$$\|x\|^2 \geq \sum_{k=1}^{\infty} (x_n, x)^2. \tag{4.3}$$

*4.2.6 Separability*    A set $S$ is said to be **denumerable or countable** if there exists a function, say $f$, mapping $N$ onto $S$. In other words there exists a sequence $p_n = f(n)$ whose values exhaust the set $S$. Of course all finite sets are countable and so is $N$ itself. The set of rational numbers is countable, too, but $\mathbb{R}$ is not. We also note that a union of two countable sets is countable, and, more generally, countable union (i.e. a union of a countable number) of countable sets is countable. A metric space is said to be **separable** if there exists a countable set that is dense in this space. We say that a Banach space (Hilbert space) is separable if it is separable as a metric space with the metric $d(x, y) = \|x - y\|$.

*4.2.7 Separability of $C[a, b]$*    The space $C[0, 1]$ is separable. Indeed, the set of polynomials with rational coefficients is countable, any polynomial can be approximated by polynomials with rational coefficients, and the set of polynomials is dense in $C[0, 1]$ by the Weierstrass Theorem (see 2.3.29). Since $C[0, 1]$ is isometrically isomorphic to $C[a, b]$ where $-\infty \leq a < b \leq \infty$, $C[a, b]$ is separable also.

*4.2.8 The space $L^p[a, b]$ is separable*    Let $-\infty < a < b < \infty$, and $1 \leq p < \infty$ be real numbers. From 4.2.7 and 2.2.44 it follows that $L^p[a, b]$ is separable. The spaces $L^p(-\infty, \infty)$, $L^p(-\infty, a)$ and $L^p(a, \infty)$ are separable also. We will show that the first of these spaces is separable, leaving the reader the opportunity to show the remaining claims. To do that, we note that $L^p[a, b]$ may be identified with a subspace of $L^p(-\infty, \infty)$, composed of (equivalence classes with respect to the equivalence relation $x \sim y \equiv \int_a^b |x - y| \, dleb$ of) functions in $L^p(-\infty, \infty)$ that vanish outside of $[a, b]$. With this identification in mind, we may consider a union $D$ of countable subsets $D_n$ that are dense in $L^p[-n, n]$. $D$ is certainly countable, and for any $\epsilon > 0$ and $x \in L^p(-\infty, \infty)$, one may find an $n$ such that the restriction $x_n$ of $x$ to $[-n, n]$ is within $\frac{\epsilon}{2}$ distance from $x$, and a $y \in D_n$ within the same distance from $x_n$. Thus, $D$ is dense in $L^p(-\infty, \infty)$.

**4.2.9 Exercise**  Suppose that a measure space $(\Omega, \mathcal{F}, \mu)$ is separable, i.e. that there exists a sequence $A_n$ of elements of $\mathcal{F}$ such that for every $A \in \mathcal{F}$ and $\epsilon > 0$ there exists an $n \in \mathbb{N}$ such that $\mu(A \div A_n) < \epsilon$. Prove that $L^2(\Omega, \mathcal{F}, \mu)$ is separable.

**4.2.10 Theorem**  If a Banach space is separable then there exists a (possibly finite) countable set composed of linearly independent vectors that is linearly dense in this space.

*Proof*  The argument is somewhat similar to the Gram–Schmidt orthonormalization procedure. Let $\{x_k, k \geq 1\}$ be dense in a Banach space $\mathbb{X}$. Using an induction argument we will construct a (possibly finite) sequence of linearly independent vectors $y_k$ such that

$$span\,\{x_k, 1 \leq k \leq n\} \subset span\,\{y_k, 1 \leq k \leq n\}; \qquad (4.4)$$

this will imply that $cl\,span\,\{x_n, n \geq 1\} = cl\,span\,\{y_n, n \geq 1\}$, and the theorem will follow. We use the following convention: if the sequence $y_n$ is finite, having say $n_0$ elements, then by definition $span\,\{y_k, 1 \leq k \leq n\} = span\,\{y_k, 1 \leq k \leq n_0\} = span\,\{y_n, n \geq 1\}$ for $n \geq n_0$.

For $n = 1$ we let $y_1 = x_1$. For the induction step we suppose that linearly independent vectors $y_1, ..., y_n$ have already been defined in such a way that (4.4) is satisfied. If

$$span\,\{y_k, 1 \leq k \leq n\} = \mathbb{X},$$

we are done; otherwise there exists at least one natural $j$ such that $x_j \notin span\,\{y_k, 1 \leq k \leq n\}$, for if $\{x_j, j \geq 1\}$ is included in $span\,\{y_k, 1 \leq k \leq n\}$ so is its closure. Certainly, $j \geq n+1$. We take the minimal $j$ with this property, and put $y_{n+1} = x_j$. By construction $\{x_1, ..., x_{n+1}\} \subset \{x_1, ..., x_j\} \subset \{y_1, ..., y_{n+1}\}$. $\qquad \square$

**4.2.11 Complete orthonormal sequences**  We will say that an orthogonal (orthonormal) sequence in Hilbert space is **complete** if its linear span is dense in this space. 4.2.2 and 4.2.10 above show that in any separable Hilbert space there exists a complete orthonormal sequence. We say that a Hilbert space is **infinite-dimensional** if there is no finite orthonormal set that is complete in this space.

**4.2.12 Theorem**    Let $\mathbb{H}$ be a Hilbert space and let $(x_n)_{n\geq 1}$ be an orthonormal sequence. The following are equivalent:

(a) $(x_n)_{n\geq 1}$ is complete,
(b) for all $x$ in $\mathbb{H}$, $x = \lim_{n\to\infty} \sum_{i=1}^{n}(x_i,x)x_i$,
(c) for all $x$ and $y$ in $\mathbb{H}$, $(x,y) = \sum_{n=1}^{\infty}(x_n,x)(x_n,y)$,
(d) for all $x$ in $\mathbb{H}$, $\|x\|^2 = \sum_{n=1}^{\infty}(x_n,x)^2$,
(e) for all $x$ in $\mathbb{H}$, $\|x\|^2 \leq \sum_{n=1}^{\infty}(x_n,x)^2$,
(f) for all $x$ in $\mathbb{H}$, $(x,x_n) = 0$ for all $n \geq 1$ implies $x = 0$.

*Proof* If (a) holds, then $\mathbb{H}_n = span\{x_i, 1 \leq i \leq n\}$ is an non-decreasing sequence of subspaces of $\mathbb{H}$ with $\mathbb{H} = cl(\bigcup_{n=1}^{\infty}\mathbb{H}_n)$. By 3.6.8, projections $P_n$ on $\mathbb{H}_n$ converge strongly to the identity operator $I : \lim_{n\to\infty} P_n x = x, x \in \mathbb{H}$. Since $P_n x = \sum_{i=1}^{n}(x_i,x)x_i$, (see 4.2.3 above) (b) follows. Implications

$$(b) \Rightarrow (c) \Rightarrow (d) \Rightarrow (e) \Rightarrow (f)$$

are trivial. Finally, if $(a)$ does not hold, then there exists an $x \in \mathbb{H}$, perpendicular to the closure of the linear span of $(x_n)_{n\geq 1}$, contradicting (f).    □

**4.2.13 Exercise**    Prove that $x_k = (\delta_{k,n})_{n\geq 1}, n \geq 1$ is a complete orthonormal sequence in $l^2$.

**4.2.14 Example**    Let $x_{2^n+k} = 2^{\frac{n}{2}} z_{2^n+k}$ where $0 \leq k < 2^n$ and $z_m$ where defined in 2.1.32. The random variables $x_m$ and $z_m$ and are centered $(E\, x_m = E\, z_n = 0)$. For any $n$ and $0 \leq k \neq k' < 2^n$, $z_{2^n+k}$ and $z_{2^n+k'}$ have disjoint support so that their product is zero and so is their covariance. Also, if $m > n$ then for $0 \leq k < 2^n$ and $0 \leq l < 2^m$ $z_{2^n+k}z_{2^m+l} = z_{2^m+k}$ provided $2^{m-n}k \leq l < 2^{m-n}(k+1)$, and $z_{2^n+k}z_{2^m+l} = 0$ otherwise. Since $E\, x_{2^n+k}^2 = 2^n\frac{1}{2^n} = 1, 0 \leq k < 2^n, n \geq 0$, $z_m$ are orthonormal. Furthermore, if $x \in L^2[0,1]$ is orthogonal to all $x_n$, then by 2.1.32 it is orthogonal to all $1_{[\frac{k}{2^n},\frac{k+1}{2^n})}, 0 \leq k < 2^n, n \geq 0$. Therefore, if a $y \in L^2[0,1]$ is continuous, by the Lebesgue Dominated Convergence Theorem

$$\int_0^1 x(s)y(s)\,ds = \lim_{n\to\infty}\sum_{k=0}^{2^n-1} y\left(\frac{k}{2^n}\right)\int_{\frac{k}{2^n}}^{\frac{k+1}{2^n}} x(s)\,ds = 0.$$

This proves that $x$ is orthogonal to all continuous functions on $[0,1]$, so that by 2.2.19 it is orthogonal to all $y \in L^2[0,1]$, and in particular to itself. Thus $x = 0$ and we have proven that $(x_n)_{n \geq 1}$ is a complete orthonormal sequence in $L^2[0, 1]$.

**4.2.15 Exercise**    Let $z_n, n \geq 1$, be an orthonormal sequence in a Hilbert space $\mathbb{H}$, and let $(a_n)_{n \geq 1} \in l^2$. Prove that the series $\sum_{n=1}^{\infty} a_n z_n$ converges.

**4.2.16** *Any two separable infinite-dimensional Hilbert spaces are isometrically isomorphic*    If $\mathbb{H}$ and $\mathbb{H}_1$ are infinite-dimensional separable Hilbert spaces, then there exists an operator $A$ mapping $\mathbb{H}$ onto $\mathbb{H}_1$ such that

$$(Ax, Ay)_{\mathbb{H}_1} = (x, y)_{\mathbb{H}} \qquad \text{for all } x, y \in \mathbb{H}.$$

(In particular on putting $x = y$ we obtain $\|Ax\|_{\mathbb{H}} = \|x\|_{\mathbb{H}_1}$.)

*Proof* Let $w_n, n \geq 1$ and $z_n, n \geq 1$ be complete orthonormal sequences in $\mathbb{H}$ and $\mathbb{H}_1$, respectively. Let $x \in \mathbb{H}$. The series $\sum_{n=1}^{\infty} (x, w_n)_{\mathbb{H}}^2$ converges, so we may define $Ax = \sum_{n=1}^{\infty} (x, w_n)_{\mathbb{H}} z_n$ (see 4.2.15). Note that $(Ax, Ay)_{\mathbb{H}_1} = \sum_{n=1}^{\infty} (x, w_n)_{\mathbb{H}} (y, w_n)_{\mathbb{H}} = (x, y)_{\mathbb{H}}$. Moreover, $A$ is onto, for if $y$ is in $\mathbb{H}_1$, then $\sum_{n=1}^{\infty} (y, z_n)_{\mathbb{H}_1}^2$ is finite, and by 4.2.15 we may consider $x = \sum_{n=1}^{\infty} (y, z_n)_{\mathbb{H}_1} w_n$. We check that $(x, w_i)_{\mathbb{H}} = (y, z_i)_{\mathbb{H}_1}$ so that $Ax = y$.    □

**4.2.17 Remark**    If $\mathbb{H}$ is not infinite-dimensional the theorem remains the same except that the operator $A$ does not have to be "onto".

**4.2.18 Corollary**    All infinite-dimensional separable Hilbert spaces are isometrically isomorphic to $l^2$.

**4.2.19 Corollary**    If $\mathbb{H}$ is a separable Hilbert space, then there exists a probability space $(\Omega, \mathcal{F}, \mathbb{P})$ and a linear operator $A : \mathbb{H} \to L^2(\Omega, \mathcal{F}, \mathbb{P})$ such that for any $x$ and $y$ in $\mathbb{H}$, $Ax$ is a centered Gaussian random variable, and $(Ax, Ay)_{L^2(\Omega, \mathcal{F}, \mathbb{P})} = cov(Ax, Ay) = (x, y)_{\mathbb{H}}$.

*Proof* Let $(\Omega, \mathcal{F}, \mathbb{P})$ be a probability space where a sequence $X_n$ of independent standard Gaussian random variables is defined. Note that $X_n \in L^2(\Omega, \mathcal{F}, \mathbb{P})$ since $\|X_n\|_{L^2(\Omega, \mathcal{F}, \mathbb{P})}^2 = \sigma^2(X_n) = 1$. Certainly, $X_n$ are orthonormal. Let $\mathbb{H}_1$ be the subspace of $L^2(\Omega, \mathcal{F}, \mathbb{P})$ spanned by $X_n$ ( i.e. the closure of the linear span of $X_n, n \geq 1$). Since $\mathbb{H}_1$ is separable, there exists an operator $A : \mathbb{H} \to \mathbb{H}_1$ described in 4.2.16. By 4.1.5, for any $x \in \mathbb{H}$, $Ax$ is Gaussian as a limit of Gaussian random variables. It also must be centered, for by the Hölder inequality expectation is a bounded linear functional in $L^2(\Omega, \mathcal{F}, \mathbb{P})$.    □

## 4.3 Construction and basic properties of Brownian motion

**4.3.1** *Construction: first step* There exists a process $\{w_t, t \geq 0\}$ on a probability space $(\Omega, \mathcal{F}, \mathbb{P})$ satisfying (a)–(b) of the definition 4.1.6.

*Proof* Let $\mathbb{H} = L^2(\mathbb{R}^+)$, and let $A$ be the operator described in 4.2.19. Let $w_t = A(1_{[0,t)})$. Any vector $(w_{t_1}, ..., w_{t_n})$ is Gaussian, because for any scalars $\alpha_i$, the random variable

$$\sum_{i=1}^n \alpha_i w_i = A\left(\sum_{i=1}^n \alpha_i 1_{[0,t_i)}\right)$$

is Gaussian. Moreover, $E\,w_t = 0$, and $E\,w_t w_s = (1_{[0,t)}, 1_{[0,s)})_{L^2(\mathbb{R}^+)} = \int_0^\infty 1_{[0,t)} 1_{[0,s)} \,\mathrm{d}leb = s \wedge t$. $\square$

**4.3.2** *Existence of Brownian motion on* $[0,1]$ In general it is hard, if possible at all, to check if the process constructed above has continuous paths. We may achieve our goal, however, if we consider a specific orthonormal system (other ways of dealing with this difficulty may be found in [5, 61, 100, 79]†). We will construct a Brownian motion on $[0,1]$ using the system $x_n$ from 4.2.14. As in 4.2.19, we define

$$Ax = \sum_{n=0}^\infty (x_n, x) Y_n \qquad (4.5)$$

where $Y_n$ is a sequence of standard independent random variables. Let

$$w_t(\omega) = (A1_{[0,t)})(\omega) = \sum_{n=0}^\infty (x_n, 1_{[0,t)}) Y_n(\omega) = \sum_{n=0}^\infty y_n(t) Y_n(\omega), \quad t \in [0,1].$$

$$(4.6)$$

The argument presented in 4.3.1 shows that $w_t$ satisfy the first two conditions of the definition of Brownian motion, and it is only the question of continuity of paths that has to be settled.

Note that $y_n(t) = (x_n, 1_{[0,t)}) = \int_0^t x_n(s)\,\mathrm{d}s$ is a continuous function, so that for any $\omega$, the partial sums of the series in (4.6) are (linear combinations of) continuous functions. We will prove that the series converges absolutely and uniformly in $t \in [0,1]$ for almost all $\omega \in \Omega$. To this end write

$$w_t(\omega) = \sum_{n=0}^\infty \sum_{k=0}^{2^n-1} y_{2^n+k}(t) Y_{2^n+k}(\omega) \qquad (4.7)$$

† Kwapień's approach [79] described also in [59] is of particular interest because it involves a very elegant functional-analytic argument and shows directly that Brownian paths are Hölder continuous with any parameter $\alpha \in (0, \frac{1}{2})$.

and

$$a_n(\omega) = \sup_{0 \le t \le 1} \sum_{k=0}^{2^n-1} |y_{2^n+k}(t) Y_{2^n+k}(\omega)|. \qquad (4.8)$$

Certainly, if $\sum_{n=0}^{\infty} a_n(\omega) < \infty$ a.s., we are done. Observe that for any $n \in \mathbb{N}$, functions $y_{2^n+k}$, $0 \le k \le 2^n - 1$, have disjoint supports and

$$\sup_{0 \le t \le 1} |y_{2^n+k}(t)| = \sup_{0 \le t \le 1} |y_{2^n}(t)| = y_{2^n}\left(2^{-(n+1)}\right)$$
$$= 2^{\frac{n}{2}} 2^{-(n+1)} = 2^{-(\frac{n}{2}+1)},$$

so that

$$a_n(\omega) \le 2^{-(\frac{n}{2}+1)} \sup_{0 \le k \le 2^n-1} |Y_{2^n+k}(\omega)|.$$

We want to find a convergent series $\sum_{n=1}^{\infty} b_n$ of non-negative numbers such that $a_n(\omega) \le b_n$ for all but a finite number of indexes $n$ (this number might be different for different $\omega$), except maybe for a set of measure zero. In other words we want to show that the set of $\omega$ such that $a_n(\omega) \ge b_n$ for infinite number of indexes has measure zero. The probability of this event equals $\lim_{n\to\infty} \mathbb{P}(\bigcup_{k \ge n} \{a_n(\omega) \ge b_n\}) \le \lim_{n\to\infty} \sum_{k=n}^{\infty} \mathbb{P}\{a_n(\omega) \ge b_n\}$. Thus it suffices to show that $\sum_{n=1}^{\infty} \mathbb{P}\{a_n(\omega) \ge b_n\} < \infty$. (In other words we are using the easy part of the Borel–Cantelli lemma.) Write $c_n = b_n 2^{\frac{n}{2}}$, and

$$\begin{aligned}
\mathbb{P}\{a_n(\omega) \ge b_n\} &\le \mathbb{P}\{2^{-(\frac{n}{2}+1)} \sup_{0 \le k \le 2^n-1} |Y_{2^n+k}| \ge b_n\} \\
&= \mathbb{P}\{\sup_{0 \le k \le 2^n-1} |Y_{2^n+k}| \ge 2c_n\} \\
&\le \sum_{k=0}^{2^n-1} \mathbb{P}\{|Y_{2^n+k}| \ge 2c_n\} \\
&= \frac{2^{n+1}}{\sqrt{2\pi}} \int_{2c_n}^{\infty} e^{-\frac{s^2}{2}} \, ds \le \frac{2^n}{\sqrt{2\pi} c_n} \int_{2c_n}^{\infty} s e^{-\frac{s^2}{2}} \, ds \\
&= \frac{2^n}{\sqrt{2\pi} c_n} e^{-2c_n^2}.
\end{aligned}$$

To make the last series converge we may try $2c_n^2 \ge n$ or at least $c_n = \sqrt{\frac{n}{2}}$. Then $\sum_{n=1}^{\infty} \mathbb{P}\{a_n(\omega) \ge b_n\} \le \sum_{n=1}^{\infty} \left(\frac{2}{e}\right)^n \frac{1}{\sqrt{\pi n}} < \infty$, and the series $\sum_{n=1}^{\infty} b_n = \sum_{n=1}^{\infty} \sqrt{\frac{n}{2}} 2^{-\frac{n}{2}}$ converges also. This completes the proof.

4.3.3 *Brownian Motion on* $\mathbb{R}^+$     To construct a Brownian motion on $\mathbb{R}^+$, note that we may start with a doubly-infinite matrix $Y_{n,m}$ $n, m \ge 0$, of independent Gaussian variables and construct a sequence of independent copies of Brownian motion on $[0, 1]$ : $w_t^k = \sum_{n=0}^{\infty} y_n(t) Y_{n,k}$. In other words, for any $k \geq 2$, and $s_1, \ldots, s_k \in [0, 1]$ and distinct integers $n_1, \ldots, n_k$, random variables $w_{s_k}^{n_k}$ are mutually independent. Then we define

$$w_t = \sum_{n=0}^{[t]-1} w_1^n + w_{t-[t]}^{[t]}, \qquad (4.9)$$

or, which is the same,

$$w_t = \begin{cases} w_t^0, & t \in [0, 1], \\ w_1^0 + w_{t-1}^1, & t \in [1, 2], \\ w_1^0 + w_1^1 + w_{t-2}^2, & t \in [2, 3], \\ \text{so on} & \ldots \end{cases} \qquad (4.10)$$

In other words, $w_t$ is defined so that for any $\omega \in \Omega$, $w_t(\omega)$ in the interval $[n, n+1]$ is a path that develops as a Brownian motion on $[0, 1]$, but starts at $w_n$ and is independent of the past (to be more specific: $w_t - w_n$ is independent of the past).

**4.3.4 Exercise**   Check that (4.10) defines a Brownian motion, i.e. that conditions (a)–(c) of definition 4.1.6 are satisfied.

**4.3.5** *A more direct construction*   A less intuitive, but more elegant way of constructing a Brownian motion $w_t$ on $\mathbb{R}^+$ is to put

$$w_t = (1 + t) \left( w_{\frac{t}{1+t}}^0 - \frac{t}{1+t} w_1^0 \right), \qquad (4.11)$$

where $w_t^0, t \in [0, 1]$ is a Brownian motion on $[0, 1]$. Still another way may be found in [62].

**4.3.6 Exercise**   Show that (4.11) defines Brownian motion.

**4.3.7** *Properties of Brownian motion*   Let $w_t$ be a Brownian motion on $\mathbb{R}^+$. Then,

(i) for any $a > 0$, $\frac{1}{a} w_{a^2 t}$ is a Brownian motion,
(ii) for any $s > 0$, $w_t - w_s$ is a Brownian motion, independent of $w_u, u \leq s$,
(iii) $-w_t$ is a Brownian motion,
(iv) $t w_{\frac{1}{t}}$ is a Brownian motion.

Property (i) is called the scaling property, and property (iv) is called the time-reversal property.

*Proof* We need to check that conditions (a)–(c) of definition 4.1.6 are satisfied. The proof of (a)–(b) is very similar in all cases. For example, in proving (iv) we note that for any $t_1, ..., t_n$ the vector $(t_1 w_{\frac{1}{t_1}}, t_2 w_{\frac{1}{t_2}}, ..., t_n w_{\frac{1}{t_n}})$ is Gaussian as an image of the Gaussian vector $(w_{\frac{1}{t_1}}, w_{\frac{1}{t_2}}, ..., w_{\frac{1}{t_n}})$ via the affine transformation given by the matrix with $t_1, ..., t_n$ on the diagonal, and the remaining entries zero. Moreover, since for any $a > 0$, $a(s \wedge t) = (as) \wedge (at)$, $E \, tw_{\frac{1}{t}} sw_{\frac{1}{s}} = st(\frac{1}{s}) \wedge (\frac{1}{t}) = t \wedge s$.

Continuity of paths is not a problem either; the non-trivial point in the whole theorem is that $\mathbb{P}\{\lim_{t \to 0} tw_{\frac{1}{t}} = 0\}$. A hint that it is really so is that $E(tw_{\frac{1}{t}})^2 = t$, so that $tw_{\frac{1}{t}}$ tends to zero in $L^2(\Omega, \mathcal{F}, \mathbb{P})$, and, by the Markov inequality 1.2.36, we have $\lim_{t \to 0} \mathbb{P}(|tw_{\frac{1}{t}}| \geq \epsilon) = 0$, for any $\epsilon > 0$. To prove that $\mathbb{P}\{\lim_{t \to 0} tw_{\frac{1}{t}} = 0\}$, we need to be more careful. Observe that, since the paths of $tw_{\frac{1}{t}}$ are continuous for $t > 0$,

$$\{\lim_{t \to 0} tw_{\frac{1}{t}} = 0\} = \bigcap_{n \in \mathbb{N}} \left[ \bigcup_{m \in \mathbb{N}} \bigcap_{s < \frac{1}{m}, s \in \mathbb{Q}} \{|sw_{\frac{1}{s}}| \leq \frac{1}{n}\} \right]. \qquad (4.12)$$

Note that were the paths not continuous we would be able to write merely

$$\{\lim_{t \to 0} tw_{\frac{1}{t}} = 0\} = \bigcap_{n \in \mathbb{N}} \left[ \bigcup_{m \in \mathbb{N}} \bigcap_{s < \frac{1}{m}} \{|sw_{\frac{1}{s}}| \leq \frac{1}{n}\} \right].$$

In (4.12), the sequence $A_n$ of sets in brackets is decreasing: $A_{n+1} \subset A_n$, so we can write

$$\mathbb{P}\{\lim_{t \to 0} tw_{\frac{1}{t}} = 0\} = \lim_{n \to \infty} \mathbb{P}\left\{ \bigcup_{m \in \mathbb{N}} \left[ \bigcap_{s < \frac{1}{m}, s \in \mathbb{Q}} \{|sw_{\frac{1}{s}}| \leq \frac{1}{n}\} \right] \right\}.$$

The sequence $A_{n,m}$ appearing in brackets now is increasing in $m$: for any $n$, $A_{n,m} \subset A_{n,m+1}$, and we may write

$$\mathbb{P}\{\lim_{t \to 0} tw_{\frac{1}{t}} = 0\} = \lim_{n \to \infty} \lim_{m \to \infty} \mathbb{P}\left\{ \left[ \bigcap_{s < \frac{1}{m}, s \in \mathbb{Q}} \{|sw_{\frac{1}{s}}| \leq \frac{1}{n}\} \right] \right\}.$$

Now, for any $m$, there exists a sequence $s_j(m)$ that admits and exhausts all values $s < \frac{1}{m}, s \in \mathbb{Q}$. Thus,

$$\mathbb{P}\{\lim_{t \to 0} tw_{\frac{1}{t}} = 0\} \tag{4.13}$$

$$= \lim_{n \to \infty} \lim_{m \to \infty} \lim_{j \to \infty} \mathbb{P}\{|s_1(m)w_{\frac{1}{s_1(m)}}| \leq \frac{1}{n}, ..., |s_j(m)w_{\frac{1}{s_j(m)}}| \leq \frac{1}{n}\}.$$

Moreover, both vectors

$$(s_1 w_{\frac{1}{s_1(m)}}, ..., s_j(m)w_{\frac{1}{s_j(m)}}) \text{ and } (w_{s_1(m)}, ..., w_{s_j(m)})$$

are normal and have the same covariance matrix and the same vector of expected values; in other words their distributions are the same. Thus,

$$\mathbb{P}\{|s_1(m)w_{\frac{1}{s_1(m)}}| \leq \frac{1}{n}, ..., |s_j(m)w_{\frac{1}{s_j(m)}}| \leq \frac{1}{n}\} \tag{4.14}$$

$$= \mathbb{P}\{|w_{s_1(m)}| \leq \frac{1}{n}, ..., |w_{s_j(m)}| \leq \frac{1}{n}\}.$$

But, repeating the argument presented above, we obtain

$$\mathbb{P}\{\lim_{t \to 0} w_t = 0\} = \lim_{n \to \infty} \lim_{m \to \infty} \lim_{j \to \infty} \mathbb{P}\{|w_{s_1(m)}| \leq \frac{1}{n}, ..., |w_{s_j(m)}| \leq \frac{1}{n}\}. \tag{4.15}$$

Combining (4.13)–(4.15), we see that

$$\mathbb{P}\{\lim_{t \to 0} tw_{\frac{1}{t}} = 0\} = \mathbb{P}\{\lim_{t \to 0} w_t = 0\} = 1.$$

Finally, a comment is needed on independence of $w_t - w_s$, from the $\sigma$-field $\sigma(w_u, 0 \leq u \leq s)$ – this follows by condition **2** in 4.1.7 and 1.4.12.

$\square$

**4.3.8 Remark**  Note that the argument used in the proof of the fact that $\mathbb{P}\{\lim_{t \to 0} tw_{\frac{1}{t}} = 0\} = 1$ applies to a much more general situation. In particular, we have never used the assumption that distributions are normal. The situation we were actually dealing with was that we had two stochastic processes $w_t$ and $z_t$ (in our case $z_t$ was equal to $tw_{\frac{1}{t}}$), both with continuous paths in $\mathbb{R}_*^+$, and the same joint distributions. To be more specific: for any $n \in \mathbb{N}$ and $t_1 \leq t_n \leq .... \leq t_n$ the joint distribution of $w_{t_1}, ..., w_{t_n}$ and $z_{t_1}, ..., z_{t_n}$ was the same. Using these assumptions we were able to show that existence of the limit $\lim_{t \to 0} w_t$ (a.s.) implies existence of the limit $\lim_{t \to 0} z_t$ (a.s.).

The relation to be proven in the following exercise will be needed in 4.3.10.

**4.3.9 Exercise**  Prove that $E|w(t)| = \sqrt{\frac{2t}{\pi}}$.

**4.3.10** *Brownian paths have unbounded variation*   We will prove that sample paths of the Brownian motion are a.s. not of bounded variation on any interval. Without loss of generality we will restrict our attention to the variation of the path on $[0,1]$. Let us fix $\omega$ and define $v_n(\omega) = \sum_{k=1}^{2^n} \left| w_{\frac{k}{2^n}}(\omega) - w_{\frac{k-1}{2^n}}(\omega) \right|$. Certainly, $v_n \leq v_{n+1}$ and we may define $v(\omega) = \lim_{n\to\infty} v_n(\omega)$. It suffices to show that $v = \infty$ a.s., or that $E\,e^{-v} = 0$. Notice that the random variables $\left| w_{\frac{k}{2^n}}(\omega) - w_{\frac{k-1}{2^n}}(\omega) \right|$, $0 \leq k < 2^n$, are independent and have the same distribution. Thus, by the Lebesgue Dominated Convergence Theorem, $E\,e^{-v}$ equals

$$\lim_{n\to\infty} E\,e^{-v_n} = \lim_{n\to\infty} E \prod_{k=1}^{2^n} e^{-\left| w_{\frac{k}{2^n}} - w_{\frac{k-1}{2^n}} \right|} = \lim_{n\to\infty} \left( E\,e^{-\left| w_{\frac{1}{2^n}} \right|} \right)^{2^n}.$$

Let $t = \frac{1}{2^n}$. Since

$$
\begin{aligned}
E\,e^{-|w_t|} &= \frac{1}{\sqrt{2\pi t}} \int_{-\infty}^{\infty} e^{-|s|} e^{-\frac{s^2}{2t}}\,ds = \sqrt{\frac{2}{\pi t}} \int_0^{\infty} e^{-s} e^{-\frac{s^2}{2t}}\,ds \\
&= \sqrt{\frac{2}{\pi}} \int_0^{\infty} e^{-\sqrt{t}s} e^{-\frac{s^2}{2}}\,ds = \sqrt{\frac{2}{\pi}} e^{\frac{t}{2}} \int_{\sqrt{t}}^{\infty} e^{-\frac{u^2}{2}}\,du
\end{aligned}
$$

$(u = s + \sqrt{t})$ then

$$
\begin{aligned}
E\,e^{-v} &= \lim_{t\to 0+} e^{\frac{1}{2}} \left( \sqrt{\frac{2}{\pi}} \int_{\sqrt{t}}^{\infty} e^{-\frac{u^2}{2}}\,du \right)^{\frac{1}{t}} \\
&= e^{\frac{1}{2}} \exp\left\{ \lim_{t\to 0+} \frac{1}{t} \ln\left( \sqrt{\frac{2}{\pi}} \int_{\sqrt{t}}^{\infty} e^{-\frac{u^2}{2}}\,du \right) \right\} \\
&= e^{\frac{1}{2}} \exp\left\{ -\lim_{t\to 0+} \left( \sqrt{\frac{2}{\pi}} \int_{\sqrt{t}}^{\infty} e^{-\frac{u^2}{2}}\,du \right)^{-1} \sqrt{\frac{2}{\pi}} e^{-\frac{t}{2}} \frac{1}{2\sqrt{t}} \right\} = 0,
\end{aligned}
$$

where we have used de l'Hospital's rule.

Brownian paths are not differentiable at any point either; for a direct proof see e.g. [5]

**4.3.11** *Brownian motion is a time-continuous martingale*   Let $(\Omega, \mathcal{F}, \mathbb{P})$ be a probability space. An increasing family $\mathcal{F}_t, t \geq 0$ of $\sigma$-algebras of measurable sets: $\mathcal{F}_t \subset \mathcal{F}_{t+h} \subset \mathcal{F}$; for all $t, h \geq 0$, is called **a (time-continuous) filtration**. A stochastic process $X_t, t \geq 0$, is termed **a (time-continuous) martingale** with respect to filtration $\mathcal{F}_t, t \geq 0$ iff $X_t$ is $\mathcal{F}_t$ measurable, $X_t \in L^1(\Omega, \mathcal{F}_t, \mathbb{P})$ and $\mathbb{E}(X_{t+h}|\mathcal{F}_t) = X_t, t, h \geq 0$.

To be more specific, we should say that $X_t, t \geq 0$ is a martingale with respect to $\mathcal{F}_t, t \geq 0$ and $\mathbb{P}$. If filtration is not clear from the context, a "martingale $X_t, t \geq 0$" is a stochastic process that is a martingale with respect to the **natural filtration** $\mathcal{F}_t = \sigma(X_s, s \leq t)$.

Brownian motion is a time-continuous martingale. Indeed, $\mathbb{E}\,(w(t + h)|\mathcal{F}_t) = \mathbb{E}\,(w(t + h) - w(t)|\mathcal{F}_t) + \mathbb{E}\,(w(t)|\mathcal{F}_t) = 0 + w(t)$, for $w(t + h) - w(t)$ is independent of $\mathcal{F}_t = \sigma(w(s), s \leq t)$ and $w(t)$ is $\mathcal{F}_t$ measurable. Similarly, $X_t = w^2(t) - t$ is a martingale. To this end it suffices to show that $\mathbb{E}\,(w^2(t+h) - w^2(t)|\mathcal{F}_t) = h$. Writing $w^2(t+h) - w^2(t)$ as $[w(t+h) - w(t)]^2 + 2w(t)[w(t+h) - w(t)]$, since $[w(t+h) - w(t)]^2$ is independent of $\mathcal{F}_t$ and $w(t)$ is $\mathcal{F}_t$ measurable, this conditional expectation equals $E\,[w(t+h) - w(t)]^2$ plus $w(t)\mathbb{E}\,(w(t + h) - w(t)|\mathcal{F}_t) = w(t)E\,(w(t + h) - w(t))$. The former quantity equals $h$ and the latter 0, as desired.

The celebrated **Lévy Theorem** (see e.g. [102]), a striking converse to our simple calculation, states that

*a continuous-time martingale $w(t), t \geq 0$, with continuous paths and $X(0) = 0$ such that $w^2(t) - t$ is a martingale, is a Brownian motion.*

**4.3.12 Exercise** Show that for any real $a$, $X_t = e^{aw(t) - \frac{a^2}{2}t}$ is a martingale.

## 4.4 Stochastic integrals

Let us consider the following hazard game related to a Brownian motion $w(t), t \geq 0$. Suppose that at time $t_0$ we place an amount $x(t_0)$ as a bet, to have $x(t_0)[w(t_0 + h) - w(t_0)]$ at time $t_0 + h$. More generally, suppose that we place amounts $x(t_i)$ at times $t_i$ to have

$$\sum_{i=1}^{n-1} x(t_i)[w(t_{i+1}) - w(t_i)] \tag{4.16}$$

at time $t_n$ where $0 \leq a = t_0 < t_1 < \cdots < t_n = b < \infty$. If we imagine that we may change our bets in a continuous manner, we are led to considering the limit of such sums as partitions refine to infinitesimal level. Such a limit, a random variable, would be denoted $\int_a^b x(t)\,dw(t)$. The problem is, however, whether such a limit exists and if it enjoys properties that we are used to associating with integrals. The problem is not a trivial one, for as we know from 4.3.10, $t \to w(t)$ is not of bounded variation in $[a, b]$, so that this integral may not be a Riemann–Stieltjes integral. This new type of integral was introduced and extensively studied by K. Itô, and is now known as an **Itô integral**.

The first thing the reader must keep in mind to understand this notion is that we will not try to define this integral for every path separately; instead we think of $w(t)$ as an element of the space of square integrable functions, and the integral is to be defined as an element of the same space. But the mere change of the point of view and the space where we are to operate does not suffice. As a function $t \mapsto w(t) \in L^2(\Omega, \mathcal{F}, \mathbb{P})$ where $(\Omega, \mathcal{F}, \mathbb{P})$ is the probability space where $w(t), t \geq 0$ are defined, the Brownian motion is not of bounded variation either. To see that consider a uniform partition $t_i = a + \frac{i}{n}(b - a), i = 0, .., n$ of $[a, b]$. Since $E\left[w(t_{i+1}) - w(t_i)\right]^2 = \frac{1}{n}(b - a)$, the supremum of sums $\sum_{i=1}^{n-1} \|w(t_{i+1}) - w(t_i)\|_{L^2(\Omega, \mathcal{F}, \mathbb{P})}$ over partitions of the interval $[a, b]$ is at least $\sum_{i=1}^{n-1} \sqrt{\frac{b-a}{n}} = \sqrt{n}\sqrt{b - a}$, as claimed.

We will still be able to establish existence of the limit of sums (4.16) for quite a large class of stochastic processes $x(t)$, and will see that $x \mapsto \int_a^b x \, dw$ is an isometry of two appropriate Hilbert spaces. (Note that at present it is not even clear yet whether such sums are members of $L^2(\Omega, \mathcal{F}, \mathbb{P})$.) In fact the tool that we are going to use is (2.3.33), where we proved that once we define a bounded operator on a linearly dense subset of a Banach space, this operator may be in a unique way extended to the whole of this space. Our linearly dense set will be the set of so-called simple processes, which correspond to sums (4.16).

Because $w(t)$ is not of bounded variation, properties of the Itô integral are going to be different from the Riemann–Stieltjes integral (cf. 1.3.7). To see one of the differences, let us recall that the approximating sums $S(\mathcal{T}, \Xi, x, y) = \sum_{i=0}^{n-1} x(\xi_i)[y(t_{i+1}) - y(t_i)]$ of the Riemann–Stieltjes integral $\int x \, dy$ involved both partitions $\mathcal{T} : a = t_0 < t_1 < \cdots < t_n = b$ and midpoints $\Xi : t_{i-1} \leq \xi_{i-1} \leq t_i$; and the limit as $\Delta(\mathcal{T})$ tends to zero was not to depend on $\Xi$. The fact that the choice of midpoints does not matter for the Riemann–Stieltjes integral is very much related to the fact that $y$ is of bounded variation. Indeed, if $x$ is continuous and $\Xi$ and $\Xi'$ are two sets of midpoints corresponding to one partition $\mathcal{T}$ (we may always combine two partitions to obtain a new one that will be more refined than the two original ones), then

$$|S(\mathcal{T}, \Xi, x, y) - S(\mathcal{T}, \Xi', x, y)| = \left| \sum_{i=0}^{n-1} [x(\xi_i) - x(\xi_i')][y(t_{i+1}) - y(t_i)] \right|$$

$$\leq \epsilon \sum_{i=0}^{n-1} |y(t_{i+1}) - y(t_i)| \leq \epsilon \, var[y, a, b],$$

provided $\Delta(\mathcal{T}) < \delta$ where $\delta$ is so small that $|x(t) - x(s)| < \epsilon$ when $|t - s| < \delta$.

Therefore, since the Brownian paths have unbounded variation, we should expect that in the Itô integral the choice of midpoints in the approximating sum does matter. This is really so, as we shall see shortly. Note that this is actually something that we should have predicted on account of interpretation of the integral in terms of our hazard game. Indeed, our betting strategy $x(t), a \leq t \leq b$, is in general a stochastic process, for it will most likely depend on the behavior of Brownian motion $w(t)$. In fact, it is natural to assume that $x(t)$ will be $\mathcal{F}_t$ measurable where $\mathcal{F}_t, t \geq 0$ is a natural filtration of the Brownian motion; in other words, by the Doob–Dynkin Lemma, $x(t)$ will be a function of the Brownian motion up to time $t$. Such processes are said to be **adapted**. Now, an approximating sum with $\xi_i > t_i$ would describe a strategy of a person playing the game at time $t_i$ with some foreknowledge of the future behavior of the Brownian motion, however short $\xi_i - t_i$ may be. Of course we should not expect that such a strategy will result in the same gain as the strategy of a player who does not posses this knowledge.

4.4.1 *The choice of midpoints: an example*  We will consider two approximating sums:

$$S_n = \sum_{i=1}^{n-1} w(t_{i,n})[w(t_{i+1,n}) - w(t_{i,n})] \quad \text{and}$$

$$S'_n = \sum_{i=1}^{n-1} w(t_{i+1,n})[w(t_{i+1,n}) - w(t_{i,n})],$$

where $t_{i,n} = a + \frac{i}{n}(b - a), 0 \leq i \leq n$, and show that $S_n$ converges in $L^2(\Omega, \mathcal{F}, \mathbb{P})$ to $\frac{1}{2}[w^2(b) - w^2(a)] - \frac{1}{2}(b - a)$ while $S'_n$ converges to $\frac{1}{2}[w^2(b) - w^2(a)] + \frac{1}{2}(b - a)$. To this end, note first that by the relation $\alpha(\beta - \alpha) = \frac{1}{2}(\beta^2 - \alpha^2) - \frac{1}{2}(\alpha - \beta)^2$,

$$S_n = \frac{1}{2} \sum_{i=1}^{n-1} [w^2(t_{i+1,n}) - w^2(t_{i,n})] - \frac{1}{2} \sum_{i=1}^{n-1} [w(t_{i+1,n}) - w(t_{i,n})]^2$$

$$= \frac{1}{2}[w^2(b) - w^2(a)] - \frac{1}{2} \sum_{i=1}^{n-1} [w(t_{i+1,n}) - w(t_{i,n})]^2.$$

In a similar way, by $\beta(\beta - \alpha) = \frac{1}{2}(\beta^2 - \alpha^2) + \frac{1}{2}(\alpha - \beta)^2$,

$$S'_n = \frac{1}{2}[w^2(b) - w^2(a)] + \frac{1}{2}\sum_{i=1}^{n-1}[w(t_{i+1,n}) - w(t_{i,n})]^2.$$

Hence, it remains to show that

$$\lim_{n\to\infty} Z_n = \lim_{n\to\infty} \frac{1}{2}\sum_{i=1}^{n-1}[w(t_{i+1,n}) - w(t_{i,n})]^2 = \frac{1}{2}(b-a)$$

(in $L^2(\Omega, \mathcal{F}, \mathbb{P})$). An elegant way of doing that was shown to me by Alex Renwick (for a different method see e.g. [20]). Note specifically that random variables $w(t_{i+1,n}) - w(t_{i,n})$, $0 \le i \le n-1$, are independent and have distribution $N(0, \frac{1}{n}(b-a))$. By 1.2.32, their squares have the gamma distribution with parameters $\frac{1}{2}$ and $\frac{1}{2}\frac{n}{b-a}$ and, by 1.2.33, the sum of the squares has the gamma distribution with parameters $\frac{n}{2}$ and $\frac{1}{2}\frac{n}{b-a}$. In particular, by 1.2.35, $EZ_n = b - a$ and $\|Z_n - \frac{1}{2}(b-a)\|_{L^2(\Omega,\mathcal{F},\mathbb{P})} = \sigma_{Z_n}^2 = 2\frac{(b-a)^2}{n}$, which proves our claim. $\qquad\square$

In what follows we will focus our attention on the Itô integral, which corresponds to the natural choice $\xi_i = t_i$. Note however, that other choices are of interest as well. In particular, taking the midpoint $\xi_i = \frac{1}{2}(t_i + t_{i+1})$ leads to the so-called **Stratonovich integral**.

We fix a Brownian motion $w(t), t \ge 0$ on $(\Omega, \mathcal{F}, \mathbb{P})$ and the natural filtration $\mathcal{F}_t = \sigma(w(s), s \le t)$.

4.4.2 *The integrands* The integrands in the Itô integral will be square integrable processes $x = x(t, \omega)$ that "do not depend on the future". First of all we require that $x$ is jointly measurable; i.e. that it is measurable as a function from $([a, b] \times \Omega, \mathcal{M}([a, b]) \times \mathcal{F})$ to $(\mathbb{R}, \mathcal{M}(\mathbb{R}))$. Secondly, we require that $\int_a^b \int_\Omega x^2(t, \omega)\,d\mathbb{P}\,dt$ is finite. Thirdly, to avoid the possibility of strategies dictated by knowledge of a future, we will assume that $x$ is **progressively measurable** which means that for every $t \in [a, b]$, $x1_{\Gamma(t)}$ is $\mathcal{M}([a, t]) \times \mathcal{F}_t$ measurable, where $\Gamma(t) = [0, t] \times \Omega$.

A set $A \in \mathcal{M}([a, b]) \times \mathcal{F}$ is said to be **progressively measurable** if $A \cap \Gamma(t) \in \mathcal{M}([a, t]) \times \mathcal{F}_t$. It is easy to check that the collection $\mathcal{I}$ of progressively measurable sets forms a $\sigma$-algebra. For example, to check that a countable union of elements of $\mathcal{I}$ belongs to $\mathcal{I}$ we use the formula $(\bigcup_{n=1}^\infty A_n) \cap \Gamma(t) = \bigcup_{n=1}^\infty (A_n \cap \Gamma(t))$.

Furthermore, a process $x$ is progressively measurable iff it is measurable with respect to $\mathcal{I}$. Indeed, by definition $x$ is progressively measurable iff for all $t \in [a, b]$, $x1_{\Gamma(t)}$ is $\mathcal{M}([a,b]) \times \mathcal{F}_t$ measurable. This means that

$$\forall_{A \in \mathcal{M}(\mathbb{R})} \forall_{t \in [a,b]} (x1_{\Gamma(t)})^{-1}(A) \in \mathcal{M}([a,b]) \times \mathcal{F}_t$$
$$\Longleftrightarrow \forall_{A \in \mathcal{M}(\mathbb{R})} \forall_{t \in [a,b]} x^{-1}(A) \cap 1_{\Gamma(t)} \in \mathcal{M}([a,b]) \times \mathcal{F}_t,$$
$$\Longleftrightarrow \forall_{A \in \mathcal{M}(\mathbb{R})} x^{-1}(A) \in \mathcal{I},$$

and this last relation means that $x$ is $\mathcal{I}$ measurable.

An important corollary to this is that the space of square integrable progressively measurable processes is a Hilbert space, or that it forms a closed subset of square integrable functions on $[a, b] \times \Omega$. The former space will be denoted $L_p^2 = L_p^2[a, b] = L_p^2([a,b], \Omega, \mathcal{F}, \mathbb{P}, \mathcal{F}_t)$.

The following lemma will be needed in 4.4.4, below.

4.4.3 **Lemma**    Consider the space $L^2[a,b]$ of real-valued, square integrable functions on $[a, b]$, and the operators $T_n, n \geq 1$ in this space given by $T_n x(t) = \sum_{i=0}^{n-2} \alpha_i(x) 1_{[t_{i+1}, t_{i+2})}(t)$, where $\alpha_i(x) = \frac{n}{b-a} \int_{t_i}^{t_{i+1}} x(s) \, ds$ and $t_i = a + \frac{i}{n}(b-a)$. Then $\|T_n\| = 1$ and $\lim_{n \to \infty} T_n x = x$.

*Proof* We have

$$\|T_n x\|^2 = \sum_{i=0}^{n-2} \alpha_i^2(x) \frac{b-a}{n} = \sum_{i=0}^{n-2} \frac{n}{b-a} \left( \int_{t_i}^{t_{i+1}} x(s) \, ds \right)^2$$
$$\leq \sum_{i=0}^{n-2} \frac{n}{b-a} \int_{t_i}^{t_{i+1}} x^2(s) \, ds \int_{t_i}^{t_{i+1}} 1^2 \, ds \quad \text{(Cauchy–Schwartz)}$$
$$= \sum_{i=0}^{n-2} \int_{t_i}^{t_{i+1}} x^2(s) \, ds \leq \|x\|^2.$$

Hence, $\|T_n\| \leq 1$. For equality take $x_n(t) = 1_{[0, t_{n-1})}(t)$.

To prove the rest note that if $x = 1_{(c,d)}$ where $a \leq c < d \leq b$ then $T_n x(t) = x(t)$ for all $n$ greater than $n_0 = n_0(t)$, except perhaps for $t = c$ and $t = d$. Moreover, $T_n x(t) \leq 1_{[a,b]}$, and so by the Lebesgue Dominated Convergence Theorem $T_n x$ converges to $x$, as $n \to \infty$. Since such $x$ form a linearly dense subset of $L^2[a, b]$, we are done.    □

4.4.4 *A dense subset of $L_p^2$: the collection of simple processes*    A process is called **simple** if there exists an $n \geq 1$, a partition $a = t_0 < t_1 < \cdots < t_n = b$ and $x_{t_i} \in L^2(\Omega, \mathcal{F}_{t_i}, \mathbb{P})$ such that

$$x(t, \omega) = \sum_{i=0}^{n-1} x_{t_i}(\omega) 1_{[t_i, t_{i+1})}(t) + x_{t_{n-1}}(\omega) 1_{\{b\}}. \tag{4.17}$$

We note that a simple process is of necessity progressively measurable as a sum of progressively measurable processes. Moreover, it is square integrable. Indeed, $\int_\Omega x^2(t,\omega)\,d\mathbb{P}(\omega)$ equals $\sum_{i=0}^{n-1}\|x_{t_i}\|^2_{L^2(\Omega,\mathcal{F},\mathbb{P})}1_{[t_i,t_{i+1})}(t)+\|x_{t_{n-1}}\|^2_{L^2(\Omega,\mathcal{F},\mathbb{P})}1_{\{b\}}$, and $\int_a^b\int_\Omega x^2\,d\mathbb{P}\,dleb = \sum_{i=0}^{n-1}\|x_{t_i}\|^2_{L^2(\Omega,\mathcal{F},\mathbb{P})}(t_{i+1}-t_i)$.

To prove that simple processes form a dense set in $L_p^2$, take an $x$ from this space and note that for almost all $\omega$, $x_\omega(t) = x(t,\omega)$ belongs to $L^2[a,b]$. Hence, we may use the operators $T_n$ from 4.4.3 to define a simple process $x_n(t,\omega)$ as $T_n x_\omega(t)$ for all such $\omega$ and put zero otherwise. Now,

$$\|x_n - x\|_{L_p^2} = \int_\Omega \int_a^b [T_n x(t,\omega) - x(t,\omega)]^2\,dt\,d\mathbb{P}(\omega)$$

$$= \int_\Omega \|T_n x_\omega - x_\omega\|^2_{L^2[a,b]}\,d\mathbb{P}(\omega). \qquad (4.18)$$

By Lemma 4.4.3, the integrand converges to zero and is bounded above by $4\|x_\omega\|^2_{L^2[a,b]}$. Furthermore, $\int_\Omega 4\|x_\omega\|^2_{L^2[a,b]}\,d\mathbb{P}(\omega) = 4\|x\|_{L_p^2} < \infty$, so that the integral in (4.18) converges to zero, as desired.

4.4.5 *The Itô isometry*    The simple process (4.17) may be thought of as a betting strategy in which we bet the amount $x_{t_i}$ at times $t_i$ to have

$$I(x) = \sum_{i=0}^{n-1} x_{t_i}[w(t_{i+1}) - w(t_i)]$$

at time $b$. We will show that $I(x)$ belongs to $L^2(\Omega,\mathcal{F},\mathbb{P})$ and that $\|I(x)\|_{L^2(\Omega,\mathcal{F},\mathbb{P})} = \|x\|_{L_p^2}$. This is the **Itô isometry**. For reasons to be explained in 4.4.8 below, all we will use in the proof is that $w(t), t \geq 0$ and $w^2(t) - t, t \geq 0$ are martingales.

First, writing $(w(t)-w(s))^2$ as $w^2(t)-w^2(s)-2w(t)w(s)+2w^2(s)$, since $w(t)$ is a martingale, $\mathbb{E}\left([w(t) - w(s)]^2|\mathcal{F}_s\right) = \mathbb{E}\left(w^2(t) - w^2(s)|\mathcal{F}_s\right), t \geq s$. This in turn equals $t - s$ since $w^2(t) - t, t \geq 0$ is a martingale. Next, we show that $x_{t_i}\delta_i$, where $\delta_i = w(t_{i+1}) - w(t_i)$, is square integrable and $\|x_{t_i}\delta_i\|^2_{L^2(\Omega,\mathcal{F},\mathbb{P})} = (t_{i+1} - t_i)\|x_{t_i}\|^2_{L^2(\Omega,\mathcal{F},\mathbb{P})}$. To this end, we note that $x_{t_i}^2 1_{\{x_{t_i}^2 \leq k\}}\delta_i^2 \leq k\delta_i^2$, $k \in \mathbb{N}$ is integrable and $\mathcal{F}_{t_i}$ measurable, and calculate

$$E\,x_{t_i}^2 1_{\{x_{t_i}^2 \leq k\}}\delta_i^2 = E\,\mathbb{E}\left(x_{t_i}^2 1_{\{x_{t_i}^2 \leq k\}}\delta_i^2|\mathcal{F}_{t_i}\right) = E\,x_{t_i}^2 1_{\{x_{t_i}^2 \leq k\}}\mathbb{E}\left(\delta_i^2|\mathcal{F}_{t_i}\right)$$

$$= (t_{i+1} - t_i)E\,x_{t_i}^2 1_{\{x_{t_i}^2 \leq k\}},$$

from which our claim follows by letting $k \to \infty$. Now,

$$\|I(x)\|_{L^2(\Omega,\mathcal{F},\mathbb{P})} = \sum_{i=0}^{n-1} E\, x_{t_i}^2 \delta_i^2 + 2 \sum_{0 \le i < j \le n-1} E\, x_{t_i} x_{t_j} \delta_i \delta_j,$$

and since the first sum equals $\sum_{i=0}^{n-1} (t_{i+1} - t_i) E\, x_{t_i}^2 = \|x\|_{L_p^2}$ it remains to show that $E\, [x_{t_i}\delta_i][x_{t_j}\delta_j] = 0$. Note that this last expectation exists, for variables in brackets are square integrable. Moreover, $\mathbb{E}\left(\delta_j | \mathcal{F}_{t_j}\right)$ equals zero, $w(t), t \ge 0$ being a martingale. Hence

$$E\, x_{t_i}\delta_i\, x_{t_j}\delta_j = E\, \mathbb{E}\left(x_{t_i}\delta_i\, x_{t_j}\delta_j | \mathcal{F}_{t_j}\right) = E\, x_{t_i}\delta_i\, x_{t_j} \mathbb{E}\left(\delta_j | \mathcal{F}_{t_j}\right) = 0,$$

for $i < j$, as desired.

**4.4.6 Definition**    By 4.4.4 and 4.4.5 there exists a linear isometry between $L_p^2$ and $L^2(\Omega, \mathcal{F}, \mathbb{P})$. This isometry is called **the Itô integral** and denoted $I(x) = \int_a^b x \, dw$.

**4.4.7** *Itô integral as a martingale*    It is not hard to see that taking $a < b < c$ and $x \in L_p^2[a,c]$ we have $\int_a^b x \, dw = \int_a^c x 1_{[a,b] \times \Omega} \, dw$. Hence, by linearity $\int_a^c x \, dw = \int_a^b x \, dw + \int_b^c x \, dw$. Also, we have $E \int_a^c x \, dw = 0$ for simple, and hence all processes $x$ in $L_p^2[a,c]$, since $E$ is a bounded linear functional on $L^2(\Omega, \mathcal{F}, \mathbb{P})$. Finally, $\mathbb{E}\left(\int_b^c x \, dw | \mathcal{F}_b\right) = 0$ where as before $\mathcal{F}_b = \sigma(w(s), 0 \le s \le b)$. Indeed, if $x$ is a simple process (4.17) (with $b$ replaced by $c$ and $a$ replaced by $b$), then $\mathbb{E}\left(x_{t_i}\delta_i | \mathcal{F}_{t_i}\right) = x_{t_i} \mathbb{E}\left(\delta_i | \mathcal{F}_{t_i}\right) = 0$, for all $0 \le i \le n-1$, whence by the tower property $\mathbb{E}\left(x_{t_i}\delta_i | \mathcal{F}_b\right) = 0$ as well, and our formula follows.

Now, assume that $x \in L_p^2[0,t]$ for all $t > 0$. Then we may define $y(t) = \int_0^t x \, dw$. The process $y(t), t \ge 0$, is a time-continuous martingale with respect to the filtration $\mathcal{F}_t, t \ge 0$, inherited from the Brownian motion. Indeed, $\mathbb{E}\left(y(t) | \mathcal{F}_s\right) = \mathbb{E}\left(\int_0^s x \, dw | \mathcal{F}_s\right) + \mathbb{E}\left(\int_s^t x \, dw | \mathcal{F}_s\right) = \int_0^s x \, dw + 0 = y(s)$, because $\int_0^s x \, dw$ is $\mathcal{F}_s$ measurable as a limit of $\mathcal{F}_s$ measurable functions.

**4.4.8** *Information about stochastic integrals with respect to square integrable martingales*    There are a number of ways to generalize the notion of Itô integral. For example, one may relax measurability and integrability conditions and obtain limits of integrals of simple processes in a weaker sense (e.g. in probability and not in $L^2$). The most important fact, however, seems to be that one may define integrals with respect to processes other than Brownian motion. The most general and yet plausible integrators are so-called **continuous local martingales**, but in this note we restrict ourselves to **continuous square integrable martingales**. These are time-continuous martingales $y(t), t \geq 0$ with $E\, y^2(t) < \infty, t \geq 0$, and almost all trajectories continuous. Certainly, Brownian motion is an example of such a process. It may be proven that for any such martingale there exists an adapted, non-decreasing process $a(t)$ such that $y^2(t) - a(t)$ is a martingale. A **non-decreasing process** is one such that almost all paths are non-decreasing. For a Brownian motion, $a(t)$ does not depend on $\omega$ and equals $t$. Now, the point is again that one may prove that the space $L_p^2[a, b] = L_p^2[a, b, y]$ of progressively measurable processes $x$ such that $E \int_a^b x^2(s)\, da(s)$ is finite is isometrically isomorphic to $L^2(\Omega, \mathcal{F}, \mathbb{P})$. To establish this fact one needs to show that simple processes form a linearly dense set in $L_p^2[a, b, y]$ and define the Itô integral for simple processes as $I(x) = \sum_{i=0}^{n-1} x_{t_i}[y(t_{i+1}) - y(t_i)]$. Again, the crucial step is establishing Itô isometry, and the reader now appreciates the way we established it in 4.4.5.

4.4.9 **Exercise**     Make necessary changes in the argument presented in 4.4.5 to show the Itô isometry in the case of a square integrable martingale.

# 5

# Dual spaces and convergence of probability measures

Limit theorems of probability theory constitute an integral, and beautiful, part of this theory and of mathematics as a whole. They involve, of course, the notion of convergence of random variables and the reader has already noticed there are many modes of convergence, including almost sure convergence, convergence in $L^1$, and convergence in probability. By far the most important mode of convergence is so-called weak convergence. Strictly speaking, this is not a mode of convergence of random variables themselves but of their distributions, i.e. measures on $\mathbb{R}$. The famous Riesz Theorem, to be discussed in 5.2.9, says that the space $\mathbb{BM}(S)$ of Borel measures on a locally compact topological space $S$ is isometrically isomorphic to the dual of $C_0(S)$. This gives natural ways of defining new topologies in $\mathbb{BM}(S)$ (see Section 5.3). It is almost magical, though in fact not accidental at all, that one of these topologies is exactly "what the doctor prescribes" and what is needed in probability. This particular topology is, furthermore, very interesting in itself. As one of the treats, the reader will probably enjoy looking at Helly's principle, so important in probability, from the broader perspective of Alaoglu's Theorem.

We start this chapter by learning more on linear functionals. An important step in this direction is the famous Hahn–Banach Theorem on extending linear functionals; as an application we will introduce the notion of a Banach limit. Then, in Section 5.2 we will study examples of dual spaces, and in Section 5.3 some topologies in the dual of a Banach space. Finally, we will study compact sets in the weak topology and approach the problem of existence of Brownian motion from this perspective.

## 5.1 The Hahn–Banach Theorem

**5.1.1 Definition**    If $X$ is a linear normed space, then the space of linear maps from $X$ to $\mathbb{R}$ is called the **space of linear functionals**. Its algebraic subspace composed of bounded linear functionals is termed the **dual space** and denoted $X^*$. The elements of $X^*$ will be denoted $F, G$, etc. The value of a functional $F$ on a vector $x$ will be denoted $Fx$ or $\langle F, x \rangle$. In some contexts, the letter notation is especially useful showing the duality between $x$ and $F$ (see below).

Let us recall that boundedness of a linear functional $F$ means that there exists an $M > 0$ such that

$$|Fx| \le M\|x\|, \qquad x \in X.$$

Note that $Fx$ is a number, so that we write $|Fx|$ and not $\|Fx\|$.

**5.1.2 Theorem**    Let $F$ be a linear functional in a normed space $X$, and let $L = \{x \in X : Fx = 0\}$. The following are equivalent:

(a)  $F$ is bounded,
(b)  $F$ is continuous,
(c)  $L$ is closed,
(d)  either $L = X$ or there exists a $y \in X$ and a number $r > 0$ such that $Fx \ne 0$ whenever $\|x - y\| < r$.

*Proof* Implications $(a) \Rightarrow (b) \Rightarrow (c)$ are immediate (see 2.3.3). If $L \ne X$, then there exists a $y \in X \setminus L$, and if (c) holds then $X \setminus L$ is open, so that (d) holds also.

To prove that (d) implies (a), let $B(y, r) = \{x : \|x - y\| < r\}$ and note that the sign of $Fx$ is the same for all $x \in B(y, r)$. Indeed, if $Fx < 0$ and $Fx' > 0$ for some $x, x' \in B(y, r)$, then the convex combination $x_c = \frac{Fx'}{Fx' - Fx}x + \frac{-Fx}{Fx' - Fx}x'$ satisfies $Fx_c = \frac{Fx'Fx - FxFx'}{Fx' - Fx} = 0$, contrary to our assumption (note that $B(y, r)$ is convex). Hence, without loss of generality we may assume that $Fx > 0$ for all $x \in B(y, r)$. Let $z \ne 0$ be an arbitrary element of $X$, and set $x_+ = y + \frac{r}{\|z\|}z \in B(y, r)$ and $x_- = y - \frac{r}{\|z\|}z \in B(y, r)$. Since $Fx_+ > 0$, $-Fz < \frac{Fy}{r}\|z\|$. Analogously, $Fx_- > 0$ implies $Fz < \frac{Fy}{r}\|z\|$. Thus, (a) follows with $M = \frac{Fy}{r}$.     $\square$

**5.1.3** *Examples of linear functionals*    If $(\Omega, \mathcal{F}, \mathbb{P})$ is a probability space then $FX = EX$ is a bounded linear functional on $X = L^1(\Omega, \mathcal{F}, \mathbb{P})$. More generally, one may consider any measure space $(\Omega, \mathcal{F}, \mu)$ and a functional $Fx = \int x \, d\mu$ on $L^1(\Omega, \mathcal{F}, \mu)$. If $\mu$ is finite, the same formula defines a linear functional on the space $BM(\Omega)$ of bounded measurable functions (endowed with the supremum norm). If $\mathbb{H}$ is a Hilbert space and $y \in \mathbb{H}$, then $Fx = (x, y)$ is a bounded linear functional in $\mathbb{H}$. In 3.1.28 we have seen that all functionals in $\mathbb{H}$ are of this form.

Let $(\Omega, \mathcal{F}, \mu)$ be a measure space, and let $p > 1$ be a number. Assume that $y \in L^q(\Omega, \mathcal{F}, \mu)$, where $\frac{1}{p} + \frac{1}{q} = 1$. By the Hölder inequality the absolute value of $\int_\Omega xy \, d\mu$ is no greater than $\|x\|_{L^p(\Omega, \mathcal{F}, \mu)}$ times $\|y\|_{L^q(\Omega, \mathcal{F}, \mu)}$, and therefore is finite for all $x \in L^p(\Omega, \mathcal{F}, \mu)$. Linearity of the map $Fx = \int_\Omega xy \, d\mu$ on $\mathbb{X} = L^p(\Omega, \mathcal{F}, \mu)$ is obvious, and another application of the Hölder inequality shows that $F$ is bounded. In a similar way one proves that if $(\Omega, \mathcal{F}, \mu)$ is a measure space and $y$ is essentially bounded, then $Fx = \int xy \, d\mu$ is a linear functional on $L^1(\Omega, \mathcal{F}, \mu)$.

5.1.4 *Duality*   Let $\mathbb{X}$ be a Banach space and $\mathbb{X}^*$ be its dual. Fix $x \in \mathbb{X}$, and consider a linear map $\mathbb{X}^* \ni F \to F(x)$. This is clearly a bounded linear functional on $\mathbb{X}^*$ for $|F(x)| \leq \|x\|_{\mathbb{X}} \|F\|_{\mathbb{X}^*}$, so that its norm is no greater than $\|x\|_{\mathbb{X}}$. In fact we will be able to prove later that these norms are equal, see 5.1.15 below. The main point to remember, however, is that sometimes it is profitable to view $\mathbb{X}$ as a subset of $\mathbb{X}^{**}$. Equivalently, one should remember about the duality between $\mathbb{X}$ and $\mathbb{X}^*$ which is amply expressed in the notation $F(x) = \langle F, x \rangle$. In this notation, depending on needs, one may either interpret $F$ as fixed and $x$ as arguments, or the opposite: $x$ as fixed and $F$ as arguments.

5.1.5 **Lemma**   Suppose that $\mathbb{X}$ is a normed space, $\mathbb{Y}$ is its algebraic subspace and $x \notin \mathbb{Y}$. Set

$$\mathbb{Z} = \{z \in \mathbb{X} : z = y + tx \text{ for some } y \in \mathbb{Y}, t \in \mathbb{R}\}.$$

Then, $\mathbb{Z}$ is an algebraic subspace of $\mathbb{X}$, and it is closed whenever $\mathbb{Y}$ is closed. The representation $z = y + tx$ of an element of $\mathbb{Z}$ is unique.

*Proof*  For uniqueness of representation, note that if $z = y + tx = y' + t'x$ where $y, y' \in \mathbb{Y}$ and $t, t' \in \mathbb{R}$ and $(t, x) \neq (t', y')$ then we must have $t \neq t'$, and consequently $x = \frac{1}{t-t'}(y' - y) \in \mathbb{Y}$, a contradiction.

$\mathbb{Z}$ is certainly an algebraic subspace of $\mathbb{X}$; we need to prove it is closed if $\mathbb{Y}$ is closed. Suppose that $y_n \in \mathbb{Y}$ and $t_n \in \mathbb{R}$ and that $z_n = y_n + t_n x$ converges. We claim first that $t_n$ is bounded; indeed if this is not the case, then for some sequence $n_k$, $\frac{1}{t_{n_k}} y_{n_k} + x = \frac{1}{t_{n_k}} z_{n_k}$ would converge to zero, for $z_n$ is bounded. This, however, is impossible since vectors $\frac{1}{t_{n_k}} y_{n_k}$ belong to a closed set $\mathbb{Y}$ and $-x \notin \mathbb{Y}$. Secondly, $t_n$ may not have two accumulation points, i.e. there are no two subsequences, say $t_{n_k}$ and $t_{n'_k}$, such that $t_{n_k}$ converges to $t$ and $t_{n'_k}$ converges to $t'$ where $t \neq t'$. Indeed, if this was the case, $y_{n_k}$ would converge to some $y \in \mathbb{Y}$ and $y_{n'_k}$ would converge to some $y' \in \mathbb{Y}$. However, $z_n$ converges, and so $y_{n_k} + t_{n_k} x$ would have to have the same limit as $y_{n'_k} + t_{n'_k} x$. Therefore, we would have $y + tx = y' + t'x$, which we know is impossible.

Since $t_n$ is bounded and has at most one accumulation point, it converges to, say, $t$. It implies that $y_n$ converges to, say, $y$. Since $\mathbb{Y}$ is closed, $y$ belongs to $\mathbb{Y}$ and so $z_n$ converges to $y + tx \in \mathbb{Z}$. $\qquad\square$

5.1.6 **Lemma**     Under notations of 5.1.5, suppose that there exists a linear functional $F \in \mathbb{Y}^*$ and a number $M > 0$ such that $|Fy| \leq M\|y\|, y \in \mathbb{Y}$. Then there exists an $\tilde{F} \in \mathbb{Z}^*$ such that $\tilde{F}y = Fy, y \in \mathbb{Y}$ and $|Fz| \leq M\|z\|, z \in \mathbb{Z}$.

*Proof* Note that we *do not assume that $\mathbb{Y}$ is closed*. If a linear functional $\tilde{F}$ extends $F$ to $\mathbb{Z}$, then for $z = y + tx$ we must have $\tilde{F}z = \tilde{F}y + t\tilde{F}x = Fy + t\tilde{F}x$. Thus, by Lemma 5.1.5, $\tilde{F}$ is uniquely determined by a number $a = \tilde{F}x$. The lemma reduces thus to saying that one may choose an $a$ such that

$$-M\|y + tx\| \leq Fy + ta \leq M\|y + tx\|, \qquad \text{for all } y \in \mathbb{Y}. \tag{5.1}$$

This is trivial if $t = 0$, and in the other case, dividing by $t$, we see after some easy algebra that this is equivalent to

$$-M\|y + x\| \leq Fy + a \leq M\|y + x\|, \qquad \text{for all } y \in \mathbb{Y}. \tag{5.2}$$

(Beware the case $t < 0$!) The existence of the $a$ we are looking for is thus equivalent to

$$\sup_{y \in \mathbb{Y}}\{-Fy - M\|y + x\|\} \leq \inf_{y \in \mathbb{Y}}\{-Fy + M\|y + x\|\} \tag{5.3}$$

or, which is the same,

$$-Fy - M\|y + x\| \leq -Fy' + M\|y' + x\|, \qquad \text{for all } y, y' \in \mathbb{Y}. \tag{5.4}$$

Since $F$ is bounded on $\mathbb{Y}$, however, we have:

$$\begin{aligned} -Fy + Fy' &= F(y' - y) \leq M\|y' - y\| \leq M\left[\|y' + x\| + \| - x - y\|\right] \\ &= M\|y' + x\| + M\|x + y\|, \end{aligned}$$

which proves our claim. Note that the inequality in (5.3) may be strict: we may not claim that the extension of the functional $F$ is unique. $\qquad\square$

**5.1.7 Exercise**    In the situation of the preceding lemma, assume additionally that $\mathbb{Y}$ is closed and $F(y) = 0$ for $y \in \mathbb{Y}$. Take $M = 1$ and check that $a = \inf_{y \in \mathbb{Y}} \|x - y\| > 0$ does the job in the proof.

**5.1.8 *Partially ordered sets***    A set $S$ is said to be **partially ordered** if there exists a relation $R$ in $S$ (a relation in $S$ is a subset of $S \times S$), such that (a) $(p_1, p_2) \in R$ and $(p_2, p_1) \in R$ implies $p_1 = p_2$, and (b) $(p_1, p_2) \in R$ and $(p_2, p_3) \in R$ implies $(p_1, p_3) \in R$. Instead of $(p_1, p_2) \in R$ one writes then $p_1 \leq p_2$. We say that partially ordered set $S$ is **linearly ordered** if for all $p_1, p_2 \in S$ either $p_1 \leq p_2$ or $p_2 \leq p_1$.

An element $p \in S$ is said to be an upper bound for a set $S' \subset S$, if $p' \leq p$ for all $p' \in S'$. Note that $p$ does not have to belong to $S'$.

An element $p_m \in S' \subset S$ is said to be maximal in $S'$ if for all $p' \in S'$, $p_m \leq p'$ implies $p_m = p'$.

**5.1.9 Exercise**    Prove by example that a maximal element may be not unique.

**5.1.10 *Kuratowski–Zorn Lemma***    If $S$ is partially ordered and for any linearly ordered subset of $S$ there exists its upper bound, then there exists a maximal element of $S$. We omit the proof of this famous result.

**5.1.11 Exercise**    Let $\Omega$ be a non-empty set, and let $\mathcal{F}$ be a family of subsets of $\Omega$. Suppose that unions of elements of $\mathcal{F}$ belong to $\mathcal{F}$. An example of such a family is the family of all sets that contain a fixed element $p_0 \in S$. Prove that there exists a set $A_m \in \mathcal{F}$ such that $A_m \subset A$ implies $A_m = A$ for all $A \in \mathcal{F}$.

**5.1.12 *The Hahn–Banach Theorem***    Let $M > 0$ and $\mathbb{Y}$ be an algebraic subspace of a normed space $\mathbb{X}$. Suppose that $F$ is a linear functional on $\mathbb{Y}$ such that $|Fy| \leq M\|y\|$, for $y \in \mathbb{Y}$. Then, there exists a linear functional $\tilde{F}$ on $\mathbb{X}$ such that $\tilde{F}y = Fy, y \in \mathbb{Y}$ and $|Fx| \leq M\|x\|, x \in \mathbb{X}$.

*Proof*    We may assume that $\mathbb{Y} \neq \mathbb{X}$, for otherwise the theorem is trivial. Consider the family $S$ of pairs $(\mathbb{Z}, F_{\mathbb{Z}})$ of subspaces $\mathbb{Z}$ and functionals $F_{\mathbb{Z}}$ on $\mathbb{Z}$ such that

(a)  $\mathbb{Y} \subset \mathbb{Z}$,
(b)  $F_{\mathbb{Z}}(y) = F(y)$, for all $y \in \mathbb{Y}$,
(c)  $|F_{\mathbb{Z}}z| \leq M\|z\|$, for $z \in \mathbb{Z}$.

By Lemma 5.1.6, $S$ is non-empty. We will write $(\mathbb{Z}, F_{\mathbb{Z}}) \leq (\mathbb{Z}', F_{\mathbb{Z}'})$ if $\mathbb{Z} \subset \mathbb{Z}'$, and $F_{\mathbb{Z}} z = F_{\mathbb{Z}'} z$, for $z \in \mathbb{Z}$. It is easy to see that $S$ with this relation is a partially ordered set.

Suppose for the time being that we may prove that there exists a maximal element $(\mathbb{Z}_m, F_m)$ of $S$. Then we would have $\mathbb{Z}_m = \mathbb{X}$, for otherwise by 5.1.6 we could extend $F_m$ to a subspace containing $\mathbb{Z}_m$ as a proper subset, contrary to the maximality of $(\mathbb{Z}_m, F_m)$.

Thus, by the Kuratowski–Zorn Lemma it remains to prove that every linearly ordered subset $S'$ of $S$ has an upper bound. Let the elements of $S'$ be indexed by an abstract set $\mathcal{U}$ : $S' = \{(\mathbb{Z}_u, F_u), u \in \mathcal{U}\}$; note that we write $F_u$ instead of $F_{\mathbb{Z}_u}$. Let $\mathbb{Z}_b = \bigcup_{u \in \mathcal{U}} \mathbb{Z}_u$ ("b" is for "bound"). If $x \in \mathbb{Z}_b$ and $y \in \mathbb{Z}_b$ then there exist $u, v \in \mathcal{U}$ such that $x \in \mathbb{Z}_u$ and $y \in \mathbb{Z}_v$. Since $S'$ is linearly ordered, we either have $\mathbb{Z}_u \subset \mathbb{Z}_v$ or $\mathbb{Z}_v \subset \mathbb{Z}_u$. Thus, both $x$ and $y$ belong to either $\mathbb{Z}_u$ or $\mathbb{Z}_v$, and so does their linear combination. Consequently, a linear combination of elements of $\mathbb{Z}_b$ belongs to $\mathbb{Z}_b$; $\mathbb{Z}_b$ is an algebraic subspace of $\mathbb{X}$.

Similarly one proves that if $z \in \mathbb{Z}_u \cap \mathbb{Z}_v$ for some $u, v \in \mathcal{U}$, then $F_u z = F_v z$. This allows us to define a functional $F_b$ on $\mathbb{Z}_b$ by the formula:

$$F_b(z) = F_u(z), \qquad \text{whenever } z \in \mathbb{Z}_u,$$

for the definition does not depend on the choice of $u$. Arguing as above one proves that $F_b$ is linear. Of course, the pair $(\mathbb{Z}_b, F_b)$ satisfies (a)–(c) and is an upper bound for $S'$. □

**5.1.13** *Separating vectors from subspaces*     The first application of the Hahn–Banach Theorem is that one may separate vectors from subspaces. To be more specific: let $\mathbb{Y}$ be an algebraic subspace of a Banach space, and $x \notin \mathbb{Y}$. There exists a bounded linear functional $F$ on $\mathbb{X}$ such that $Fx \neq 0$, and $Fy = 0, y \in \mathbb{Y}$. To this end, one defines first an $F$ on $\mathbb{Z}$ from 5.1.5 by $F(y + tx) = t d(x, \mathbb{Y}) = t \inf_{y \in \mathbb{Y}} \|x - y\|$. By 5.1.7 this is a bounded linear functional on $\mathbb{Z}$ such that $|Fz| \leq \|z\|$, for $z \in \mathbb{Z}$, and $Fx \neq 0$. By the Hahn–Banach Theorem this functional may be extended to the whole of $\mathbb{X}$ in such a way that $\|F\|_{\mathbb{X}*} \leq 1$.

**5.1.14 Exercise**     A map $p : \mathbb{X} \to \mathbb{R}^+$ is called a **Banach functional** if $p(x + y) \leq p(x) + p(y)$ and $p(tx) = t p(x)$ for $x$ and $y$ in $\mathbb{X}$ and $t \in \mathbb{R}^+$. An example of a Banach functional is $p(x) = M\|x\|$, $M > 0$. Repeat the argument from 5.1.6 to show the following form of the Hahn–Banach Theorem. If $p$ is a Banach functional on $\mathbb{X}$, $\mathbb{Y}$ is a subspace of $\mathbb{X}$ and $F$ is a linear functional on $\mathbb{Y}$ such that $|F(y)| \leq p(y), y \in \mathbb{Y}$, then there exists a linear functional $\tilde{F}$ on $\mathbb{X}$ such that $\tilde{F}(y) = F(y), y \in \mathbb{Y}$ and $|F(x)| \le p(x), x \in \mathbb{X}$.

**5.1.15** *More on duality* Let $x \ne 0$ be a vector in a Banach space. There exists a functional $F \in \mathbb{X}^*$ such that $\|F\| = 1$, and $Fx = \|x\|$. Indeed, let $\mathbb{Y}$ be the subspace of $\mathbb{X}$ defined by $\mathbb{Y} = \{y \in \mathbb{X}; y = tx, t \in \mathbb{R}\}$. The functional $F$ on $y$, given by $F(tx) = t\|x\|$, satisfies $Fx = \|x\|$, and $\|F\|_{\mathbb{Y}^*} = \max F(|\frac{x}{\|x\|}|, |\frac{-x}{\|x\|}|) = 1$. Therefore, the extension of $F$, that exists by the Hahn–Banach Theorem, satisfies all the desired properties. In particular we obtain

$$\|x\| = \sup_{\|F\|_{\mathbb{X}^*}=1} |Fx| = \sup_{\|F\|_{\mathbb{X}^*}=1} |\langle F, x \rangle|. \tag{5.5}$$

This should be compared to the fact that by definition

$$\|F\|_{\mathbb{X}^*} = \sup_{\|x\|=1} |Fx| = \sup_{\|x\|=1} |\langle F, x \rangle|, \tag{5.6}$$

as this again shows the duality between $\mathbb{X}$ and $\mathbb{X}^*$. As a by-product we obtain the fact that the norms considered in 5.1.4 are equal. In other words, one may consider $\mathbb{X}$ as a subspace of $\mathbb{X}^{**}$, and the norms of $x$ as an element of $\mathbb{X}$ and as a functional on $\mathbb{X}^*$ are the same.

There is, however, an important difference between (5.5) and (5.6). Indeed, while it is easy to show by example that the supremum in the latter equality does not have to be attained at a vector $x \in \mathbb{X}$, we have constructed the functional $F$ for which the supremum in the former equality is attained. This suggests the following problem. We know that continuous functions attain their suprema on compact sets; is there a topology in $\mathbb{X}^*$ under which the function $F \mapsto \langle F, x \rangle$ is continuous for all $x$, and the set of functionals with norm one is compact? The answer is affirmative, and it turns out that this topology is simply the weakest topology such that the functions $F \mapsto \langle F, x \rangle$ are continuous for all $x \in \mathbb{X}$, called the weak* topology. This is the subject of Alaoglu's Theorem 5.7.5, below. This topology is exactly the topology that this chapter is devoted to.

Before closing this subsection, we introduce the notion of a Banach limit, a tool of great usefulness in functional analysis, ergodic theory, etc. See e.g. 5.3.6 for a simple application.

**5.1.16** *Banach limit* As in 2.3.9, let $L$ be the left translation $L(\xi_n)_{n \ge 1} = (\xi_{n+1})_{n \ge 1}$ in $l^\infty$, and let $e$ be the member of $l^\infty$ with all coordinates equal to 1. There exists a functional $B$ on $l^\infty$ such that $\|B\| = 1$,

(a)  $BL = B$, and $Be = 1$,
(b)  $B(\xi_n)_{n\geq 1} \geq 0$ provided $\xi_n \geq 0, n \geq 1$,
(c)  $\liminf_{n\to\infty} \xi_n \leq B(\xi_n)_{n\geq 1} \leq \limsup_{n\to\infty} \xi_n$,
(d)  $B(\xi_n)_{n\geq 1} = \lim_{n\to\infty} \xi_n$ if this limit exists.

Such a functional is called a **Banach limit**.

*Proof* Let $\mathbb{Y} = Range(L - I) \subset l^\infty$, where $I$ is the identity operator. Consider

$$a = \inf_{y \in cl\, \mathbb{Y}} \|e - y\| = \inf_{y \in \mathbb{Y}} \|e - y\| = \inf_{x \in l^\infty} \|e - (Lx - x)\|.$$

Taking $x = 0$, we have $a \leq 1$. Also, if we had $\|e - (Lx - x)\| < 1$ for some $x \in l^\infty$, then all the coordinates of $Lx - x = (\xi_{n+1} - \xi_n)_{n\geq 1}$ would need to be positive, so that $x = (\xi_n)_{n\geq 1}$ would be increasing. Since $x \in l^\infty$, $(\xi_n)_{n\geq 1}$ would need to converge and we would have $\lim_{n\to\infty}(\xi_{n+1} - \xi_n) = 0$ and, hence, $\|e - (Lx - x)\| = 1$. This contradiction shows that $a = 1$.

By 5.1.7 and the Hahn–Banach Theorem there exists a functional $B$ such that $\|B\| = 1, Be = 1$ and $By = 0, y \in cl\,\mathbb{Y}$. In particular $(a)$ holds.

Suppose that $Bx < 0$ even though $x \neq 0$ has non-negative coordinates. Then $\|e - \frac{1}{\|x\|}x\| \leq 1$ and yet $B(e - \frac{1}{\|x\|}x) > 1$. This contradicts $\|B\| = 1$. Hence, (b) follows.

We are left with proving (c), condition (d) following directly from (c). We will show that $B(\xi_n)_{n\geq 1} \geq l := \limsup_{n\to\infty} \xi_n$; the other inequality will follow from this by noting that $\limsup_{n\to\infty}(-\xi_n) = -\liminf_{n\to\infty} \xi_n$. Take $\epsilon > 0$ and choose an $n_0$ such that $\xi_n \geq l - \epsilon, n \geq n_0$. By (b), $B(\xi_n)_{n\geq 1} = BL^{n_0}(\xi_n)_{n\geq 1} = B(\xi_{n_0+n})_{n\geq 1}$. Hence by (b) we obtain $B(\xi_{n+n_0} - (l - \epsilon))_{n\geq 1} \geq 0$, so that $B(\xi_n)_{n\geq 1} = B(\xi_{n_0+n})_{n\geq 1} \geq l - \epsilon$. Since $\epsilon$ is arbitrary, we are done. $\square$

**5.1.17 Exercise**   Prove (d) directly without referring to (c), by noting that $c_0 \subset cl\,\mathbb{Y}$.

## 5.2 Form of linear functionals in specific Banach spaces

In 3.1.28 we saw that all bounded linear functionals on a Hilbert space $\mathbb{H}$ are of the form $Fx = (x, y)$ where $y$ is an element of $\mathbb{H}$. In this section we will provide forms of linear functionals in some other Banach spaces. It will be convenient to agree that from now on the phrase "a linear functional" means "a bounded linear functional", unless stated otherwise.

5.2.1 **Theorem**    Let $X = c_0$ be the space of sequences $x = (\xi_n)_{n\geq 1}$ such that $\lim_{n\to\infty} \xi_n = 0$, equipped with the supremum norm. $F$ is a functional on $X$ if and only if there exists a unique sequence $(\alpha_n)_{n\geq 1} \in l^1$ such that

$$Fx = \sum_{n=1}^{\infty} \xi_n \alpha_n \tag{5.7}$$

where the last series converges uniformly. Also, $\|F\|_{c_0^*} = \|(\alpha_n)_{n\geq 1}\|_{l^1}$. In words: $c_0^*$ is isometrically isomorphic to $l^1$.

*Proof*  Define $e_i = (\delta_{i,n})_{n\geq 1}$. Since $\|\sum_{i=1}^{n} \xi_i e_i - x\| = \sup_{i\geq n+1} |\xi_i|$ which tends to zero as $n \to \infty$, we may write $x = \lim_{n\to\infty} \sum_{i=1}^{n} \xi_i e_i = \sum_{i=1}^{\infty} \xi_i e_i$. In particular, $(e_n)_{n\geq 1}$ is linearly dense in $c_0$. This is crucial for the proof.

If $(\alpha_n)_{n\geq 1}$ belongs to $l^1$, then

$$\sum_{n=1}^{\infty} |\xi_n \alpha_n| \leq \|x\|_{c_0} \sum_{n=1}^{\infty} |\alpha_n| = \|x\|_{c_0} \|(\alpha_n)_{n\geq 1}\|_{l^1} \tag{5.8}$$

and the formula (5.7) defines a bounded linear functional on $c_0$.

Conversely, suppose that $F$ is a linear functional on $c_0$. Define $\alpha_n = Fe_n$, and $x_n = \sum_{i=1}^{n} (\operatorname{sgn} \alpha_i) e_i \in c_0$. We have $\|x_n\|_{c_0} \leq 1$, and $Fx_n = \sum_{i=1}^{n} |\alpha_i|$. Since $|Fx| \leq \|F\|$, if $\|x\| \leq 1$, $(\alpha_n)_{n\geq 1}$ belongs to $l^1$ and its norm in this space is does not exceed $\|F\|$. Using continuity and linearity of $F$, for any $x \in c_0$,

$$Fx = F \lim_{n\to\infty} \sum_{i=1}^{n} \xi_i e_i = \lim_{n\to\infty} F \sum_{i=1}^{n} \xi_i e_i$$

$$= \lim_{n\to\infty} \sum_{i=1}^{n} \xi_i F e_i = \lim_{n\to\infty} \sum_{i=1}^{n} \xi_i \alpha_i = \sum_{n=1}^{\infty} \xi_i \alpha_i. \tag{5.9}$$

Estimate (5.8) proves that the last series converges absolutely and that $\|F\| \leq \|(\alpha_n)_{n\geq 1}\|_{l^1}$. Combining this with (5.8) we obtain $\|F\|_{c_0^*} = \|(\alpha_n)_{n\geq 1}\|_{l^1}$.    $\square$

5.2.2 **Exercise**    State and prove an analogous theorem for the space $c$ of convergent sequences.

**5.2.3 Theorem**    Let $\mathbb{X} = l^1$. $F$ is a functional on $\mathbb{X}$ if and only if there exists a unique sequence $(\alpha_n)_{n \geq 1} \in l^\infty$ such that

$$Fx = \sum_{n=1}^{\infty} \xi_n \alpha_n, \qquad x = (\xi_n)_{n \geq 1} \in l^1 \qquad (5.10)$$

where the last series converges uniformly. Also, we have $\|F\|_{(l^1)^*} = \|(\alpha_n)_{n \geq 1}\|_{l^\infty}$. In other words $(l^1)^*$ is isometrically isomorphic to $l^\infty$.

*Proof*  If $(\alpha_n)_{n \geq 1}$ belongs to $l^\infty$ then the series (5.10) converges absolutely, and the series of absolute values is no greater than $\|x\|_{l^1}$ times $\|(\alpha_n)_{n \geq 1}\|_{l^\infty}$. In particular, $\|F\| \leq \|(\alpha_n)_{n \geq 1}\|_{l^\infty}$.

Conversely, suppose that $F$ is a linear functional on $l^1$, and define $\alpha_n = Fe_n$ where as before $e_i = (\delta_{i,n})_{n \geq 1}$. Since $\|e_n\|_{l^1} = 1$, $\sup_n |\alpha_n| \leq \|F\|$, proving that $(\alpha_n)_{n \geq 1}$ belongs to $l^\infty$, and that its norm in $l^\infty$ does not exceed $\|F\|$. Now, $\|x - \sum_{i=1}^n \xi_i e_i\|_{l^1} = \sum_{i=n+1}^{\infty} |\xi_i|$ so that $x = \sum_{i=1}^{\infty} \xi_i e_i$. Therefore, we may argue as in (5.9) to obtain (5.10). This in turn implies that $\|F\| \leq \|(\alpha_n)_{n \geq 1}\|_{l^\infty}$, and so the two quantities are equal.    □

**5.2.4 Remark**    The theorems proven above illustrate in particular the fact that $\mathbb{X}$ may be viewed as a subspace of $\mathbb{X}^{**}$. In our case we have $c_0 \subset l^1$. It is also worth noting how duality is expressed in formulae (5.7) and (5.10).

**5.2.5 Exercise**    Let $l_r^1, r > 0$ be the space of sequences $(\xi_n)_{n \geq 1}$ such that $\sum_{n=1}^{\infty} |\xi_n| r^n < \infty$. When equipped with the norm $\|(\xi_n)_{n \geq 1}\| = \sum_{n=1}^{\infty} |\xi_n| r^n$, $l_r^1$ is a Banach space. Prove that this space is isomorphic to $l^1$ and use this result to find the form of a linear functional on $l_r^1$.

**5.2.6 Theorem**    Let $\mathbb{X}$ be the space $C[0,1]$ of continuous functions on the interval $[0,1]$. $F$ is a linear functional on $\mathbb{X}$ iff there exists a (unique) signed Borel measure $\mu$ on $[0,1]$ such that

$$Fx = \int_0^1 x \, \mathrm{d}\mu, \qquad (5.11)$$

and $\|F\| = \|\mu\|_{\mathrm{BM}[0,1]}$.

*Proof*  Certainly, if $\mu$ is a signed measure then (5.11) defines a linear functional on $C[0,1]$ and $\|F\| \leq \|\mu\|$. To complete the proof we show that for a given functional $F$ there exists a $\mu$ such that (5.11) holds and $\|\mu\| \leq \|F\|$, and then prove uniqueness.

The space $C[0,1]$ is a subspace of the space $BM[0,1]$ of bounded measurable functions on this interval. Let $F$ be an extension of our functional to $BM[0,1]$, which exists by the Hahn–Banach Theorem. (This extension may perhaps be not unique but this will not concern us.) Define

$$y(t) = \begin{cases} 0, & t < 0, \\ F1_{[0,t]}, & 0 \le t \le 1, \\ F1_{[0,1]}, & t \ge 1. \end{cases} \tag{5.12}$$

If $a < 0$ and $b > 1$ and $a = t_0 \le t_1 \le ... \le t_n = b$, where $t_i < 0 \le t_{i+1}$, and $t_j \le 1 < t_{j+1}$, for some $1 \le i \le j < n$, then

$$\sum_{k=1}^{n} |y(t_k) - y(t_{k-1})|$$

$$= |y(t_{i+1})| + \sum_{k=i+2}^{j} |y(t_k) - y(t_{k-1})| + |y(t_{j+1}) - y(t_j)|$$

$$= |F1_{[0,t_{i+1}]}| + \sum_{k=i+2}^{j} |F1_{(t_{k-1},t_k]}| + |F1_{(t_j,1]}|$$

$$= F\left(\beta_{i+1}1_{[0,t_{i+1}]} + \sum_{k=i+2}^{j} \beta_k 1_{(t_{k-1},t_k]} + \beta_{j+1}1_{(t_j,1]}\right)$$

where $\beta_{i+1} = \operatorname{sgn} F1_{[0,t_{i+1}]}$, $\beta_k = \operatorname{sgn} F1_{(t_{k-1},t_k]}, k = i+2, ..., j$, and $\beta_{j+1} = \operatorname{sgn} F1_{(t_k,t_1]}$. Above, the argument of $F$ is a function with norm 1 in $BM[0,1]$, hence, the whole expression is bounded by $\|F\|$. This shows that $var[y, a, b] \le \|F\|$ and so $var[y, -\infty, \infty] \le \|F\|$. Moreover, condition (1.22) is satisfied. Let $\mu$ be the unique Borel measure on $\mathbb{R}$ corresponding to the regularization of $y$. Even though $\mu$ is a measure on $\mathbb{R}$, it is concentrated on $[0,1]$ for $y$ is constant outside of this interval. Hence, $\mu$ may be identified with a measure on $[0,1]$. Also, $\mu(\{0\}) = y(0) - y(0-) = F1_{\{0\}}$, and $\|\mu\| = var[y, -\infty, \infty] \le \|F\|$.

For $x \in C[0,1]$ define $x_n = \sum_{i=1}^{n} x(\frac{i}{n})1_{(\frac{i-1}{n}, \frac{i}{n}]}$ on $(0,1]$ and $x_n(0) = x(\frac{1}{n})$. By uniform continuity of $x$, $x_n$ tends to $x$ in $BM[0,1]$. Therefore, $Fx = \lim_{n \to \infty} Fx_n$. On the other hand, $Fx_n = \sum_{i=1}^{n} x(\frac{i}{n})[y(\frac{i}{n}) - y(\frac{i-1}{n})] + x(\frac{1}{n})1_{\{0\}}$. The sum here is an approximating sum of the Riemann–Stieltjes integral of $x$ with respect to $y$. By 1.3.8, this sum converges to the same limit as the corresponding approximating sum of the Riemann–Stieltjes integral of $x$ with respect to the regularization $y_r$ of $y$, for $y(0+) = y(0)$. Since $\mu(-\infty, t] = y_r(t), t \in \mathbb{R}$, $Fx$ is the limit of

$\sum_{i=1}^{n} x(\frac{i}{n})\mu(\frac{i-1}{n}, \frac{i}{n}] + x(\frac{1}{n})\mu(\{0\})$, which by the Lebesgue Dominated Convergence Theorem equals $\int_{[0,1]} x \, d\mu$.

To prove uniqueness of representation (5.11), assume that $\mu$ and $\nu$ are two measures such that this formula holds. Let $\mu = \mu^+ - \mu^-$ and $\nu = \nu^+ - \nu^-$, for some positive measures $\mu^+, \mu^-, \nu^+$, and $\nu^-$ on $[0,1]$. Then $\int x \, d(\mu^+ + \nu^-) = \int x \, d(\nu^+ + \mu^-)$, for all $x \in C[0,1]$, which by 1.2.20 implies $\mu^+ + \nu^- = \nu^+ + \mu^-$, and so $\mu = \nu$.     $\square$

**5.2.7 Exercise**     In Subsection 1.2.20 we dealt with measures on $\mathbb{R}$ and in the concluding step of the proof above we needed to apply 1.2.20 to two measures on $[0,1]$. Fill out the necessary details.

**5.2.8 Remark**     It is *not* true that for any functional $F$ on $C[0,1]$ there exists a regular function $y$ of bounded variation on $[0,1]$ such that

$$Fx = \int x \, dy, \qquad \text{for all } x \in C[0,1]. \qquad (5.13)$$

The assumption that $y$ is right-continuous does not allow us to express $Fx = x(0)$ in this form. However, one may prove that for any $F$ on $C[0,1]$ there exists a unique function of bounded variation on $[0,1]$, right-continuous in $(0,1)$, such that $y(0) = 0$ and (5.13) holds. In a similar way, there is no one-to-one correspondence between measures on $\{-\infty\} \cup \mathbb{R} \cup \{\infty\}$ and functions of bounded variation on $\mathbb{R}$ such that (1.22) holds, for the measure of $\{-\infty\}$ may possibly be non-zero.

**5.2.9 *Riesz Theorem***     In Theorem 5.2.6, the assumption that we are dealing with functions at $[0,1]$ is inessential. In particular, we could consider any interval $[a,b]$, half-axis $(-\infty, a]$ or $[a, +\infty)$, with or without endpoints, or the whole of $\mathbb{R}$; we could also include one or both points at infinity (see below). From the topological point of view it is the assumption that we consider functions defined on a locally compact topological space that is important. A famous **Theorem of Riesz** says what follows.

*Let $S$ be a locally compact topological space, and let $C_0(S)$ be the space of continuous functions vanishing at infinity. $F$ is a functional on $C_0(S)$ iff there exist a finite (signed) measure on $S$ such that $Fx = \int_S x \, d\mu$. Moreover, $\|F\|_{C_0^*(S)}$ is equal to the total variation of $\mu$.*

We note that this theorem includes not only 5.2.6 but also 5.2.1 and 5.2.2 as special cases (see 2.2.37). Corollary 5.2.10 and Exercise 5.2.11 below are also instances of this result, but we will derive them from 5.2.6 without referring to the Riesz Theorem. The proof of the Riesz Theorem may be found e.g. in [103]. It is important to note that the measure in the Riesz Theorem is non-negative iff $F$ maps the **non-negative cone** $C_0^+(S)$ of all non-negative numbers into $\mathbb{R}^+$ ([103]). Moreover, it is **inner regular,** i.e. for any $\epsilon > 0$ and any open set $G$ there exists its compact subset $K$ of $G$ such that $\mu(G \setminus K) < \epsilon$.

**5.2.10 Corollary** For any linear functional on $C[-\infty, \infty]$ there exists a unique Borel measure $\mu$ on $\mathbb{R}$ and two unique real numbers $a$ and $b$ such that

$$Fx = \int x \, \mathrm{d}\mu + ax(+\infty) + bx(-\infty). \tag{5.14}$$

*Proof* By 2.2.31, $C[-\infty, \infty]$ is isomorphic to $C[0, 1]$ and the isomorphism $I : C[0, 1] \to C[-\infty, \infty]$ is given by $Ix(\tau) = x(\frac{1}{\pi} \arctan \tau + \frac{1}{2})$. If $F$ is a functional on $C[-\infty, \infty]$, then $F \circ I$ is a functional on $C[0, 1]$, and thus we have $F \circ Ix = \int x \, \mathrm{d}\nu$ for some measure $\nu$ on $[0, 1]$. Now, for $y \in C[-\infty, \infty]$,

$$Fy = F \circ I \circ I^{-1}y = \int I^{-1}y \, \mathrm{d}\nu = \int y \circ f \, \mathrm{d}\nu,$$

where $f(\varsigma) = \tan(\pi\varsigma - \frac{\pi}{2})$ is a map from $[0, 1]$ to $[-\infty, \infty]$. Let $\nu = \nu^+ - \nu^-$ be a representation of $\nu$ and let $\nu_f^+$ and $\nu_f^-$ be the transports of measures $\nu^+$ and $\nu^-$ via $f$. Note that $\nu^+$ and $\nu^-$ are measures on $\{-\infty\} \cup \mathbb{R} \cup \{\infty\}$. Let the measures $\mu^+$ and $\mu^-$ on $\mathbb{R}$ be restrictions of $\nu^+$ and $\nu^-$ to $\mathbb{R}$, respectively. Then

$$Fy = \int_{[0,1]} y \circ f \, \mathrm{d}\nu^+ - \int_{[0,1]} y \circ f \, \mathrm{d}\nu^-$$

$$= \int_{\{-\infty\} \cup \mathbb{R} \cup \{\infty\}} y \, \mathrm{d}\nu_f^+ - \int_{\{-\infty\} \cup \mathbb{R} \cup \{\infty\}} y \, \mathrm{d}\nu_f^-$$

$$= \int_{\mathbb{R}} y \, \mathrm{d}\mu + ay(+\infty) + by(-\infty)$$

where $\mu = \mu^+ - \mu^-$, $a = \nu_f^+(\{+\infty\}) - \nu_f^-(\{+\infty\})$, and $b = \nu_f^+(\{-\infty\}) - \nu_f^-(\{-\infty\})$. Uniqueness is proven as in 5.2.6. $\qquad\square$

**5.2.11 Exercise** Using 5.2.10 find the form of linear functionals on $C_0(\mathbb{R})$, and on the space of functions that vanish at infinity and have a finite limit at $-\infty$.

**5.2.12 Exercise** Find the form of a linear functional on $C_0(\mathbb{G})$ where $\mathbb{G}$ is the Kisyński group.

5.2.13 **Corollary**      If $T$ is an operator on $C_0(\mathbb{R})$ that commutes with all translations $T_t, t \geq 0$, given by $T_t x(\tau) = x(\tau + t)$, then there exists a unique (signed) Borel measure $\mu$ on $\mathbb{R}$ such that $Tx(\tau) = \int x(\tau + \varsigma)\mu(d\varsigma)$.

*Proof* Define $Fx = Tx(0)$. $F$ is a bounded linear functional on $C_0(\mathbb{R})$. Hence (see 5.2.11) there exists a charge $\mu$ such that $Tx(0) = \int x(\varsigma)\mu(d\varsigma)$. We have

$$Tx(\tau) = T_\tau Tx(0) = TT_\tau x(0) = \int T_\tau x(\varsigma)\mu(d\varsigma) = \int x(\tau + \varsigma)\mu(d\varsigma).$$

Uniqueness is obvious.                                                                           $\square$

5.2.14 **Remark**      By the remark at the end of 5.2.9, the measure in the corollary above is non-negative iff $T$ maps the non-negative cone $C_0^+(\mathbb{R})$ into itself.

Our next goal is to find the form of a functional on $L^1(\Omega, \mathcal{F}, \mu)$. Before we will do that, however, we need a lemma.

5.2.15 **Lemma**      Suppose that $(\Omega, \mathcal{F}, \mu)$ is a measure space and that $F$ is a linear functional on $L^1(\Omega, \mathcal{F}, \mu)$. There exist two functionals $F^+$ and $F^-$ such that $F = F^+ - F^-$, and $F^+x$ and $F^-x$ are non-negative whenever $x$ is (a.e.).

*Proof* If $x \geq 0$, define $F^+x = \sup_{0 \leq y \leq x} Fy$. This is a finite number, since for $0 \leq y \leq x$, $|Fy| \leq \|F\|\|y\| \leq \|F\|\|x\|$. Moreover,

$$F^+x \geq 0 \quad \text{and} \quad \|F^+x\| \leq \|F\|\|x\|. \tag{5.15}$$

We have (a) $F^+(x_1 + x_2) = F^+(x_1) + F^+(x_2)$, for $x_1, x_2 \geq 0$ and (b) $F^+(\alpha x) = \alpha F^+(x)$ for $x \geq 0$ and non-negative number $\alpha$. The proof of (b) is immediate. To prove (a), for any $\epsilon > 0$ choose $0 \leq y_i \leq x_i$ such that $Fy_i > F^+x_i - \epsilon$, $i = 1, 2$. Then, $0 \leq y_1 + y_2 \leq x_1 + x_2$, and $F^+(x_1 + x_2) \geq F(y_1 + y_2) \geq F^+x_1 + F^+x_2 - 2\epsilon$. Thus $F^+(x_1 + x_2) \geq F^+x_1 + F^+x_2$. For the other inequality, we note that for every $\epsilon > 0$ there exists a $0 \leq y \leq x_1 + x_2$ such $Fy \geq F^+(x_1 + x_2) - \epsilon$. Then $y_1 = \min(y, x_1)$, satisfies $0 \leq y_1 \leq x_1$. Moreover, $y_2 = y - y_1$ equals either $0$ or $y - x_1$ and so is no greater than $x_2$. Since we have $y_1 + y_2 = y$, $F^+x_1 + F^+x_2 \geq Fy_1 + Fy_2 = F(y_1 + y_2) \geq F^+(x_1 + x_2) - \epsilon$. This completes the proof of (a).

Now, one checks that the functional $F^+x = F^+x_+ - F^+x_-$ where

$x_+ = \max(x, 0)$ and $x_- = \max(-x, 0)$ is linear, and $F^+x$ is non-negative whenever $x \geq 0$. $F^+$ is also bounded since by (5.15),

$$\|F^+x\| = \|F^+x_+ - F^+x_-\| \leq \|F^+x_+\| + \|F^+x_-\|$$
$$\leq \|F\|[\|x_+\| + \|x_-\|] = \|F\|\,\|x\|.$$

It remains to check that a bounded linear functional $F^-x = F^+x - Fx$ maps non-negative functions into non-negative numbers, but this is obvious by definition of $F^+$. $\qquad\square$

**5.2.16 Theorem**  Let $(\Omega, \mathcal{F}, \mu)$ be a $\sigma$-finite measure space. $F$ is a bounded linear functional on $\mathbb{X} = L^1(\Omega, \mathcal{F}, \mu)$ iff there exists a function $y \in L^\infty(\Omega, \mathcal{F}, \mu)$ such that

$$Fx = \int_\Omega xy \, d\mu. \tag{5.16}$$

In such a case, $\|F\| = \|y\|_{L^\infty(\Omega, \mathcal{F}, \mu)}$.

*Proof* The "if part" is immediate and we will restrict ourselves to proving the converse. Also, to prove the last statement of the theorem it suffices to show $\|y\| \leq \|F\|$ since the other inequality is immediate from (5.16).

1 We will show that if the theorem is true for finite $\mu$, then it also holds for $\sigma$-finite $\mu$. To see that let $\Omega = \bigcup_{n \geq 1} \Omega_n$, where $\Omega_n$ are disjoint and $\mu(\Omega_n) < \infty$. For all $n \geq 1$ let $\mathbb{X}_n = L^1(\Omega_n, \mathcal{F}_n, \mu_n)$ where $\mathcal{F}_n$ is the $\sigma$-algebra of measurable subsets of $\Omega_n$, and $\mu_n$ is the restriction of $\mu$ to this $\sigma$-algebra. Consider the restriction $F_n$ of $F$ to $\mathbb{X}_n$. Then $\|F_n\|_{\mathbb{X}_n^*} \leq \|F\|_{\mathbb{X}^*}$ and hence by assumption there exists a function $y_n \in L^\infty(\Omega_n, \mathcal{F}_n, \mu_n)$ such that for all $x \in \mathbb{X}_n$, $F_n x = \int xy_n \, d\mu_n = \int xy_n \, d\mu$, and $\|y_n\| \leq \|F_n\|$. Let us extend each $y_n$ to the whole of $\Omega$ by putting $y_n(\omega) = 0$ for $\omega \notin \Omega_n$. Define also $y = \sum_{n=1}^\infty y_n$, which amounts to saying that $y(\omega) = y_n(\omega)$ for $\omega \in \Omega_n$. Take an $x \in \mathbb{X}$ and define $x_n = x1_{\Omega_n}$. Then $\|\sum_{i=1}^n x_i - x\| = \int_{\Omega \setminus \bigcup_{i=1}^n \Omega_i} |x| \, d\mu$ tends to zero, which means that $\lim_{n \to \infty} \sum_{i=1}^n x_i = x$ in the norm in $\mathbb{X}$. On the other hand, $x_n$ may be identified with an element of $\mathbb{X}_n$. Therefore,

$$Fx = \lim_{n \to \infty} \sum_{i=1}^n Fx_i = \lim_{n \to \infty} \sum_{i=1}^n F_i x_i = \lim_{n \to \infty} \sum_{i=1}^n \int_{\Omega_i} y_i x_i \, d\mu_i$$
$$= \lim_{n \to \infty} \sum_{i=1}^n \int_{\Omega_i} y_i x \, d\mu = \lim_{n \to \infty} \int_\Omega \Big(\sum_{i=1}^n y_i\Big) x \, d\mu = \int_\Omega yx \, d\mu,$$

the last step following by the Lebesgue Dominated Convergence Theorem. Moreover, since $\|y_n\| \leq \|F_n\| \leq \|F\|$, we have $\|y\| \leq \|F\|$.

**2** Using Lemma 5.2.15, one may argue that we may further restrict ourselves to the case where $Fx \geq$ for $x \geq 0$.

**3** Assume therefore that $\mu(\Omega) < \infty$, and $Fx \geq$ for $x \geq 0$. Define a set-function $\nu$ on $(\Omega, \mathcal{F})$ by $\nu(A) = F1_A$. This is a (positive) bounded, measure, since (a) $\nu(\Omega) = F1_\Omega \leq \|F\|\|1_\Omega\| = \|F\|\mu(\Omega)$, (b) finite additivity follows from linearity of $F$, and (c) if $A_n$ are disjoint sets, then $\|1_{\bigcup_{i=1}^n A_i} - 1_{\bigcup_{i=1}^\infty A_i}\| = \|1_{\bigcup_{i=n+1}^\infty A_i}\| = \mu(\bigcup_{i=n+1}^\infty A_i)$ tends to zero, and thus $\nu(\bigcup_{i=1}^n A_i) = F1_{\bigcup_{i=1}^n A_i}$ tends to $F1_{\bigcup_{i=1}^\infty A_i} = \nu(\bigcup_{i=1}^\infty A_i)$. Moreover, $\nu(A) = F1_A \leq \|F\|\|1_A\| = \|F\|\mu(A)$, and in particular, $\mu(A) = 0$ implies $\nu(A) = 0$. By the Radon–Nikodym Theorem there exists a non-negative $y$ such that $F1_A = \nu(A) = \int_A y \, d\mu = \int_\Omega 1_A y \, d\mu$. Hence, by linearity, for any simple function $x$, $Fx = \int_\Omega xy \, d\mu$. Furthermore, for non-negative $x$ we may use an increasing sequence $x_n$ of simple functions converging almost everywhere to $x$. By the Monotone Convergence Theorem such functions converge to $x$ in the sense of the norm in $L^1(\Omega, \mathcal{F}, \mu)$. Hence, $Fx_n$ tends to $Fx$. On the other hand, by the same theorem $\int_\Omega x_n y \, d\mu$ converges to $\int_\Omega xy \, d\mu$. Therefore, $Fx = \int xy \, d\mu$ for non-negative $x$, and hence for all $x \in L^1(\Omega, \mathcal{F}, \mu)$.

It remains to show that $\|y\| \leq \|F\|$. Suppose that on the contrary, $\|y\| > \|F\|$. This means that the measure $\mu(A)$ of a measurable set $A = \{|y| > \|F\|\}$ is positive. Let $x = 1_A$. This is an element of $L^1(\Omega, \mathcal{F}, \mu)$, and $\|x\| = \mu(A)$. Note that $Fx = \int_A y \, d\mu > \|F\|\mu(A) = \|F\|\|x\|$. This, however, contradicts the definition of $\|F\|$. $\square$

**5.2.17 Exercise**     Prove claim **2** made above.

**5.2.18 Exercise**     Show that 5.2.16 implies 5.2.3.

**5.2.19 Exercise**     Let $(\Omega, \mathcal{F}, \mu)$ be a $\sigma$-finite measure space. Show that $L^p(\Omega, \mathcal{F}, \mu), p > 1$, is isomorphic to $L^q(\Omega, \mathcal{F}, \mu)$, where $\frac{1}{p} + \frac{1}{q} = 1$.

## 5.3 The dual of an operator

**5.3.1** *The dual operator*     Let $\mathbb{X}$ and $\mathbb{Y}$ be Banach spaces, and let $A$ be an operator $A \in \mathcal{L}(\mathbb{X}, \mathbb{Y})$. For any functional $F$ on $\mathbb{Y}$, the map $F \circ A$ is a linear functional on $\mathbb{X}$. Since $\|F \circ A\|_{\mathbb{X}^*} \leq \|F\|_{\mathbb{Y}^*}\|A\|_{\mathcal{L}(\mathbb{X}, \mathbb{Y})}$, the map $F \mapsto F \circ A$, denoted $A^*$, is a bounded linear map from $\mathbb{Y}^*$ to $\mathbb{X}^*$

and $\|A^*\| \leq \|A\|$. The operator $A^*$ is called the **dual operator** or the **adjoint operator** of $A$. Since for any $x \in \mathbb{X}$,

$$\|Ax\| = \sup_{F \in \mathbb{Y}^*, \|F\|=1} |FAx| = \sup_{F \in \mathbb{Y}^*, \|F\|=1} |A^*Fx| \leq \|A^*\|\|x\|,$$

we see that $\|A\| \leq \|A^*\|$, and so $\|A\| = \|A^*\|$. Furthermore, we see that $A^*$ is adjoint to $A$ if for any $x \in \mathbb{X}$ and $F \in \mathbb{Y}^*$,

$$\langle F, Ax \rangle = \langle A^*F, x \rangle.$$

By Definition 3.1.19 a linear operator in a Hilbert space is self-adjoint if $A = A^*$. More examples follow.

**5.3.2 Exercise** Let $\mathbb{X} = \mathbb{R}^n$. A linear operator on $\mathbb{X}$ may be identified with a matrix $A(n \times n)$. Check that its adjoint is identified with the transpose of $A$.

**5.3.3 Example** Now we can clarify the relation between the operators $T_\mu$ and $S_\mu$ introduced in 2.3.17 and 2.3.23. For any bounded measurable function $x$ on $\mathbb{R}$, and any measure $\nu$ on $\mathbb{R}$,

$$\langle S_\mu \nu, x \rangle = \int x \, dS_\mu \nu = \int x \, d\mu * \nu = \int \int x(\tau + \varsigma) \mu(d\tau) \nu(d\varsigma)$$
$$= \int T_\mu x \, d\nu = \langle \nu, T_\mu x \rangle. \tag{5.17}$$

Since this relation holds in particular for $x \in C_0(\mathbb{R})$ this formula proves that $S_\mu$ is dual to $T_\mu$. Similarly, considering the operators $S_\mu$ and $T_\mu$ introduced in 2.3.25 we see that for any $x \in BM(\mathbb{G})$ and in particular for $x \in C_0(\mathbb{G})$,

$$\langle S_\mu \nu, x \rangle = \int x \, dS_\mu \nu = \int x \, d\mu * \nu = \int \int x(gh) \mu(dh) \nu(dg)$$
$$= \int T_\mu x \, d\nu = \langle \nu, T_\mu x \rangle$$

which proves that $S_\mu$ is dual to $T_\mu$. Analogously, $\tilde{S}_\mu$ is dual to $\tilde{T}_\mu$. The reader should examine Examples 2.3.26 and 2.3.27 in the light of the above discussion.

**5.3.4 Exercise** Let $\mathbb{X}, \mathbb{Y}$ and $\mathbb{Z}$ be Banach spaces and suppose that $A \in \mathcal{L}(\mathbb{X}, \mathbb{Y})$ and $B \in \mathcal{L}(\mathbb{Y}, \mathbb{Z})$. Show that $(BA)^* = A^*B^* \in \mathcal{L}(\mathbb{Z}^*, \mathbb{X}^*)$. Use this result to show equivalence of (a) and (c) in 3.1.24.

5.3.5 **Exercise**   Let $R$ be the right shift in $l^1$ : $R\left(\xi_n\right)_{n\geq 1} = \left(\xi_{n-1}\right)_{n\geq 1}$ where we set $\xi_0 = 0$. Check that $R^*$ in $l^\infty$ is the left shift $R^*\left(\alpha_n\right)_{n\geq 1} = L\left(\alpha_n\right)_{n\geq 1} = \left(\alpha_{n+1}\right)_{n\geq 1}$.

5.3.6 *An application: invariant measures*    Both stochastic and deterministic phenomena (such as forming of fractals, for instance) are often described by specifying the evolution of the distribution of their particular characteristic in time. If the time is discrete, this is simply a sequence $\mu_n, n \geq 0$ of measures on a space $S$. Such a sequence naturally depends on the initial distribution $\mu_0$ and this dependence is often described by means of a single operator $P$, for we have $\mu_n = P^n \mu_0$. Typically, $S$ is a compact topological space and $P$ is a linear map from $\mathbb{BM}(S)$, the space of Borel measures on $S$, to itself. Because of the interpretation, we also assume that $P\mu$ is a probability measure whenever $\mu$ is. Such operators are also called **Markov operators** (of course this is a more general class of operators than that described in 2.3.37).

One of the questions that a mathematician may ask is that of existence of an **invariant measure** of a Markov operator $P$, i.e. of such a probability measure $\mu_\diamond$ that $P\mu_\diamond = \mu_\diamond$. If $\mu_\diamond$ is a distribution of our process at time 0, this distribution does not change in time.

As we will see shortly, the search for an invariant measure may be facilitated by a dual operator. In general, though, the dual $P^*$ of $P$ is defined in $[\mathbb{BM}(S)]^*$ which in the first place is difficult to describe. A situation that can be easily handled is that where $P^*$ leaves the subspace $C(S)$ of $[\mathbb{BM}(S)]^*$ invariant. (Well, $C(S)$ is not a subspace of $[\mathbb{BM}(S)]^*$ but it is isometrically isomorphic to a subspace of $[\mathbb{BM}(S)]^*$, which in a sense is quite the same as being a subspace.) In such a case $P$ is said to be a **Feller operator**.

Quite often we actually start with a linear operator $U$ that maps $C(S)$ into itself, such that $U1_S = 1_S$, and $Ux \geq 0$ provided $x \geq 0$. Then, the dual $P$ of $U$ is a Markov operator in $\mathbb{BM}(S)$. Indeed, for any non-negative $x$ and probability measure $\mu$, $\int_S x \, dP\mu = \int_S Ux \, d\mu \geq 0$ so that $P\mu$ is a non-negative measure, and the calculation $P\mu(S) = \int_S 1_S \, dP\mu = \int_S U1_S \, d\mu = \int_S 1_S \, d\mu = \mu(S) = 1$ shows that $P\mu$ is a probability measure.

To find an invariant measure of such an operator $P$ it suffices to find a non-negative invariant functional $F$ on $C(S)$, i.e. a linear functional such that $F1_S = 1$, $FUx = Fx, x \in C(S)$, and $Fx \geq 0$ whenever $x \geq 0$. Indeed, by the Riesz Theorem, with such a linear functional $F$ we have a non-negative measure $\mu_\diamond$ such that $\int_S x \, d\mu_\diamond = Fx$; this is a probability measure since $\mu_\diamond(S) = \int_S 1_S \, d\mu_\diamond = F1_S = 1$. This measure is invariant for $P$ since for any $x \in C(S)$ we have $\int_S x \, dP\mu_\diamond = \int_S Ux \, d\mu_\diamond = FUx = Fx = \int_S x \, d\mu_\diamond$, which implies $P\mu_\diamond = \mu_\diamond$.

If $S$ is not compact, however, this scheme fails to work. The problem lies in the fact that in general the operator $U^*$ acts in $C(S)^*$ and not in $\mathbb{BM}(S)$. For example take a locally compact $S = \mathbb{N}$. We have $C(S) = l^\infty$, the space of bounded sequences, $C_0(S) = c_0$, the space of sequences converging to zero, and $\mathbb{BM}(S) = l^1$, where "$=$" means "is isometrically isomorphic to". Let us take a closer look at functionals on $C(S)$. For any $F \in C(S)^*$ we may define a functional $F_0$ on $C_0(S)$ given by $F_0 x = Fx, x \in C_0(S)$. By the Riesz Theorem, there exists a measure $\mu$ such that $\int_S x \, d\mu = F_0 x, x \in C_0(S)$. This formula may be used to extend $F_0$ to the whole of $C(S)$. The functionals $F_0$ and $F$ agree only on a subspace of $C(S)$. Moreover, if $F$ is non-negative, then $F_0 x \le Fx, x \in C(S)$. Indeed, for a natural $k$, $F_0(x1_{\{n \le k\}}) = F(x1_{\{n \le k\}}) \le Fx$. Hence, by the Lebesgue Dominated Convergence Theorem, $F_0 x = \int_S x \, d\mu = \lim_{k \to \infty} \int_S x1_{\{n \le k\}} \, d\mu \le Fx$, which also may be proved using 5.2.1. Therefore we conclude that any positive $F \in C(S)^*$ can be expressed as a sum $F_0 + F_1$, where $F_0$ is its measure-part defined above, and $F_1$ is its positive "singular" part, $F_1 = F - F_0$. An example of such a singular functional is a Banach limit.

Now, given a non-negative $U$ in $C(S)$ with $U1_S = 1_S$ and a non-negative $\mu \in \mathbb{BM}(S)$, it is natural to define $P\mu$ as $(U^*\mu)_0$, the measure-part of $U^*\mu$, (and extend this definition in a natural way to all $\mu \in \mathbb{BM}(S)$). Then, $P\mu$ is a non-negative measure, since for non-negative $x \in C_0$, $\int_S x \, dP\mu = \langle U^*\mu, x \rangle = \int_S x \, d\mu \ge 0$. However, in general $P$ is not a Markov operator, since for any probability measure $\mu$ and $x \in C(S)$ we have

$$P\mu(S) = (U^*\mu)_0 1_S \le \langle U^*\mu, 1_S \rangle = \langle \mu, U1_S \rangle = \langle \mu, 1_S \rangle = \mu(S) = 1,$$

and the inequality may be strict. In particular, if we find a non-negative functional $F$ that is also invariant, i.e.

$$F1_S = 1, \quad U1_S = 1_S, \quad \text{and} \quad FUx = Fx, \tag{5.18}$$

the corresponding measure $\mu$ may happen not to be invariant. In the following example, due to R. Rudnicki and taken from [81], we have $P\mu = \frac{1}{2}\mu$. We consider the functional $Fx = F(\xi_n)_{n \ge 1} = \frac{1}{2}Bx + \sum_{n=1}^\infty \frac{1}{2^{n+1}}\xi_n$, where $B$ is a Banach limit, and the operator $Ux = Fx1_S$. Clearly, (5.18) is satisfied, since $FUx = F(Fx1_S) = FxF1_S = Fx$. It is easy to see that the measure $\mu$ on $S$ that corresponds to $F_0$ is given by $\mu(A) =$

$\sum_{n \in S} \frac{1}{2^{n+1}}$. In particular, $F_0$ has a natural extension to $l^\infty$ given by $F_0 (\xi_n)_{n \geq 1} = \sum_{n=1}^\infty \frac{1}{2^{n+1}} \xi_n$ (which agrees with $F$ only on the kernel of $B$ which is a proper subspace of $l^\infty$). Now, for any $x \in c_0$,

$$\langle P\mu, x \rangle = \langle U^*\mu, x \rangle = \langle \mu, Ux \rangle = \langle \mu, Fx 1_S \rangle = Fx \langle \mu, 1_S \rangle$$
$$= Fx \, \mu(S) = \frac{1}{2} Fx = \langle \frac{1}{2}\mu, x \rangle.$$

as claimed.

### 5.3.7 *Von Neumann's Ergodic Theorem*

A bounded operator $U$ in a Hilbert space $\mathbb{H}$ is said to be **unitary** if its inverse (both left and right) exists and equals $U^*$. In other words we have $UU^* = U^*U = I$. Note that $U$ is unitary iff $(Ux, Uy) = (x, y)$ for all $x$ and $y$ in $\mathbb{H}$. In particular unitary operators are isometric isomorphisms of $\mathbb{H}$.

The famous von Neumann's Ergodic Theorem says that if $U$ is unitary, then $\lim_{n \to \infty} \frac{1}{n} \sum_{k=1}^n U^k x = Px, x \in \mathbb{H}$, where $P$ is the projection on the subspace $\mathbb{H}_1 = \{x | Ux = x\} = Ker(U - I)$. Observe that $x = Ux$ iff $U^*x = U^{-1}x = x$. Hence $\mathbb{H}_1 = Ker(U^* - I)$. The main step in the proof is establishing that $\mathbb{H}_1^\perp$ is the closure of $Range(U - I)$ or, which is the same, that $\{cl \, Range(U - I)\}^\perp = \mathbb{H}_1$. To this end we note that, by continuity of the scalar product, $x$ is perpendicular to $cl \, Range(U-I)$ iff it is perpendicular to $Range(U - I)$. Since for all $y \in \mathbb{H}_1$, $(x, Uy - y) = (U^*x - x, y)$, then our claim is proven. Now, for $x \in \mathbb{H}_1$ we find a $y$ such that $x = Uy - y$ and then $\frac{1}{n} \sum_{k=1}^n U^k x = \frac{1}{n} \|U^{k+1}y - Uy\| \leq \frac{2}{n} \|y\| \to 0$, as $n \to \infty$. Also, if $x = Ux$ then $\frac{1}{n} \sum_{k=1}^n U^k x = x$. This completes the proof by 3.1.16.

An important example is the case where $\mathbb{H} = L^2(\Omega, \mathcal{F}, \mu)$ where $\mu$ is a finite measure, and $U$ is given by $Ux = x \circ f$ for some measure-preserving map $f$. A map $f$ is said to be **measure-preserving** if it is measurable and $\mu(f^{-1}(B)) = \mu(B)$ for all measurable sets $B \subset \Omega$. In other words the transport $\mu_f$ of the measure $\mu$ via $f$ is the same as $\mu$. To prove that such a $U$ is indeed unitary we calculate as follows:

$$(Ux, Uy) = \int_\Omega x(f(\omega)) y(f(\omega)) \, d\mu(\omega) = \int_\Omega xy \, d\mu_f = \int_\Omega xy \, d\mu = (x, y).$$

## 5.4 Weak and weak* topologies

Distributions of random variables, i.e. probability measures on $\mathbb{R}$, are functionals on $C_0(\mathbb{R})$ and we have a well-defined metric in $C_0(\mathbb{R})^*$ which may be used in studying asymptotic behavior of these distributions and hence of these random variables. However, there are very few limit theorems of probability that can be expressed in the language provided by this topology. In fact I am aware of only one interesting case, namely the Poisson approximation to binomial (see 5.8.4). This topology, the strong topology in $C_0(\mathbb{R})^*$, is simply too strong to capture such delicate phenomena like the Central Limit Theorem. In this section we give a functional analytic view on other possible choices of topology. First, the weak topology and then the weak\* topology are discussed.

5.4.1 *Convergence determining sets*    Let $\mathbb{Y}$ be a Banach space. A set $\Lambda \subset \mathbb{Y}^*$ is said to be a **convergence determining set** if $\Lambda$ separates points in $\mathbb{Y}$, i.e. if for all $y_1$ and $y_2$ in $\mathbb{Y}$ there exists a functional $F \in \Lambda$ such that $Fy_1 \neq Fy_2$. Let $\mathcal{U}_\Lambda$ be the smallest topology in $\mathbb{Y}$ under which all $F \in \Lambda$ are continuous. The family of subsets of $\mathbb{Y}$ of the form

$$U(y_0, F, \epsilon) = \{y \in \mathbb{Y}; |Fy - Fy_0| < \epsilon\}$$

where $y_0 \in \mathbb{Y}, F \in \mathbb{Y}$· and $\epsilon > 0$ are given, is a subbase of this topology. Note that $\mathcal{U}_\Lambda$ is a Hausdorff topology for $\Lambda$ separates points of $\mathbb{Y}$. By definition, $\mathcal{U}_\Lambda$ is smaller than the strong topology in $\mathbb{Y}$ (generated by the open balls). Note also that for any $U(y_0, F, \epsilon)$ there exists a $\delta$ such that the open ball $B(y_0, \epsilon) \subset U(y_0, F, \epsilon)$.

A sequence $y_n$ converges to a $y \in \mathbb{Y}$ in the $\mathcal{U}_\Lambda$ topology iff

$$\lim_{n \to \infty} Fy_n = Fy$$

for all $F \in \Lambda$. Of course, the same can be said of a net. Since $\Lambda$ separates points of $\mathbb{Y}$, there may be only one $y$ like that. The fact that the strong topology is stronger than $\mathcal{U}_\Lambda$ is now expressed in the implication:

$$\lim_{n \to \infty} y_n = y \text{ (strongly)} \Rightarrow \lim_{n \to \infty} y_n = y \text{ (in } \mathcal{U}_\Lambda),$$

which can also be verified as follows: for all $F \in \mathcal{U}_\Lambda$, $|Fy_n - Fy| \leq \|F\|\|y_n - y\|$.

All such topologies are termed **weak topologies**. From among many weak topologies we will discuss two of (probably) greatest importance: the weak topology and the weak\* topology (or the weak-star topology).

5.4.2 *The weak topology*    By the Hahn–Banach Theorem, $\mathbb{Y}^*$ is a convergence determining set. The resulting topology $\mathcal{U}_\Lambda$ in $\mathbb{Y}$ is called **the weak topology.**

In general, weakly convergent sequences are not strongly convergent.

In fact, the only known infinite-dimensional Banach space where weakly convergent sequences are strongly convergent is $l^1$ (see 5.4.9). As an example consider the sequence $e_k = (\delta_{n,k})_{n \geq 1}$ in $c_0$. Certainly, $\|e_k - e_l\| = 1$ for $k \neq l$, and so $e_k$ does not converge in the strong topology, as $k \to \infty$. However, any $F \in c_0^*$ is of the form (5.7). Hence, $Fe_k = \alpha_k$ tends to 0 as $k \to \infty$, which proves that $e_k$ converges weakly to the zero functional.

**5.4.3 Exercise** Let $y_n \in L^1(\mathbb{R})$ be represented by $y_n(\tau) = \frac{1}{\sqrt{2\pi n}} e^{-\frac{\tau^2}{2n}}$. Show that $x_n$ converges weakly, but not strongly, to 0.

**5.4.4 Example** Suppose that for any $F \in \mathbb{Y}^*$, $Fy_n$ converges. Does it imply that $y_n$ converges weakly? The answer is in the negative; in particular $\mathbb{Y}$ equipped with the weak topology is not complete. To see that consider an example of the space $C(S)$ of continuous functions on a compact space, and assume that one may construct a sequence of equibounded functions $y_n \in C(S)$, $\sup_{n \geq 1} \|y_n\| < \infty$, that converges pointwise to a function $y \notin C(S)$ (as in 1.2.20 or 2.2.44 for instance; one may also take $S = [0,1]$ and $y_n(s) = s^n$). For any $F \in C(S)^*$ there exists a measure $\mu$ on $S$ such that $Fy_n = \int_S y_n \, d\mu$ which converges to $\int_S y \, d\mu$ by the Lebesgue Dominated Convergence Theorem. On the other hand, $y_n$ may not converge to a $y_0$ in $C(S)$ because this would imply $\int y \, d\mu = \int y_0 \, d\mu$ for all $\mu \in C(S)^*$, and, consequently, taking $\mu = \delta_p$, $y(p) = y_0(p)$ for all $p \in S$, which we know is impossible.

**5.4.5 Exercise** Let $y_n = \sum_{k=1}^n e_k$, $e_k = (\delta_{n,k})_{n \geq 1} \in c_0$. Show that $Fy_n$ converges for all $F \in c_0^*$, and yet $y_n$ does not converge weakly in $c_0$.

Let us continue with examples of criteria for weak convergence; one is general; the other one relates to the space $L^p, p > 1$. In both cases we assume that a sequence $y_n$ to be proven weakly convergent is bounded. This is a natural assumption, for in 7.1.8 we prove that weakly convergent sequences are bounded.

**5.4.6 Proposition** Let $y_n$ be a bounded sequence in a Banach space $\mathbb{Y}$, that $\mathbb{Y}_0^* \subset \mathbb{Y}^*$ is linearly dense in $\mathbb{Y}^*$, and that $\lim_{n \to \infty} Fy_n = Fy$, for $F \in \mathbb{Y}_0^*$. Then $y_n$ converges weakly to $y$.

*Proof* Let $F \in \mathbb{Y}^*$. Fix $\epsilon > 0$. There exists a linear combination $G = \sum_{i=1}^k \alpha_i F_i$ of elements of $\mathbb{Y}_0^*$ such that $\|F - G\|_{\mathbb{Y}^*} < \frac{\epsilon}{3M}$ where $M =$

$\sup_{n\geq 1} \|y_n\| \vee \|y\|$. Of course, $\lim_{n\to\infty} Gy_n = Gy$. Let $n_0$ be large enough so that $|G(y_n - y)| < \frac{\epsilon}{3}$, for $n \geq n_0$. Then

$$|Fy_n - Fy| \leq |(F - G)y_n| + |G(y_n - y)| + |(G - F)y|$$
$$\leq \frac{\epsilon}{3M}\|y_n\| + \frac{\epsilon}{3} + \frac{\epsilon}{3M}\|y\| \leq \epsilon.$$

$\square$

**5.4.7 Example**    Assume that $(\Omega, \mathcal{F}, \mu)$ is a $\sigma$-finite measure space. A sequence $y_n$ of elements of $\in L^p(\Omega, \mathcal{F}, \mu), p > 1$ is weakly convergent iff (a) $y_n$ is bounded and (b) the numerical sequence $\int_A y_n \, d\mu$ converges for all measurable sets $A$ with finite measure.

*Proof*  Necessity of (a) was discussed above, and necessity of (b) follows from the fact that for any $A$ with finite measure $Fy = \int_A y \, d\mu$ is a bounded linear functional on $L^p(\Omega, \mathcal{F}, \mu)$, because by the Hölder inequality

$$\|Fy\|_{L^p} \leq \left(\int 1_A \, d\mu\right)^{\frac{1}{q}} \|y\|_{L^q} = \mu(A)^{\frac{1}{q}} \|y\|_{L^q}.$$

To prove sufficiency recall that the set of indicator functions $1_A$ where $A$ is of finite measure is linearly dense in $L^q(\Omega, \mathcal{F}, \mu)$ (see 2.2.39). Hence, arguing as in 5.4.6 one may show that the sequence $\int xy_n \, d\mu$ converges for all $x \in L^q(\Omega, \mathcal{F}, \mu)$, by showing that $\int xy_n \, d\mu$ is a Cauchy sequence. Let $Hx = \lim_{n\to\infty} \int xy_n \, d\mu$. Certainly, $H$ is linear and

$$|Hx| \leq \|x\|_{L^q(\Omega, \mathcal{F}, \mu)} \sup_{n\in\mathbb{N}} \|y_n\|_{L^p(\Omega, \mathcal{F}, \mu)}.$$

Hence, $H \in (L^p)^*$, and by 5.2.19 there exists a $y \in L^q$ such that $Hx = \int xy \, d\mu = \lim_{n\to\infty} \int xy_n \, d\mu$. $\square$

**5.4.8 Exercise**    The argument from 5.4.7 proves that $L^p(\Omega, \mathcal{F}, \mu)$ has a property that if for some bounded sequence $y_n$ and all linear functionals $F$ on this space, the numerical sequence $Fy_n$ converges, then $y_n$ converges weakly to some $y$. The property of $L^p$ that makes the proof work is that it is reflexive. A Banach space $\mathbb{X}$ is said to be reflexive iff it is isometrically isomorphic to its second dual, i.e. if for any functional $x^{**}$ on $\mathbb{X}^*$ there exists an $x \in \mathbb{X}$ such that $x^{**}(F) = F(x)$ for all $F \in \mathbb{X}^*$. State and prove the appropriate result on weak convergence in reflexive Banach spaces.

**5.4.9 Weak and strong topologies are equivalent in $l^1$**     By linearity, it
suffices to show that if a sequence $x_k, k \geq 1$ of elements of $l^1$ converges
weakly to 0, then it converges strongly, as well. Suppose that it is not
so. Then, $\|x_k\|$ does not converge to 0. Any sequence of non-negative
numbers that does not converge to 0 contains a positive subsequence
converging to a positive number. Moreover, a subsequence of a weakly
convergent sequence converges weakly. Hence, without loss of generality,
we may assume that $\lim_{k\to\infty} \|x_k\| = r > 0$, and that $\|x_k\| \neq 0$. Taking
$y_k = \frac{1}{\|x_k\|} x_k$ we obtain a sequence $y_k = (\eta_{k,n})_{n\geq 1}$ converging weakly to
zero and such that $\|y_k\| = \sum_{n=1}^{\infty} |\eta_{k,n}| = 1$. We will show that such a
sequence may not exist.

By 5.2.3, for any bounded $(\alpha_n)_{n\geq 1}$ we have $\lim_{k\to\infty} \sum_{n=1}^{\infty} \alpha_n \eta_{k,n} = 0$.
In particular, taking $\alpha_n = \delta_{l,n}, n \geq 1$ for $l \geq 1$ we see that

$$\lim_{k\to\infty} \eta_{k,l} = 0, \quad l \geq 1. \tag{5.19}$$

We will define two sequences $k_i, i \geq 1$, and $n_i, i \geq 1$, of integers induc-
tively. First we put $k_1 = 1$ and choose $n_1$ so that $\sum_{n=1}^{n_1} |\eta_{1,n}| \geq \frac{3}{5}$. By
(5.19), having chosen $k_i$ and $n_i$ we may choose $k_{i+1}$ large enough to have
$k_{i+1} > k_i$ and $\sum_{n=1}^{n_i} |\eta_{k_{i+1},n}| < \frac{1}{5}$ and then, since $\|y_k\| = 1$ for all $k \geq 1$,
we may choose an $n_{i+1}$ so that $\sum_{n=n_i+1}^{n_{i+1}} |\eta_{k_{i+1},n}| > \frac{3}{5}$.

Now, define

$$\alpha_n = \text{sgn}\eta_{k_i,n}, \quad \text{for} \quad n \in A_i := \{n_{i-1}+1, ..., n_i\}$$

where $n_0 := 0$, and let $F$ be a continuous linear functional on $l^\infty$ related
to this bounded sequence. Then,

$$Fx_{k_i} = \sum_{n\in A_i} \alpha_n \eta_{k_i,n} + \sum_{n\notin A_i} \alpha_n \eta_{k_i,n} = \sum_{n\in A_i} |\eta_{k_i,n}| + \sum_{n\notin A_i} \alpha_n \eta_{k_i,n}$$

$$\geq \sum_{n\in A_i} |\eta_{k_i,n}| - \sum_{n\notin A_i} |\eta_{k_i,n}| = 2\sum_{n\in A_i} |\eta_{k_i,n}| - 1 > \frac{1}{5},$$

contrary to the fact that $x_{k_i}$ converges weakly to 0, as $i \to \infty$.

**5.4.10 Weak\* topology**     Another important example of a convergence
determining set arises if $\mathbb{Y}$ itself is a dual space of a Banach space, say $\mathbb{X}$.
In such a case, all elements of $\mathbb{Y}$ are functionals on $\mathbb{X}$ and we may consider
the set of functionals on $\mathbb{Y}$ that are of the form $y \to y(x)$ for some fixed
$x \in \mathbb{X}$. By the Hahn–Banach Theorem this is a convergence determining
set. The resulting topology in $\mathbb{Y}$ is called the **weak\* topology**. Note
that the sets $U_{y_0,x,\epsilon} = \{y \in \mathbb{Y} | |y(x) - y_0(x)| < \epsilon\}$ form the subbase
of weak\* topology. A sequence $y_n$ converges to $y$ in this topology iff

$\lim_{n\to\infty} y_n(x) = y(x)$ for all $x \in \mathbb{X}$. The most important example of such a topology is the case where $\mathbb{X} = C(S)$ for some compact space $S$. Then $\mathbb{Y} = \mathbb{X}^*$ is the space of Borel measures on $S$ and $\mu_n \in \mathbb{Y}$ converges to $\mu$ iff

$$\lim_{n\to\infty} \int x \, d\mu_n = \lim_{n\to\infty} \int x \, d\mu,$$

for all $x \in C(S)$. We note here that weak\* topology is weaker than the weak topology – see the examples below.

**5.4.11 Example** Let $S = [0,1]$. If $S \ni p_n \to p$, as $n \to \infty$, then $\delta_{p_n}$ converges to $\delta_p$ in the weak\* topology but not in the strong or weak topology.

**5.4.12 *Weak\* convergence to the Dirac measure at a point*** Establishing weak\* convergence of, say, a sequence $\mu_n$ of probability measures is particularly simple if the limit measure is concentrated in a single point, say $s_0$, for in such a case it is enough to show that for any neighborhood $V$ of $s_0$, $\lim_{n\to\infty} \mu_n(V^{\complement}) = 0$. Indeed, for an arbitrary continuous $x$ on $S$, given $\epsilon > 0$ we choose a neighborhood $V = V(\epsilon)$ of $p_0$ such that $|x(p) - x(p_0)| < \frac{\epsilon}{2}$ for $p \in V$. Next, for $n$ sufficiently large, we have $\mu_n(V^{\complement}) \leq \frac{\epsilon}{4\|x\|}$. Since for all $n$, $\int x \, d\delta_{p_0} = x(p_0) = \int x(p_0)\mu_n(\,dp)$, we have

$$\left| \int x \, d\mu_n - \int x \, d\delta_{s_0} \right| \leq (\int_V + \int_{V^{\complement}})|x(p) - x(p_0)|\mu_n(\,dp)$$

$$\leq \int_V |x(p) - x(p_0)|\mu_n(\,dp) + 2\|x\|\mu_n(V^{\complement}),$$

which for $V$ and $n$ described above is less than $\frac{\epsilon}{2} + \frac{\epsilon}{2}$, as desired.

Certainly, this result remains true for nets as well. For example, the probability measures $\mu_r$ on the unit circle $\mathbf{C} = \{z \in \mathbb{C}; |z| = 1\}$ with densities being Poisson kernels $p_r, 0 \leq r < 1$ (see 1.2.29) converge, as $r \to 1$, to $\delta_1$ in the weak\* topology. To show this, we note that for any $\delta > 0$, setting $V_\delta = \{e^{i\alpha} \in \mathbf{C}; |\alpha| < \delta\}$, we have

$$\mu_r(V_\delta^{\complement}) = \frac{1}{\pi} \int_\delta^\pi \frac{1 - r^2}{1 - 2r\cos\alpha + r^2} \, d\alpha \leq \frac{\pi - \delta}{\pi} \frac{1 - r^2}{1 - 2r\cos\delta + r^2} \xrightarrow[r\to 1]{} 0.$$

This implies that the same is true for any open neighborhood $V$ of 1 and proves our claim by the previous remarks.

As another example note that in 2.3.29 we have actually proven that the distributions of variables $X_n/n$ tend in weak\* topology to $\delta_s$.

**5.4.13 Example**    In Example 1.4.14, we have actually proved that for any $n$ the distributions $\mu_{n,k}$ converge, as $k \to \infty$, in the weak* topology to the Lebesgue measure on $[0, 1]$. Taking $f(s) = 0$ if $s$ is rational and 1 otherwise we see, however, that $\int f \, d\mu_{n,k}$ does not converge to $\int f \, d leb$, since measures $\mu_{n,k}$ are concentrated on rational numbers. Thus, $\mu_{n,k}$ do not converge weakly or strongly to $leb$.

**5.4.14 Example**    Let $x_k = (\delta_{k,n})_{n \geq 1} \in l^1, k \geq 1$. By 5.2.1, $x_k$ converges in the weak* topology to 0. Since $\|x_k\| = 1, k \geq 1$, however, it cannot converge to zero strongly, and hence does not converge strongly to anything at all. By 5.4.9, this shows in particular that in $l^1$ the weak* topology is strictly weaker than the weak topology. This can be seen from 5.2.3, as well.

**5.4.15** *Measures escaping to infinity*    If $S$ is locally compact, the space $C_0(S)$ is often not the best choice of test functions to study the weak* convergence of measures on $S$. It is more convenient to treat measures on $S$ as measures on a (say, one-point) compactification $\overline{S}$ of $S$ and take $C(\overline{S})$ as the set of test functions. The reason is that it may happen that some mass of involved measures escapes "to infinity", as in the example where $S = \mathbb{R}^+$ and $\mu_n = \frac{1}{2}\delta_0 + \frac{1}{2}\delta_n$. In this case, for any $x \in C_0(\mathbb{R})$, $\int x \, d\mu_n$ converges to $\frac{1}{2}x(0)$, so that $\mu_n$ as functionals on $C_0(\mathbb{R}^+)$ converge in the weak* topology to an improper distribution $\frac{1}{2}\delta_0$. In this approach it is unclear what happened with the missing mass. Taking an $x \in C(\overline{\mathbb{R}^+}) = C([0, \infty])$ clarifies the situation, because for such an $x$ we see that $\int x \, d\mu_n$ converges to $\frac{1}{2}x(0) + \frac{1}{2}x(\infty)$, and so $\mu_n$ converges to $\frac{1}{2}\delta_0 + \frac{1}{2}\delta_\infty$. Working with compactification of $S$ instead of $S$ itself helps avoid misunderstandings especially when it is not so clear whether and how much measure escapes to infinity. If we work only with probability measures, an equivalent approach is to check that the limiting measure is a probability measure and in the case it is not, to find out what happened with the missing mass. (See also 5.7.12.)

**5.4.16 Example**    Let $X_n$ be geometric random variables with parameters $p_n$ respectively. If $p_n \to 0$ and $np_n \to a > 0$, as $n \to \infty$, then the distribution $\mu_n$ of $\frac{1}{n}X_n$ converges in the weak* topology to the exponential distribution.

*Proof* Let $x$ be a member of $C(\overline{\mathbb{R}^+}) = C([0, \infty])$. Then

$$\int_{\mathbb{R}^+} x \, d\mu_n = \sum_{k=0}^{\infty} x \left( \frac{k}{n} \right) p_n q_n^k = n p_n \int_0^{\infty} x \left( \frac{[nt]}{n} \right) q_n^{[nt]} \, dt$$

where $[nt]$ denotes the integer part of $nt$. Since

$$(1 - p_n)^{nt} \leq (1 - p_n)^{[nt]} \leq (1 - p_n)^{nt-1} \tag{5.20}$$

and extreme terms in this inequality converge to $e^{-at}$ (since $(1 - p_n)^{\frac{1}{p_n}}$ converges to $e^{-1}$), so does the $(1 - p_n)^{[nt]}$. Similarly, $\frac{[nt]}{n} \to t$, and so the integrand converges pointwise to $ae^{-at}x(t)$. Using the right inequality in (5.20) one proves that the convergence is actually dominated and the Lebesgue theorem applies to show that $\int_{\mathbb{R}^+} x \, d\mu_n$ converges to $\int_0^{\infty} ae^{-at}x(t) \, dt$. $\qquad \square$

**5.4.17** *The role of dense sets*    The weak\* convergence of functionals may be viewed as a special case of strong convergence of operators where operators have scalar values. Therefore, all theorems concerning strong convergence of operators apply to weak\* convergence. In particular, one may use 2.3.34, especially if we deal with weak\* convergence of probability measures, for then the assumption of equiboundedness is automatically satisfied. As an example let us consider the measures from 5.4.16. By 2.3.31 the functions $e_\lambda(\tau) = e^{-\lambda\tau}$, $\lambda \geq 0$, form a linearly dense subset of $C(\overline{\mathbb{R}^+})$ and so to prove 5.4.16 it suffices to show that $\int_0^{\infty} e_\lambda \, d\mu_n$ converges to $a \int_0^{\infty} e^{-\lambda t}e^{-at} \, dt = \frac{a}{\lambda+a}$ for all $\lambda \geq 0$. On the other hand,

$$\int_0^{\infty} e_\lambda \, d\mu_n = \sum_{k=0}^{\infty} p_n q_n^k e^{-\frac{\lambda k}{n}} = \frac{p_n}{1 - e^{-\frac{\lambda}{n}}(1 - p_n)} \frac{n}{n}$$

which converges to the desired limit since $n(1 - e^{-\frac{\lambda}{n}})$ converges to $\lambda$.

The following lemma serves as a very useful tool in proving weak\* convergence of measures. We will use it in particular in establishing the Central Limit Theorem, to be presented in the next section.

**5.4.18 Lemma**    *A sequence $\mu_n$ of probability measures on $\mathbb{R}$ converges to a probability measure $\mu$ in the weak\* topology iff the corresponding operators $T_{\mu_n}$ in $C[-\infty, \infty]$ converge strongly to $T_\mu$.*

*Proof* The "if" part of this lemma is immediate. To prove the "only if" part note that by assumption, for any $x \in C[-\infty, \infty]$ any any $\tau \in$

$(-\infty, \infty)$, there exists the limit of $y_n(\tau) = \int_\mathbb{R} x(\tau - \sigma)\mu(\,d\sigma)$, as $n \to \infty$, and equals $y(\tau) = \int_\mathbb{R} x(\tau - \sigma)\mu(\,d\sigma)$. Moreover, $\lim_{n\to\infty} y_n(\pm\infty) = x(\pm\infty) = y(\pm\infty)$. We will prove that $y_n$ converges to $y$ uniformly, i.e. strongly in $C[-\infty, \infty]$.

We claim that the family $y_n$ is bounded (by $\|x\|$) and equicontinuous on $[-\infty, \infty]$, so that the assumptions of the well-known Arzela–Ascoli Theorem (see e.g. [22] or 5.7.17, below) are satisfied. To see that this implies our result assume that $y_n$ does not converge to $y$ strongly, and choose a subsequence that stays at some distance $\epsilon > 0$ from $y$. By the Arzela–Ascoli Theorem, there exists a subsequence of our subsequence that converges uniformly to some $z \in C[-\infty, \infty]$. Being chosen from $y_n$ this subsequence must also converge (pointwise) to $y$, implying that $z = y$, a contradiction.

It remains to prove the claim. For a given $\epsilon > 0$, a $\delta > 0$ may be chosen so that $|x(\sigma) - x(\sigma')| < \epsilon$ provided $|\sigma - \sigma'| < \delta, \sigma, \sigma' \in \mathbb{R}$. Hence, for any $\tau \in \mathbb{R}$ and $|h| < \delta$ we also have $|y_n(\tau + h) - y_n(\tau)| \leq \int_\mathbb{R} |x(\tau + h - \sigma) - x(\tau - \sigma)|\mu(\,d\sigma) < \epsilon$, proving that $y_n, n \geq 1$ is equicontinuous at $\tau \in \mathbb{R}$. To prove that it is equicontinuous at $\infty$ we first take a $T > 0$ and define $x_k \in C[-\infty, \infty], k \geq 1$ as $x_k(\tau) = \frac{1}{1 + k\max\{T - \tau, 0\}}$. Then, $\lim_{k\to\infty} x_k(\tau) = 1_{[T,\infty)}(\tau), \tau \in \mathbb{R}$. Hence

$$\limsup_{n\to\infty} \mu_n[T, \infty) \leq \lim_{n\to\infty} \int_\mathbb{R} x_k \,d\mu_n = \int_\mathbb{R} x_k \,d\mu_n \underset{k\to\infty}{\longrightarrow} \mu[T, \infty).$$

This implies that given an $\epsilon > 0$ we may choose a $T > 0$ so that $\mu_n[T, \infty) < \epsilon$, for sufficiently large $n$. Since such a $T$ may be chosen for each $n \geq 1$ individually, as well, and $x$ belongs to $C[-\infty, \infty]$, we may choose a $T$ so that $\mu_n[T, \infty) < \epsilon$ for all $n \geq 1$ and $|x(\tau) - x(\infty)| < \epsilon$, for $\tau > T$. Now, for $\tau > 2T$,

$$|y_n(\infty) - y_n(\tau)| \leq \int_{\sigma \leq T} + \int_{\sigma > T} |x(\infty) - x(\tau - \sigma)|\mu_n(\,d\sigma) \leq \epsilon + 2\|x\|\epsilon$$

proving that $y_n$ are equicontinuous at $\infty$. The case of $-\infty$ is treated in the same way. □

**5.4.19 Remark**   In the above proof, to use the Arzela–Ascoli Theorem, it was crucial to show that $y_n$ are uniformly continuous on $[-\infty, \infty]$ and not just on $\mathbb{R}$. Note that for a given $x \in C[-\infty, \infty]$, the functions $y_n(\tau) = x(n + \tau)$ are equicontinuous in $\mathbb{R}$ but not in $[-\infty, \infty]$. As a result, the Arzela–Ascoli Theorem does not apply, even though $y_n$ are also bounded. In fact, taking non-zero $x$ with support contained in $[0, 1]$, we have $\|y_n - y_m\| = \|x\|, n \neq m$, so that $y_n, n \geq 1$ cannot have a converging subsequence.

**5.4.20 Corollary**    Let $\mu_n, n \geq 1$ and $\nu_n, n \geq 1$ be two sequences of probability measures on $\mathbb{R}$ converging weakly to probability measures $\mu$ and $\nu$, respectively. Then, the measures $\mu_n * \nu_n$ converge weakly to $\mu * \nu$.

*Proof*  This follows directly from 5.4.18, $\|T_{\mu_n}\| = \|T_\nu\| = 1$ and the triangle inequality applied to $T_{\mu_n * \nu_n} x - T_{\mu * \nu} x = T_{\mu_n}(T_{\nu_n} - T_\nu)x + T_\nu(T_{\mu_n} - T_\mu)x, x \in C[-\infty, \infty]$.    $\square$

**5.4.21 Remark**    In probability theory one rarely considers convergence of measures in the weak topology. Although it may sound strange, the reason for this is that *the weak topology is still too strong!* On the other hand, weak* topology is used quite often, but for historical reasons, convergence in this topology is termed the *weak convergence*. (Sometimes *narrow* convergence, from the French *étroite*.) **In what follows we will adhere to this custom.** This should not lead to misunderstandings as the "real" weak convergence will not concern us any more.

## 5.5 The Central Limit Theorem

By far the most important example of weak convergence is the Central Limit Theorem. For its proof we need the following lemma, in which the lack on dependence of the limit on $X$ is of greatest interest; the fact that the second derivative in the limit points out to the normal distribution will become clear in Chapters 7 and 8 (see 8.4.18 in particular).

**5.5.1 Lemma**    Let $X$ be square integrable with $E\,X = 0$ and $E\,X^2 = 1$. Also, let $a_n, n \geq 1$ be a sequence of positive numbers such that $\lim_{n \to \infty} a_n = 0$. Then, for any $x \in \mathcal{D}$, the set of twice differentiable functions $x \in C[-\infty, \infty]$ with $x'' \in C[-\infty, \infty]$, the limit of $\frac{1}{a_n^2}(T_{a_n X} x - x)$ exists and *does not depend on* $X$. In fact it equals $\frac{1}{2} x''$.

*Proof*  By the Taylor formula, for a twice differentiable $x$, and numbers $\tau$ and $\varsigma$,

$$x(\tau + \varsigma) = x(\tau) + \varsigma x'(\tau) + \frac{\varsigma^2}{2} x''(\tau + \theta \varsigma), \qquad (5.21)$$

where $0 \leq \theta \leq 1$ depends on $\tau$ and $\varsigma$ (and $x$). Thus,

$$
\frac{1}{a_n^2} [T_{a_n X} x(\tau) - x(\tau)] = \frac{1}{a_n^2} E\left[x(\tau + a_n X) - x(\tau)\right]
$$
$$
= \frac{1}{a_n} x'(\tau) E X + \frac{1}{2} E\left[X^2 x''(\tau + \theta a_n X)\right]
$$
$$
= \frac{1}{2} E\left[X^2 x''(\tau + \theta a_n X)\right], \qquad (5.22)
$$

for $E X = 0$.† Since $E X^2 = 1$,

$$
\left| \frac{1}{a_n^2} (T_{a_n X} x - x)(\tau) - \frac{1}{2} x''(\tau) \right| = \left| \frac{1}{2} E X^2 \left(x''(\tau + \theta a_n X) - x''(\tau)\right) \right|.
$$

For $x \in \mathcal{D}$, and $\epsilon > 0$, one may choose a $\delta$ such that $|x''(\tau + \varsigma) - x''(\tau)| < \epsilon$, provided $|\varsigma| < \delta$. Calculating the last expectation on the set where $|X| \geq \frac{\delta}{a_n}$ and its complement separately we get the estimate

$$
\left\| \frac{1}{a_n^2} (T_{a_n X} x - x) - \frac{1}{2} x'' \right\| \leq \frac{1}{2} \|x''\| E X^2 1_{\{|X| \geq \frac{\delta}{a_n}\}} + \frac{1}{2}\epsilon. \qquad (5.23)
$$

Since $\mathbb{P}\{|X| \geq \frac{\delta}{a_n}\} \to 0$ as $n \to \infty$ we are done by the Lebesgue Dominated Convergence Theorem. $\qquad\square$

5.5.2 *The Central Limit Theorem*     The Central Limit Theorem in its classical form says that

*if $X_n, n \geq 1$ is a sequence of i.i.d. (independent, identically distributed) random variables with expected value $m$ and variance $\sigma^2 > 0$, then*

$$
\frac{1}{\sqrt{n\sigma^2}} \sum_{k=1}^{n} (X_k - m)
$$

*converges weakly to the standard normal distribution.*

*Proof (of CLT)* Without loss of generality we may assume that $m = 0$ and $\sigma^2 = 1$, since the general case may be reduced to this one. Let $T_n = T_{\frac{1}{\sqrt{n}} X}$ where $X$ is any of the variables $X_n, n \geq 1$. By the independence assumption, $T_{\frac{1}{\sqrt{n}} \sum_{k=1}^{n} X_k} = T_n^n$ and we need to show that $T_n^n$ converges strongly to $T_Z$ where $Z$ is a standard normal variable. The set $\mathcal{D}$ of twice

---

† Let us observe here that although it is not obvious that the map $\omega \mapsto \theta(\tau, \frac{1}{\sqrt{n}} X)$ is measurable, measurability of the function $\omega \mapsto X^2 x''[\tau + \theta(\tau, \frac{1}{\sqrt{n}} X) \frac{1}{\sqrt{n}} X]$ is assured by the fact that this function equals $x(\tau + \frac{1}{\sqrt{n}} X) - x(\tau) - \frac{1}{\sqrt{n}} X x'(\tau)$.

differentiable functions $x$ with $x'' \in C[-\infty, \infty]$ is dense in $C[-\infty, \infty]$; hence it suffices to show convergence for $x \in \mathcal{D}$. Now, by (2.10),

$$\|T_n^n x - T_Z x\| = \|T_n^n x - T_{\frac{1}{\sqrt{n}} Z}^n x\| \le n \|T_n x - T_{\frac{1}{\sqrt{n}} Z} x\|$$

$$\le \|n(T_{\frac{1}{\sqrt{n}} X} x - x) - n(T_{\frac{1}{\sqrt{n}} Z} x - x)\|,$$

which by Lemma 5.5.1 converges to 0, as $n \to \infty$. $\qquad\square$

5.5.3 *The Lindeberg condition* A sequence $X_n, n \ge 1$ of independent (not necessarily identically distributed) square integrable random variables is said to satisfy **the Lindeberg condition** iff for every $\delta > 0$,

$$\frac{1}{s_n^2} \sum_{k=1}^n E\left(X_k - \mu_k\right)^2 1_{\{|X_k - \mu_k| > \delta s_n\}}$$

tends to 0, as $n \to \infty$, where $\mu_k = E X_k$ and $s_n^2 = \sum_{k=1}^n \sigma_k^2$, $\sigma_k^2 = D^2 X_k > 0$. In what follows we will use $E \mathcal{X}_n^2 1_{\{|\mathcal{X}_n| > \delta s_n\}}$ as a shorthand for the sum above. Note that i.i.d. variables satisfy the Lindeberg condition; for such variables we have $\frac{1}{s_n^2} E \mathcal{X}_n^2 1_{\{|\mathcal{X}_n| > \delta s_n\}} = \frac{1}{\sigma_1^2} E (X_1 - \mu)^2 1_{\{|X_1 - \mu_1| > \sqrt{n} \sigma_1 \delta\}}$ which certainly converges to zero, as $n \to \infty$.

The celebrated Lindeberg–Feller Theorem says that *the Lindeberg condition holds iff* $\lim_{n \to \infty} \frac{\max(\sigma_1^2, \dots, \sigma_n^2)}{s_n^2} = 0$ *and the sequence* $\frac{1}{s_n} \sum_{k=1}^n X_k$ *converges weakly to the standard normal distribution.*

We will prove merely the "only if" part which is perhaps less remarkable but more applicable.

*Proof* The proof is a modification of the proof of 5.5.2. As before, we assume without loss of generality that $E X_n = 0, n \ge 0$. By the independence assumption $T_{\frac{1}{s_n} \sum_{k=1}^n X_k} = T_{\frac{1}{s_n} X_n} \dots T_{\frac{1}{s_n} X_1}$. Analogously, we may write $T_Z$ where $Z$ is standard normal as $T_{\frac{1}{s_n} Z_n} \dots T_{\frac{1}{s_n} Z_1}$ where $Z_k$ is normal with zero mean and variance $\sigma_k^2$. Now, using (2.9), $\|T_{\frac{1}{s_n} \sum_{k=1}^n X_k} - T_Z\| \le \sum_{k=1}^n \|T_{\frac{1}{s_n} X_k} - T_{\frac{1}{s_n} Z_k}\|$ and our task reduces to showing that for $x \in \mathcal{D}$,

$$\sum_{k=1}^n \|T_{\frac{1}{s_n} X_k} x - x - \frac{1}{2} \frac{\sigma_k^2}{s_n^2} x''\| + \sum_{k=1}^n \|T_{\frac{1}{s_n} Z_k} x - x - \frac{1}{2} \frac{\sigma_k^2}{s_n^2} x''\| \qquad (5.24)$$

converges to 0, as $n \to \infty$. Arguing as in (5.22) and (5.23), we obtain

$$\left\|\frac{s_n^2}{\sigma_k^2} [T_{\frac{1}{s_n} X_k} x - x] - \frac{1}{2} x''\right\| \le \frac{1}{2} \epsilon + \frac{1}{\sigma_k^2} \|x''\| E X_k^2 1_{\{|X_k| > s_n \delta\}},$$

where we chose $\delta$ in such a way that $|x''(\tau + \varsigma) - x''(\tau)| < \epsilon$, provided $|\varsigma| < \delta$. Multiplying both sides by $\frac{\sigma_k^2}{s_n^2}$ and summing from $k = 1$ to $k = n$,

$$\sum_{k=1}^{n} \left\| T_{\frac{1}{s_n} X_k} x - x - \frac{1}{2} \frac{\sigma_k^2}{s_n^2} x'' \right\| \le \frac{1}{2}\epsilon + \frac{1}{s_n^2} \|x''\| E \, \mathcal{X}_n^2 1_{\{|\mathcal{X}_n| > \delta s_n\}}.$$

This proves by the Lindeberg condition that the first sum in (5.24) converges to zero. Since the second sum has the same form as the first sum it suffices to show that $Z_n, n \ge 1$, satisfies the Lindeberg condition. Noting that $E \, Z_n^4 = 3\sigma_n^4$,

$$\frac{1}{s_n^2} \sum_{k=1}^{n} E \, Z_k^2 1_{\{|Z_k| > \delta s_n\}} \le \frac{1}{s_n^4 \epsilon^2} \sum_{k=1}^{n} E \, Z_k^4 \le \frac{3}{\epsilon^2} \frac{\max(\sigma_1^2, ..., \sigma_n^2)}{s_n^2}.$$

In Exercise 5.5.4 the reader will check that this last quantity converges to zero.  □

**5.5.4 Exercise**   Complete the proof above by showing that the Lindeberg condition implies $\lim_{n \to \infty} \frac{\max(\sigma_1^2, ..., \sigma_n^2)}{s_n^2} = 0$.

**5.5.5 Exercise**   Show that the Lyapunov condition

$$\lim_{n \to \infty} \frac{1}{s_n^{2+\alpha}} \sum_{k=1}^{n} E \, |X_k - \mu_k|^{2+\alpha} = 0$$

where $\alpha > 0$, and $s_n$ is defined as before, implies the Lindeberg condition.

## 5.6 Weak convergence in metric spaces

The assumption that the space $S$ where our probability measures are defined is compact (or locally compact) is quite restrictive and is not fulfilled in many important cases of interest. On the other hand, assuming just that $S$ is a topological space leads to an unnecessarily general class. The golden mean for probability seems to lie in **separable metric spaces**, or perhaps, **Polish spaces**. A Polish space is by definition a separable, complete metric space. We start with general metric spaces to specialize to Polish spaces later when needed. As an application of the theory developed here, in the next section we will give another proof of the existence of Brownian motion.

**5.6.1 Definition**　Let $(S, d)$ be a metric space, and let $BC(S)$ be the space of continuous (with respect to the metric $d$, of course) functions on $S$. A sequence $\mathbb{P}_n$ of Borel probability measures on $S$ is said to **converge weakly** to a Borel probability measure $\mathbb{P}$ on $S$ iff, for all $x \in BC(S)$,

$$\lim_{n \to \infty} \int_S x \, d\mathbb{P}_n = \int_S x \, d\mathbb{P}. \tag{5.25}$$

It is clear that this definition agrees with the one introduced in the previous section, as in the case where $S$ is both metric and compact, $BC(S)$ coincides with $C(S)$.

We will sometimes write $E_n x$ for $\int_S x \, d\mathbb{P}_n$ and $Ex$ for $\int_S x \, d\mathbb{P}$.

**5.6.2 Corollary**　Suppose $\mathbb{P}_n, n \geq 1$ is a sequence of Borel probability measures on $(S, d)$ and $f : S \to S'$, where $(S', d')$ is another metric space, is a continuous map. Then the transport measures $(\mathbb{P}_n)_f, n \geq 1$ on $S'$ converge weakly. The proof is immediate by the change of variables formula (1.6).

**5.6.3 *Portmanteau Theorem*** 　Let $\mathbb{P}$ and $\mathbb{P}_n, n \geq 1$ be probability measures on a metric space $(S, d)$. The following are equivalent:

(a)　$\mathbb{P}_n$ converge weakly to $\mathbb{P}$,
(b)　condition (5.25) holds for Lipschitz continuous $x$ with values in $[0, 1]$,
(c)　$\limsup_{n \to \infty} \mathbb{P}_n(F) \leq \mathbb{P}(F)$, for closed $F \subset S$,
(d)　$\liminf_{n \to \infty} \mathbb{P}_n(G) \geq \mathbb{P}(G)$, for open $G \subset S$,
(e)　$\lim_{n \to \infty} \mathbb{P}_n(B) = \mathbb{P}(B)$, for Borel $B$ with $\mu(\partial B) = 0$.

*Proof*　Recall that $\partial B = cl B \cap cl(S \setminus B)$.

Implication (a)$\Rightarrow$(b) is obvious. Assume (b) and for a closed $F$ and $s \in S$ define $d(p, F) := \inf_{q \in F} d(p, q)$. Note that $|d(p, F) - d(p', F)| \leq d(p, p')$ so that functions $x_k(p) = (1 + kd(p, F))^{-1}, k \geq 1$, are Lipschitz continuous (with Lipschitz constant $k$). Also, $\lim_{k \to \infty} x_k(p) = 1$ or $0$ according as $p \in F$ or $p \notin F$. This gives (c) by

$$\limsup_{n \to \infty} \mathbb{P}_n(F) \leq \limsup_{n \to \infty} E_n x_k = E \, x_k \xrightarrow[k \to \infty]{} \mathbb{P}(F),$$

the last relation following by the Monotone Convergence Theorem.

Taking complements we establish equivalence of (c) and (d). Next, that (c) and (d) imply (e) can be seen from

$$\limsup_{n \to \infty} \mathbb{P}_n(B) \leq \limsup_{n \to \infty} \mathbb{P}(cl\,B) \leq \mathbb{P}(cl\,B) = \mathbb{P}(\partial B) + \mathbb{P}(B^\circ) = \mathbb{P}(B^\circ)$$

$$\leq \liminf_{n \to \infty} \mathbb{P}_n(B^\circ) \leq \liminf_{n \to \infty} \mathbb{P}(B)$$

where the set $B^\circ := S \setminus cl(S \setminus B) = clB \setminus \partial B$ is open.

Finally, in proving that (e) implies (a), by linearity and the assumption that $\mathbb{P}$ and $\mathbb{P}_n$ are probability measures, it suffices to show 5.25 for continuous functions with values in $[0,1]$. Since $x$ is continuous, $cl\{x > t\} \subset \{x = t\}, t \geq 0$, and the sets $\{x = t\}$ are disjoint. Hence, the set $\{t \geq 0; \mathbb{P}\{x = t\}\}$, being countable, has Lebesgue measure zero. By (1.18) with $\beta = 1$, and the Dominated Convergence Theorem,

$$E_n x = \int_0^\infty \mathbb{P}_n\{x > t\}\,dt = \int_0^1 \mathbb{P}_n\{x > t\}\,dt$$

$$\underset{n \to \infty}{\longrightarrow} \int_0^1 \mathbb{P}\{x > t\}\,dt = \int_0^\infty \mathbb{P}\{x > t\}\,dt = E\,x,$$

as desired.    □

**5.6.4 Remark**    Note that we *assume a priori* that the limit measure in the above theorem is a probability measure.

**5.6.5** *The space of measures as a metric space*    If $S$ is separable, one may introduce a metric $D$ in the space $\mathbb{PM}(S)$ of probability measures on $S$ in such a way that $\lim_{n \to \infty} D(\mathbb{P}_n, \mathbb{P}) = 0$ iff $\mathbb{P}_n$ converges weakly to $\mathbb{P}$. In particular, in discussing convergence of measures it is justified to restrict our attention to sequences of measures (as opposed to general nets). The most famous metric of such a type is the **Prohorov–Lévy metric** $D_{\mathrm{PL}}$ defined as follows: $D_{\mathrm{PL}}(\mathbb{P}, \mathbb{P}^\sharp)$ is the infimum of those positive $\epsilon$ for which both $\mathbb{P}(A) \leq \mathbb{P}^\sharp(A^\epsilon) + \epsilon$, as well as $\mathbb{P}^\sharp(A) \leq \mathbb{P}(A^\epsilon) + \epsilon$, for any Borel subset $A$ of $S$. Here $A^\epsilon$ is the set of those $p \in S$ that lie within the $\epsilon$ distance from $A$, i.e. such that there is a $p' \in A$ such that $d(p, p') < \epsilon$. It turns out that if $S$ is a Polish space, then so is $(\mathbb{PM}(S), D_{\mathrm{PL}})$. This result is not only of special beauty, but also of importance, especially in the theory of point processes [24].

Another example of such a metric is the **Fortet–Mourier metric** $D_{\mathrm{FM}} : D_{\mathrm{FM}}(\mathbb{P}, \mathbb{P}^\sharp) = \sup |\int x\,d\mathbb{P} - \int x\,d\mathbb{P}^\sharp|$ where the supremum is taken over all $x \in BC(S)$ such that $|x(p) - x(p')| \leq d(p, p'), p, p' \in S$ and $\sup_{p \in S} |x(p)| \leq 1$.

These results are discussed in detail in many monographs, see e.g. Billingsley [5], Edgar [36], Ethier and Kurtz [38], Shiryaev [106]. A rich source of further information is Dudley [31] and Zolotarev [116].

**5.6.6** *Weak convergence in* $\mathbb{R}$ ...    For measures on $\mathbb{R}$ it suffices to check condition (e) of 5.6.3 for $B$ of the form $B = (-\infty, t], t \in \mathbb{R}$. In other words, it is enough to check convergence of cumulative distribution functions, $F_n(t) = \mathbb{P}_n(-\infty, t]$ to $F(t) = \mathbb{P}(-\infty, t]$ at every point $t$ where $\mathbb{P}\{t\} = 0$, i.e. at every point of continuity of $F$. To this end note first that our assumption implies obviously that (e) holds for all intervals $(a, b]$ with $a$ and $b$ being points of continuity of $F$. For the proof we will need the following properties of the class $\mathcal{I}$ of such intervals:

(i) $\mathcal{I}$ is a $\pi$-system;
(ii) condition (e) holds for all finite unions of elements of $\mathcal{I}$;
(iii) for every $s \in \mathbb{R}$, and $\epsilon > 0$ there is an interval $(a, b] \in \mathcal{I}$ such that $s \in (a, b)$, and $b - a < \epsilon$;

and the fact that $\mathbb{R}$ is separable, and hence satisfies the following

**Lindelöf property**: *any open cover of a subset of $\mathbb{R}$ contains a denumerable subcover.*

Condition (i) is obvious, (ii) follows by (i), induction argument and $\mathbb{P}_n(A \cup B) = \mathbb{P}_n(A) + \mathbb{P}_n(B) - \mathbb{P}_n(A \cap B)$. (iii) is true, since the set of $c \in \mathbb{R}$, with $\mathbb{P}\{c\} > 0$, is countable.

As for the Lindelöf property, consider open balls with centers at rational numbers and radii $\frac{1}{n}$. There are countably many balls like that and we can arrange them in a sequence $B_n, n \geq 1$. Since rational numbers form a dense set in $\mathbb{R}$, any $s \in \mathbb{R}$ belongs to at least one of $B_n, n \geq 1$. Now, let $U_\gamma, \gamma \in \Gamma$ be a cover of a set $A \subset \mathbb{R}$. To a $B_n$ assign one of the $U_\gamma, \gamma \in \Gamma$ containing it, if such a $U_\gamma$ exists. Since there are countably many balls, there are countably many sets $U_\gamma$ chosen in this process. We will show that their union covers $G$. Let $s$ belong to $G$; then there is a $\gamma$ such that $s \in U_\gamma$, and since $U_\gamma$ is open, there is an $n_0$ such that $s \in B_{n_0} \subset G$. The set of $\gamma$ such that $B_{n_0} \subset U_\gamma$ is non-empty and there is a $U_{\gamma_0}$ assigned to this $n_0$. We have $s \in B_{n_0} \subset U_{\gamma_0}$, which implies our claim.

Now, by (iii), any open $G$ is a union of intervals $(a, b)$, such that $(a, b] \in \mathcal{I}$ and $(a, b] \subset G$. By the Lindelöf property, we have $G = \bigcup_{k \geq 1}(a_k, b_k) = \bigcup_{k \geq 1}(a_k, b_k]$, for some $(a_k, b_k] \in \mathcal{I}, k \geq 1$. By assumption, for any integer $l$, $\liminf_{n \to \infty} \mathbb{P}_n(G) \geq \lim_{n \to \infty} \mathbb{P}_n(\bigcup_{k=1}^{l}(a_k, b_k]) = \mathbb{P}(\bigcup_{k=1}^{l}(a_k, b_k])$. Since for any $\epsilon > 0$ one may choose an $l$ such that $\mathbb{P}(\bigcup_{k=1}^{l}(a_k, b_k]) > \mathbb{P}(G) - \epsilon$, we have $\liminf_{n \to \infty} \mathbb{P}_n(G) > \mathbb{P}(G) - \epsilon$. But $\epsilon$ was arbitrary, and we conclude that (d) in 5.6.3 holds, as desired.

**5.6.7 Example** Here is the alternative proof of 5.4.16 using 5.6.6. If $X$ is exponential with parameter $a$ then for any $t \geq 0$, $\mathbb{P}\{X > t\} = \mathrm{e}^{-at}$. Taking complements we see that it suffices to show that $\mathbb{P}\{\frac{1}{n}X_n > t\}$ converges to $\mathrm{e}^{-at}$. Now, $\frac{1}{n}X_n > t$ iff $X_n > [nt]$. Therefore, the involved probability equals $\sum_{k=[nt]+1}^{\infty} p_n q_n^k = q_n^{[nt]+1}$. Arguing as in (5.20), we get our claim.

**5.6.8 ... and in $\mathbb{R}^k$** For probability measures $\mathbb{P}_n$ in $\mathbb{R}^k$, we have a result analogous to 5.6.6. That is, $\mathbb{P}_n$ converge weakly to a $\mathbb{P}$ if $F_n(\underline{a}) = \mathbb{P}_n\left(\prod_{i=1}^{k}(-\infty, a_i]\right)$ converges to $F(\underline{a}) = \mathbb{P}\left(\prod_{i=1}^{k}(-\infty, a_i]\right)$ for all $\underline{a} = (a_1, ..., a_k) \in \mathbb{R}^k$ with $\mathbb{P}\{\underline{a}\} = 0$. The proof is analogous to 5.6.6; first we show that our assumption implies that (e) in 5.6.3 holds for all **rectangles** $(\underline{a}, \underline{b}]$, (i.e. sets of $\underline{s} = (s_1, ..., s_k) \in \mathbb{R}^k$, such that $a_i < s_i \leq b_i$, $i = 1, ..., k$, where $\underline{a} = (a_1, ..., a_k)$ and similarly for $\underline{b}$), with $\mathbb{P}\{\underline{a}\} = \mathbb{P}\{\underline{b}\} = 0$. Then we show that the class $\mathcal{I}$ of such rectangles satisfies conditions (i)–(ii) of 5.6.6, and the following version of (iii): for any $\underline{s} \in \mathbb{R}^k$, and $\epsilon > 0$ there exist $(\underline{a}, \underline{b}]$ in this class such that the Euclidian distance between $\underline{a}$ and $\underline{b}$ is less than $\epsilon$. Since $\mathbb{R}^k$ is separable, the Lindelöf property completes the proof.

**5.6.9 Remark** We know from 1.2.20 that the values of a measure on the sets of the form $(-\infty, t]$ determine this measure. In 5.6.6 we proved that such sets also determine convergence of measures. We should not expect, however, that in general a collection of sets that determines a measure must also determine convergence – see [6]. In the same book it is shown that conditions (i)–(iii) can be generalized to give a nice criterion for convergence of probability measures in a (separable) metric space.

**5.6.10 Example** Probability measures on $\mathbb{R}^k$ may of course escape to infinity; and this may happen in various ways. In studying such phenomena we need to be careful to keep track of how much of the mass escapes and where it is being accumulated. The criterion given in 5.6.8 can be of assistance, if we use it in an intelligent way.

The following example originates from population genetics and describes the limit distribution of a pair $(X_t, Y_t)$, $t > 0$, of random variables, in which the first coordinate may be interpreted as a time (measured backwards) to the first common ancestor of two individuals taken from a large population (see [16, 17, 18]). The so-called effective population size $2N(\cdot)$ is supposed to be known as a function of time (the factor 2 is here for genetical reasons: individuals are in fact interpreted as chromosomes and chromosomes come in pairs). The larger $2N(\cdot)$ is, the longer time it takes to find the ancestor. The variable $X_t$ is truncated at a $t > 0$, and the second coordinate is defined as $Y_t = t - X_t$. Formally, $X_t$ is a random variable taking values in $[0, t]$ with distribution determined by $2N(\cdot)$ according to the formula $\mathbb{P}[X_t \geq u] = e^{-\int_{t-u}^{t} \frac{dv}{2N(v)}}$ (in particular: $\mathbb{P}[X = t] = e^{-\int_{0}^{t} \frac{dv}{2N(v)}}$). Let $\mathbb{P}_t$ be the distribution of the pair $(X_t, Y_t)$ in $\mathbb{R}^+ \times \mathbb{R}^+$. We are interested in the limit of $\mathbb{P}_t$ as $t \to \infty$ and consider the following cases:

(a) $\lim_{t \to \infty} N(t) = 0$,
(b) $\lim_{t \to \infty} N(t) = N, 0 < N < \infty$,
(c) $\lim_{t \to \infty} N(t) = \infty$, and $\int_{0}^{\infty} \frac{dt}{2N(t)} = \infty$,
(d) $\lim_{t \to \infty} N(t) = \infty$, and $\int_{0}^{\infty} \frac{dt}{2N(t)} < \infty$.

We will show that $\mathbb{P}_t$ converges weakly to a measure on $[0, \infty]^2$. In the cases (a) and (c), this measure is the Dirac measure at $\{0\} \times \{\infty\}$ and $\{\infty\} \times \{\infty\}$, respectively. In (b) and (d), it is the measure identified with the functionals on $C([0, \infty]^2)$ given by

$$x \to \frac{1}{2N} \int_{0}^{\infty} e^{-\frac{t}{2N}} x(t, \infty)\, dt,$$

$$x \to e^{-\int_{0}^{\infty} \frac{du}{2N(u)}} x(\infty, 0) + \int_{0}^{\infty} \frac{1}{2N(u)} e^{-\int_{u}^{\infty} \frac{dv}{2N(v)}} x(\infty, u)\, du,$$

respectively. These claims may be summarized in the form of the following table.

Table 5.1

| behavior of $N(t)$ | variable $X = X_\infty$ | variable $Y = Y_\infty$ |
|---|---|---|
| $\lim_{t \to \infty} N(t) = 0$ | $0$ | $\infty$ |
| $\lim_{t \to \infty} N(t) = N,$ $0 < N < \infty$ | exponential with parameter $2N$ | $\infty$ |
| $\lim_{t \to \infty} N(t) = \infty,$ $\int_{0}^{\infty} \frac{du}{2N(u)} = \infty$ | $\infty$ | $\infty$ |
| $\lim_{t \to \infty} N(t) = \infty,$ $\int_{0}^{\infty} \frac{du}{2N(u)} < \infty$ | $\infty$ | finite, $\mathbb{P}(Y > w) =$ $1 - e^{-\int_{w}^{\infty} \frac{du}{2N(u)}}$ $\mathbb{P}(Y = 0) = e^{-\int_{0}^{\infty} \frac{du}{2N(u)}}$ |

For the proof, note that for any $v, w \geq 0$ and $t$ so large that $t - w \geq v$,

$$\mathbb{P}_t[(v, \infty) \times (w, \infty)] = \mathbb{P}[v < X_t < t - w] = e^{-\int_{t-v}^{t} \frac{du}{2N(u)}} - e^{-\int_{w}^{t} \frac{du}{2N(u)}}.$$

Hence, $\lim_{t\to\infty} \mathbb{P}_t[(v,\infty)\times(w,\infty)]$ equals 0 in the case (a), for $v > 0$. It equals 1 in the case (a), for $v = 0$, and in the case (c). It equals $e^{-\frac{v}{2N}}$ in the case (b), and $1 - e^{-\int_u^\infty \frac{dv}{2N(v)}}$ in the case (d). This proves our claim in cases (b)–(d); to treat (a) we need to note additionally that $\lim_{t\to\infty} \mathbb{P}[X \le v, t - X > w] = \lim_{t\to\infty} \mathbb{P}[X \le v] = 1, v \ge 0$.

## 5.7 Compactness everywhere

I believe saying that the notion of compactness is one of the most important ones in topology and the whole of mathematics is not an exaggeration. Therefore, it is not surprising that it comes into play in a crucial way in a number of theorems of probability theory as well (see e.g. 5.4.18 or 6.6.12). To be sure, Helly's principle, so familiar to all students of probability, is simply saying that any sequence of probability measures on $\mathbb{R}$ is relatively compact; in functional analysis this theorem finds its important generalization in Alaoglu's Theorem. We will discuss compactness of probability measures on separable metric spaces, as well (Prohorov's Theorem), and apply the results to give another proof of existence of Brownian motion (Donsker's Theorem). On our way to Brownian motion we will prove the Arzela–Ascoli Theorem, too.

We start by looking once again at the results of Section 3.7, to continue with Alexandrov's Lemma and Tichonov's Theorem that will lead directly to Alaoglu's Theorem mentioned above.

5.7.1 *Compactness and convergence of martingales*     As we have seen in 3.7.7, a martingale converges in $L^1$ iff it is uniformly integrable. Moreover, in 3.7.15 we proved that a martingale converges in $L^p, p > 1$ iff it is bounded. Consulting [32] p. 294 we see that uniform integrability is necessary and sufficient for a sequence to be relatively compact in the weak topology of $L^1$. Similarly, in [32] p. 289 it is shown that a sequence in $L^p, p > 1$ is weakly relatively compact iff it is bounded. Hence, the results of 3.7.7 and 3.7.15 may be summarized by saying that *a martingale in $L^p, p \ge 1$, converges iff it is weakly relatively compact*. However, my attempts to give a universal proof that would work in both cases covered in 3.7.7 and 3.7.15 have failed. I was not able to find such a proof in the literature, either.

5.7.2 **Definition**     We say that an open infinite cover of a topological space $S$ is **truly infinite** iff it does not contain a finite subcover.

5.7.3 *Alexandrov's Lemma*    Let $\mathcal{V}$ be a subbase in $S$. If there is an open truly infinite cover of $S$, then there also is a truly infinite cover of $S$ build with elements of $\mathcal{V}$.

*Proof* (i) The set of truly infinite subcovers of $S$ is non-empty and partially ordered by the relation of inclusion. Moreover, any linearly ordered subset, say $C_t, t \in \mathbb{T}$ (where $\mathbb{T}$ is a non-empty index set) of this set has its upper bound; indeed $C_b = \bigcup_{t \in \mathbb{T}} C_t$ is such a bound. To prove this assume that $C_b$ is not truly infinite; it covers $S$ and is infinite as it contains at least one cover $C_t$. Then there exists an integer $n$ and elements $U_1, U_2, ..., U_n$ of $C_b$ that cover $S$. Since $C_t, t \in \mathbb{T}$, is linearly ordered, there exists a $t$ such that all $U_i \in C_t$ which contradicts the fact that $C_t$ is truly infinite, thus proving that $C_b$ is truly infinite. By the Kuratowski–Zorn Lemma, there exists a maximal element of the set of truly infinite covers. Let $C_m$ be such a cover.

(ii) Suppose an open set $G$ does not belong to $C_m$. Then there exist an $n \in \mathbb{N}$ and members $U_1, U_2, ..., U_n$ of $C_m$ such that

$$G \cup \bigcup_{i=1}^{n} U_i = S, \tag{5.26}$$

because $\{G\} \cup C_m$ contains $C_m$ as a proper subset, and hence cannot be a truly infinite cover. Conversely, for no member $G$ of $C_m$ may we find $n \in \mathbb{N}$ and $U_i$ in $C_m$ so that (5.26) holds, for $C_m$ is truly infinite. Hence, the possibility of writing (5.26) fully characterizes open sets that do not belong to $C_m$. It follows immediately that if $G_1 \subset G_2$ are open sets and $G_1 \notin C_m$ then $G_2 \notin C_m$. Moreover, if $G_1, G_2 \notin C_m$, then $G_1 \cap G_2 \notin C_m$, either.

(iii) We will show that $\mathcal{V}' = \mathcal{V} \cap C_m$ is a cover of $S$. This will imply that it is a truly infinite cover, as it is a subset of a truly infinite cover. Take a $p \in S$ and its open neighborhood $U \in C_m$. By definition, there exists a $k \in \mathbb{N}$ and members $V_1, ..., V_k$ of $\mathcal{V}$ such that $p \in \bigcap_{i=1}^{k} V_k \subset U$. Now, by (ii), one of $V_i$ must belong to $C_m$, for otherwise their intersection would not belong to $C_m$, and neither would $U$. This however, shows that $\mathcal{V}'$ covers $S$. $\qquad \square$

5.7.4 *Tichonov's Theorem*    Let $\mathbb{T}$ be a non-empty set and let, for each $t \in \mathbb{T}$, $S_t$ be a topological space. Let $\prod_{t \in \mathbb{T}} S_t$ be the set of functions $f : \mathbb{T} \to \bigcup_{t \in \mathbb{T}} S_t$ such that $f(t)$ belongs to $S_t$ for all $t \in \mathbb{T}$. Let us introduce a topology in $\prod_{t \in \mathbb{T}} S_t$ by defining its subbase to be formed of sets of the form $V_{t,U} = \{f \in \prod_{t \in \mathbb{T}} S_t | f(t) \in U\}$ where $t \in \mathbb{T}$ and

$U \subset S_t$ is open. This is the weakest topology making all the maps $\prod_{t \in \mathbb{T}} S_t \ni f \mapsto f(t) \in S_t$, $t \in \mathbb{T}$ continuous. Tichonov's Theorem says that *if the $S_t$ are compact then so is $\prod_{t \in \mathbb{T}} S_t$.*

*Proof* Suppose that $\prod_{t \in \mathbb{T}} S_t$ is not compact.

(i) By Alexandrov's Lemma, there exists a family $\mathcal{V}'$ of subsets $V_{t,U}$ that is a truly infinite cover of $\prod_{t \in \mathbb{T}} S_t$. Now, fix a $t \in \mathbb{T}$ and consider the family $\mathcal{U}_t$ of open sets $U$ in $S_t$ such that $V_{t,U}$ belongs to $\mathcal{V}'$. Then, none of $\mathcal{U}_t$ is a cover of $S_t$. Indeed, if it were, there would exist an $n \in \mathbb{N}$ and sets $U_{1,t}, ..., U_{n,t} \in \mathcal{U}_t$ such that $S_t = \bigcup_{i=1}^n U_{n,t}$. Consequently, we would have $\prod_{t \in \mathbb{T}} S_t = \{f \in \prod_{t \in \mathbb{T}} S_t | f(t) \in S_t\} \subset \bigcup_{i=1}^n \{f \in \prod_{t \in \mathbb{T}} S_t | f(t) \in U_i\} = \bigcup_{i=1}^n V_{t,U_i}$, contradicting the fact that $\mathcal{V}'$ is a truly infinite cover.

(ii) By (i), for any $t \in \mathbb{T}$ there exists a $p = f(t) \in S_t$ such that $p \notin V_{t,U}$ for $U \in \mathcal{U}_t$. On the other hand, thus defined $f$ is a member of $\prod_{t \in \mathbb{T}} S_t$, and $\mathcal{V}'$ is a cover of this space. Hence, there exists a $t$ and an open set $U \subset S_t$, $U \in \mathcal{U}_t$ such that $f \in V_{t,U}$, i.e. $f(t) \in U$. This contradiction shows that $\prod_{t \in \mathbb{T}} S_t$ must be compact. □

5.7.5 *Alaoglu's Theorem*     Let $\mathbb{X}$ be a Banach space. The unit (closed) ball $B = \{F \in \mathbb{X}^* | \|F\| \leq 1\}$ in $\mathbb{X}^*$ is weak* compact.

*Proof* By Tichonov's Theorem, all bounded functionals of norm not exceeding 1 are members of the compact space $\prod_{x \in \mathbb{X}} S_x$ where $S_x$ are compact intervals $[-\|x\|, \|x\|]$. Moreover, the weak* topology in $B$ is the topology inherited from $\prod_{x \in \mathbb{X}} S_x$. Hence, it suffices to show that $B$ is closed in $\prod_{x \in \mathbb{X}} S_x$.

To this end, let us assume that an $f \in \prod_{x \in \mathbb{X}} S_x$ belongs to the closure of $B$. We need to show that $f$ is a linear functional with the norm not exceeding 1. Let us take $x$ and $y \in \mathbb{X}$ and the neighborhoods $V_{f,x,\epsilon}, V_{f,y,\epsilon}$ and $V_{f,x+y,\epsilon}$, where $V_{f,z,\epsilon} = \{g \in \prod_{x \in \mathbb{X}} S_x | |g(z) - f(z)| < \epsilon\}$. There is an $F \in B$ that belongs to the intersection of these three neighborhoods. Hence $|f(x+y) - f(x) - f(y)| = |f(x+y) - f(x) - f(y) - F(x+y) + F(x) + F(y)| \leq |f(x+y) - F(x+y)| + |F(x) - f(x)| + |f(y) - F(y)| < 3\epsilon$. Since $\epsilon > 0$ can be chosen arbitrarily, $f(x+y) = f(x) + f(y)$. Similarly one shows that $f(\alpha x) = \alpha f(x)$. Finally, $|f(x)| \leq \|x\|$ by definition of $S_x$ in $\prod_{x \in \mathbb{X}} S_x$. □

5.7.6 **Corollary**     Let $S$ be a compact topological space, and let $\mathbb{P}_n$, $n \geq 1$, be a sequence of probability measures in $S$. By Alaoglu's Theorem and the Riesz Theorem there exists a subsequence of $\mathbb{P}_n, n \geq 1$, converging to a linear functional $F$ on $C(S)$, with $\|F\| \leq 1$. Moreover, $F$ is non-negative and $F(1_S) = 1$. This implies that $F$ corresponds to a probability measure $\mathbb{P}$ on $S$. Hence, for any sequence of probability measures on $S$ there exists a subsequence converging to a probability measure on $S$.

**5.7.7** *Helly's principle*   Helly's principle says (see e.g. [41]) that any sequence $\mathbb{P}_n, n \geq 1$, of probability measures on $\mathbb{R}$ has a subsequence converging to some measure $\mu$; yet in general the inequality $\mu(\mathbb{R}) \leq 1$ may be strict. The reason for this last complication is probably already clear for the reader: $\mathbb{R}$ is not compact. If we consider $\mathbb{P}_n, n \geq 1$ as measures on the one-point compactification, or the natural two-point compactification of $\mathbb{R}$, and use 5.7.6 it may happen that the limit probability measure has some mass at one of the adjoint points.

In a similar way, Helly's principle applies to measures on any locally compact space.

**5.7.8** **Exercise**   Prove 5.4.20 without alluding to 5.4.18, and using 5.7.7 instead (applied to $\mu_n \otimes \nu_n$ on $\mathbb{R}^2$) – cf. 6.5.6.

**5.7.9** *Tightness of measures*   Let $S$ be a separable metric space. Then, a sequence of probability measures $\mathbb{P}_n, n \geq 1$, does not have to have a converging subsequence. Well, as we shall see in the proof of 5.7.12, it does have to, but the support of the limit measure may be partly or totally outside of $S$. To make sure the limit measure is concentrated on $S$ we require the measures $\mathbb{P}_n$ to "hold on tight" to $S$. By definition, a family of probability measures on $S$ is said to be tight if for every $\epsilon > 0$ there exists a compact set $K$ such that $\mathbb{P}(K) > 1 - \epsilon$ for all $\mathbb{P}$ in this family.

**5.7.10** *Urysohn's Theorem*   The universal space where the supports of limit measures "live" is the Hilbert cube $\mathcal{H} = [0, 1]^{\mathbb{N}} = \prod_{i=1}^{\infty} S_i$ where all $S_i = [0, 1]$. The topology in $\mathcal{H}$ is introduced in the general way described in 5.7.4, but in this special case we may go further and introduce the norm $d_{\mathcal{H}}$ in $\mathcal{H}$ by

$$d_{\mathcal{H}}(f, g) = \sum_{i=1}^{\infty} \frac{1}{2^i} |f(i) - g(i)|.$$

It is clear that if $d_{\mathcal{H}}(f_n, f)$ converges to 0 then $f_n(i)$ converges to $f(i)$. The converse statement follows by the Lebesgue Dominated Convergence

Theorem. As a corollary, the topology induced in $\mathcal{H}$ by the metric $d_{\mathcal{H}}$ is the same as the topology introduced in 5.7.4.

Urysohn's Theorem states that *any separable metric space $S$ is home-omorphic to a subset of $\mathcal{H}$.*

*Proof* We need to show that there is an injective map $\Phi : S \to \mathcal{H}$ such that for any members $p$ and $p_n, n \geq 1$, of $S$, $d(p_n, p)$ converges to 0 iff $d_{\mathcal{H}}(\Phi(p_n), \Phi(p))$ does. Without loss of generality, we may assume that the metric $d$ in $S$ is bounded by 1, i.e. that $d(p, p') \leq 1$ for all $p$ and $p'$ in $S$. Indeed, in the general case we introduce an equivalent metric $d'$ in $S$ by $d' = \min(d, 1)$. Let $e_i, i \geq 1$, be dense in $S$, and let $\Phi(p) = f$ be a function $f : \mathbb{N} \to [0, 1]$ given by $f(i) = d(p, e_i)$. By continuity of metric, $\lim_{n \to \infty} d(p_n, p) = 0$ implies $\lim_{n \to \infty} d(p_n, e_i) = d(p, e_i)$ for all $i \in \mathbb{N}$, and hence $d_{\mathcal{H}}(\Phi(p_n), \phi(p))$ converges to 0. Conversely, if $d(p, e_i)$ converges to 0 for all $i$ then, since $\{e_i, i \geq 1\}$ is dense, given $\epsilon > 0$ we may find an $e_i$ such that $d(p, e_i) < \frac{\epsilon}{3}$. Next, we may find an $n_0$ such that $|d(p_n, e_i) - d(p, e_i)| < \frac{\epsilon}{3}$ for $n \geq n_0$. For such an $n$, $d(p_n, p) \leq d(p_n, e_i) + d(p, e_i) < \epsilon$, as desired. Finally, the same argument shows that $d(p, e_i) = d(p', e_i), i \in \mathbb{N}$, implies $d(p, p') = 0$ and hence $p = p'$. □

**5.7.11 Corollary** The transport of a measure on $S$ via $\Phi$ together with the transport of the measure on $\Phi(S)$ via $\Phi^{-1}$ establishes a one-to-one correspondence between Borel measures on $S$ and on $S' = \Phi(S)$. Moreover, by the change of variables formula (1.6), a sequence of probability measures $\mathbb{P}_n, n \geq 1$, on $S$ converges weakly to a $\mathbb{P}$ iff a corresponding sequence $(\mathbb{P}_n)_\Phi$ converges weakly to $\mathbb{P}_\Phi$. Furthermore, $\Phi$ and $\Phi^{-1}$ map compact sets into compact sets. Hence, $\mathbb{P}_n, n \geq 1$, is tight iff $(\mathbb{P}_n)_\Phi, n \geq 1$, is.

**5.7.12 *Prohorov's Theorem*** Suppose a sequence of Borel probability measures on a separable metric space $S$ is tight. Then, it is relatively compact.

*Proof* By Urysohn's Theorem and 5.7.11, it suffices to show that a tight sequence $\mathbb{P}_n, n \geq 1$, of Borel measures on a subset $S$ of $\mathcal{H}$ is relatively compact. Of course the idea is to reduce this situation to that described in 5.7.6. We note that in general Borel subsets of $S$ are not Borel in $\mathcal{H}$, unless $S$ is Borel itself. We may, however, use characterization (1.3) and given a Borel measure $\mathbb{P}$ on $S$ define a Borel measure $\mathbb{P}^\sharp$ on $\mathcal{H}$ by $\mathbb{P}^\sharp(A) = \mathbb{P}(S \cap A)$. Now, the sequence $\mathbb{P}_n^\sharp, n \geq 1$, has a subsequence

$\mathbb{P}^{\sharp}_{n_i}, i \geq 1$, converging to a probability measure $\mu$ on $\mathcal{H}$. By 5.6.3,

$$\limsup_{i \to \infty} \mathbb{P}^{\sharp}_{n_i}(A) \leq \mu(A), \qquad \text{for any closed subset } A \text{ of } \mathcal{H}. \qquad (5.27)$$

Since $\mathbb{P}_n, n \geq 1$, is tight, there are sets $K_k, k \geq 1$, that are compact in $S$ (hence, compact in $\mathcal{H}$ as well) such that $\mathbb{P}_n(K_k) \geq 1 - \frac{1}{k}, n \geq 1$. Moreover, $K_k$ are Borel subset of $\mathcal{H}$ and $\mathbb{P}^{\sharp}_n(K_k) = \mathbb{P}_n(K_k), n \geq 1$. Hence, $\mu(K_n) \geq 1 - \frac{1}{k}$. If we let $K = \bigcup_{k \geq 1} K_k$, then $\mu(K) = 1$.

Let $B \in \mathcal{B}(S)$. If $A_1$ and $A_2$ are Borel in $\mathcal{H}$ and $B = A_1 \cap S = A_2 \cap S$, then the symmetric difference of $A_1$ and $A_2$ is contained in $\mathcal{H} \backslash S \subset \mathcal{H} \backslash K$. Therefore, the quantity $\mathbb{P}(B) := \mu(A)$ where $A \in \mathcal{B}(S)$ and $B = A \cap S$ is well defined. We check that $\mathbb{P}$ is a Borel measure on $S$ and $\mathbb{P}^{\sharp} = \mu$.

Finally, if $B \subset S$ is closed in $S$, then there is an $A \subset \mathcal{H}$ that is closed in $\mathcal{H}$ such that $B = S \cap A$. Since $\mathbb{P}^{\sharp}_n(A) = \mathbb{P}_n(B)$ and $\mathbb{P}(B) = \mu(A)$, (5.27) shows that $\mathbb{P}_{n_i}, i \geq 1$, converge weakly to $\mathbb{P}$.  □

**5.7.13 Exercise** Complete the above proof by showing that if $B \subset S$ is compact in $S$, then it is compact in $\mathcal{H}$, too. Note that an analogous statement about closed sets is in general not true.

**5.7.14 Remark** The converse to Prohorov's Theorem is true under the assumption that $S$ is complete. However, we will neither use nor prove this result here.

**5.7.15 *Brownian motion as a measure*** Let $C(\mathbb{R}^+)$ be the space of continuous functions $x, y, \dots$ mapping $\mathbb{R}^+$ into $\mathbb{R}$ and such that $x(0) = 0$. When equipped with the metric

$$d(x, y) = \sum_{n=1}^{\infty} \frac{1}{2^n} \min\{1, \sup_{s \in [0,n]} |x(s) - y(s)|\},$$

$C(\mathbb{R}^+)$ is a metric space. Moreover, polynomials with rational coefficients form a dense set in $C(\mathbb{R}^+)$, i.e. $C(\mathbb{R}^+)$ is separable. Indeed, given $x \in C(\mathbb{R}^+)$ and $1 > \epsilon > 0$, we may choose an $n_0 \in \mathbb{N}$ so that $\frac{1}{2^{n_0 - 1}} < \epsilon$ and a polynomial $y$ with rational coefficients such that $\sup_{s \in [0, n_0]} |x(s) - y(s)| < \frac{\epsilon}{2}$. Then, $d(x, y) \leq \sum_{n=1}^{n_0} \frac{1}{2^n} \frac{\epsilon}{2} + \sum_{n=n_0+1}^{\infty} \frac{1}{2^n} < \frac{\epsilon}{2} + \frac{1}{2^{n_0}} < \epsilon$, as claimed. A similar argument shows that $\lim_{n \to \infty} d(x_n, x) = 0$ iff functions $x_n \in C(\mathbb{R}^+)$ converge to $x \in C(\mathbb{R}^+)$ uniformly on compact subintervals of $\mathbb{R}^+$, and that $C(\mathbb{R}^+)$ is complete.

We will show that the Borel $\sigma$-algebra $\mathcal{B}(C(\mathbb{R}^+))$ is the $\sigma$-algebra generated by the maps $C(\mathbb{R}^+) \ni x \mapsto \pi_t(x) := x(t) \in \mathbb{R}, t > 0$. To this end we note that all these maps are continuous, hence Borel measurable, and it suffices to show that $\mathcal{B}(C(\mathbb{R}^+))$ is a subset of the $\sigma$-algebra just mentioned. By the characterization of convergence in $C(\mathbb{R}^+)$ given at the end of the previous paragraph, for $y \in C(\mathbb{R}^+)$ and positive $T$, the map $C(\mathbb{R}^+) \ni x \mapsto f_{y,T}(x) = \sup_{t \in [0,T]} |x(t) - y(t)|$ is continuous. Hence, the sets $V_{y,T,\epsilon} := f_{y,T}^{-1}(-\epsilon, \epsilon)$, where $\epsilon > 0$, are open in $C(\mathbb{R}^+)$. Moreover, these sets form a subbase of the topology in $C(\mathbb{R}^+)$ and belong to the $\sigma$-algebra generated by the $\pi_t, t > 0$, as may be seen from

$$V_{y,T,\epsilon} = \bigcup_{n \geq 1} \bigcap_{t \in \mathbb{Q}} A(y, t, \frac{n}{n+1}\epsilon)$$

where $A(y, t, \epsilon) := \{x \in C(\mathbb{R}^+) | \,|x(t) - y(t)| \leq \epsilon\} = \Phi_t^{-1}[-\epsilon, \epsilon]$.

Now, suppose that $(\Omega, \mathcal{F}, \mathbb{P})$ is a probability space where a Brownian motion process is defined. Without loss of generality we may assume that all trajectories $\omega \mapsto w(t, \omega)$ are continuous and $w(0, \omega) = 0$. The map $W : \Omega \to C(\mathbb{R}^+)$ that assigns a trajectory to an element of $\Omega$ is measurable because $W^{-1} B_{t,A} = \{\omega \in \Omega | x(t, \omega) \in A\} \in \mathcal{F}$ where $B_{t,A} = \{x \in C(\mathbb{R}^+) | x(t) \in A\}$ and $A$ is a Borel subset of $\mathbb{R}$, and the sets $B_{t,A}$ generate $\mathcal{B}(C(\mathbb{R}^+))$. Hence, given a Brownian motion on $(\Omega, \mathcal{F}, \mathbb{P})$ we may construct a measure $\mathbb{P}_W$ on $C(\mathbb{R}^+)$, called the Wiener measure, as the transport of $\mathbb{P}$ via $W$. Note that finite intersections of the sets $B_{t,A}$ where $t > 0$ and $B \in \mathcal{B}(\mathbb{R})$ form a $\pi$-system. Hence the Wiener measure is determined by its values on such intersections. In other words, it is uniquely determined by the condition that $(\mathbb{P}_W)_{\pi_{t_1, t_2, \ldots, t_n}}$ (called a **finite-dimensional distribution** of $\mathbb{P}_W$) is a Gaussian measure with covariance matrix $(t_i \wedge t_j)_{i,j=1,\ldots,n}$ (see 4.1.7), where $\pi_{t_1, \ldots, t_n} : C(\mathbb{R}^+) \to \mathbb{R}^n$ is given by $\pi_{t_1, \ldots, t_n} x = (x(t_1), \ldots, x(t_n))$ and $t_1 < t_2 < \ldots < t_n$ are positive numbers.

On the other hand, if we could construct a measure $\mathbb{P}_W$ on the space $(C(\mathbb{R}^+), \mathcal{B}(C(\mathbb{R}^+)))$ possessing the properties listed above, the family of random variables $\pi_t, t \geq 0$, would be a Brownian motion. In the following subsections we will prove existence of $\mathbb{P}_W$ without alluding to Chapter 4. We start by presenting the Arzela–Ascoli Theorem which plays a role in the proof. However, the reader who wants to be sure he understands this subsection well should not skip the following exercise.

**5.7.16 Exercise**    Check that the sets $V_{y,T,\epsilon}$ form a subbase of the topology in $C(\mathbb{R}^+)$.

**5.7.17 *Arzela–Ascoli Theorem*** The **Arzela–Ascoli Theorem**, the famous criterion for compactness of a set $A \subset C(S)$ where $(S, d)$ is a compact metric space, turns out to be another direct consequence of Ti-chonov's Theorem. The former theorem says that the set $A$ is relatively compact iff it is composed of equicontinuous functions and there exists an $M > 0$ such that

$$\|x\|_{C(S)} \leq M, \text{ for all } x \in A. \tag{5.28}$$

Let us recall that $A$ is said to be composed of equicontinuous functions iff

$$\forall_{\epsilon>0}\forall_{p\in S}\exists_{\delta>0}\forall_{p'\in S}\forall_{x\in A} \; d(p,p') < \delta \Rightarrow |x(p) - x(p')| \leq \epsilon. \tag{5.29}$$

For the proof of the Arzela–Ascoli Theorem we need to recall two lem-mas.

**5.7.18 Lemma** Let us suppose $A \subset C(S)$ is composed of equicontin-uous functions and (5.28) is satisfied. Let us consider $A$ as a subset of $\prod_{p\in S} S_p$ where $S_p = [-M, M]$ for all $p \in S$. Then, the closure of $A$ in this space is composed of equicontinuous functions, and (5.29) holds for all $x \in cl\, A$, too. In particular, the limit points of $A$ are continuous.

*Proof* Let us fix $y \in cl\, A$ and $\epsilon > 0$, and choose a $\delta > 0$ so that (5.29) holds. Let $p'$ be such that $d(p,p') < \delta$. For all $n \geq 1$, there exists an $x_n \in V_{p,y,\frac{1}{n}} \cap V_{p',y,\frac{1}{n}} \cap A$ where $V_{p,y,\frac{1}{n}}$ is a neighborhood of $y$ composed of $x \in C(S)$ such that $|x(p) - y(p)| < \frac{1}{n}$. Hence,

$$|y(p) - y(p')| \leq |y(p) - x_n(p)| + |x_n(p) - x_n(p')| + |x_n(p') - y(p')| \leq \epsilon + \frac{2}{n},$$

for all $n \geq 1$. Therefore, $|y(p) - y(p')| \leq \epsilon$. $\qquad\square$

**5.7.19 Lemma** Suppose $x_n, n \geq 1$, is a sequence of equicontinuous functions on a compact metric space $S$, and $\lim_{n\to\infty} x_n(p)$ exists for all $p \in S$. Then, the convergence is in fact uniform, i.e. $\lim_{n\to\infty} \|x_n - x\|_{C(S)} = 0$. (Note that $x \in C(S)$ by the previous lemma.)

*Proof* Suppose that this is not so. Then, there exists a $c > 0$ such that $\|x_n - x\| \geq c$ for infinitely many $n \geq 1$. On the other hand, for any $n \geq 1$, there exists a $p_n \in S$ such that $\|x_n - x\| = |x_n(p_n) - x(p_n)|$. Hence, there are infinitely many $n$ such that $|x_n(p_n) - x(p_n)| \geq c$. Since $S$ is compact, there exists a further subsequence of $p_n, n \geq 1$, converging to a $p_0 \in S$. Without loss of generality, to simplify notation, we assume that $\lim_{n\to\infty} p_n = p_0$. Then, for any $\epsilon > 0$ and $n$ large enough, $d(p_n, p) < \delta$

where $\delta$ is chosen as in (5.29). Moreover, by assumption, for $n$ large enough $|x_n(p_0) - x(p_0)| \le \epsilon$. Therefore, for such $n$,

$$|x_n(p_n) - x(p_n)| \le |x_n(p_n) - x_n(p_0)| + |x_n(p_0) - x(p_0)| + |x(p_0) - x(p_n)|$$

does not exceed $3\epsilon$ (we use the previous lemma here!), which contradicts the way the $p_n$ were chosen. □

*Proof (of the "if" part of the Arzela–Ascoli Theorem)* Let $x_n, n \ge 1$, be a sequence of elements of $A$. By the first lemma, the closure of $A$ in the space $\prod_{p \in S} S_p$ is composed of equicontinuous functions. Since $\prod_{p \in S} S_p$, is compact, so is $cl A$. Moreover, the topology in this space is the topology of pointwise convergence. Hence, there exists a subsequence of $x_n, n \ge 1$, converging to a continuous $x$. By the second lemma, the convergence is in fact uniform. □

**5.7.20 Exercise**     Show the "only if" part of the Arzela–Ascoli Theorem.

**5.7.21 Remark**     Suppose $S = [0, t]$ where $t > 0$, and $A \subset C[0, t]$ is composed of equicontinuous functions $x$ such that $|x(0)| \le M$ for some constant $M$ independent of the choice of $x \in A$. Then $A$ is relatively compact.

*Proof* We may take an $n > 0$ such that $|x(s) - x(s')| < 1$ provided $|s - s'| \le \frac{1}{n}$. Then, for all $x \in A$ and $s \in [0, t]$, $|x(s)| \le |x(0)| + \sum_{k=1}^{[ns]} |x\left(\frac{k}{n}\right) - x\left(\frac{k-1}{n}\right)| + |x(s) - x\left(\frac{[ns]}{n}\right)| \le M + [ns] + 1 \le M + [nt] + 1$. This shows that (5.28) holds with $M$ replaced by $M + [nt] + 1$. □

**5.7.22 *Compact sets in $C(\mathbb{R}^+)$***     A set $A \subset C(\mathbb{R}^+)$ is relatively compact iff it is composed of functions that are equicontinuous at every subinterval of $\mathbb{R}^+$.

*Proof* Necessity is obvious by the Arzela–Ascoli Theorem. To show sufficiency, for $i \in \mathbb{N}$, let $A_i$ be the set of $y \in C[0, i]$ such that there is an $x \in A$ such that $y = x_{|[0,i]}$ (the restriction of $x$ to $[0, i]$). Let us recall that $x(0) = 0$ for all $x \in C(\mathbb{R}^+)$. Therefore, by 5.7.21, the sets $A_i$ are compact and so is $\prod_{i \in \mathbb{N}} A_i$.

There is a one-to-one correspondence between elements of $A$ and sequences $(y_i)_{i \ge 1} \in \prod_{i \in \mathbb{N}} A_i$ such that $(y_{i+1})_{|[0,i]} = y_i, i \ge 1$. Moreover, a sequence $x_n, n \geq 1$ of elements of $C(\mathbb{R}^+)$ converges in $C(\mathbb{R}^+)$ iff the corresponding $(y_{n,i})_{i \geq 1}, n \geq 1$, converges in $\prod_{i \in \mathbb{N}} A_i$.

If a $x_n \in C(\mathbb{R}^+), n \geq 1$, does not have a converging subsequence, then there is no converging subsequence of the corresponding $(y_{n,i})_{i \geq 1}, n \geq 1$, contradicting compactness of $\prod_{i \in \mathbb{N}} A_i$. $\square$

**5.7.23 *Donsker's Theorem*** The theorem we are to discuss may be called a "$C(\mathbb{R}^+)$-version of the Central Limit Theorem". We suppose that $X_i, i \geq 1$, is a sequence of independent, identically distributed random variables with mean zero and variance 1 defined on $(\Omega, \mathcal{F}, \mathbb{P})$. Let $S_0 = 0$, and $S_n = \sum_{i=1}^n X_i, n \geq 1$. For any $\omega \in \Omega$ and $n \in \mathbb{N}$ we define a continuous function $\mathcal{X}_n(t) = \mathcal{X}_n(t, \omega)$ of argument $t \in \mathbb{R}^+$ by letting $\mathcal{X}_n\left(\frac{k}{n}\right) = \frac{1}{\sqrt{n}} S_k$ and requiring $\mathcal{X}_n(t)$ to be linear in between these points. In other words,

$$\sqrt{n}\mathcal{X}_n(t) = (1 - \alpha(t))S_{[nt]} + \alpha(t)S_{[nt]+1} = S_{[nt]} + \alpha(t)X_{[nt]+1} \quad (5.30)$$

where $\leq \alpha(t) = nt - [nt] \leq 1 \in [0,1]$. As in 5.7.15, we prove that $\mathcal{X}_n$ is a measurable map from $\Omega$ to $C(\mathbb{R}^+)$. Let $\mathbb{P}_n, n \geq 1$, denote the transport measures on $C(\mathbb{R}^+)$ related to these maps. The **Donsker's Theorem** says that

*the measures $\mathbb{P}_n, n \geq 1$ converge weakly and the limit measure is the Wiener measure $\mathbb{P}_W$ on $C(\mathbb{R}^+)$.*

In particular, the existence of $\mathbb{P}_W$ is a part of the theorem. For the proof we need two lemmas, a hard one and an easy one. We start with the former.

**5.7.24 Lemma** The sequence $\mathbb{P}_n, n \geq 1$ is tight in $C(\mathbb{R}^+)$.

*Proof* For $x \in C(\mathbb{R}^+)$ and $t > 0$ let us define the functions $m_t(h) = m_t(x, h)$ of argument $h \geq 0$ by $m_t(h) = \sup_{0 \leq s \leq t} |x(s) - x(s + h)|$. By 5.7.22, it is clear that we need to make sure that large values of $m_t(\mathcal{X}_n, h)$ do not show up too often. In other words, we want to have a control over $\sup_n \mathbb{P}_n\{m_t(x, h) > \epsilon\}, \epsilon > 0$. We will gain it in several steps.

**1** Fix $0 \leq s < t$. Since maximum of a polygonal function is attained at one of its vertices,

$$\sup_{s \leq u \leq t} |\mathcal{X}_n(u) - \mathcal{X}_n(s)| \leq \max_{\lceil ns \rceil \leq k \leq \lceil nt \rceil} \left| \frac{1}{\sqrt{n}} S_k - \mathcal{X}_n(s) \right|$$

where $\lceil s \rceil$ denotes the smallest integer $k$ such that $k \geq s$. Let $\lfloor s \rfloor (= [s])$ denote the largest integer dominated by $s$. Note that $\frac{1}{\sqrt{n}} S_{\lceil ns \rceil} \geq \mathcal{X}_n(s) \geq$

$\frac{1}{\sqrt{n}}S_{\lfloor ns \rfloor}$ provided $X_{\lfloor ns \rfloor} \geq 0$, and both inequalities reverse in the other case. Hence, our supremum does not exceed

$$\max_{\lceil ns \rceil \leq k \leq \lceil nt \rceil} \left| \frac{1}{\sqrt{n}}S_k - \frac{1}{\sqrt{n}}S_{\lfloor ns \rfloor} \right| \vee \max_{\lceil ns \rceil \leq k \leq \lceil nt \rceil} \left| \frac{1}{\sqrt{n}}S_k - \frac{1}{\sqrt{n}}S_{\lceil ns \rceil} \right|$$

so that, for any $\epsilon > 0$, $\mathbb{P}\{\sup_{s \leq u \leq t}|\mathcal{X}_n(u) - \mathcal{X}_n(s)| > \epsilon\}$ is no greater than $\mathbb{P}\left\{ \max_{\lceil ns \rceil \leq k \leq \lceil nt \rceil} \left| \frac{1}{\sqrt{n}} \sum_{i=\lfloor ns \rfloor+1}^{k} X_i \right| > \epsilon \right\}$ plus a similar probability with $\lfloor ns \rfloor$ replaced by $\lceil ns \rceil$. Since random variables $X_i$ are independent and identically distributed, both probabilities are no greater than $\mathbb{P}\left\{ \max_{1 \leq k \leq \lceil nt \rceil - \lceil ns \rceil} \left| \frac{1}{\sqrt{n}}S_k \right| > \epsilon \right\}$. Moreover, for $s >$ and $h > 0$, $\lceil n(s+h) \rceil - \lceil ns \rceil \leq \lceil nh \rceil + 1$. Thus,

$$\sup_{s} \mathbb{P}\left\{ \sup_{s \leq u \leq s+h} |\mathcal{X}_n(u) - \mathcal{X}_n(s)| \right\} \leq 2\mathbb{P}\{ \max_{1 \leq k \leq \lceil nh \rceil+1} \frac{1}{\sqrt{n}}|S_k| > \epsilon\}. \tag{5.31}$$

For $m \geq 1$, let $S_m^* = \max_{1 \leq k \leq m}|S_k|$ and for $a > 0$ let $\tau_a = \min\{k \in \mathbb{N} \mid |S_k| > a\}$. Note that $\tau_a$ is a Markov time and $S_\tau$ is a well-defined random variable. Moreover, the probability on the right-hand side of (5.31) equals $\mathbb{P}\{S_m > \sqrt{n}\epsilon\} = \mathbb{P}\{\tau_{\sqrt{n}\epsilon} \leq m\}$ where $m = m(n,h) = \lceil nh \rceil + 1$.

**2** We have the following **maximal inequality of Ottaviani:**

$$\mathbb{P}(S_m^* \geq 2\sqrt{m}r) \leq \frac{2\mathbb{P}\{S_m \geq \sqrt{m}r\}}{1 - r^{-2}}, \quad r > 1. \tag{5.32}$$

For its proof we note that

$$\mathbb{P}\{|S_m| > r\sqrt{m}\} \geq \mathbb{P}\{|S_m| > r\sqrt{m}, S_m^* > 2r\sqrt{m}\}$$
$$\geq \mathbb{P}\{\tau \leq m, |S_m - S_\tau| < r\sqrt{m}\}$$
$$= \sum_{k=1}^{m} \mathbb{P}\{\tau = k, |S_m - S_k| < r\sqrt{m}\},$$

where $\tau = \tau_{2r\sqrt{m}}$ for simplicity. Since $S_m - S_k$ is independent of the variables $X_1, ..., X_k$ and has the same distribution as $S_{m-k}$, and $\{\tau = k\} \in \sigma(X_1, ..., X_k)$, the last sum equals

$$\sum_{k=1}^{m} \mathbb{P}\{\tau = k\}\mathbb{P}\{|S_{m-k}| < r\sqrt{m}\}$$
$$\geq \min_{1 \leq k \leq m} \mathbb{P}\{|S_{m-k}| < r\sqrt{m}\} \sum_{k=1}^{m} \mathbb{P}\{\tau = k\}$$
$$\geq \min_{1 \leq k \leq m} \mathbb{P}\{|S_k| < r\sqrt{m}\}\mathbb{P}\{\tau \leq m\}.$$

Moreover, by Chebyshev's inequality

$$\min_{1 \le k \le m} \mathbb{P}\{|S_k| < r\sqrt{m}\} = 1 - \max_{1 \le k \le m} \mathbb{P}\{|S_k| \ge r\sqrt{m}\}$$

$$\ge 1 - \max_{1 \le k \le m} \frac{k}{r^2 m} = 1 - \frac{1}{r^2}.$$

Hence, $\mathbb{P}\{|S_m| > r\sqrt{m}\} \ge (1 - \frac{1}{r^2})\mathbb{P}\{S_m^* > 2r\sqrt{m}\}$ and (5.32) follows by dividing by $1 - \frac{1}{r^2}$ (this is where we need $r > 1$).

**3** Let $r_n = \frac{1}{2}\sqrt{\frac{n}{m}}\epsilon$. Since $\lim_{n\to\infty} r_n = \frac{\epsilon}{2}\frac{1}{\sqrt{h}}$, for sufficiently small $h$ and sufficiently large $n$ we have $r_n > 1$. Moreover, by **1**, **2** and the Central Limit Theorem,

$$\limsup_{n\to\infty}\sup_{s\ge 0}\mathbb{P}\left\{\sup_{s\le u\le s+h}|\mathcal{X}_n(u) - \mathcal{X}_n(s)|\right\} \le \lim_{n\to\infty}\frac{2\mathbb{P}\{|S_m| > \sqrt{m}r_n\}}{1 - r_n^{-2}}$$

$$= \left(1 - \frac{4h}{\epsilon^2}\right)^{-1}\frac{4}{\sqrt{2\pi}}\int_{\frac{\epsilon}{2\sqrt{h}}}^{\infty} e^{-u^2/2}\,du.$$

Since $\lim_{h\to 0+}\frac{1}{h}\frac{4}{\sqrt{2\pi}}\int_{\frac{\epsilon}{2\sqrt{h}}}^{\infty} e^{-u^2/2}\,du = 0$,

$$\limsup_{h\to 0+}\frac{1}{h}\limsup_{n\to\infty}\sup_{s\ge 0}\mathbb{P}\left\{\sup_{s\le u\le s+h}|\mathcal{X}_n(u) - \mathcal{X}_n(s)| > \epsilon\right\} = 0. \quad (5.33)$$

The first "lim sup" here may be replaced by "lim" for we are dealing with limits of non-negative functions.

**4** Relation (5.33) implies that, for any $t > 0$,

$$\limsup_{h\to 0+}\limsup_{n\to\infty}\mathbb{P}\left\{\sup_{0\le s\le t}\sup_{s\le u\le s+h}|\mathcal{X}_n(u) - \mathcal{X}_n(s)| > \epsilon\right\} = 0. \quad (5.34)$$

To prove this, we take a $\delta > 0$ and choose $h_\delta < 1$ small enough to have

$$\frac{1}{h}\limsup_{n\to\infty}\sup_{s\ge 0}\mathbb{P}\left\{\sup_{s\le u\le s+h}|\mathcal{X}_n(u) - \mathcal{X}_n(s)| > \frac{\epsilon}{3}\right\} < \frac{\delta}{t}, \quad \text{for } h < h_\delta.$$

Next, we take $l = \lfloor t/h \rfloor$ and divide the interval $[0, t]$ into $l + 1$ subintervals $[s_i, s_{i+1}]$, $i = 0, ..., l$ where $s_0 = 0$ and $s_{l+1} = t$, such that $\max_{i=0,...,l}|s_{i+1} - s_i| < h$. For $s < u \le t$ with $u - s < h$, there exists an $s_i$ such that $|s_i - u| \wedge |s_i - s| < h$ and we either have $s_i \le s < u$ or $i \ge 2$ and $s \le s_i \le u$. In the former case, $|\mathcal{X}_n(u) - \mathcal{X}_n(s)|$ does not exceed $|\mathcal{X}_n(u) - \mathcal{X}_n(s_i)| + |\mathcal{X}_n(u) - \mathcal{X}_n(s_i)|$ and in the latter it does not exceed $|\mathcal{X}_n(u) - \mathcal{X}_n(s_i)| + |\mathcal{X}_n(u) - \mathcal{X}_n(s_{i-1})| + |\mathcal{X}_n(s_i) - \mathcal{X}_n(s_{i-1})|$. Therefore, $\{\sup_{0\le s\le t}\sup_{s\le u\le s+h}|\mathcal{X}_n(u) - \mathcal{X}_n(s)| > \epsilon\}$ is contained in the union $\bigcup_{i=0}^{l}\{\sup_{s_i\le u\le s_i+h}|\mathcal{X}_n(u) - \mathcal{X}_n(s)| > \frac{\epsilon}{3}\}$, and its probability is no greater than the sum of probabilities of the involved events, which by assumption does not exceed $(\lfloor t/h \rfloor + 1) h \frac{\delta}{t+1} \leq \frac{t+h}{t+1} \delta < \delta$. Since $\delta$ was arbitrary, we are done.

Again, the first "lim sup" in (5.34) may be replaced by "lim". Hence, by definition of $m_t$ and $\mathbb{P}_n$, we have

$$\lim_{h \to 0+} \limsup_{n \to \infty} \mathbb{P}\{m_t(\mathcal{X}_n, h) > \epsilon\} = \lim_{h \to 0+} \limsup_{n \to \infty} \mathbb{P}_n\{m_t(x, h) > \epsilon\} = 0.$$

(5.35)

Finally, $\lim_{h \to 0+} m_t(\mathcal{X}_n(\omega), h)$ for every $\omega \in \Omega$ and $n \in \mathbb{N}$, the function $\mathcal{X}_n(\omega)$ being continuous. Therefore, by the Dominated Convergence Theorem,

$$\lim_{h \to 0+} \mathbb{P}\{m_t(\mathcal{X}_n, h) > 0\} = \lim_{h \to 0+} E 1_{\{m_t(\mathcal{X}_n, h) > \epsilon\}} = 0, \qquad n \in \mathbb{N}.$$

(5.36)

This proves that the "lim sup" in (5.35) may be replaced by "sup".

**5**   Fix $\epsilon > 0$ and $t > 0$. For any $k \in \mathbb{N}$, there exists an $h_k$ such that $\sup_n \mathbb{P}_n\{m_{kt}(x, h_k) > \frac{1}{k}\} \leq \frac{\epsilon}{2^{k+1}}$. Let $K = \bigcap_{k \geq 1}\{m_{kt}(x, h_k) \leq \frac{1}{k}\}$. By 5.7.22, $K$ is compact (it is closed by 5.7.18), and

$$\min_{n \geq 1} \mathbb{P}_n(K) = 1 - \sup_{n \geq 1} \mathbb{P}_n(K^{\mathsf{C}}) \geq 1 - \sup_{n \geq 1} \sum_{k=1}^{\infty} \mathbb{P}_n \left\{ m_{kt}(x, h_k) > \frac{1}{k} \right\}$$

$$\geq 1 - \sum_{k=1}^{\infty} \frac{\epsilon}{2^{k+1}} = 1 - \epsilon.$$

$\square$

**5.7.25 Lemma**   For any $0 \leq t_1 < t_2 < \ldots < t_k$, the distributions $(\mathbb{P}_n)_{\pi_{t_1, \ldots, t_k}}$ of $(\mathcal{X}_n(t_1), \ldots, \mathcal{X}_n(t_k))$ converge weakly to the normal distribution with covariance matrix $(t_i \vee t_j)_{i,j=1,\ldots,k}$.

*Proof*   By 5.6.2 and 4.1.7, it suffices to show that

$$\underline{\mathcal{X}_n} = \left( \frac{\mathcal{X}_n(t_1)}{\sqrt{t_1}}, \frac{\mathcal{X}_n(t_2) - \mathcal{X}_n(t_1)}{\sqrt{t_2 - t_1}}, \ldots, \frac{\mathcal{X}_n(t_k) - \mathcal{X}_n(t_{k-1})}{\sqrt{t_k - t_{k-1}}} \right)$$

converges weakly to the standard normal distribution. To this end, we show that

$$\underline{S_n} = \frac{1}{\sqrt{n}} \left( \frac{S_{[nt_1]}}{\sqrt{t_1}}, \frac{S_{[nt_2]} - S_{[nt_1]}}{\sqrt{t_2 - t_1}}, \ldots, \frac{S_{[nt_k]} - S_{[nt_{k-1}]}}{\sqrt{t_2 - t_1}} \right)$$

converges to the standard normal distribution, and that

$$\lim_{n \to \infty} \left| E x \circ \underline{\mathcal{X}_n} - E x \circ \underline{S_n} \right| = 0,$$

for any continuous $x$ on $\mathbb{R}^k$ with compact support. The reasons why we may restrict our attention to continuous functions with compact support are (a) continuous functions with compact support form a dense set in $C_0(\mathbb{R}^k)$ and (b) by the first claim we know that the the limit measure is a probability measure, so that no mass escapes to infinity.

To prove the first claim we note that the coordinates of $\underline{S}_n$ are independent and that the $i$th coordinate has the same distribution as $\frac{1}{\sqrt{n}} \frac{1}{\sqrt{t_i - t_{i-1}}} S_{[nt_i] - [nt_{i-1}]}$ (we put $t_0 = 0$) which by the Central Limit Theorem and $\lim_{n \to \infty} \frac{[nt_i] - [nt_{i-1}]}{n} = t_i - t_{i-1}$ converges weakly to the $N(0,1)$ distribution. Hence, for any $a_i \in \mathbb{R}, i = 1, ..., k$,

$$\mathbb{P}\left\{ \underline{S}_n \in \prod_{i=1}^{k} (-\infty, a_i] \right\} = \prod_{i=1}^{k} \mathbb{P}\left\{ \frac{S_{[nt_i]} - S_{[nt_{i-1}]}}{\sqrt{t_2 - t_1}} \le a_i \right\}$$

converges to $\prod_{i=1}^{k} \frac{1}{\sqrt{2\pi}} \int_{a_i}^{\infty} e^{-u^2/2}\,du$, as desired.

To prove the second claim we argue similarly as in 4.1.4. Let $\underline{R}_n := \underline{X}_n - \underline{S}_n$. By definition (5.30), the $i$th coordinate of $\underline{R}_n$ equals

$$\frac{1}{\sqrt{n}} \frac{\alpha(t_i) X_{[nt_i]+1} - \alpha(t_{i-1}) X_{[nt_{i-1}]+1}}{\sqrt{t_i - t_{i-1}}}$$

and the variance of this coordinate is no greater than $\frac{2}{n(t_i - t_{i-1})}$. Let $|\underline{R}_n|$ denote the sum of absolute values of coordinates of $\underline{R}_n$. For $\delta > 0$, the probability $\mathbb{P}\left\{ |\underline{R}_n| \ge \delta \right\}$ does not exceed

$$\sum_{i=1}^{k} \mathbb{P}\left\{ \frac{1}{\sqrt{n}} \frac{\alpha(t_i) X_{[nt_i]+1} - \alpha(t_{i-1}) X_{[nt_{i-1}]+1}}{\sqrt{t_i - t_{i-1}}} \ge \frac{\delta}{k} \right\}$$

which, by Chebyshev's inequality, is dominated by $\frac{\delta^2}{n} k^2 \sum_{i=1}^{k} \frac{2}{(t_i - t_{i-1})}$.

Given $\epsilon > 0$ we may find a $\delta$ such that $|x(\underline{t}) - x(\underline{s})| < \frac{\epsilon}{2}$ provided $|\underline{t} - \underline{s}| < \delta, \underline{t}, \underline{s} \in \mathbb{R}^k$. Moreover, we may find an $n$ such that $\mathbb{P}\{|\underline{R}_n| \ge \delta\} < \frac{\epsilon}{4\|x\|}$. Calculating $E\left|x \circ \underline{S}_n - x \circ \underline{X}_n\right|$ on the set where $|\underline{R}_n| < \delta$ and its complement, we see that this expectation is no greater than $\frac{\epsilon}{2} + 2\|x\| \frac{\epsilon}{4\|x\|} = \epsilon$. $\square$

*Proof (of Donsker's Theorem)* This is a typical argument using compactness – compare e.g. 6.6.12.

The first lemma shows that there exists a subsequence of $\mathbb{P}_n, n \ge 1$ that converges weakly to a probability measure on $C(\mathbb{R}^+)$. By 5.6.2 and the second lemma, the finite-dimensional distributions of this measure are normal with appropriate covariance matrix. Hence, this limit measure is the Wiener measure $\mathbb{P}_W$. It remains to prove that the measures $\mathbb{P}_n$ converge to $\mathbb{P}_W$.

Suppose that this is not so. Then there exists a subsequence of our sequence that stays at some distance $\epsilon > 0$ away from $\mathbb{P}_W$. On the other hand, this subsequence has a further subsequence that converges to some probability measure. Using 5.6.2 and the second lemma again, we see that this limit measure must be the Wiener measure – a contradiction. $\qquad\square$

## 5.8 Notes on other modes of convergence

5.8.1 *Modes of convergence of random variables*      Throughout the book, we have encountered various modes of convergence of random variables. To list the most prominent ones, we have the following definitions:

(a) $X_n$ converges to $X$ a.s. if $\mathbb{P}\{\omega|\lim_{n\to\infty} X_n(\omega) = X(\omega)\} = 1$,

(b) $X_n$ converges to $X$ in $L^1$ norm if $\lim_{n\to\infty}\|X_n - X\|_{L^1} = 0$,

(c) $X_n$ converges to $X$ in probability if $\lim_{n\to\infty}\mathbb{P}\{|X_n - X| > \epsilon\} = 0$,

(d) $X_n$ converges to $X$ weakly iff $\mathbb{P}_{X_n}$ converges weakly to $\mathbb{P}_X$.

Note that in the first three cases we need to assume that $X_n$ and $X$ are random variables defined on the same probability space $(\Omega, \mathcal{F})$; additionally, in the third case these random variables need to be absolutely integrable. In the fourth case such an assumption is not needed.

It is easy to see that a.s. convergence does not imply convergence in $L^1$ norm, even if all the involved variables are absolutely integrable. Conversely, convergence in $L^1$ norm does not imply a.s. convergence (see 3.6.15).

If the random variables $X_n$ and $X$ are defined on the same probability space $(\Omega, \mathcal{F}, \mathbb{P})$, and are absolutely integrable, and $X_n$ converge to $X$ in $L^1(\Omega, \mathcal{F}, \mathbb{P})$, then they converge to $X$ in probability also – see (3.36). Similarly, a.s. convergence implies convergence in probability. Indeed, $\mathbb{P}\{|X_n - X| > \epsilon\} = E\mathbf{1}_{\{|X_n-X|>\epsilon\}}$ and the claim follows by the Dominated Convergence Theorem (compare (5.36)).

Finally, if $X_n, n \geq 1$, converge to $X$ in probability then they converge to $X$ weakly. This has been proved in 4.1.4, but we offer another proof here. To this end we note first that if $a$ is a point of continuity of the cumulative distribution function of $X$ then for any $\epsilon > 0$ one may find a

$\delta$ such that the difference between $\mathbb{P}\{X \leq a\}$ and $\mathbb{P}\{X \leq a - \delta\}$ is less than $\epsilon$. Therefore, $|\mathbb{P}\{X_n \leq a\} - \mathbb{P}\{X \leq a\}|$ is less than

$$\mathbb{P}\{X_n > a \wedge X \leq a\} + \mathbb{P}\{X_n \leq a \wedge X > a\}$$
$$\leq \mathbb{P}\{X_n > a \wedge X \leq a - \delta\} + \mathbb{P}\{X_n \leq a \wedge X > a + \delta\} + \epsilon$$
$$\leq 2\mathbb{P}\{|X_n - X| \geq \delta\} + \epsilon.$$

This implies that the upper limit of $|\mathbb{P}\{X_n \leq a\} - \mathbb{P}\{X \leq a\}|$ is less than $\epsilon$ and thus proves our claim.

In general, weak convergence does not imply convergence in probability even if all involved variables are defined on the same probability space. However, if they converge weakly to a constant, then they converge in probability also; the reader should prove it.

**5.8.2 Exercise**    Prove the last claim.

**5.8.3** *Scheffé's Theorem*    In general a.s. convergence does not imply $L^1$ convergence. We have, however, the following **Scheffé's Theorem.** Let $(\Omega, \mathcal{F}, \mu)$ be a measure space, and let $\phi$ and $\phi_n$ $n \geq 1$ be non-negative functions on $\Omega$ such that $\phi_n(\omega)$ converges to $\phi(\omega)$ a.s. and $c_n = \int \phi_n \, d\mu$ converges to $c = \int \phi \, d\mu$. Then $\int |\phi_n - \phi| \, d\mu$ converges to 0, i.e. the measures $\mu_n$ with densities $\phi_n$ converge strongly to the measure with density $\phi$.

*Proof*  Let $A_n = \{\phi \geq \phi_n\}$. We have

$$\int |\phi_n - \phi| \, d\mu = \int_{A_n} |\phi_n - \phi| \, d\mu + \int_{\Omega \setminus A_n} |\phi_n - \phi| \, d\mu$$
$$= 2 \int_{A_n} |\phi_n - \phi| \, d\mu + c_n - c.$$

Since $(\phi - \phi_n)1_{A_n} \leq \phi$, the theorem follows by the Lebesgue Dominated Convergence Theorem. $\qquad\square$

**5.8.4** *The Poisson Theorem as an example of strong convergence*    By 1.2.31 and Scheffé's Theorem, binomial distributions with parameters $n$ and $p_n$ converge strongly to the Poisson distribution with parameter $\lambda$ provided $np_n$ converges to $\lambda$, as $n \to \infty$.

**5.8.5 Example**    Let $\Omega = N, \phi_n(i) = \delta_{in}, \phi(i) = 0$ and $\mu$ be the counting measure. The requirements of Scheffé's Theorem are satisfied except for the one concerning integrals. We see that Dirac measures $\delta_n$ of which $\phi_n$ are densities do not converge to zero measure strongly.

**5.8.6 Corollary**      Let $\mu$ and $\mu_n, n \geq 1$, be probability measures on a set $S$ with countably many elements. The following are equivalent:

(a)  $\mu_n$ converges strongly to $\mu$,
(b)  $\mu_n$ converges weakly to $\mu$,
(c)  $\mu_n$ converges to $\mu$ in weak* topology,
(d)  $\mu_n(\{p\})$ converges to $\mu(\{p\})$ for all $p \in S$.

*Proof* Implications, (a) $\Rightarrow$(b)$\Rightarrow$(c) are obvious. To show that (c) implies (d) we note that the function $x_{p_0}(p) = 1$ for $p = p_0$ and zero otherwise is continuous on $S$ (with discrete topology). (d) implies (a) by Scheffé's Theorem.                                                                                    $\square$

# 6

# The Gelfand transform and its applications

## 6.1 Banach algebras

6.1.1 *Motivating examples*    Let $S$ be a locally compact Hausdorff topological space, and $C_0(S)$ be the space of continuous functions on $S$ that vanish at infinity, equipped with the supremum norm. Until now, we treated $C_0(S)$ as a merely Banach space. This space, however, unlike general Banach spaces, has an additional algebraic structure: two functions on $S$ may not only be added, they may be multiplied. The product of two functions, say $x$ and $y$, is another function, a member of $C_0(S)$ given by

$$(xy)(p) = x(p)y(p). \tag{6.1}$$

The operation so defined is associative and enjoys the following properties that relate it to the algebraic and topological structure introduced before:

(a)  $\|xy\| \le \|x\|\|y\|$,
(b)  $(\alpha x)y = \alpha(xy) = x(\alpha y), \alpha \in \mathbb{R}$,
(c)  $x(y_1 + y_2) = xy_1 + xy_2$,
(d)  $xy = yx$.

Moreover, if $S$ is compact, then $C_0(S) = C(S)$ has a unit, an element $u = 1_S$ such that $ux = xu = x$ for all $x \in C(S)$. We have $\|u\| = 1$.

For another example, let $\mathbb{X}$ be a Banach space and let $\mathcal{L}(\mathbb{X})$ be the space of bounded linear operators on $\mathbb{X}$. If we define $xy$ as the composition of two linear maps $x$ and $y \in \mathbb{X}$, it will be easy to see that conditions (a)–(c) above are satisfied (use 2.3.11), and that the identity operator is a unit. Such multiplication does not, however, satisfy condition (d).

Yet another example is the space $l^1(\mathbb{Z})$ of absolutely summable sequences $(\xi_n)_{n \in \mathbb{Z}}$. If we define the product of two sequences $x = (\xi_n)_{n \in \mathbb{Z}}$

and $y = (\eta_n)_{n \in \mathbb{Z}}$ as their convolution

$$\left( \sum_{i=-\infty}^{\infty} \xi_{n-i} \eta_i \right)_{n \in \mathbb{Z}}, \tag{6.2}$$

then conditions $(a) - (d)$ are satisfied. Also, the sequence $e_0 = (\delta_{n,0})_{n \geq 0}$ plays the role of the unit.

The final, very important example is the space $\mathbb{BM}(\mathbb{R})$ of (signed) Borel measures on $\mathbb{R}$ with convolution (2.14) as multiplication.

As a generalization of these examples, let us introduce the following definition.

6.1.2 **Definition**　　A **Banach algebra** $\mathbb{A}$ is a Banach space, equipped with an additional associative operation $\mathbb{A} \times \mathbb{A} \ni (x, y) \mapsto xy \in \mathbb{A}$, such that conditions (a)–(c) above hold, and additionally

(c′)　$(y_1 + y_2)x = y_1 x + y_2 x.$

6.1.3 *Various comments and more definitions*　　If (d) holds, $\mathbb{A}$ is said to be a **commutative Banach algebra** and condition (c′) is superfluous. If there is a unit $u$ satisfying the properties listed above for $u = 1_S$, $\mathbb{A}$ is said to be a **Banach algebra with unit**. All Banach algebras described above have units, except perhaps for $C_0(S)$, which has a unit iff $S$ is compact.

We will say that a subspace $\mathbb{B}$ of a Banach algebra $\mathbb{A}$ is a subalgebra of $\mathbb{A}$ if $\mathbb{B}$ is a Banach algebra itself, or that, in other words, the product of two elements of $\mathbb{B}$ lies in $\mathbb{B}$. As an example one may take $\mathbb{A} = \mathbb{BM}(\mathbb{R})$ and $\mathbb{B} = l^1(\mathbb{Z})$; the elements of $l^1(\mathbb{Z})$ may be viewed as charges on $\mathbb{R}$ that are concentrated on integers $\mathbb{Z}$; it is almost obvious that for two such charges their convolution is concentrated on $\mathbb{Z}$ as well, and that (2.14) becomes (6.2).

A bounded linear map $H$ from an algebra $\mathbb{A}$ to an algebra $\mathbb{B}$ is said to be a homomorphism if $H(xy) = (Hx)(Hy)$, for all $x$ and $y$ in $\mathbb{A}$. If $\mathbb{B}$ is the algebra of bounded linear operators on a Banach space $\mathbb{X}$, we say that $H$ is a **representation** of $\mathbb{A}$ in $\mathcal{L}(\mathbb{X})$.

Two Banach algebras $\mathbb{A}$ and $\mathbb{B}$ are said to be (isometrically) isomorphic if there exists a map $J : \mathbb{A} \to \mathbb{B}$ that is an (isometric) isomorphism between them as Banach spaces and, additionally, for all elements $x$ and $y$ of $\mathbb{A}$, $J(xy) = J(x)J(y)$.

We will illustrate these notions with examples and exercises given below. Before we do that, though, let us note that Exercises 2.3.12,

2.3.13, 2.3.14 and 2.3.15 are actually ones on Banach algebras. In other words, instead of operators $A, B \in \mathcal{L}(\mathbb{X})$ one may consider elements $a, b$ of an abstract Banach algebra with a unit $u$, define $\exp(a)$ and $\exp(b)$ as sums of the appropriate series and prove that $\exp(a)\exp(b) = \exp(a+b)$, provided $a$ commutes with $b$. Similarly, if $\|u - a\| < 1$, one may define $\log a$ and check that $\exp(\log a) = a$. We start by providing two more examples of this type.

**6.1.4 Exercise** Let $\mathbb{A}$ be a Banach algebra. Suppose that $\|x\| < 1$. Prove that the series $\sum_{n=1}^{\infty} x^n$ converges to an element, say $y$, of $\mathbb{A}$, and that $xy + x - y = yx + x - y = 0$.

**6.1.5 Exercise** Let $\mathbb{A}$ be a Banach algebra with unit $u$. Suppose that $\|x\| < 1$. Prove that the series $\sum_{n=0}^{\infty} x^n$, where $x^0 = u$ converges to an element, say $y$, of $\mathbb{A}$, and that $(u - x)y = y(u - x) = u$.

**6.1.6 Example** The left canonical representation of a Banach algebra $\mathbb{A}$ in the space $\mathcal{L}(\mathbb{A})$ of bounded linear operators in $\mathbb{A}$ is the map $\mathbb{A} \ni a \mapsto L_a \in \mathcal{L}(\mathbb{A})$ where $L_a b = ab$ for all $b \in \mathbb{A}$; certainly $L_a L_b = L_{ab}$. The right canonical representation is a similar map $a \mapsto R_a$ with $R_a$ given by $R_a b = ba$, and we have $R_a R_b = R_{ba}$. The reader will note that operators $S_\mu$ and $\tilde{S}_\mu$ from 2.3.25 are the left and right canonical representations of the algebra $\mathbb{BM}(\mathbb{R})$, respectively.

The inequality $\|L_a\| \leq \|a\|$ holds always, but to make sure that it is not strict it suffices to assume that $\mathbb{A}$ has a unit, or that it has a right approximate unit bounded by 1, i.e. that there exists a sequence $u_n \in \mathbb{A}, \|u_n\| \leq 1, n \geq 1$ such that $\lim_{n\to\infty} a u_n = a, a \in \mathbb{X}$. (Similarly, $\|R_a\| = \|a\|$ if $\mathbb{A}$ has a left approximate unit bounded by 1, i.e. that there exists a sequence $u_n \in \mathbb{A}, \|u_n\| \leq 1, n \geq 1$ such that $\lim_{n\to\infty} u_n a = a, a \in \mathbb{X}$.)

**6.1.7 Exercise** Let $\mathbb{A}$ be a Banach algebra without unit. Check that the set $\mathbb{A}_u = L^1(\mathbb{R}^+) \times \mathbb{R}$ equipped with multiplication $(a, \alpha)(b, \beta) = (\alpha b + \beta a + ab, \alpha\beta)$ and the norm $\|(a, \alpha)\| = \|a\| + |\alpha|$ is a Banach algebra with unit $(0, 1)$. It is commutative iff $\mathbb{A}$ is commutative. Moreover, the map $a \mapsto (a, 0)$ is an isometric isomorphism of $\mathbb{A}$ and the subalgebra $\mathbb{A} \times \{0\}$ of $\mathbb{A}_u$.

**6.1.8 Example** Let $L^1(\mathbb{R})$ be the Banach space of (equivalence classes of) absolutely integrable functions on $\mathbb{R}$ with the usual norm. As the reader will easily check, $L^1(\mathbb{R})$ is a commutative Banach algebra with convolution as multiplication:

$$x * y(\tau) = \int_{-\infty}^{\infty} x(\tau - \varsigma)y(\varsigma)\,d\varsigma. \tag{6.3}$$

This algebra may be viewed as a subalgebra of $\mathbb{BM}(\mathbb{R})$ composed of signed measures that are absolutely continuous with respect to the Lebesgue measure. Here "may be viewed" actually means that there exists an isometric isomorphism between $L^1(\mathbb{R})$ and the said subalgebra of measures (see Exercise 1.2.18). This isomorphism maps a measure into its density with respect to Lebesgue measure.

From the same exercise we see that if $x$ and $y$ are representatives of two elements of $L^1(\mathbb{R})$ such that $x(\tau) = y(\tau) = 0$, for $\tau < 0$, then $x * y(\tau) = 0$ for $\tau < 0$ and

$$x * y(\tau) = \int_0^\tau x(\tau - \varsigma)y(\varsigma)\,d\varsigma, \qquad \tau > 0. \tag{6.4}$$

This proves that the set of (equivalence classes) of integrable functions that vanish on the left half-axis is a subalgebra of $L^1(\mathbb{R})$, for obviously it is also a subspace of $L^1(\mathbb{R})$. This subalgebra is isometrically isomorphic to the algebra $L^1(\mathbb{R}^+)$ of (classes of) absolutely integrable functions on $\mathbb{R}^+$ with convolution given by (6.4). The isomorphism maps (a class of) a function from $L^1(\mathbb{R}^+)$ into (a class) of its natural extension $\tilde{x}$ to the whole axis given by $\tilde{x}(\tau) = x(\tau), \tau \geq 0$, $\tilde{x}(\tau) = 0, \tau < 0$.

In a similar fashion one proves that the space $L_e^1(\mathbb{R})$ of (equivalence classes of) even functions that are absolutely integrable on $\mathbb{R}$ is a subalgebra of $L^1(\mathbb{R})$. To this end it is enough to show that if $x$ and $y$ are even then $x * y$ defined in (6.3) is even also.

6.1.9 **Exercise** Prove that the subspace of $l^1(\mathbb{Z})$ formed by absolutely summable sequences $(\xi_n)_{n \in \mathbb{Z}}$ such that $\xi_n = 0$ for $n < 0$ is a subalgebra of $l^1(\mathbb{Z})$. Show that this subalgebra is isometrically isomorphic to the algebra $l^1(\mathbb{N}_0)$ of absolutely summable sequences $(\xi_n)_{n \geq 0}$ with convolution

$$(\xi_n)_{n \geq 0} * (\eta_n)_{n \geq 0} = \left( \sum_{i=0}^{n} \xi_{n-i}\eta_i \right)_{n \geq 0} \tag{6.5}$$

as multiplication.

**6.1.10 Example**    Consider the elements $e_\lambda, \lambda > 0$, of $L^1(\mathbb{R}^+)$ defined by their representatives $e_\lambda(\tau) = e^{-\lambda\tau}$. One checks that the following **Hilbert equation** is satisfied:

$$(\lambda - \mu)e_\lambda * e_\mu = e_\mu - e_\lambda, \qquad \lambda, \mu > 0. \tag{6.6}$$

Now, let us note that as a Banach space, $L^1(\mathbb{R}^+)$ is isometrically isomorphic to the space $L_e^1(\mathbb{R})$ of (equivalence classes of) even functions that are absolutely integrable on $\mathbb{R}$. The isomorphism $J : L_e^1(\mathbb{R}) \to L^1(\mathbb{R}^+)$ is given by $Jx(\tau) = 2x(\tau), \tau \in \mathbb{R}^+$, and $J^{-1}y(\tau) = \frac{1}{2}y(|\tau|)$ for $y \in L^1(\mathbb{R}^+)$ and $\tau \in \mathbb{R}$. Even though Banach algebras $L^1(\mathbb{R}^+)$ and $L_e^1(\mathbb{R})$ are isometrically isomorphic as Banach spaces, it turns out that as Banach algebras they are quite different. In particular, one checks that $\tilde{e}_\lambda = Je_\lambda \in L_e^1(\mathbb{R}), \lambda > 0$, instead of satisfying (6.6), satisfy

$$(\mu^2 - \lambda^2)\tilde{e}_\lambda * \tilde{e}_\mu = \mu\tilde{e}_\lambda - \lambda\tilde{e}_\mu, \qquad \lambda, \mu > 0. \tag{6.7}$$

(Note, though, that we *did not* prove that $L^1(\mathbb{R}^+)$ and $L_e^1(\mathbb{R})$ are not isomorphic; relations (6.6)–(6.7) show merely that the natural isomorphism of Banach spaces introduced above is not an isomorphism of Banach algebras.)

**6.1.11 Exercise**    Show that the space $l_e^1(\mathbb{Z})$ of even absolutely summable sequences $(\xi_n)_{n\in\mathbb{Z}}$ forms a subalgebra of $l^1(\mathbb{Z})$. Show that as a Banach space $l_e^1(\mathbb{Z})$ is isomorphic to $l^1(\mathbb{N}_0)$, but that the natural isomorphism of these Banach spaces is not an isometric isomorphism of Banach algebras.

**6.1.12 Exercise**    Prove that the family $u_t = 1_{[0,t)} \in L^(\mathbb{R}^+), t > 0$, satisfies the so-called "integrated semigroup equation"

$$u_t * u_s = \int_0^{t+s} u_r \, dr - \int_0^t u_r \, dr - \int_0^s u_r \, dr,$$

and that $\tilde{u}_t = Ju_t = \frac{1}{2}1_{(-t,t)} \in L_e^1(\mathbb{R})$ satisfies the so-called "sine function equation":

$$\tilde{u}_t\tilde{u}_s = \int_0^{t+s} \tilde{u}_r \, dr - \int_0^{|t-s|} \tilde{u}_r \, dr.$$

**6.1.13 Exercise**    Show that $e_n(\tau) = ne^{-n\tau}$ (a.s.), $n \geq 1$, is an approximate unit in $L^1(\mathbb{R}^+)$.

## 6.2 The Gelfand transform

6.2.1 *Multiplicative functionals*    As we have seen in the previous chapter, one may successfully study Banach spaces with the help of bounded linear functionals. However, to study Banach algebras, linear functionals will not suffice for they do not reflect the structure of multiplication in such algebras. We need to use *linear multiplicative functionals*. By definition, a linear multiplicative functional on a Banach algebra is a linear functional $F$, not necessarily bounded, such that, $F(xy) = Fx\,Fy$. In other words $F$ is a homomorphism of $\mathbb{A}$ and the algebra $\mathbb{R}$ equipped with the usual multiplication. To avoid trivial examples, we assume that $F$ is not identically equal to 0.

6.2.2 **Exercise**    Prove that if $F$ is a multiplicative functional on a Banach algebra with unit $u$ then $Fu = 1$.

6.2.3 **Lemma**    Linear multiplicative functionals are bounded; their norms never exceed 1.

*Proof* Suppose that for some $x_0$, $|Fx_0| > \|x_0\|$. Then for $x = \frac{\mathrm{sgn}Fx_0}{|Fx_0|}x_0 \in$ $\mathbb{A}$ we have that $\|x\| < 1$ and $Fx = 1$. Let $y$ be defined as in 6.1.4. We have $F(xy + x - y) = 0$, but on the other hand this equals $F(xy) + Fx - Fy = (Fx)(Fy) + Fx - Fy = 1$, a contradiction.    □

6.2.4 **Remark**    The idea of the above proof becomes even more clear if $\mathbb{A}$ has a unit. We construct $x$ as before but then take $y$ defined in 6.1.5 to arrive at the contradiction:

$$1 = Fu = Fy(Fu - Fx) = Fy(1 - 1) = 0.$$

6.2.5 **Exercise**    Suppose that an algebra $\mathbb{A}$ has a right or left approximate unit. Show that multiplicative functionals on $\mathbb{A}$ have norm 1.

6.2.6 **Example**    Let $C(S)$ be the algebra of continuous functions on a compact topological space. A linear functional on $C(S)$ is multiplicative iff there exists a point $p \in S$ such that $Fx = x(p)$ for all $x \in C(S)$.

*Proof* As always in such cases, the non-trivial part is the "only if" part. Let $F$ be a multiplicative functional. We claim first that there exists a $p \in S$ such that for all $x \in C(S)$, $Fx = 0$ implies $x(p) = 0$. Suppose that this is not so. Then for all $p \in S$, there exist a neighborhood $U_p$ of $p$ and a continuous function $x_p \in C(S)$ such that $Fx_p = 0$ but $x_p$ is non-zero in $U_p$. Consequently, $y_p = x_p^2$ is positive in $U_p$ and $Fy_p = 0$. Since $S$ is compact there exist $p_1, ..., p_n$ such that $S \subset \bigcup_{i=1}^{n} U_{p_i}$. Defining $z = \sum_{i=1}^{n} y_i$, and $z' = \frac{1}{z}$ (recall that $z > 0$), we obtain a contradiction: $1 = F(1_S) = F(z)F(z') = 0$.

Now, let $p \in S$ be such that $Fx = 0$ implies $x(p) = 0$. For any $x \in C(S)$, we have $F(x - F(x)1_S) = 0$. Hence, $x(p) = F(x)F(1_S) = F(x)$ as claimed. □

6.2.7 *The Gelfand transform*   Let us look once again at the definition of a multiplicative functional using the duality notation: a multiplicative functional on a Banach algebra is a bounded linear functional satisfying:

$$\langle F, xy \rangle = \langle F, x \rangle \langle F, y \rangle, \qquad \text{for all } x, y \in \mathbb{A}. \qquad (6.8)$$

In this definition, $F$ is fixed, while $x$ and $y$ vary. We will think differently now, however: we fix $x$ and $y$, and vary $F$. As we have seen, multiplicative functionals form a subset $\mathcal{M}$ of a unit ball in $\mathbb{A}^*$, which is compact in the weak* topology. One may prove (Exercise 6.2.8) that $\mathcal{M} \cup \{0\}$ where $0$ is a zero functional on $\mathbb{A}$ is closed in this topology, and thus $\mathcal{M}$ itself is locally compact ($0$ plays the role of "infinity" here). Hence, in (6.8) the elements $x$ and $y$ may be viewed as *functions defined on the locally compact set of multiplicative functionals*. In such an interpretation, (6.8) may be written down in a still more clear way:

$$xy(F) = x(F)y(F);$$

which is just (6.1) with $p$ replaced by $F$! So, if we forget about the fact that $F$s are "really" functionals and treat them as points of a topological space, we will be able to view our abstract algebra $\mathbb{A}$ in a much simpler way. In other words, we have succeeded in transforming an abstract Banach algebra into a subset of the algebra $C_0(\mathcal{M})$. (We need to talk about a subset of $C_0(\mathcal{M})$ for there might be some functions in $C_0(\mathcal{M})$ that do not correspond to any element of $\mathbb{A}$; see e.g. 6.3.1 further on.) A map described above, transforming elements $x \in \mathbb{A}$ into functions $x(F)$ defined on $\mathcal{M}$, is called *the Gelfand transform*. Certainly, the Gelfand transform is a homomorphism of Banach algebras $\mathbb{A}$ and $C_0(\mathcal{M})$. As we will see shortly by looking at a number of examples, the set $\mathcal{M}$, though defined in an abstract way, in many cases forms a very nice and simple topological space. This method of viewing Banach algebras has proven useful in many situations; see e.g. [27, 51, 82, 103, 104, 112, 115].

Unfortunately, in our presentation we did not take care of one important fact. Specifically, in contrast to the rich spaces of linear functionals that allowed us to describe a Banach space in such a pleasant way, the space of multiplicative functionals may be too small to describe the Banach algebra well. In fact, there are Banach algebras with *no multiplicative functionals*, and the only general condition for existence of a rich space of multiplicative functionals is that the Banach algebra is commutative, with identity and – complex. This means in particular that multiplication of vectors by complex scalars is allowed. Technically, the situation is not much different from the real case; all that happens is that in all the involved definitions the real field $\mathbb{R}$ must be replaced by a complex field $\mathbb{C}$. An example of a complex Banach space (and a complex algebra) is the space of complex-valued functions on a compact topological space. On the other hand, properties of real and complex spaces differ considerably. In particular, as we have said, a complex, commutative Banach algebra has a rich space of multiplicative functionals, while a real Banach algebra may have none, even if it is commutative and has a unit.

Nevertheless, there are a good number of real Banach algebras for which the space of multiplicative functionals is rich enough for the Gelfand transform to preserve the most important properties of the element of this algebra. The examples presented in the next section, all of them often used in probability, explain this idea in more detail. As we shall see, a function mapping a measure on $\mathbb{N}$ into the related probability generating function is an example of the Gelfand transform, and so are the Fourier and Laplace transforms of (densities of) measures on the real axis and positive half-axis, respectively. The reader should not, however, jump to the examples before solving the problems presented below.

**6.2.8 Exercise**　　Let $\mathbb{A}$ be a Banach algebra. Prove that $\mathcal{M} \cup \{0\}$ is closed in the weak* topology of $\mathbb{A}^*$. Moreover, if $\mathbb{A}$ has a unit then $\mathcal{M}$ is closed.

**6.2.9 Exercise**　　Let $\mathbb{A}$ be a Banach algebra with a unit $u$ and let $x \in \mathbb{A}$. Show that for any multiplicative functional $F$ on $\mathbb{A}$, $F(\exp x) = \exp(Fx)$. In other words, the Gelfand transform of $\exp x$ equals $\exp \hat{x}$, where $\hat{x}$ is the Gelfand transform of $x$.

### 6.3 Examples of Gelfand transform

6.3.1 *Probability generating function*　　The space $l^1 = l^1(\mathbb{N}_0)$ of absolutely summable sequences $x = (\xi_n)_{n \geq 0}$ is a Banach algebra with convolution (6.5) as multiplication. We already know (see 5.2.3 or 5.2.16) that a linear functional on $l^1$ has to be of the form

$$Fx = \sum_{n=0}^{\infty} \xi_n \alpha_n \tag{6.9}$$

where $\alpha_i = Fe_i = F(\delta_{i,n})_{n \geq 0}$, and $\sup_{n \geq 0} |\alpha_n| < \infty$. To determine the form of a multiplicative functional, observe first that $e_2 = e_1 * e_1$ and, more generally, $e_n = e_1^{n*}, n \geq 0$ where the last symbol denotes the $n$th convolution power of $e_1$. Also, $Fe_0 = 1$. Hence $\alpha_n = \alpha^n, n \geq 0$, where $\alpha = \alpha_1$. Therefore, if $F$ is multiplicative, then there exists a real $\alpha$ such that $Fx = \sum_{n=0}^{\infty} \alpha^n \xi_n$. This $\alpha$ must belong to the interval $[-1, 1]$ for otherwise the sequence $\alpha_n$ would be unbounded. Conversely, any functional of this form is multiplicative.

This proves that a multiplicative functional on $l^1$ may be identified with a number $\alpha \in [-1, 1]$. Moreover, one may check that the weak* topology restricted to the set of multiplicative functionals is just the usual topology in the interval $[-1, 1]$. In other words, the image via the Gelfand transform of the vector $x \in l^1$ is a function $\hat{x}$ on $[-1, 1]$ given by $\hat{x}(\alpha) = \sum_{i=0}^{\infty} \alpha^i \xi_i$. However, in this case, the space of multiplicative functionals is "too rich"; the information provided by $\alpha \in [-1, 0)$ is superfluous. Therefore, in probability theory it is customary to restrict the domain of $\hat{x}$ to the interval $[0, 1]$.

We note that the interval $[-1, 1]$ is compact, and that this is related to the fact that $l^1$ has a unit. It is also worth noting that the image via Gelfand transform of a sequence in $l^1$ is not just "any" continuous function on $[-1, 1]$; this function must be expandable into a series $\hat{x}(\alpha) = \sum_{n=0}^{\infty} \xi_n \alpha^n$ and in particular be infinitely differentiable and analytic. This illustrates the fact that the image of a Banach algebra via the Gelfand transform is usually a subset of the algebra $C_0(\mathcal{M})$.

**6.3.2 Exercise**    Consider the space $l^1(\mathbb{N}_0 \times \mathbb{N}_0)$ of infinite matrices $x = (\xi_{i,j})_{i,j \geq 0}$ such that $\sum_{i=0}^{\infty} \sum_{j=0}^{\infty} |\xi_{i,j}| < \infty$ with the norm $\|x\| = \sum_{i=0}^{\infty} \sum_{j=0}^{\infty} |\xi_{i,j}|$. Prove that this is a Banach space and a Banach algebra with multiplication defined as

$$(\xi_{i,j})_{i,j \geq 0} * (\eta_{i,j})_{i,j \geq 0} = \left( \sum_{k=0}^{i} \sum_{l=0}^{j} \xi_{k,l} \eta_{i-k,j-l} \right)_{i,j \geq 0}.$$

Prove that all multiplicative functionals on this space are of the form $Fx = \sum_{i=0}^{\infty} \sum_{j=0}^{\infty} \alpha^i \beta^j \xi_{i,j}$, where $\alpha$ and $\beta$ are in $[-1, 1]$. In other words, the topological space $\mathcal{M}$ of multiplicative functionals may be identified (i.e. it is homeomorphic) with $[-1, 1] \times [-1, 1]$. As in 6.3.1 we note that in probability theory one usually uses only non-negative values of $\alpha$ and $\beta$. If $x$ is a distribution, the function

$$g(\alpha, \beta) = \sum_{i=0}^{\infty} \sum_{j=0}^{\infty} \alpha^i \beta^j \xi_{i,j}, \qquad \alpha, \beta \in [0, 1]$$

is termed the **joint probability generating function**.

**6.3.3** *The Laplace transform*     Let $L^1(\mathbb{R}^+)$ be the Banach algebra of absolutely integrable functions on $\mathbb{R}^+$ with the usual norm and convolution as multiplication. From 5.2.16 we know that if $F$ is a linear functional on $L^1(\mathbb{R}^+)$, then there exists a bounded function, say $\alpha(\cdot)$, on $\mathbb{R}^+$ such that

$$Fx = \int_0^\infty x\alpha \, dleb, \qquad x \in L^1(\mathbb{R}^+). \tag{6.10}$$

Assume that $F$ is multiplicative. We have

$$\begin{aligned}
F(x * y) &= \int_0^\infty \alpha(\tau) \, x * y(\tau) \, d\tau \\
&= \int_0^\infty \alpha(\tau) \int_0^\tau x(\tau - \varsigma) y(\varsigma) \, d\varsigma \, d\tau \\
&= \int_0^\infty \int_\varsigma^\infty \alpha(\tau) x(\tau - \varsigma) y(\varsigma) \, d\tau \, d\varsigma \\
&= \int_0^\infty \int_0^\infty \alpha(\tau + \varsigma) x(\tau) y(\varsigma) \, d\tau \, d\varsigma \tag{6.11}
\end{aligned}$$

and

$$FxFy = \int_0^\infty \int_0^\infty \alpha(\tau) \alpha(\varsigma) x(\tau) y(\varsigma) \, d\tau \, d\varsigma. \tag{6.12}$$

First we show that $\alpha$ may be chosen to be continuous, i.e. that there is a continuous function in the equivalence class of $\alpha$. By 2.2.44, there exists a continuous function $x_0$ with compact support such that $Fx_0 = \int_0^\infty \alpha x_0 \, dleb \neq 0$. Using (6.11) and (6.12),

$$\int_0^\infty y(\varsigma) \alpha(\varsigma) \int_0^\infty \alpha(\tau) x_0(\tau) \, d\tau \, d\varsigma = \int_0^\infty y(\varsigma) \int_0^\infty \alpha(\tau + \varsigma) x_0(\tau) \, d\tau \, d\varsigma.$$

Since $y \in L^1(\mathbb{R}^+)$ is arbitrary, $\alpha(\varsigma) = \frac{\int_0^\infty \alpha(\tau + \varsigma) x_0(\tau) \, d\tau}{Fx_0}$, for almost all $\varsigma \in \mathbb{R}^+$. Hence it suffices to show that $\beta(\varsigma) = \int_0^\infty \alpha(\tau + \varsigma) x_0(\tau) \, d\tau =$

$\int_{\varsigma}^{\infty} \alpha(\tau) x_0(\tau - \varsigma) \, d\tau$ is continuous. To this end, note that for $\varsigma, h > 0$ we have

$$|\beta(\varsigma + h) - \beta(\varsigma)| \leq \int_{\varsigma+h}^{\infty} |\alpha(\tau)| \, |x_0(\tau - \varsigma - h) - x_0(\tau - \varsigma)| \, d\varsigma$$
$$+ \int_{\varsigma}^{\varsigma+h} |\alpha(\tau)| \, |x_0(\tau - \varsigma)| \, d\tau,$$

that the second integral tends to zero as $h \to 0$, and that the first may be estimated by $\|\alpha\|_{L^{\infty}(\mathbb{R}^+)} \sup_{\sigma \in \mathbb{R}^+} |x_0(\sigma+h) - x_0(\sigma)|$ which also converges to zero by virtue of uniform continuity of $x_0$.

Let us assume thus that $\alpha$ is continuous. Using (6.11) and (6.12), we see now that a linear functional $F$ is multiplicative if $\alpha(\tau+\varsigma) = \alpha(\tau)\alpha(\varsigma)$, for almost all $\tau, \varsigma \in \mathbb{R}^+$. Since $\alpha$ is continuous this relation holds for all $\tau$ and $\varsigma$ in $\mathbb{R}^+$. In particular, taking $\tau = \varsigma = \frac{1}{2}\sigma$ we see that for any $\sigma > 0$, $\alpha(\sigma) = \alpha(\sigma)^2 \geq 0$. Moreover, we may not have $\alpha(\sigma) = 0$ for some $\sigma \in \mathbb{R}^+$ for this would imply $\alpha(\tau) = 0$ for all $\tau \in \mathbb{R}^+$ and $Fx = 0$ for all $x \in L^1(\mathbb{R}^+)$. Thus we may consider $\beta(\tau) = \ln \alpha(\tau)$ (ln is the natural logarithm). We have $\beta(\tau + \varsigma) = \beta(\tau) + \beta(\varsigma)$. In Section 1.6 we have proven that this implies $\beta(\tau) = \mu\tau$ for some $\mu \in \mathbb{R}$. Thus, $\alpha(\tau) = e^{\mu\tau}$. Since $\alpha$ must be bounded, we must have $\mu = -\lambda, \lambda \geq 0$.

We have proven that all multiplicative functionals on $L^1(\mathbb{R}^+)$ are of the form

$$Fx = \int_{0}^{\infty} e^{-\lambda\tau} x(\tau) \, d\tau;$$

in other words, the set $\mathcal{M}$ of multiplicative functionals on $L^1(\mathbb{R}^+)$ may be identified with $\mathbb{R}^+$. One still needs to check that topologies in $\mathbb{R}^+$ and in the space $\mathcal{M}$ agree, but this is not difficult. We remark that the fact that $L^1(\mathbb{R}^+)$ does not have a unit has its reflection in the fact that $\mathbb{R}^+$ is not compact; the point at infinity corresponds to the zero functional on $L^1(\mathbb{R}^+)$.

6.3.4 *The Fourier cosine series*   What is the general form a multiplicative functional on the algebra $l_e^1(\mathbb{Z})$ defined in Exercise 6.1.11? We know that as a Banach space $l_e^1(\mathbb{Z})$ is isometrically isomorphic to $l^1(\mathbb{N}_0)$ and the isomorphism is given by

$$Ix = I(\xi_n)_{n \geq 0} = y = (\eta_n)_{n \in \mathbb{Z}}, \quad \text{where } \eta_n = \frac{1}{2}\xi_{|n|}, n \neq 0, \eta_0 = \xi_0$$

and

$$I^{-1}y = I^{-1}(\eta_n)_{n \in \mathbb{Z}} = x = (\xi_n)_{n \geq 0}, \quad \text{where } \xi_n = 2\eta_n, n > 0, \xi_0 = \eta_0.$$

If $F$ is a linear functional on $l_e^1(\mathbb{Z})$, then $FI$ is a linear functional on $l^1(\mathbb{N}_0)$ which must be of the form (6.9). Hence $FIx = \sum_{n=0}^{\infty} \alpha_n \xi_n$ for all $x = (\xi_n)_{n \geq 0}$ in $x \in l^1(\mathbb{N}_0)$, and

$$Fy = FIx = \sum_{n=0}^{\infty} \alpha_n \xi_n = \alpha_0 \xi_0 + 2 \sum_{n=1}^{\infty} \alpha_n \eta_n$$

$$= \alpha_0 \xi_0 + \sum_{n \in \mathbb{Z}, n \neq 0} \alpha_n \eta_n = \sum_{n \in \mathbb{Z}} \alpha_n \eta_n. \qquad (6.13)$$

Here we put $\alpha_{-n} = \alpha_n$ for $n \geq 1$, making $\alpha_n$ an even sequence.

Now, assume that $F$ is multiplicative. Since $e_0$ is a unit for our algebra, $\alpha_0 = F(e_0) = 1$. Note that $e_n, n \neq 0$, do not belong to $l_e^1(\mathbb{Z})$ but $\frac{1}{2}(e_{-n} + e_n)$ do. Also, $\alpha_n = F\frac{1}{2}(e_{-n} + e_n)$. If $m$ and $n$ are both odd or both even, then

$$2\alpha_{\frac{n-m}{2}} \alpha_{\frac{n+m}{2}} = \frac{1}{2}F(e_{-\frac{1}{2}(n-m)} + e_{\frac{1}{2}(n-m)})(e_{-\frac{1}{2}(n+m)} + e_{\frac{1}{2}(n+m)})$$

$$= \frac{1}{2}F(e_{-n} + e_n + e_{-m} + e_m) = \alpha_n + \alpha_m.$$

Let $\alpha = \alpha_1$. Since $\|\frac{1}{2}(e_{-1} + e_1)\| = 1$, $\alpha$ belongs to $[-1, 1]$, and one may choose a $t \in [0, 2\pi)$ such that $\alpha = \cos t$. We claim that $\alpha_n = \cos nt, n \in \mathbb{Z}$. To prove this note first that there is only one even sequence $\alpha_n, n \in \mathbb{Z}$, such that

$$2\alpha_{\frac{n-m}{2}} \alpha_{\frac{n+m}{2}} = \alpha_n + \alpha_m, \qquad \text{and } \alpha_0 = 1, \alpha_1 = \cos t \qquad (6.14)$$

where $n$ and $m$ are either both even or both odd. In other words the sequence $\alpha_n, n \in \mathbb{Z}$ is uniquely determined by the functional equation and initial conditions (6.14). This fact is left as an exercise for the reader. Secondly, using the well-known trigonometric identity we check that $\alpha_n = \cos nt$ satisfies (6.14). As a result, invoking (6.13) we see that all multiplicative functionals on $l_e^1(\mathbb{Z})$ are of the form

$$F(\eta_n)_{n \in \mathbb{Z}} = \sum_{n=-\infty}^{\infty} \eta_n \cos nt, \qquad t \in [0, 2\pi).$$

The reader may notice, however, that the topological space $\mathcal{M}$ is not isomorphic to $[0, 2\pi)$; it may not be since $[0, 2\pi)$ is not compact, while $\mathcal{M}$ is, for $l_e^1(\mathbb{Z})$ has a unit. Moreover, if a sequence $t_k \in [0, 2\pi)$ tends to $2\pi$ then the corresponding functionals $F_{t_k} = \sum_{n=-\infty}^{\infty} \eta_n \cos nt_k$ tend in the weak* topology to $F_0 = \sum_{n=-\infty}^{\infty} \eta_n$. This suggests that instead of the interval $[0, 2\pi)$ we should think of a unit circle in the complex plane

$\mathbf{C} = \{z; z = e^{it}, t \in [0, 2\pi)\}$, which, for the one thing, is compact. The reader will check that $\mathcal{M}$ may indeed be identified with this set.

**6.3.5 Exercise**   Show that (6.14) determines the sequence $\alpha_n, n \in \mathbb{Z}$ uniquely.

**6.3.6** *The Fourier series*   Let us consider the algebra $l^1(\mathbb{Z})$. By 5.2.16, all linear functionals on this space have to be of the form

$$F(\xi_n)_{n \in \mathbb{Z}} = \sum_{n=-\infty}^{\infty} \alpha_n \xi_n \qquad (6.15)$$

where $\alpha_n, n \in \mathbb{Z}$ is bounded. As in 6.3.1, we see that $\alpha_n = Fe_n = (Fe_1)^n = \alpha^n$; this time however this relation holds for all $n \in \mathbb{Z}$. Hence, the only multiplicative functionals in $l^1(\mathbb{Z})$ are

$$F(\xi_n)_{n \in \mathbb{Z}} = \sum_{n=-\infty}^{\infty} \xi_n, \qquad \text{and} \qquad F(\xi_n)_{n \in \mathbb{Z}} = \sum_{n=-\infty}^{\infty} (-1)^n \xi_n.$$

The situation changes if we allow sequences in $l^1(\mathbb{Z})$ to have complex values. Repeating the argument from 5.2.3 one proves that linear functionals on $l^1(\mathbb{Z})$ (that are complex-valued now) have to be of the form (6.15) with a bounded, complex sequence $\alpha_n$. In such case, there are infinitely many bounded solutions to $\alpha_n = \alpha^n$, all of them of the form $\alpha_n = e^{itn}, t \in [0, 2\pi)$. In other words, the image of the element of the complex $l^1(\mathbb{Z})$ via the Gelfand transform is a function $\hat{x}$ on $[0, 2\pi)$ given by

$$\hat{x}(t) = \sum_{n=-\infty}^{\infty} \xi_n e^{itn}, \quad t \in [0, 2\pi),$$

called the Fourier series of $x$. As in the previous subsection we note, however, that the topological space $\mathcal{M}$ of multiplicative functionals may not be identified with the interval $[0, 2\pi)$ but rather with the unit circle.

**6.3.7** *The Fourier transform*   Let $L^1(\mathbb{R})$ be the Banach algebra of absolutely integrable functions on $\mathbb{R}^+$ with the usual norm and convolution as multiplication. Arguing as in 6.3.3 we obtain that all multiplicative functionals on this algebra have to be of the form

$$Fx = \int_{-\infty}^{\infty} \alpha \, x \, dleb \qquad (6.16)$$

where $\alpha$ is continuous and satisfies

$$\alpha(\tau + \varsigma) = \alpha(\tau)\alpha(\varsigma), \qquad \text{for all } \tau, \varsigma \in \mathbb{R}. \tag{6.17}$$

This implies that $\alpha(\tau) = a^\tau$ where $a = \alpha(1)$. Since $\alpha(1) = \alpha(1/2)\alpha(1/2)$ is non-negative we may choose a real $\mu$ such that $\alpha(\tau) = e^{\mu\tau}$. The requirement that $\alpha$ be bounded forces now $\mu = 0$, so that the only multiplicative functional on $L^1(\mathbb{R})$ is

$$Fx = \int_{-\infty}^{\infty} x \, dleb.$$

If we allow the members of $L^1(\mathbb{R})$ to have complex values, the analysis presented above does not change except that $\alpha(\tau)$ turns out to be of the form $\alpha(\tau) = e^{\mu\tau}$ where $\mu$ is complex. The boundedness condition leads now to $\mu = it$ where $t \in \mathbb{R}$. As a result, all multiplicative functionals on the complex space $L^1(\mathbb{R})$ are of the form:

$$Fx = \int_{-\infty}^{\infty} e^{it\tau} x(\tau) \, d\tau, \quad t \in \mathbb{R}.$$

One proves that $\mathcal{M}$ is indeed topologically isomorphic to $\mathbb{R}$. The image $\hat{x}$ of an element $x$ of the complex $L^1(\mathbb{R})$ is its Fourier transform. Note that $\mathbb{R}$ is not compact and that $L^1(\mathbb{R})$ does not have a unit. As a bonus we obtain the fact that the Fourier transform of an absolutely integrable function belongs to $C_0(\mathbb{R})$.

6.3.8 *The Fourier cosine transform*   Let $L_e^1(\mathbb{R})$ be the Banach algebra of even, absolutely integrable functions on $\mathbb{R}$ with the usual norm and convolution as multiplication. As a Banach space $L_e^1(\mathbb{R})$ is isometrically isomorphic to $L^1(\mathbb{R}^+)$. Hence arguing as in 6.3.4, one proves that all linear functionals on $L_e^1(\mathbb{R})$ are of the form

$$Fx = \int_{-\infty}^{\infty} x(\tau)\alpha(\tau) \, d\tau = 2\int_0^{\infty} x(\tau)\alpha(\tau) \, d\tau, \tag{6.18}$$

where $\alpha$ is a bounded, even function. Calculating as in (6.11) and (6.12), we obtain

$$F(x * y) = \int_{-\infty}^{\infty}\int_{-\infty}^{\infty} \alpha(\tau + \varsigma)x(\tau)y(\varsigma) \, d\tau \, d\varsigma \tag{6.19}$$

and

$$Fx \, Fy = \int_{-\infty}^{\infty}\int_{-\infty}^{\infty} \alpha(\tau)\alpha(\varsigma)x(\tau)y(\varsigma) \, d\tau \, d\varsigma. \tag{6.20}$$

Even though these two equations are formally the same as (6.11) and (6.12), we may not claim that they imply that $\alpha(\tau+\varsigma) = \alpha(\tau)\alpha(\varsigma)$ since $x$ and $y$ are not arbitrary: they are even. Nevertheless, rewriting the right-hand side of (6.19) as

$$\int_{-\infty}^{\infty}\int_{-\infty}^{\infty}\alpha(\tau+\varsigma)x(\tau)y(-\varsigma)\,\mathrm{d}\tau\,\mathrm{d}\varsigma = \int_{-\infty}^{\infty}\int_{-\infty}^{\infty}\alpha(\tau-\varsigma)x(\tau)y(\varsigma)\,\mathrm{d}\tau\,\mathrm{d}\varsigma$$

we obtain

$$F(x*y) = \int_{-\infty}^{\infty}\int_{-\infty}^{\infty}\frac{1}{2}[\alpha(\tau+\varsigma)+\alpha(\tau-\varsigma)]x(\tau)y(\varsigma)\,\mathrm{d}\tau\,\mathrm{d}\varsigma. \quad (6.21)$$

Observe that

$$\alpha(\tau+\varsigma)+\alpha(\tau-\varsigma) = \alpha(\tau+\varsigma)+\alpha(-\tau+\varsigma) = \alpha(-\tau-\varsigma)+\alpha(-\tau+\varsigma)$$
$$= \alpha(-\tau-\varsigma)+\alpha(\tau-\varsigma)$$

so that (6.21) equals

$$2\int_{0}^{\infty}\int_{0}^{\infty}[\alpha(\tau+\varsigma)+\alpha(\tau-\varsigma)]x(\tau)y(\varsigma)\,\mathrm{d}\tau\,\mathrm{d}\varsigma.$$

Using (6.18),

$$FxFy = 4\int_{0}^{\infty}\int_{0}^{\infty}\alpha(\tau)\alpha(\varsigma)x(\tau)y(\varsigma)\,\mathrm{d}\tau\,\mathrm{d}\varsigma.$$

Since $x$ and $y$ restricted to $\mathbb{R}$ may be arbitrary members of $L^1(\mathbb{R})$ we infer that if $F$ is a multiplicative functional then

$$\alpha(\tau+\varsigma)+\alpha(\tau-\varsigma) = 2\alpha(\tau)\alpha(\varsigma) \quad (6.22)$$

for almost all $\tau$ and $\varsigma$ in $\mathbb{R}^+$. This is the famous cosine equation, a continuous equivalent of the equation which we have encountered in 6.3.4. Arguing as in 6.2.6 we prove that $\alpha$ may be chosen to be continuous, and thus that (6.22) holds for all $\tau$ and $\varsigma$ in $\mathbb{R}^+$. The reader will not be surprised to learn that all continuous solutions to (6.22) are of the form $\alpha(\tau) = \cos(t\tau), t \in \mathbb{R}$. Hence, all multiplicative functionals on $L_e^1(\mathbb{R})$ are of the form

$$Fx = \int_{-\infty}^{\infty}\cos(t\tau)\,x(\tau)\,\mathrm{d}\tau.$$

Also, the locally compact topological space $\mathcal{M}$ may be identified with $\mathbb{R}$. The image

$$\hat{x}(t) = \int_{-\infty}^{\infty}\cos(t\tau)\,x(\tau)\,\mathrm{d}\tau$$

of a member of $L_e^1(\mathbb{R})$ is termed the Fourier cosine transform.

**6.3.9** *The Gelfand transform on the Klein group*    Consider the space $L^1(\mathbb{G})$ of functions on the Klein group that are absolutely summable with respect to the counting measure. This space coincides with the space of all signed measures on the Klein group. The convolution in this space was discussed in 2.2.19. The space $L^1(\mathbb{G})$ with multiplication thus defined is a Banach algebra with unit identified with a sequence $(1, 0, 0, 0)$; which is just the Dirac measure on the unit of the Klein group. (The elements of this space will be written as column vectors or row vectors.)

Let us find all multiplicative functionals on $L^1(\mathbb{G})$. To this end, note first that by 5.2.16 all linear functionals on $L^1(\mathbb{G})$ are of the form $F(a_i)_{i=1,2,3,4} = \sum_{i=1}^{4} \alpha_i a_i$ where $\alpha_i$ are real numbers. If $F$ is to be multiplicative, we must have

$$\sum_{i=1}^{4} \alpha_i c_i = \left[ \sum_{i=1}^{4} \alpha_i a_i \right] \left[ \sum_{i=1}^{4} \alpha_i b_i \right]$$

for all $(a_i)_{i=1,2,3,4}$ and $(b_i)_{i=1,2,3,4}$ where $c_i$ are defined by (1.12). A direct calculation shows that the possible choices for $\alpha_i$ are

$$F_1 = (+1, +1, +1, +1),$$
$$F_2 = (+1, -1, -1, +1),$$
$$F_3 = (+1, -1, +1, -1),$$
$$F_4 = (+1, +1, -1, -1).$$

We could write this as follows:

$$\begin{bmatrix} F_1\mu \\ F_2\mu \\ F_3\mu \\ F_4\mu \end{bmatrix} = G\mu = \begin{bmatrix} +1 & +1 & +1 & +1 \\ +1 & -1 & -1 & +1 \\ +1 & -1 & +1 & -1 \\ +1 & +1 & -1 & -1 \end{bmatrix} \begin{bmatrix} a_1 \\ a_2 \\ a_3 \\ a_4 \end{bmatrix}, \qquad \text{where } \mu = \begin{bmatrix} a_1 \\ a_2 \\ a_3 \\ a_4 \end{bmatrix}.$$
(6.23)

In other words, $\mathcal{M} = \{F_1, F_2, F_3, F_4\}$ is a finite set, and the Gelfand transform of an element of $\mu$ of $L^1(\mathbb{G})$ may be identified with

$$(\mu(F_i))_{i=1,2,3,4} = (F_i\mu)_{i=1,2,3,4},$$

treated as members of the algebra $\mathbb{R}^4$ with coordinatewise multiplication.

Note that the matrix $G$, appearing in (6.23), is invertible and that $G^{-1} = \frac{1}{4} G$.

We will apply this result in 7.8.4 below.

**6.3.10 Remark**   As a simple calculation shows, the functions $\tau \to e^{it\tau}$, $t \in \mathbb{R}$, generate multiplicative functionals on the algebra $\mathbb{BM}(\mathbb{R})$ and $\tau \to e^{-\lambda\tau}$ generate multiplicative functionals on the algebra $\mathbb{BM}(\mathbb{R}^+)$. Finding the general form of multiplicative functionals on the algebra $\mathbb{BM}(\mathbb{R})$ or $\mathbb{BM}(\mathbb{R}^+)$ is a more difficult task. We will not pursue this subject because fortunately *we do not need* all multiplicative functionals to describe members of these algebras. All we need is the Fourier transform of a member of $\mathbb{BM}(\mathbb{R})$ and the Laplace transform of a member of $\mathbb{BM}(\mathbb{R}^+)$ (of course it requires a proof, see Section 6.6). In a sense, the situation is similar to that in 6.3.1 where we remarked that although the space of multiplicative functionals on $l^1(\mathbb{N}_0)$ is isomorphic to $[-1, 1]$ it suffices to consider the functionals corresponding to $\alpha \in [0, 1]$. This is not only fortunate but also very useful, and we will have plenty of examples that illustrate this.

## 6.4 Examples of explicit calculations of Gelfand transform

**6.4.1 Exercise**   Let $Z_n$ be the Galton–Watson process described in 3.5.5 and let $f$ be the common probability generating function of the (distribution of) random variables $X_n^k$ used in the construction of $Z_n$. Show that the probability generating function of the (distribution of) $Z_n$ is the $n$th composition $f^{\circ n}$ of $f$, where $f^{\circ(n+1)} = f(f^{\circ n})$, and $f^{\circ 1} = f$.

**6.4.2 Exercise**   Find the probability generating function of (a) the binomial distribution, (b) the Poisson distribution, and (c) the geometric distribution.

**6.4.3 Exercise**   Let $X$ and $Y$ be two independent random variables with values in the set of non-negative integers, with probability generating functions $f_X$ and $f_Y$, respectively. Let $f_{X,Y}$ be the joint probability generating function of the (distribution of the) pair $(X, Y)$. Show that $f_{X,Y}(\alpha, \beta) = f_X(\alpha) f_Y(\beta)$.

**6.4.4 Exercise**   Let $f_X$ be the probability generating function of a random variable $X$ with values in non-negative integers. Show that the joint probability generating function of the pair $(X, X)$ is $f_{X,X}(\alpha, \beta) = f_X(\alpha + \beta)$.

**6.4.5 Example**     Let us recall that for any $0 \le \alpha < 1$ and a positive number $r$,

$$\sum_{n=0}^{\infty} \binom{n+r-1}{n} \alpha^n = \frac{1}{(1-\alpha)^r} \qquad (6.24)$$

where $\binom{n+r-1}{n} = \frac{r(r+1)\cdots(r+n-1)}{n!}$. This may be checked by expanding the right-hand side into Maclaurin's series. In particular, if $0 < p \le 1$, and $r = k$ is a positive integer, then $p_n = \binom{n-1}{k-1}q^{n-k}p^k, q = 1-p, n = k, k+1, \ldots$ is a distribution of a random variable, because

$$\sum_{n=k}^{\infty} \binom{n-1}{k-1} q^{n-k} p^k = p^k \sum_{n=0}^{\infty} \binom{n-1+k}{n} q^n = \frac{p^k}{(1-q)^k} = 1.$$

The quantity $p_n$ is the probability that in a series of independent Bernoulli trials with parameter $p$ the $k$th success will occur at the $n$th trial [40]. In other words $(p_n)_{n \ge k}$ is the distribution of the time $T$ to the $k$th success. The distribution $(p_n)_{n \ge 0}$ where $p_n = 0$ for $n < k$ is termed the **negative binomial** distribution with parameters $p$ and $k$. The probability generating function of this distribution is given by

$$f(\alpha) = p^k \sum_{n=k}^{\infty} \binom{n-1}{k-1} q^{n-k} \alpha^n = p^k \alpha^k \sum_{n=0}^{\infty} \binom{n-1+k}{k-1} q^n \alpha^n$$

$$= \frac{(\alpha p)^k}{(1-\alpha q)^k}.$$

This result becomes obvious if we note that $T$ is the sum of $k$ independent times $T_i$ to the first success, and that $T_i - 1$ is geometrically distributed: $\mathbb{P}(T_i = n) = pq^{n-1}, n \ge 1$. Indeed, this implies that the probability generating function of $T$ is the $k$th power of the probability generating function of $T_1$, which equals $\frac{p\alpha}{1-\alpha q}$.

**6.4.6 Exercise**     Let $X_m, n \ge 1$, be independent Bernoulli random variables with parameter $p$. Define a random walk $S_n, n \ge 0$, by $S_0 = 0$ and $S_n = \sum_{i=1}^{n} X_i, n \ge 1$. Let $p_n = \mathbb{P}(S_n = 0)$, and let $r_n = \mathbb{P}(S_1 \ne 0, \ldots, S_{n-1} \ne 0, S_n = 0)$ be the probability that at time $n$ the random walk returns to 0 for the first time. Note that a priori we may not exclude the case $\sum_{n=1}^{\infty} p_n = \infty$ or $\sum_{n=1}^{\infty} r_n < 1$. Show that

(a) $p_{2n} = \binom{2n}{n} p^n q^n, p_{2n+1} = 0, n \ge 0$,

(b) $f(\alpha) = \sum_{n=1}^{\infty} p_{2n} \alpha^n$ equals $(1 - 4\alpha pq)^{-\frac{1}{2}}$ for $\alpha \in [0,1]$ if $p \ne q$; if $p = q = \frac{1}{2}$ the formula holds for $\alpha \in [0,1)$,

(c) $\sum_{n=1}^{\infty} p_n = \frac{1}{|p-q|}$ so that $(p_n)_{n \ge 0}$ belongs to $l^1$ iff $p \ne q$,

(d) $p_n = \sum_{k=1}^{n} r_k p_{n-k}, n \geq 1$ (this is almost $(p_n)_{n \geq 0} = (p_n)_{n \geq 0} * (r_n)_{n \geq 0}$, but not quite!), and so $g(\alpha) = \sum_{n=1}^{\infty} r_n \alpha^n$ satisfies $f(\alpha) = 1 + f(\alpha)g(\alpha)$,

(e) $g(\alpha) = 1 - (1 - 4\alpha pq)^{\frac{1}{2}}$,

(f) $\sum_{n=1}^{\infty} r_n = 1 - |p - q|$.

**6.4.7 Exercise**   For $\tau \in \mathbb{R}$ and $t > 0$ let $g(\tau, t) = \frac{1}{\sqrt{2\pi t}} e^{-\frac{\tau^2}{2t}}$ and $g(\tau, 0) = \lim_{t \to 0} g(\tau, t) = 0$. Show that the Laplace transform of $g(\tau, t)$,

$$G(\tau, \lambda) = \int_0^\infty e^{-\lambda t} g(\tau, t) \, dt$$

equals

$$\frac{1}{\sqrt{2\lambda}} e^{-\sqrt{2\lambda}|\tau|}, \qquad \lambda > 0.$$

Hint: check that $G$ is differentiable in $\tau$ and $\frac{d^2 G}{d\tau^2} = 2\lambda G$.

**6.4.8 Exercise**   Calculate the Fourier transform (the characteristic function) $\phi(\alpha) = E\, e^{i\alpha X}$ of a standard normal random variable $X$, by checking that it satisfies the following differential equation:

$$\frac{d}{d\alpha}\phi(\alpha) = -\alpha\phi(\alpha), \quad \phi(0) = 1. \tag{6.25}$$

**6.4.9 Campbell's Theorem**   Let us come back to the random variable $(Px)(\omega) = \sum_{n=1}^{\infty} x(S_n(\omega))$ discussed in 2.3.45, where $x \in L^1(\mathbb{R}^+, a \times \text{leb})$, and try to find its characteristic function. Assume first that $x$ is continuous with support in $[0, K]$, say, and define $x_n(\tau) = x(\frac{jK}{n})$ for $\frac{jK}{n} < \tau \leq \frac{(j+1)K}{n}$, $1 \leq j \leq n - 1$, $x(0) = x(\frac{K}{n})$. Then, $x_n$ converges to $x$ pointwise (even uniformly) and so for any $\omega$, $Px_n(\omega)$ converges to $Px(\omega)$. By the Lebesgue Dominated Convergence Theorem characteristic functions $\phi_n(\alpha)$ of $Px_n$ converge to that of $Px$. On the other hand,

$$Px_n = \sum_{j=0}^{n-1} x\left(\frac{jK}{n}\right) Y_{n,j}$$

where $Y_{n,j}$ is a number of $S_l$ in the interval $(\frac{jK}{n}, \frac{(j+1)K}{n}]$. In Exercise 7.5.7 in Chapter 7 the reader will show that $Y_{n,j}$ are independent, Poisson distributed random variables with parameter $\lambda = \frac{aK}{n}$. Hence

$$\phi_n(\alpha) = \prod_{j=0}^{n-1} \exp\{\frac{aK}{n}(e^{i\alpha x(\frac{jK}{n})} - 1)\} = \exp\{\frac{aK}{n} \sum_{j=0}^{n-1} (e^{i\alpha x(\frac{jK}{n})} - 1)\}.$$

The sum in the exponent above is an approximating sum of the Riemann integral

$$a \int_0^K [e^{i\alpha x(u)} - 1]\,du = a \int_0^\infty [e^{i\alpha x(u)} - 1]\,du,$$

and thus the characteristic function of $Px$ equals

$$\exp\{a \int_0^\infty [e^{i\alpha x(u)} - 1]\,du\}. \tag{6.26}$$

This result may be extended to all $x \in L^1(\mathbb{R}^+, a \times leb)$. To this end, one approximates $x$ in the strong topology of $L^1(\mathbb{R}^+, a \times leb)$ by a sequence $x_n$ of continuous functions with compact support (see 2.2.44). Since $P$ is a Markov operator, $Px_n$ converge to $Px$ in $L^1(\Omega, \mathcal{F}, \mathbb{P})$, and hence weakly also (see 5.8.1). Therefore, the characteristic functions of $Px_n$ converge to that of $Px$. Our claim follows now by the Lebesgue Dominated Convergence Theorem and the estimate $|1 - e^{i\alpha x(u)}| \le |\alpha x(u)|$ (see the solution to 6.4.8).

Note that $\alpha$ in equation (6.26) is redundant in the sense that it appears only together with $x$, while $x$ is an arbitrary element of $L^1(\mathbb{R}^+, a \times leb)$ anyway. In other words we may restate our (actually, Campbell's) theorem by saying that

$$Ee^{iPx} = \exp\{a \int_0^\infty [e^{ix(u)} - 1]\,du\} \tag{6.27}$$

and if necessary recover (6.26) by substituting $\alpha x$ for $x$. Formula (6.27) is a particular case of the so-called **characteristic functional** of a **point process**. In this setting, $\{S_n, n \ge 1\}$ is viewed as a random subset of $\mathbb{R}^+$. The characteristic functional $x \mapsto Ee^{iPx}$ (which is not a linear functional) describes the distribution of such a random set in a quite similar way to that in which the characteristic function describes the distribution of a random variable (see [24], [66]). In the case of a single random variable $Z$ we describe its distribution by means of expected values $Ee^{iY}$ of a family $Y = \alpha Z$ of random variables indexed by $\alpha \in \mathbb{R}$. Since a random set is a more complex entity than a random variable we need more "test-variables" $Y$ to describe it and need to take, for example, $Y = Px$ indexed by $x \in L^1(\mathbb{R}^+, a \times leb)$. The random set $\{S_n, n \ge 1\}$ is called the **Poisson (point) process** with parameter $a$ (see also 7.5.5).

The characteristic functional (6.27) reveals much more than is seen at a first glance. For example, taking $x = \sum_{i=1}^n \alpha_i x_i$ where $\alpha_i \in \mathbb{R}$ and $x_i \in L^1(\mathbb{R}^+, a \times leb)$ we obtain from it the joint characteristic function of $Px_i, 0 \leq i \leq n$. In particular, taking $x_i$ to be indicators of measurable disjoint subsets $A_i$ of $\mathbb{R}^+$ with finite measure we see that the number of points of our random set in each $A_i$ is Poisson distributed with parameter $a\,leb(A_i)$ and independent of a number of such points in $A_j, j \neq i$. Moreover, if $x_1$ and $x_2$ belong to $L^2(\mathbb{R}^+)$, we differentiate the joint characteristic function

$$\phi(\alpha_1, \alpha_2) = \exp\{a \int_0^\infty [e^{i\alpha_1 x_1(u) + i\alpha_2 x_2(u)} - 1]\,du\}$$

of $Px_1$ and $Px_2$ at $(0,0)$ to obtain

$$E\,Px_1 Px_2 = a^2 \int x_1 x_2 \,dleb + a^2 \int x_1 \,dleb \int x_2 \,dleb. \tag{6.28}$$

**6.4.10 Exercise**   Let $a, b > 0$, and $P_t$ be the Markov operator related to a Poisson process with parameter $at$. There is a probability space $(\Omega, \mathcal{F}, \mathbb{P})$ where one may construct a random subset† $S$ of $\mathbb{R}^+$ such that for $x \in L^1(\mathbb{R}^+, a \times leb)$, the expected value $F(x) = E e^{iPx}$ where

$$Px(\omega) = \sum_{s \in S(\omega)} x(s)$$

is given by

$$F(x) = b \int_0^\infty e^{-bt} E\,e^{iP_t x} \,dt = \frac{b}{b + a \int_0^\infty [1 - e^{ix(u)}]\,du}.$$

Such random sets are called **geometric point processes** and are of importance in population genetics – see [18]. Show that for any bounded measurable $A$, the number of elements of this random set in $A$ is geometrically distributed with parameter $p = \frac{b}{b + a\,leb(A)}$. Also, show that $P$ is a Markov operator and that for $x_1$ and $x_2$ in $L^2(\mathbb{R}^+)$ we have

$$E\,Px_1 Px_2 = 2\frac{a^2}{b^2} \int x_1 \,dleb \int x_2 \,dleb + \frac{a^2}{b} \int x_1 x_2 \,dleb.$$

Deduce in particular that random variables $P1_A$ and $P1_B$ are not independent even if the sets $A$ and $B$ are disjoint.

**6.4.11 Exercise**   The bilateral exponential measure with parameter $a > 0$ is the probability measure with density $x(\tau) = \frac{a}{2} e^{-|\tau|a}$. Show by direct calculation that the Fourier cosine transform of the density of the bilateral exponential distribution with parameter $a$ equals $\frac{a^2}{a^2 + t^2}$. Note

---

† The phrase *random set* means here that one may introduce a $\sigma$-algebra in the set of (countable) subsets in $\mathbb{R}^+$ such that the map $\omega \to S(\omega)$ is measurable.

that since the bilateral exponential is symmetric its Fourier transform and Fourier cosine transform coincide.

**6.4.12 Exercise**     The Cauchy measure with parameter $a > 0$ is the probability measure with density $x(\tau) = \frac{1}{\pi}\frac{a}{a^2+\tau^2}$. Use the calculus of residues to show that the Fourier transform (and the Fourier cosine transform) of the density of the Cauchy measure with parameter $a$ is $e^{-|t|a}$.

## 6.5  Dense subalgebras of $C(S)$

In this short section we present the Kakutani–Krein Theorem and derive from it the Stone–Weierstrass Theorem. Of course, in functional analysis this last theorem is of great interest by itself, but the main reason to present it here is that it constitutes a tool in proving inversion theorems of the next section section (Theorem 6.6.7).

**6.5.1 Definition**     Let $S$ be a topological space. A subset of the set of continuous functions on $S$ is said to be a **lattice** if both $x \vee y$ and $x \wedge y$ belong to this subset whenever $x$ and $y$ do. It is said to **separate points** if for any $p, q \in S$, there exists a $z$ in this set with $z(p) \neq z(q)$.

**6.5.2** *The Kakutani–Krein Theorem*     Let $C(S)$ be the space of continuous functions on a compact topological space $S$ and let $\mathbb{A}$ be an algebraic subspace of $C(S)$ which contains $1_S$ and is a lattice. If $\mathbb{A}$ separates points, then the closure of $\mathbb{A}$ is the whole of $C(S)$.

*Proof* Fix $x \in C(S)$. For any $p, q \in S$, take a $z \in \mathbb{A}$ such that $z(q) \neq z(p)$. Then there is exactly one pair of numbers $\alpha, \beta$ such that $\alpha + \beta z(p) = x(p)$ and $\alpha + \beta z(q) = x(q)$. Writing $y_{p,q} = \alpha 1_S + \beta z \in \mathbb{A}$ we see that at $p$ and $q$ the values of $y_{p,q}$ and $x$ are the same.

Now, fix $\epsilon > 0$ and $p \in S$. Functions $y_{p,q}$ and $x$ being continuous, we may find an open neighborhood $U_{\epsilon,q}$ of $q$ such that $y_{p,q}(r) < x(r) + \epsilon$, $r \in U_{\epsilon,q}$. These neighborhoods cover $S$. Since $S$ is compact, one may find a finite subcover of this cover. Let $q_1, ..., q_n$ be the points defining this subcover. Note that $y_p = y_{p,q_1} \wedge y_{p,q_2} \wedge ... \wedge y_{p,q_n} \in \mathbb{A}$ satisfies $y_p < x + \epsilon 1_S$ because any $r$ belongs to $U_{q_i,\epsilon}$ for some $i = i(r)$ and for this $i$ we have $y_p(r) \leq y_{p,q_i}(r) < x(r) + \epsilon$. Also, $y_p(p) = x(p)$.

Finally, let $V_p$ be open neighborhoods of $p$ such that $y_p(r) > x(r) - \epsilon$ for $r \in V_p$. Again, $V_p, p \in S$ form an open cover of $S$. Let $p_1, ..., p_k$ be the points defining a finite subcover of this cover. Writing $y = y_{p_1} \vee y_{p_2} \vee ... \vee y_{p_n} \in \mathbb{A}$ and arguing as above we see that $y > x - \epsilon 1_S$. Since, obviously, $y < x + \epsilon 1_S$, we obtain $\|x - y\| < \epsilon$. This proves that $\mathbb{A}$ is dense in $C(S)$, as desired. □

**6.5.3** *The Stone–Weierstrass Theorem*    Let $C(S)$ be the space of real continuous functions on a compact topological space $S$, and let $\mathbb{B}$ be its subalgebra. If $\mathbb{B}$ separates points and contains $1_S$, then the closure of $\mathbb{B}$ equals $C(S)$.

*Proof* Let $\mathbb{A}$ be the closure of $\mathbb{B}$. Since $\mathbb{A}$ separates points and contains $1_S$ it suffices to show that it is a lattice, and by Exercise 6.5.4 it suffices to show that $|x|$ belongs to $\mathbb{A}$ whenever $x$ does. Without loss of generality we may assume $\|x\| \leq 1$. By the Weierstrass Theorem 2.3.29, for any $\epsilon > 0$ there exists a polynomial $p_n$ such that $\sup_{\tau \in [0,1]} |\, |\tau| - p_n(\tau)| < \epsilon$. Hence, for any $p \in S$, $|\, |x(p)| - p_n(x(p))| < \epsilon$. This shows that $|x|$ can be uniformly approximated by a polynomial in $x$. But $\mathbb{A}$ is an algebra, as the reader will easily check, and contains $1_S$ so that a polynomial in $x$ belongs to $\mathbb{A}$. Hence, $|x|$ belongs to the closure of $\mathbb{A}$, equal to $\mathbb{A}$, as claimed. □

**6.5.4 Exercise**    Show that an algebraic subspace of $C(S)$ is a lattice iff $|y|$ belongs to this subspace whenever $y$ does.

**6.5.5 Exercise**    Prove the following complex version of the Stone–Weierstrass Theorem. If $\mathbb{A}$ is a subalgebra of the space $C(S)$ of *complex* continuous functions on a compact space $S$ satisfying conditions of 6.5.3 and such that the complex conjugate $\overline{x}$ of any $x \in \mathbb{A}$ belongs to $\mathbb{A}$ then the closure of $\mathbb{A}$ equals $C(S)$.

**6.5.6 Exercise**    Let $C(\overline{\mathbb{R}^2})$ be the algebra of functions $x$ on $\mathbb{R}^2$ such that the limit $\lim_{\tau^2 + \sigma^2 \to \infty} x(\tau, \sigma)$ exists, and let $C(\overline{\mathbb{R}})$ be the algebra of functions $y$ such that the limit $\lim_{|\tau| \to \infty} y(\tau)$ exists. Show that linear combinations of functions of the form $x(\tau, \sigma) = y_1(\tau)y_2(\sigma)$ where $y_i \in C(\overline{\mathbb{R}}), i = 1, 2$, are dense in $C(\overline{\mathbb{R}^2})$. Use this to offer yet another proof of 5.4.20 (cf. 5.7.8).

**6.5.7 Exercise** (*Continuation*) Use the previous exercise to show that if $X$ and $Y$ are random variables, $X$ is independent from a $\sigma$-algebra $\mathcal{F}$ and $Y$ is $\mathcal{F}$ measurable, then $\mathbb{E}(x(X, Y)|\mathcal{F}) = \int_{\mathbb{R}} x(\tau, Y)\, \mathbb{P}_X(\, d\tau)$ for any $x \in C(\overline{\mathbb{R}^2})$.

## 6.6 Inverting the abstract Fourier transform

In probability theory, three examples of a Banach algebra seem to be of particular importance. These are: the algebra $\mathbb{BM}(\mathbb{R})$ of Borel measures on $\mathbb{R}$, the algebra $\mathbb{BM}(\mathbb{R}^+)$ of Borel measures on $\mathbb{R}^+$ and the algebra $\mathbb{BM}(\mathbb{N}_0)$ of Borel measures on $\mathbb{N}_0$ (isometrically isomorphic to $l^1(\mathbb{N}_0)$). This is because we are interested in random variables in general, but quite often our interest focuses on positive random variables or random variables with natural values. From 6.3.10 we know that the Gelfand transform in the first two of these algebras is related to the Fourier transform and the Laplace transform, respectively. The Gelfand transform in the third algebra is a probability generating function, except that it is defined on $[-1, 1]$ and not on $[0, 1]$ as customary in probability. The reader may be familiar with the fact that there are well-known inversion theorems for the Fourier transform, for the Laplace transform and for the probability generating function; all serious books in probability discuss them (or at least one of them). Such theorems assert that the values of the Fourier or Laplace transform or the probability generating function of a (probability) measure determine this measure. This section is devoted to proving these results.

At this point it is crucial to recall that the set of multiplicative functionals on a Banach algebra $\mathbb{A}$ may be empty. Hence, in general the Gelfand transform does not determine a member of this algebra and there is no "inversion theorem" for the abstract Gelfand transform. Our case is not hopeless, however, since we may restrict ourselves to abelian algebras. Moreover, we are not dealing with general Banach algebras but rather with convolution algebras of measures. Inspecting the arguments used in the previous section we discover that the notion of a multiplicative functional on the algebra of Borel measures on a locally compact commutative topological semigroup $\mathbb{G}$ with multiplication $\circ$ is closely related to that of a **semicharacter** of $\mathbb{G}$, defined to be a bounded, real or complex, continuous function $\alpha$ on $\mathbb{G}$ such that

$$\alpha(g_1 \circ g_2) = \alpha(g_1)\alpha(g_2);$$

a semicharacter on a group is called a **character**. To be more specific, we check that given a semicharacter $\alpha$ we may define a multiplicative functional on $\mathbb{BM}(\mathbb{G})$ by

$$F_\alpha \mu = \int_{\mathbb{G}} \alpha \, d\mu.$$

As a subset of a locally compact space, the set $\Lambda$ of such functionals is a locally compact space itself. Following [52], the map

$$\mathbb{BM}(S) \ni \mu \mapsto \hat{\mu} \in C(\Lambda),$$

where $\hat{\mu}(\alpha) = F_\alpha \mu$ will be called the **abstract Fourier transform**.

Hence, inversion theorems of probability theory say that *for many locally compact semigroups* $\mathbb{G}$*, a measure in* $\mathbb{BM}(\mathbb{G})$ *is determined by its abstract Fourier transform.*

Unfortunately, it is not clear whether this theorem is true for *all* locally compact commutative semigroups. It is definitely true for locally compact commutative *groups*, but the proof is based on the results of Gelfand and Raikov concerning unitary representations of a locally compact commutative group [52], and cannot be presented in this elementary treatise. Therefore, we need to take up a still more modest course. Specifically, we will prove that the theorem is true for *compact* commutative semigroups, provided that semicharacters separate their points (this will follow immediately from the Stone–Weierstrass Theorem) and then treat the remaining cases that are important from the probabilistic point of view separately.

We start, though, with examples and exercises on characters and the elementary inversion theorem for probability generating function.

**6.6.1 Example** In Section 6.3 we have showed that $\alpha(n) = \alpha^n, \alpha \in [-1, +1]$, and $\alpha(\tau) = \mathrm{e}^{-\lambda \tau}, \lambda \geq 0$, are the only real semicharacters of the semigroups $\mathbb{N}$ and $\mathbb{R}^+$, respectively. Moreover, $\alpha(\tau) = \mathrm{e}^{i\tau t}, t \in \mathbb{R}$, are the only complex characters of the group $\mathbb{R}$.

**6.6.2 Exercise** Show that semicharacters of a semigroup $\mathbb{G}$ have values in the complex unit disc (i.e. $|\alpha(g)| \leq 1, g \in \mathbb{G}$) and that characters of a group have values in a unit circle.

**6.6.3 Exercise** Show that characters of a group form a group (called the **dual group** or the **character group**). Show that the dual group to $\mathbb{Z}$ is (izomorphic to) the unit circle with complex multiplication. Conversely, the dual of this last group is $\mathbb{Z}$. This is a simple example of the general theorem of Pontryagin and Van Kampen saying that the dual of the dual group is the original group – [51] p. 378.

**6.6.4** *Inversion theorem for the probability generating function* The values of $\hat{x}(\alpha) = \sum_{n=0}^{\infty} \xi_n \alpha^n$ determine $x = (\xi_n)_{n \geq 0} \in l^1(\mathbb{N}_0)$.

*Proof* The function $\hat{x}$ is analytic with right-hand derivatives at 0 equal $\frac{d^n \hat{x}(0)}{d\alpha^n} = n! \xi_n$. □

**6.6.5 Exercise** Prove the inversion theorem for the joint probability generating function (see 6.3.2).

**6.6.6 Exercise** Prove that the values of $\hat{x}(t) = \sum_{n=-\infty}^{\infty} \xi_n e^{itn}$ determine $x = (\xi_n)_{n \in \mathbb{Z}} \in l^1(\mathbb{Z})$.

**6.6.7 Theorem** Suppose $\mathbb{G}$ is a compact topological semigroup and the set of its semicharacters separates points. Then, a Borel measure $\mu$ on $\mathbb{G}$ is determined by its abstract Fourier transform.

*Note:* as a corollary to the important theorem of Gelfand and Raikov mentioned earlier one may show that the set of characters of a commutative locally compact *group* separates points – [51] pp. 343–345.

*Proof* By the Riesz Theorem, it suffices to show that the linear span $\mathbb{Y}$ of characters is dense in $C(\mathbb{G})$. $\mathbb{Y}$ is a subalgebra of $C(\mathbb{G})$. Moreover, it separates points and contains $1_\mathbb{G}$. Hence, if all semicharacters are real, our claim follows by the Stone–Weierstrass Theorem. In the general case, we note that $\mathbb{Y}$ also has the property of being closed under complex conjugation – a function belongs to $\mathbb{Y}$ iff its complex conjugate does. Hence, the theorem follows by 6.5.5. □

**6.6.8 Example** The unit circle $\mathbf{C}$ with complex multiplication is a compact commutative group. All its characters are of the form $\alpha(z) = z^n, z \in \mathbf{C}, n \geq 0$. Obviously, the set of characters separates points of $\mathbf{C}$. Hence, a measure $\mu$ on $\mathbf{C}$ is determined by $\hat{\mu}(n) = \int_{\mathbf{C}} z^n \mu(\mathrm{d}z)$. More generally, all characters on the compact group $\mathbf{C^k}$ are of the form $\alpha(z_1, ..., z_k) = z_1^{n_1}...z_k^{n_k}, n_i \geq 0, i = 1, ..., k$, and a measure $\mu$ on $\mathbf{C^k}$ is determined by $\hat{\mu}(n_1, ..., n_k) = \int_{\mathbf{C^k}} z_1^{n_1}...z_k^{n_k} \mu(\mathrm{d}z_1...z_k)$.

**6.6.9 *Inversion theorem for the Laplace transform*** A measure $\mu \in \mathbb{BM}(\mathbb{R}^+)$ is determined by its Laplace transform, $\hat{\mu}(\lambda) = \int_{\mathbb{R}^+} e_\lambda \mathrm{d}\mu, \lambda \geq 0, e_\lambda(\tau) = e^{-\lambda\tau}, \tau \geq 0$.

*Proof* This is a direct consequence of 2.3.31. We may also derive it from 6.6.7 by noting that $\mathbb{G} = \mathbb{R}^+ \cup \{\infty\}$, with the usual addition and supplementary rule $a + \infty = \infty, a \in \mathbb{R}^+ \cup \{\infty\}$, is a compact semigroup, so that $\mu$ may be treated as an element of $\mathbb{BM}(\mathbb{G})$, with $\mu(\{\infty\}) = 0$. Moreover, all semicharacters on $\mathbb{R}^+$ may be continuously extended to the whole of $\mathbb{G}$ to form semicharacters of $\mathbb{G}$. These semicharacters separate points of $\mathbb{G}$.                                                          □

**6.6.10 Inversion theorem for the Fourier transform**    All characters of $R^k, k \geq 1$ are of the form $R^k \ni \underline{s} = (s_1, ..., s_k) \mapsto e_{\underline{t}}(\underline{s}) = e^{i\underline{s} \cdot \underline{t}}$ where $\underline{t} = (t_1, ..., t_k) \in R^k$ and "$\cdot$" denotes the scalar product $\underline{s} \cdot \underline{t} = \sum_{i=1}^{k} s_i t_i$. Hence, the multidimensional Fourier transform of a measure $\mu \in \mathbb{BM}(\mathbb{R})$ is given by $\hat{\mu}(\underline{t}) = \int_{\mathbb{R}^k} e_{\underline{t}} \, d\mu$. The inversion theorem for the Fourier transform does not follow from 6.6.7, since $\mathbb{R}^k$ is not compact. The compactification argument of 6.6.9 does not work either, since the characters of $\mathbb{R}^k$ do not have limits at infinity. To prove that *a measure $\mu \in \mathbb{BM}(\mathbb{R}^k)$ is determined by its Fourier transform*, $\hat{\mu}(\underline{t}) = \int_{\mathbb{R}^k} e_{\underline{t}} \, d\mu$, we proceed as follows.

We assume, without loss of generality, that $\mu$ is a probability measure. It suffices to show that for any $x \in C_0(\mathbb{R}^k)$ and $\|x\| > \epsilon > 0$ there exists a linear (complex) combination $y$ of $e_{\underline{t}}$ such that

$$\left| \int_{\mathbb{R}^k} x \, d\mu - \int_{\mathbb{R}^k} y \, d\mu \right| < \epsilon. \tag{6.29}$$

Let $n$ be large enough so that $\mu(S_n) > 1 - \frac{\epsilon}{6\|x\|}$ where

$$S_n = \{\underline{s} = (s_1, ..., s_k) \in \mathbb{R}^k, |s_i| \leq n, i = 1, ..., k\}.$$

The linear combinations of $\tilde{e}_{\underline{m}} = \left(e_{\pi n^{-1} \underline{m}}\right)_{|S_n}, \underline{m} \in \mathbb{Z}^k$ are, by 6.5.5, dense in $C(S_n)$. Let $y_0$ be a linear combination of $e_m$ such that $\|x_{|S_n} - y_0\|_{C(S_n)} < \frac{\epsilon}{2}$. Since the $\tilde{e}_{\underline{m}}$ are periodic with period $2n$ (in each coordinate), $y_0$ has the $2n$-periodic extension $y$, defined on $\mathbb{R}$. We have $\|y_0\|_{C(S_n)} = \|y\|_{BM(\mathbb{R}^k)}$. In particular, $\|y\|_{BM(\mathbb{R}^k)} \leq \|x\| + \epsilon \leq 2\|x\|$. Hence, integrating $|x - y|$ over $S_n$ and its complement we see that the left-hand side of (6.29) is less than $\mu(S_n)\frac{\epsilon}{2} + \frac{\epsilon}{6\|x\|}(\|x\| + 2\|x\|) < \epsilon$.

We conclude this section with applications.

**6.6.11 Distributions of linear combinations of coordinates determine the distribution of a random vector**

*Proof* If $\underline{X} = (X_1, ..., X_k)$, then for $\mu = \mathbb{P}_{\underline{X}}$,

$$\hat{\mu}(\underline{t}) = \int_{\mathbb{R}^k} e_{\underline{t}} \, d\mu = \int_{\Omega} e^{i\underline{t} \cdot \underline{X}} \, d\mathbb{P} = \int_{\mathbb{R}} x \, d\mathbb{P}_{\underline{t} \cdot \underline{X}}$$

where $x(s) = e^{is}, s \in \mathbb{R}$.                                        □

6.6.12 *Continuity theorem*   Let $\mu$ and $\mu_n, n \geq 1$, be probability measures on a locally compact semigroup $\mathbb{G}$. If $\mu_n$ converge weakly to $\mu$, then $\hat{\mu}_n$ converges pointwise to $\hat{\mu}$. Conversely, if we assume that the abstract Fourier transform determines a measure, and that $\hat{\mu}_n$ converges pointwise to $\hat{\mu}$, then $\mu_n$ converges weakly to $\mu$.

*Proof*   The first part is immediate. For the other part observe that, by Alaoglu's Theorem, any subsequence of the sequence $\mu_n, n \geq 1$, contains a further subsequence converging to some $\mu_0$ (perhaps not a probability measure, but definitely a non-negative measure). By the first part, the abstract Fourier transforms of the measures forming this sub-subsequence converge to $\hat{\mu}_0$. On the other hand, by assumption they converge to $\hat{\mu}$. In particular $\mu_0(\mathbb{G}) = \mu(\mathbb{G}) = 1$, so that $\mu_0$ is a probability measure. Since the abstract Fourier transform is assumed to determine the measure, we must have $\mu_0 = \mu$. This shows that $\mu_n$ converges to $\mu$. $\square$

6.6.13 **Corollary**   In view of 6.6.4 and 5.8.6 a sequence $x_n = (\xi_{i,n})_{i \geq 0}$, $n \geq 1$, of distributions in $l^1(\mathbb{N}_0)$ converges strongly, weakly or in the weak* topology to a density $x = (\xi_i)_{i \geq 0}$ iff $\lim_{n \to \infty} \sum_{i=0}^{\infty} \alpha^i \xi_{i,n} = \sum_{i=0}^{\infty} \alpha^i \xi_i$ for all $\alpha \in [0,1]$.

6.6.14 **Exercise**   Use the continuity theorem for the generating function to prove the Poisson approximation to binomial 1.2.34.

6.6.15 *Negative binomial approximates gamma*   Let $X_n, n \geq 1$, be negative binomial random variables with parameters $(p_n, k)$, respectively. If $p_n \to 0$ and $np_n \to a > 0$, as $n \to \infty$, then the distribution $\mu_n$ of $\frac{1}{n}X_n$ converges weakly to the gamma distribution with parameters $k$ and $a$. Indeed, all measures may be considered as elements of $\mathbb{BM}(\overline{\mathbb{R}^+})$. The Laplace transform of $\frac{1}{n}X_n$ equals

$$\phi_n(\lambda) = E \exp\{-\frac{\lambda}{n}X_n\} = E\left(\exp\{-\frac{\lambda}{n}\}\right)^{X_n},$$

which is the probability generating function of $X_n$ at $\alpha = \exp\{-\frac{\lambda}{n}\}$. Therefore, $\phi_n(\lambda) = \frac{\exp\{-\frac{k\lambda}{n}\}p_n^k}{(1-\exp\{-\frac{\lambda}{n}\}q_n)^k} \times \frac{n^k}{n^k}$ converges to $\frac{a^k}{(\lambda+a)^k}$, which is the Laplace transform of the gamma distribution with prescribed parameters:

$$\int_0^{\infty} e^{-\lambda t} e^{-at} \frac{a^k}{(k-1)!} t^{k-1} \, \mathrm{d}t = \frac{a^k}{(\lambda+a)^k}.$$

Note that we have once again established 5.4.16 (take $k = 1$).

**6.6.16 *Another proof of the Central Limit Theorem*** The continuity theorem is a powerful tool in proving weak convergence of probability measures. The point is that we need to consider merely integrals $\int \alpha \, d\mu$ with $\alpha$ continuous and having nice algebraic properties. In particular, the proof of CLT (in its classical form 5.5.2) is now an easy exercise in calculus.

Recall that for any continuously differentiable function $\phi$ on $\mathbb{R}$ we have $\phi(t) = \phi(0) + \int_0^t \phi'(s) \, ds$. Using this, if $\phi$ is twice continuously differentiable

$$\phi(t) = \phi(0) + \int_0^t \left( \phi'(0) + \int_0^s \phi(u) \, du \right) ds$$

$$= \phi(0) + t\phi'(0) + \int_0^t (t - s)\phi''(s) \, ds \qquad (6.30)$$

by Fubini's Theorem. (This is a particular case of the Taylor formula with integral remainder.) Also, one checks that for continuous function $\phi$, $\lim_{t \to 0} \frac{1}{t} \int_0^t \phi(s) \, ds = \phi(0)$. In a similar way,

$$\lim_{t \to 0} \frac{2}{t^2} \int_0^t (t - s)\phi(s) \, ds = \lim_{t \to 0} \frac{2}{t^2} \int_0^t \int_0^s \phi(u) \, du \, ds = \phi(0). \qquad (6.31)$$

Now, without loss of generality we may assume that $m = 0$ and $\sigma^2 = 1$, since the general case may be reduced to this one. Then, arguing as in 1.2.12 we see that the (common) characteristic function $\phi(t) = E \, e^{itX_n}$ of all $X_n$ is twice differentiable with $\phi'(0) = i\mu = 0$ and $\phi''(0) = -E \, X^2 = -\sigma^2 = -1$. Moreover, the characteristic function of $\frac{1}{\sqrt{n}} \sum_{k=1}^n X_k$ equals $\phi^n(\frac{t}{\sqrt{n}})$. By (6.30),

$$\phi^n \left( \frac{t}{\sqrt{n}} \right) = \left[ 1 + \int_0^{\frac{t}{\sqrt{n}}} \left( \frac{t}{\sqrt{n}} - s \right) \phi''(s) \, ds \right]^n$$

and (6.31) implies that $n \int_0^{\frac{t}{\sqrt{n}}} (\frac{t}{\sqrt{n}} - s)\phi''(s) \, ds$ tends to $-\frac{t^2}{2}$ as $n \to \infty$. Recalling $\lim_{\tau \to 0+} (1 - \tau)^{\frac{1}{\tau}} = e^{-1}$ we obtain $\phi^n(\frac{t}{\sqrt{n}}) \to e^{-\frac{t^2}{2}}$, as desired.

**6.6.17 *Central Limit Theorem with (Poisson distributed) random number of terms*** Let $X_n, n \geq 1$, be the i.i.d. random variables from the previous subsection. Let $Z_n$ be Poisson random variables with $E \, Z_n = an, a > 0$, independent of $X_n, n \geq 1$. Defining $Y_n = \sum_{k=1}^{Z_n} X_k$ we have that $\frac{Y_n - anm}{\sqrt{an\sigma^2}}$ converges weakly to the standard normal distribution.

*Proof*  Again, we may assume $m = 0$ and $\sigma^2 = 1$. Let $\mu$ and $\phi$ be the (common) distribution and characteristic function of the $X_n, n \geq 1$, respectively. The key step in the proof is that the distribution of $Y_n$ equals

$$\mathrm{e}^{-an} \sum_{k=0}^{\infty} \frac{(an)^k}{k!} \mu^{*k} = \mathrm{e}^{-an} \exp(an\mu)$$

where $*k$ denotes the $k$th convolution power and exp is the exponent in the algebra $\mathbb{BM}(\mathbb{R})$. Hence, the characteristic function of $\frac{1}{\sqrt{an}} Y_n$ is $\mathrm{e}^{-an} \exp\left[an\phi\left(\frac{t}{\sqrt{an}}\right)\right]$. By (6.30) and (6.31),

$$an\phi\left(\frac{t}{\sqrt{an}}\right) - an = an \int_0^{\frac{t}{\sqrt{an}}} \left(\frac{t}{\sqrt{an}} - s\right) \phi''(s)\, \mathrm{d}s$$

tends to $-\frac{t^2}{2}$, as claimed.                                             □

**6.6.18 Example**    In 8.4.31 we show that, for $a > 0$ and $t \geq 0$,

$$\phi_{a,t}(\tau) = \begin{cases} \mathrm{e}^{-at}\left[\cosh\sqrt{a^2 - \tau^2}t + \frac{\mathrm{i}\tau + a}{\sqrt{a^2 - \tau^2}}\cosh\sqrt{a^2 - \tau^2}t\right], & |\tau| < a, \\ \mathrm{e}^{-at}\left[\cos\sqrt{\tau^2 - a^2}t + \frac{\mathrm{i}\tau + a}{\sqrt{\tau^2 - a^2}}\cos\sqrt{\tau^2 - a^2}t\right], & |\tau| > a, \\ \mathrm{e}^{-at}(1 \pm \mathrm{i}at + at), & \tau = \pm a, \end{cases}$$
$$(6.32)$$

is the characteristic function of a random variable, say $\xi_a(t)$. (To be more specific, $\xi_a(t) = \int_0^t (-1)^{N_a(s)}\, \mathrm{d}s$ where $N_a(t), t \geq 0$, is the Poisson process to be introduced in 7.5.5, but this is of no importance for now.) We will show that, as $a \to \infty$, $\sqrt{a}\xi_a(t)$ converges to an $N(0, t)$ variable.

To this end, it suffices to prove that $\lim_{a \to \infty} \phi_{a,t}(\sqrt{a}\tau) = \mathrm{e}^{-\frac{t\tau^2}{2}}$. Let $\tau \in \mathbb{R}$ be fixed. For $a > \tau^2$, $\phi_{a,t}(\sqrt{a}\tau)$ is calculated using the first formula in the definition above, and some algebra shows that it equals

$$\exp\left\{at\left[\sqrt{1 - \frac{\tau^2}{a}} - 1\right]\right\}\left[\frac{1}{2} + \frac{\frac{\mathrm{i}\tau}{\sqrt{a}} + 1}{2\sqrt{1 - \frac{\tau^2}{\sqrt{a}}}}\right]$$

$$+ \exp\left\{at\left[-\sqrt{1 - \frac{\tau^2}{a}} - 1\right]\right\}\left[\frac{1}{2} - \frac{\frac{\mathrm{i}\tau}{\sqrt{a}} + 1}{2\sqrt{1 - \frac{\tau^2}{\sqrt{a}}}}\right].$$

Observe that the expressions in square brackets tend to 1 and 0, respectively. Next, $0 < \exp\left\{ at\left[ -\sqrt{1 - \frac{\tau^2}{a}} - 1 \right] \right\} \le 1$ , and

$$\lim_{a\to\infty} at\left[ \sqrt{1 - \frac{\tau^2}{a}} - 1 \right] = \lim_{a\to\infty} \frac{at\left[ 1 - \frac{\tau^2}{a} - 1 \right]}{\sqrt{1 - \frac{\tau^2}{a}} + 1} = \frac{-\tau^2 t}{2},$$

as desired.

## 6.7 The Factorization Theorem

This section is devoted to Cohen's Factorization Theorem, later gener-alized independently by E. Hewitt, P. C. Curtis and A. Figá-Talamanca, and S. L. Gulik, T. S. Liu and A. C. M. van Rooij, see [52] and the overview article [72]. This is one of the fundamental theorems of the theory of Banach algebras but this is not the main reason why we dis-cuss it here. Rather, we are motivated by the fact that this theorem constitutes an integral part of the structure of the Kisyński's algebraic version of the Hille–Yosida Theorem to be discussed in Chapter 8. On the other hand, Cohen's Theorem is not crucial for the *proof* of the Hille–Yosida Theorem. Hence, a casual reader may take my advice from Section 1.6.

6.7.1 *The Factorization Theorem* Let $H$ be a representation of $L^1(\mathbb{R}^+)$ by bounded linear operators in a Banach space $\mathbb{X}$, and let $\mathcal{R} \subset \mathbb{X}$ be defined as

$$\mathcal{R} = \{x \in \mathbb{X} | x = H(\phi)y \text{ for some } \phi \in L^1(\mathbb{R}^+), y \in \mathbb{X}\}.$$

Then, for every $x$ in the closed linear span $\mathbb{X}_0$ of $\mathcal{R}$ there exists a $y \in \mathbb{X}$ and a non-negative $\phi \in L^1(\mathbb{R}^+)$ such that $x = H(\phi)y$. In particular, $\mathcal{R}$ is an algebraic subspace of $\mathbb{X}$ and is closed.

*Proof*

(a) Let $A_u = L^1(\mathbb{R}^+) \times \mathbb{R}$ be the algebra with unit $u = (0, 1)$ described in 6.1.7. For notational convenience, we will write $\phi$ and $\phi\psi$ to denote $(\phi, 0)$ and $(\phi, 0)(\psi, 0) \in A_u$, respectively. In other words we identify $L^1(\mathbb{R}^+)$ with a subalgebra $L^1(\mathbb{R}^+) \times \{0\}$ of $A_u$. Define $H : A_u \to \mathcal{L}(\mathbb{X})$ by $H(\phi, \alpha) = H(\phi) + \alpha I_\mathbb{X}$. $H$ is now a representation of $\mathbb{A}_u$.

**(b)** Let $i_\lambda \in L^1(\mathbb{R}^+)$ be given by $i_\lambda(\tau) = \lambda e^{-\lambda\tau}, \tau > 0$. Note that

$$\lim_{\lambda\to\infty} i_\lambda\phi = \phi, \quad \phi \in L^1(\mathbb{R}^+), \tag{6.33}$$

$$\|i_\lambda\| = 1. \tag{6.34}$$

This implies that if $x = H(\phi)y, \phi \in L^1(\mathbb{R}^+), y \in \mathbb{X}$, then $\lim_{\lambda\to\infty} H(i_\lambda)x$ $= \lim_{\lambda\to\infty} H(i_\lambda)H(\phi)y = \lim_{\lambda\to\infty} H(i_\lambda\phi)y = H(\phi)y = x$. By linearity and continuity,

$$\lim_{\lambda\to\infty} H(i_\lambda)x = x, \tag{6.35}$$

for all $x \in \mathbb{X}_0$.

**(c)** Let $x \in \mathbb{X}_0$ and $(\lambda_n)_{n\geq 1}$ be a sequence of positive numbers. Define $b_n \in A_u$ and $y_n \in \mathbb{X}_0$ by

$$b_n = \prod_{j=1}^{n} (2u - a_j)^{-1}, \qquad y_n = H\left(\prod_{j=1}^{n} (2u - a_j)\right) x, \tag{6.36}$$

where $a_j = i_{\lambda_j}$. Note that $H(b_n)y_n = H(u)x = x$. Since (see 6.1.5)

$$(2u - a_j)^{-1} = \frac{1}{2}\left(u - \frac{1}{2}a_j\right)^{-1} = \frac{1}{2}u + \sum_{k=1}^{\infty} 2^{-(k+1)}a_j^k,$$

(6.34) implies $\|(2u - a_j)^{-1}\| \leq 1$ and $\|b_n\| \leq 1$. Moreover,

$$b_n = 2^{-n}u + \phi_n \tag{6.37}$$

for appropriate choice of $\phi_n \in L^1(\mathbb{R}^+)$, and $b_n$ converges (in $A_u$) iff $\phi_n$ converges (in $L^1(\mathbb{R}^+)$). Note that $\phi_n \geq 0$. Since $\phi_{n+1} - \phi_n$ equals

$$b_{n+1} - b_n + 2^{-(n+1)}u = \left\{(2u - a_{n+1})^{-1} - u\right\}b_n + 2^{-(n+1)}u$$

$$= \left\{(2u - a_{n+1})^{-1} - u\right\}(\phi_n + 2^{-n}u) + 2^{-(n+1)}u$$

$$= (2u - a_{n+1})^{-1}(a_{n+1} - u)\phi_n$$

$$\quad + 2^{-(n+1)}(2u - a_{n+1})^{-1}(2a_{n+1} - 2u)$$

$$\quad + 2^{-(n+1)}(2u - a_{n+1})^{-1}(2u - a_{n+1}),$$

$$= (2u - a_{n+1})^{-1}\{i_{\lambda_{n+1}}\phi_n - \phi_n\} + 2^{-(n+1)}(2u - a_{n+1})^{-1}a_{n+1},$$

we have

$$\|\phi_{n+1} - \phi_n\| = \|i_{\lambda_{n+1}} * \phi_n - \phi_n\| + 2^{-(n+1)}. \tag{6.38}$$

Also, $y_{n+1} - y_n = \left(I - H(i_{\lambda_{n+1}})\right)y_n$ so that

$$\|y_{n+1} - y_n\| \leq \|H(i_{\lambda_{n+1}})y_n - y_n\|. \tag{6.39}$$

(d) By (6.33), (6.35), (6.38) and (6.39), the sequence $(\lambda_n)_{n \geq 1}$ can be constructed inductively in such a way that both $\|y_{n+1} - y_n\|$ and $\|\phi_{n+1} - \phi_n\|$ are less than $\frac{1}{2^n}$. Consequently, the series $\sum_{n=1}^{\infty} \|y_{n+1} - y_n\|$ and $\sum_{n=1}^{\infty} \|\phi_{n+1} - \phi_n\|$ are convergent and, consequently, there exist the limits $\lim_{n \to \infty} y_n = y$ and $\lim_{n \to \infty} \phi_n = \phi$. We have $\phi \geq 0$ since $\phi_n \geq 0$. Finally,

$$x = \lim_{n \to \infty} H(b_n)y_n = \lim_{n \to \infty} \left[ \frac{1}{2^n} y_n + H(\phi_n)y_n \right] = H(\phi)y$$

as desired.                                                                                           □

**6.7.2 Corollary**     For any $\varphi \in L^1(\mathbb{R}^+)$ there exists a $\psi \in L^1(\mathbb{R}^+)$ and a nonnegative $\phi \in L^1(\mathbb{R}^+)$ such that $\varphi = \phi * \psi$.

*Proof* Take $\mathbb{X} = L^1(\mathbb{R}^+)$ and $H(\phi) = L_\phi$. By (6.33), the closure of $\mathcal{R}$ equals $L^1(\mathbb{R}^+)$.                                                                                           □

# 7

# Semigroups of operators and Lévy processes

Our plan is to prepare a functional-analytic background for the semi-group-theoretical treatment of Markov processes. We start with the Banach–Steinhaus uniform boundedness principle, the result that on its own is one of the most important theorems in functional analysis as well as one of its most powerful tools. In Section 7.2 we prove basic facts from the calculus of Banach space valued functions. Section 7.3 is crucial: as we shall see from the Hille–Yosida theorem to be presented in Chapter 8, there is a one-to-one correspondence between Markov processes and a class of linear operators – the class of generators of corresponding semigroups. In general, these operators are not bounded, but are closed. Hence, Section 7.3 presents the definition and basic properties of closed operators. Section 7.4 is devoted to the rudiments of the theory of semigroups of operators, and in 7.5 we study Lévy processes (a particular type of Markov processes) with the aid of the theory introduced in the foregoing sections. Following the example of Feller [41], we postpone the treatment of general Markov processes to the next chapter.

## 7.1 The Banach–Steinhaus Theorem

We start with the following exercise.

**7.1.1 Exercise** Let $r_n, n \geq 1$, be a non-increasing sequence of positive numbers and $x_n, n \geq 1$, a sequence of elements of a Banach space $\mathbb{X}$, and let $clB_n = clB(x_n, r_n)$ be the closed ball with radius $r_n$ and center $x_n$. Assume that $clB_n, n \geq 1$, is a decreasing sequence of sets: $clB_{n+1} \subset clB_n$. Show that $\bigcap_{n \in N} clB_n$ is non-empty.

**7.1.2 Baire's Category Theorem**    A subset, say $S$, of a normed space $\mathbb{X}$ is termed nowhere dense iff its closure does not contain any open ball. It turns out that if $S$ is nowhere dense, then any open ball $B$ contains a ball $B'$ such that $B' \cap S$ is empty. Indeed, if we suppose that any ball $B'$ that is a subset of a certain ball $B$ contains an element of $S$, then it is easy to see that every point of $B$ belongs to the closure of $S$. To this end it suffices to consider, for every point $x$ of $B$, the sequence of balls $B(x, \frac{1}{n})$ with $n$ large enough to have $B(x, \frac{1}{n}) \subset B$. This leads us to the following important statement about Banach spaces: *a Banach space may not be represented as a countable union of nowhere dense sets.* This is the famous Baire's Category Theorem. The name comes from the fact that sets that may be represented as a countable union of nowhere dense sets (e.g. countable sets) are termed sets of the first category. In this terminology, Baire's Theorem states that Banach spaces are not of the first category; they are sets of the second category. To prove this theorem, assume that we have $\mathbb{X} = \bigcup_{n \in N} S_n$ where $\mathbb{X}$ is a Banach space and $S_n$ are nowhere dense sets. Let $B_0$ be the open ball with radius 1 and center 0. Since $S_1$ is nowhere dense, $B_0$ contains an open ball $B_1$ that is disjoint with $S_1$. We may actually assume that the radius of $B_1$ is smaller than $\frac{1}{2}$, and that the closure of $B_1$ is contained in $B_0$ and disjoint with $S_1$; it is just a matter of taking a smaller radius, if necessary. This procedure may be repeated: we may find an open ball $B_2$ of radius lesser than $\frac{1}{3}$ such that its closure is contained in $B_1$ and is disjoint with $S_2$. More generally, having found an open ball $B_n$, we may find an open ball $B_{n+1}$ with radius less than $\frac{1}{n+2}$, whose closure is contained in $B_n$ and yet is disjoint with $S_{n+1}$. This, however, leads to a contradiction. Specifically, we may use 7.1.1 for the sequence of closures of balls $B_n$ to see that there exists an $x$ that belongs to all the closed balls $clB_n$. On the other hand, $clB_n$ is disjoint with $S_n$, and therefore, $x$ does not belong to the union of $S_n$, which is impossible by assumption that $\mathbb{X} = \bigcup_{n \in N} S_n$.

**7.1.3** *The Banach–Steinhaus Theorem (uniform boundedness principle)*
Suppose that $\mathbb{X}$ is a Banach space and that $A_n, n \geq 1$, is a sequence of bounded linear operators. Assume that for every $x \in \mathbb{X}$, the supremum of $\|A_n x\|$ is finite. Then $\sup_{n \in N} \|A_n\|$ is finite also.

*Proof* Let $S_n = \{x \in \mathbb{X}; \sup_{k \in N} \|A_k x\| \leq n\}$. Since the operators $A_n$ are continuous, the sets $S_n$ are closed. Our assumption states that $\mathbb{X} = \bigcup_{n \in N} S_n$. Therefore, by Baire's Category Theorem, there exists an $l \in$

$N$, such that there exists a closed ball $B$ contained in $S_l$. Let $x$ be the center of this ball and let $r > 0$ be its radius. Consider a non-zero $y \in \mathbb{X}$ and a vector $z = x + \frac{r}{\|y\|}y \in B$. We have

$$\|A_n y\| = \left\| \frac{\|y\|}{r}A_n z - \frac{\|y\|}{r}A_n x \right\| \leq \frac{\|y\|}{r}\|A_n z\| + \frac{\|y\|}{r}\|A_n x\| \leq \frac{2l}{r}\|y\|$$

(7.1)

since both $z$ and $x$ belong to $B \subset S_l$. Thus, $\sup_{n \in N}\|A_n\| \leq \frac{2l}{r}$. □

**7.1.4 Corollary**    Suppose that $A_n, n \geq 1$, is a sequence of continuous linear operators in a Banach space $\mathbb{X}$, and that the limit $\lim_{n \to \infty} A_n x$ exists for all $x \in \mathbb{X}$. Then the operator $Ax = \lim_{n \to \infty} A_n x$ is linear and bounded.

*Proof* Linearity of $A$ is obvious; it is its boundedness that is non-trivial and needs to be proven. It follows, however, from the Banach–Steinhaus Theorem. Indeed, by assumption the sequence $\|A_n x\|$ is bounded for all $x \in \mathbb{X}$, and so the sequence of norms $\|A_n\|$ must be bounded by a constant, say $K$. Therefore, $\|Ax\| = \lim_{n \to \infty}\|A_n x\| \leq K\|x\|$ for all $x \in \mathbb{X}$, as desired. □

**7.1.5 Remark**    It is worthwhile saying that the above theorem is true only for linear operators. The reader should contrast this situation with the fact that, for example, functions $x_n(\tau) = \tau^n$ converge pointwise on $[0, 1]$ but their limit is not continuous.

**7.1.6 Corollary**    Suppose that $A_t$, $t \in (0, 1]$, is a family of bounded linear operators such that for every $x \in \mathbb{X}$, the limit $\lim_{t \to 0} A_t x$ exists. Then there exists a $\delta > 0$ such that $\sup_{0 < t \leq \delta}\|A_t\|$ is finite.

*Proof* The difficulty lies in the fact that we are now dealing with an uncountable family of operators. We may argue, however, in this way: if the thesis of the theorem is not true, then for any $n$ there exists $t_n < \frac{1}{n}$ such that $\|A_{t_n}\| \geq n$. On the other hand, $\lim_{n \to \infty} A_{t_n} x = \lim_{t \to 0} A_t x$ exists for all $x \in \mathbb{X}$. This is a contradiction, by 7.1.3. □

**7.1.7 Exercise**    Following the argument from the previous subsection show that if $A_t, t \in \mathbb{T}$, is a family of bounded linear operators in a Banach space, indexed by an abstract set $\mathbb{T}$, and if $\sup_{t \in \mathbb{T}}\|A_t x\|$ is finite for any $x \in \mathbb{X}$, then $\sup_{t \in \mathbb{T}}\|A_t\|$ is finite, too.

**7.1.8 Corollary** Weakly* convergent sequences are bounded. In particular, weakly convergent sequences are bounded.

*Proof* Let $F_n \in \mathbb{X}^*$ converge in weak* topology. Then for every $x \in \mathbb{X}$, $F_n x$ converges and therefore is bounded. The uniform boundedness principle completes the proof. □

**7.1.9 Exercise** If $A_n, n \geq 1$, and $A$ are bounded linear operators in a Banach space $\mathbb{X}$, with values in a normed space $\mathbb{Y}$, and if $A_n$ converges strongly to $A$ then

$$\lim_{n \to \infty} A_n x_n = Ax$$

for each sequence $x_n \in \mathbb{X}, n \geq 1$, that converges to $x$.

**7.1.10 Corollary** Let $\mathbb{Y}$ be a compact subset of a Banach space $\mathbb{X}$. Suppose that $A$ and $A_n, n \geq 1$, are bounded linear operators in $\mathbb{X}$ with values in a normed space $\mathbb{Z}$, and that $\lim_{n \to \infty} A_n x = Ax$ for all $x \in \mathbb{X}$. Then

$$\lim_{n \to \infty} \sup_{y \in \mathbb{Y}} \|A_n y - Ay\| = 0,$$

i.e. the convergence is uniform on $\mathbb{Y}$.

*Proof* Let $\epsilon > 0$ be given. Since $\mathbb{Y}$ is compact, there exist $k \in \mathbb{N}$ and $y_1, ..., y_k \in \mathbb{Y}$ such that for any $y \in \mathbb{Y}$ there exists a $1 \leq i \leq k$ such that $\|y - y_i\| < \frac{\epsilon}{4M}$. For any $y \in \mathbb{Y}$, the norm of $A_n y - Ay$ is less than the minimal value of $\|A_n y - A_n y_i\| + \|A_n y_i - Ay_i\| + \|Ay_i - Ay\|$. By the Banach–Steinhaus Theorem $A_n$ are equibounded by, say, $M$. Hence we may estimate this quantity by

$$2M \min_{i=1,...,k} \|y - y_i\| + \max_{i=1,...,k} \|A_n y_i - Ay_i\|.$$

The first term above is less than $\frac{\epsilon}{2}$ and the second may be made that small by choosing $n$ large enough. □

**7.1.11 Corollary** Suppose that $x_t, t \in \mathbb{T}$, where $\mathbb{T}$ is an abstract index set, is a family of equicontinuous functions on $\mathbb{R}$ such that

$$\sup_{t \in \mathbb{T}} \sup_{\tau \in \mathbb{R}} |x_t(\tau)| = c$$

is finite. Suppose that $\mu_n, n \geq 1$, is a sequence of probability measures converging weakly to a probability measure $\mu$. Then, $\lim_{n \to \infty} \int_{\mathbb{R}} x_t \, d\mu_n = \int_{\mathbb{R}} x_t \, d\mu$ uniformly in $t \in \mathbb{T}$.

*Proof* By 5.7.17, for any $T > 0$, the functions $(x_t)_{|[-T,T]}$ form a compact set in $C[-T,T]$. Applying 7.1.10 to the functionals $x \mapsto \int_{[-T,T]} x \, d\mu_n$ we see that

$$\lim_{n \to \infty} \int_{[-T,T]} x_t \, d\mu_n = \int_{[-T,T]} x_t \, d\mu \qquad (7.2)$$

uniformly in $t \in \mathbb{T}$.

On the other hand, for any $\epsilon > 0$ we may find a $T > 0$ such that $\mu(\{-T,T\}) = 0$ and $\mu(-T,T)^{\complement} < \epsilon$. Then, for large $n$, $\mu_n(-T,T)^{\complement}$ is less than $\epsilon$, too. Hence, $\left| \int_{(-T,T)^{\complement}} x_t \, d\mu_n \right| \leq \epsilon c$. This, together with (7.2) completes the proof.                                                          $\square$

**7.1.12 Corollary**     A sequence $\mu_n$ of probability measures on $\mathbb{R}$ converges weakly to a probability measure $\mu$ iff the corresponding operators $T_{\mu_n}$ in $BUC(\mathbb{R})$ converge strongly to $T_{\mu}$.

*Proof* If $x \in BUC(\mathbb{R})$, then the functions $x_\sigma, \sigma \in \mathbb{R}$ defined as $x_\sigma(\tau) = x(\sigma + \tau)$ are equicontinuous on $\mathbb{R}$.                                          $\square$

## 7.2 Calculus of Banach space valued functions

7.2.1 *The derivative*     Let $a < b$ be two numbers. A function $(a,b) \to x_t$ taking values in a normed space is said to be differentiable at a point $t_0 \in (a,b)$ iff the limit $\lim_{h \to 0+} \frac{x_{t_0+h} - x_{t_0}}{h} = x'_{t_0}$ exists, and $x'_{t_0}$ is then called the derivative of $x_t$ at $t_0$. Analogously one defines the right-hand (left-hand) derivative at a point $t_0$, if the function is defined in $[t_0, t_0+h)$ $((t_0 - h, t_0])$ for some positive $h$, and the appropriate limit exists.

Many results from the calculus of real-valued functions remain valid with this definition. For example, one proves that if a function $(a,b) \ni t \to x_t$ is differentiable, and the derivative equals zero, then the function is constant. Let us carry out the proof, for it illustrates well the method of proving this sort of theorem. Let $x^* \in \mathbb{X}^*$ be a linear functional. Since $x^*$ is linear and continuous, the scalar-valued function $t \to x^*(x_t)$ is differentiable with derivative $x^*(x'_t)$, which by assumption equals zero. Thus, $x^*(x_t)$ is constant for any functional $x^*$. Choose a $t_0 \in (a,b)$. For any $t$ in this interval, $\|x_t - x_{t_0}\| = \sup_{x^* \in \mathbb{X}, \|x^*\|=1} |x^*(x_t - x_{t_0})| = 0$, as desired.

7.2.2 **Example**     Here is another example of a generalization of a classical result that is often found useful. If $t \to x_t$ is continuous in an interval $[a,b]$ (right-continuous at $a$ and left-continuous at $b$), then for any $x \in \mathbb{X}$, the function $t \rightarrow y_t = x + \int_a^t x_s\, ds$ is differentiable and $y_t' = x_t$ (with right-hand derivative at $a$ etc.).

*Proof* For any $t < b$ and $h > 0$ sufficiently small,

$$\left\| \frac{y_{t+h} - y_t}{h} - x_t \right\| = \left\| \frac{1}{h} \int_t^{t+h} (x_s - x_t)\, ds \right\| \leq \sup_{s \in [t, t+h]} \|x_s - x_t\|,$$

which tends to zero, as $h \rightarrow 0$. For $t > a$ and sufficiently small $h < 0$ the argument is similar. $\square$

**7.2.3 Corollary** A function $[a, b] \ni t \rightarrow x_t$ is continuously differentiable iff there exists a continuous function $[a, b] \ni t \rightarrow y_t$ such that

$$x_t = x_a + \int_a^t y_s\, ds, \qquad t \in [a, b]. \tag{7.3}$$

*Proof* If $t \rightarrow x_t$ is continuously differentiable in $[a, b]$, then the function $t \rightarrow z_t = x_t - x_a - \int_a^t x_s'\, ds$ is differentiable in $(a, b)$ with $z_t' = x_t' - x_t' = 0$, and thus it is constant. Since $\|z_t\| \leq \|x_t - x_a\| + (t - a) \sup_{s \in [a,b]} \|x_s'\|$, we have $\lim_{t \to a} z_t = 0$, implying $z_t = 0$ for all $t \in [a, b]$, by continuity.

On the other hand, by 7.2.2, if (7.3) holds, then $x_t$ is continuously differentiable, and $x_t' = y_t$. $\square$

**7.2.4 Example** Certainly, not all theorems from calculus are true for Banach space valued functions. For instance, the Lagrange Theorem fails, i.e. for a continuously differentiable function $x_t$ on an interval $(a, b)$, with values in a Banach space $\mathbb{X}$, continuous on $[a, b]$, there may be no $\theta \in (a, b)$ such that $(b - a)x_\theta' = x_b - x_a$. To see that consider $x_t = (\sin t, \cos t), t \in [0, \frac{\pi}{2}]$ with values in the Hilbert space $\mathbb{R}^2$. Then $x_\theta' = (\cos \theta, -\sin \theta)$ and $\|x_\theta'\| = \frac{\pi}{2}$, for all $\theta \in [0, \frac{\pi}{2}]$, while $\|x_{\frac{\pi}{2}} - x_0\| = \|(-1, 1)\| = \sqrt{2}$.

We may prove, however, the following useful estimate: if $x_t$ satisfies the above assumption and $\|x_t'\| \leq M$ for some $M > 0$ and all $t \in (a, b)$ then $\|x_b - x_a\| \leq M(b - a)$. To this end, for a functional $F$ on $\mathbb{X}$ consider the scalar-valued function $t \rightarrow Fx_t$. This function satisfies the assumptions of Lagrange's Theorem and therefore there exists a $\theta \in (a, b)$ such that $Fx_b - Fx_a = Fx_\theta'(b - a)$; note that $\theta$ depends on $F$ and that is why we may not omit the functional $F$ in this relation. Nevertheless,

$$|F(x_b - x_a)| \leq \|F\|\, \|x_\theta'\|(b - a) \leq \|F\|M(b - a),$$

and our claim follows by (5.5).

## 7.3 Closed operators

**7.3.1 Definition**    Let $\mathbb{X}$ be a Banach space. A linear operator $A$ with domain $\mathcal{D}(A) \subset \mathbb{X}$ and values in $\mathbb{X}$ is termed closed iff for any converging sequence $x_n$ of elements of its domain such that $Ax_n$ converges, the limit $x = \lim_{n\to\infty} x_n$ belongs to $\mathcal{D}(A)$ and $Ax = \lim_{n\to\infty} Ax_n$. It is clear that all continuous operators are closed; there are, however, many examples of linear operators that are closed but not continuous.

**7.3.2 Example**    Consider $C[0,1]$, the space of continuous functions on $[0,1]$ (with supremum norm). Let the operator $A$ be defined on the algebraic subspace $\mathcal{D}(A)$ of continuously differentiable functions (with right-hand derivative at $\tau = 0$ and left-hand derivative at $\tau = 1$), by $Ax = x'$. $A$ is not bounded. Indeed, first of all it is not defined on the entire space. But, since the domain of $A$ is dense in $C[0,1]$ maybe we could extend $A$ to a bounded linear operator? The answer is no. Extending $A$ to a bounded linear operator is possible only if there exists a constant $K$ such that $\|Ax\| \leq K\|x\|$ for all $x \in \mathcal{D}(A)$ (cf. 2.3.33) Defining, however, the functions $x_n(\tau) = \tau^n, n \geq 1$, we see that $\|x_n\| = 1$ and $Ax_n = nx_{n-1}$, so that $\|Ax_n\| = n$.

On the other hand, if for a converging sequence $x_n$ of elements of $\mathcal{D}(A)$, there also exists the limit $\lim_{n\to\infty} x'_n$, then $x$ is continuously differentiable and $x' = \lim_{n\to\infty} x'_n$. This is a well-known fact from calculus; to reproduce its proof note that for $\tau \in [0,1]$ and any $n \geq 1$, we have $x_n(\tau) = x_n(0) + \int_0^\tau x'_n(\sigma)\, d\sigma$. Since $x'_n$ converges to some continuous $y$ (even uniformly), the integral above converges to $\int_0^\tau y(\sigma)\, d\sigma$, while $x_n(\tau)$ converges to $x(\tau)$ and $x_n(0)$ converges to $x(0)$. Thus, $x(\tau) = x(0) + \int_0^\tau y(\sigma)\, d\sigma$. This implies our result.

**7.3.3 Exercise**    Let $A$ be a linear operator in a Banach space $\mathbb{X}$. Equip the algebraic subspace $\mathcal{D}(A)$ with the norm $\|x\|_A = \|x\| + \|Ax\|$. Prove that $\mathcal{D}(A)$ with this norm is a Banach space iff $A$ is closed.

**7.3.4 Example**    Suppose that $A$ is a closed linear operator in a Banach space $\mathbb{X}$, and that $t \to x_t \in \mathcal{D}(A)$ is a Riemann integrable function on an interval $[a, b]$. If $t \mapsto Ax_t$ is integrable also, then $\int_a^b x_s\, ds$ belongs to $\mathcal{D}(A)$, and

$$A \int_a^b x_s\, ds = \int_a^b Ax_s\, ds. \tag{7.4}$$

*Proof* Let $(\mathcal{T}_n, \Xi_n), n \geq 1$, be a sequence of pairs considered in definition 2.2.48. Since $x_t \in \mathcal{D}(A)$, for all $t \in [a, b]$, $S(\mathcal{T}_n, \Xi_n, x.)$ belongs to $\mathcal{D}(A)$ also, and we see that $AS(\mathcal{T}_n, \Xi_n, x.) = S(\mathcal{T}_n, \Xi_n, Ax.)$. If, additionally, $\lim_{n \to \infty} \Delta(\mathcal{T}_n) = 0$, then the limit of $S(\mathcal{T}_n, \Xi_n, x.)$ exists and equals $\int_a^b x_s \, ds$. Similarly, the limit of $S(\mathcal{T}_n, \Xi_n, Ax.)$ exists and equals $\int_a^b Ax_s \, ds$. Closedness of $A$ implies that $\int_a^b x_s \, ds$ belongs to $\mathcal{D}(A)$ and that (7.4) holds. □

**7.3.5 Exercise** Assume that $A$ is a closed operator in a Banach space $\mathbb{X}$, and $B \in \mathcal{L}(X)$. Prove that $C = A + B$ with domain $\mathcal{D}(C) = \mathcal{D}(A)$ is closed. In particular, a linear operator $A$ is closed iff for some $\lambda \in \mathbb{R}$, $\lambda I_{\mathbb{X}} - A$ is closed.

**7.3.6 Example** Consider the space $BUC(\mathbb{R})$ and the operator $A = \frac{d^2}{d\tau^2}$ with domain $\mathcal{D}(\frac{d^2}{d\tau^2})$ composed of all twice differentiable functions $x$ with $x'' \in BUC(\mathbb{R})$. We claim that $A$ is closed. It may seem that the way to prove this claim is simply to follow the argument given in 7.3.2. Thus, we would assume that a sequence of $x_n \in \mathcal{D}(\frac{d^2}{d\tau^2})$ converges to some $x$ and that the sequence $x_n''$ of derivatives converges to some $y$, and write

$$x_n(\tau) = x_n(0) + \tau x_n'(0) + \int_0^\tau \int_0^\sigma x_n''(\varsigma) \, d\varsigma \, d\sigma$$

$$= x_n(0) + \tau x_n'(0) + \int_0^\tau (\tau - \sigma) x_n''(\sigma) \, d\sigma. \qquad (7.5)$$

The reader already perceives, however, that there is going to be a problem with passage to the limit as $n \to \infty$. The question is: does the numerical sequence $x_n'(0)$ converge? Or, perhaps, we have not defined our operator properly? We have not assumed that $x'$ belongs to $BUC(\mathbb{R})$; maybe we should have done it? The answer is in the negative: it turns out that the assumption that $x$ and $x''$ belong to $BUC(\mathbb{R})$ implies that $x'$ does belong to $BUC(\mathbb{R})$, too. This follows from formula (7.8), below. We will prove this formula first, and then complete the proof of the fact that $A$ is closed. Note that uniform continuity of $x'$ follows from boundedness of $x''$ and that it is only boundedness of $x'$ that is in question.

Let $x$ be a twice differentiable function on the interval $[0, n]$, $n \in \mathbb{N}$. Integrating by parts twice we see that for any $\tau \in (0, n]$,

$$\int_0^\tau \frac{\sigma^{n+1}}{\tau^n} x''(\sigma) \, d\sigma = x'(\tau)\tau - (n+1)x(\tau) + n(n+1) \int_0^\tau \frac{\sigma^{n-1}}{\tau^n} x(\sigma) \, d\sigma. \qquad (7.6)$$

Therefore, for $\tau \in [0, n)$,

$$\int_\tau^n \frac{(n-\sigma)^{n+1}}{(n-\tau)^n} x''(\sigma)\, d\sigma = \int_0^{n-\tau} \frac{\sigma^{n+1}}{(n-\tau)^n} x''(n-\sigma)\, d\sigma \qquad (7.7)$$

$$= -x'(\tau)(n-\tau) - (n+1)x(\tau) + n(n+1)\int_\tau^n \frac{(n-\sigma)^{n-1}}{(n-\tau)^n} x(\sigma)\, d\sigma.$$

Subtracting (7.7) from (7.6), after some algebra,

$$x'(\tau) = \int_0^n a_n(\tau, \sigma) x''(\sigma)\, d\sigma + \int_0^n b_n(\tau, \sigma) x(\sigma)\, d\sigma, \qquad \tau \in (0, n),$$

where

$$a_n(\tau, \sigma) = \begin{cases} \dfrac{1}{n}\dfrac{\sigma^{n+1}}{\tau^n}, & \sigma \le \tau, \\[2mm] -\dfrac{1}{n}\dfrac{(n-\sigma)^{n+1}}{(n-\tau)^n}, & n \ge \sigma > \tau, \end{cases}$$

and

$$b_n(\tau, \sigma) = \begin{cases} -(n+1)\dfrac{\sigma^{n-1}}{\tau^n}, & \sigma \le \tau, \\[2mm] (n+1)\dfrac{(n-\sigma)^{n-1}}{(n-\tau)^n}, & n \ge \sigma > \tau. \end{cases}$$

Since

$$\int_0^n |a_n(\tau, \sigma)|\, d\sigma = \frac{1}{n(n+2)} \frac{\sigma^{n+2}}{\tau^n}\bigg]_{\sigma=0}^{\sigma=\tau} - \frac{1}{n(n+2)} \frac{(n-\sigma)^{n+2}}{(n-\tau)^n}\bigg]_{\sigma=\tau}^{\sigma=n}$$

$$= \frac{1}{n(n+2)}[\tau^2 + (n-\tau)^2] \le \frac{1}{2}\frac{n}{n+2}, \qquad \tau \in (0, n),$$

and

$$\int_0^n |b_n(\tau, \sigma)|\, d\sigma = \frac{n+1}{n}\frac{\sigma^n}{\tau^n}\bigg]_{\sigma=0}^{\sigma=\tau} - \frac{n+1}{n}\frac{(n-\sigma)^n}{(n-\tau)^n}\bigg]_{\sigma=\tau}^{\sigma=n} = \frac{2(n+1)}{n},$$

we have, by continuity,

$$\sup_{\tau\in[0,n]} |x'(\tau)| \le \frac{1}{2}\frac{n}{n+2} \sup_{\tau\in[0,n]} |x''(\tau)| + \frac{2(n+1)}{n} \sup_{\tau\in[0,n]} |x(\tau)|.$$

Arguing similarly on the negative half-axis, we obtain

$$\sup_{\tau\in[-n,n]} |x'(\tau)| \le \frac{1}{2}\frac{n}{n+2} \sup_{\tau\in[-n,n]} |x''(\tau)| + \frac{2(n+1)}{n} \sup_{\tau\in[-n,n]} |x(\tau)|.$$

Therefore,

$$\|x'\|_{BUC(\mathbb{R})} \le \frac{1}{2}\|x''\|_{BUC(\mathbb{R})} + 2\|x\|_{BUC(\mathbb{R})}. \qquad (7.8)$$

This implies that the first derivative of a member of $\mathcal{D}(\frac{d^2}{d\tau^2})$ belongs to $BUC(\mathbb{R})$. Moreover, it implies that if $x_n$ converges to an $x$, and $x_n''$

converges to a $z$, then $x_n'$ is a Cauchy sequence, and thus converges to a $y$. As in 7.3.2 we prove that this forces $x$ to be differentiable and $x' = y$. Now, the sequence $y_n = x_n'$ converges to $y$ and $y_n' = x_n''$ converges to $z$, thus $y$ is differentiable with $y' = z$, so that $x$ is twice differentiable with $x'' = z$.

**7.3.7** *The left inverse of a closed linear operator is closed*   Suppose that $\mathbb{X}$ is a Banach space and that $A$ is a closed linear operator in $\mathbb{X}$. Let

$$\mathcal{R} = \{y \in \mathbb{X}; y = Ax \text{ for some } x \in \mathbb{X}\}$$

denote its range, and suppose that $A$ is injective, i.e. that for any $x \in \mathcal{D}(A)$, condition $Ax = 0$ implies $x = 0$. This forces $A$ to map $\mathcal{D}(A)$ onto $\mathcal{R}$ in a one-to-one fashion. Indeed, if there are two elements of $\mathcal{D}(A)$, say $x_1$ and $x_2$, such that $Ax_1 = Ax_2$, then we have $A(x_1 - x_2) = 0$, and so $x_1$ equals $x_2$. Define the operator $A^{-1}$ (often called the left inverse of $A$) with domain $\mathcal{D}(A^{-1}) = \mathcal{R}$, and range equal to $\mathcal{D}(A)$, by

$$A^{-1}y = x \text{ iff } Ax = y, \qquad \text{for } y \in \mathcal{R}, x \in \mathcal{D}(A).$$

We claim that $A^{-1}$ is closed. To prove this, suppose that a sequence $y_n$ of elements of $\mathcal{R}$ converges to $y$ and that $A^{-1}y_n$ converges to some $x$. Then, elements $x_n = A^{-1}y_n$ belong to $\mathcal{D}(A)$ and converge to $x$. Furthermore, the sequence $Ax_n = y_n$ converges also. Since $A$ is closed, $x$ belongs to $\mathcal{D}(A)$ and $Ax = y$. This means, however, that $y$ belongs to the domain of $A^{-1}$ and that $A^{-1}y = x$, as desired.

As a corollary, we obtain that the left inverse of a bounded linear operator is closed. In general, the left inverse of a bounded linear operator may be unbounded.

**7.3.8 Exercise**   Prove that a linear operator $A : \mathbb{X} \supset \mathcal{D}(A) \to \mathbb{X}$ is closed iff its graph $G_A = \{(x, y) \in \mathbb{X} \times \mathbb{X}; x \in \mathcal{D}(A), y = Ax\}$ is closed in $\mathbb{X} \times \mathbb{X}$ equipped with any one of the norms defined in 2.2.27. Use this result to give a simple proof of the fact that the left inverse of a closed operator (if it exists) is closed.

**7.3.9 Example**   Let $C_0[0, 1]$ be the subspace of $C[0, 1]$ composed of functions $x$ such that $x(0) = 0$. For $x \in C_0[0, 1]$ define $Ax$ in $C_0[0, 1]$ by $Ax(\tau) = \int_0^\tau x(\sigma) \, d\sigma$. The operator $A$ is linear and bounded, since

$$\sup_{\tau \in [0,1]} |Ax(\tau)| \leq \sup_{\tau \in [0,1]} \tau \|x\| = \|x\|.$$

Moreover, if $Ax = 0$ then $\int_0^\tau x(\sigma)\, d\sigma = 0$, for all $\tau \in [0,1]$, and so $x(\tau) = \frac{d}{d\tau} \int_0^\tau x(\sigma)\, d\sigma = 0$. It is easy to see that the range $\mathcal{R}$ of $A$ is the set of all differentiable functions $x \in C_0[0,1]$ such that $x'$ belongs to $C_0[0,1]$. Therefore, the operator $A^{-1} = \frac{d}{d\tau}$ defined on $\mathcal{R}$ is closed.

**7.3.10 Example**    Here is another proof of the fact that the operator introduced in 7.3.6 is closed. Let $\lambda > 0$, we will show that in $BUC(\mathbb{R})$ there exists exactly one solution to the equation

$$\lambda x - \frac{1}{2}x'' = y, \tag{7.9}$$

where $y$ is any (given) member of $BUC(\mathbb{R})$. To prove uniqueness, it suffices to consider the homogeneous ODE: $\lambda x - \frac{1}{2}x'' = 0$,. Recall that solutions to this equation are of the form $x(\tau) = C_1 e^{-\tau\sqrt{2\lambda}} + C_2 e^{\tau\sqrt{2\lambda}}$, where $C_1$ and $C_2$ are arbitrary constants ($r = \pm\sqrt{2\lambda}$ are the roots of $\lambda - \frac{1}{2}r^2 = 0$). If $C_1 \neq 0$, then $\lim_{\tau \to -\infty} |x(\tau)| = \infty$, and $x$ must not belong to $BUC(\mathbb{R})$. Analogously, we exclude the case where $C_2 \neq 0$. Hence, the only solution to the homogeneous equation that belongs to $BUC(\mathbb{R})$ is trivial.

Now, we need to show that for any $y$, (7.9) has at least one solution. From the theory of ODEs we know that we should look for an $x$ of the form

$$\begin{aligned}
x(\tau) &= C_1 e^{-\tau\sqrt{2\lambda}} + C_2 e^{\tau\sqrt{2\lambda}} \\
&\quad + \frac{1}{\sqrt{2\lambda}} \int_0^\tau \left[ e^{-(\tau-\sigma)\sqrt{2\lambda}} - e^{(\tau-\sigma)\sqrt{2\lambda}} \right] y(\sigma)\, d\sigma.
\end{aligned} \tag{7.10}$$

For any constants $C_1$ and $C_2$, this function solves (7.9), but they are to be determined in such a way that $x \in BUC(\mathbb{R})$. For $\tau > 0$,

$$\left| C_1 e^{-\tau\sqrt{2\lambda}} + \frac{1}{\sqrt{2\lambda}} \int_0^\tau e^{-(\tau-\sigma)\sqrt{2\lambda}} y(\sigma)\, d\sigma \right| \leq |C_1| + \frac{1}{2\lambda} \|y\|_{BUC(\mathbb{R}^+)}.$$

Therefore, if $x$ is to be bounded,

$$\tau \to C_2 e^{\tau\sqrt{2\lambda}} - \frac{1}{\sqrt{2\lambda}} e^{\tau\sqrt{2\lambda}} \int_0^\tau e^{-\sigma\sqrt{2\lambda}} y(\sigma)\, d\sigma$$

must be bounded also. Since $\lim_{\tau \to \infty} e^{\tau\sqrt{2\lambda}} = \infty$, we must have

$$C_2 = \frac{1}{\sqrt{2\lambda}} \int_0^\infty e^{-\sigma\sqrt{2\lambda}} y(\sigma)\, d\sigma. \tag{7.11}$$

Similarly,

$$C_1 = \frac{1}{\sqrt{2\lambda}} \int_{-\infty}^0 e^{\sigma\sqrt{2\lambda}} y(\sigma)\, d\sigma.$$

Hence,

$$
\begin{aligned}
x(\tau) &= \frac{1}{\sqrt{2\lambda}} \int_{-\infty}^{\tau} e^{-(\tau-\sigma)\sqrt{2\lambda}} y(\sigma)\,d\sigma + \frac{1}{\sqrt{2\lambda}} \int_{\tau}^{\infty} e^{(\tau-\sigma)\sqrt{2\lambda}} y(\sigma)\,d\sigma \\
&= \frac{1}{\sqrt{2\lambda}} \int_{-\infty}^{\infty} e^{-|\tau-\sigma|\sqrt{2\lambda}} y(\sigma)\,d\sigma.
\end{aligned} \tag{7.12}
$$

We need to check that this $x$ belongs to $BUC(\mathbb{R})$. Certainly,

$$
\begin{aligned}
\|x\|_{BUC(\mathbb{R})} &\le \sup_{\tau \in \mathbb{R}} \frac{1}{\sqrt{2\lambda}} \int_{-\infty}^{\infty} e^{-|\tau-\sigma|\sqrt{2\lambda}}\,d\sigma \|y\|_{BUC(\mathbb{R})} \\
&= \frac{2}{\sqrt{2\lambda}} \int_{0}^{\infty} e^{-\sigma\sqrt{2\lambda}}\,d\sigma \|y\|_{BUC(\mathbb{R})} \\
&= \frac{1}{\lambda} \|y\|_{BUC(\mathbb{R})},
\end{aligned} \tag{7.13}
$$

proving boundedness of $x$. Next, note that

$$
|x(\tau) - x(\tau')| \le \frac{1}{\sqrt{2\lambda}} \int_{-\infty}^{\infty} \left| e^{-|\tau-\sigma|\sqrt{2\lambda}} - e^{-|\tau'-\sigma|\sqrt{2\lambda}} \right|\,d\sigma \|y\|_{BUC(\mathbb{R})}.
$$

Assuming, as we may, that $\tau' > \tau$, and writing the last integral as the sum of integrals over $(-\infty, \tau), [\tau, \tau']$ and $(\tau', \infty)$, we estimate it by

$$
\left(1 - e^{\sqrt{2\lambda}(\tau-\tau')}\right) \frac{1}{\sqrt{2\lambda}} + 2(\tau' - \tau) + \left(1 - e^{\sqrt{2\lambda}(\tau-\tau')}\right) \frac{1}{\sqrt{2\lambda}}
$$

which implies uniform continuity of $x$.

We have proved that for any $\lambda > 0$ the operator $\lambda I_{BUC(\mathbb{R})} - \frac{1}{2}\frac{d^2}{d\tau^2}$ is one-to-one with range equal to the whole of $BUC(\mathbb{R})$. Moreover, by (7.13), the inverse operator is bounded. It implies that $\frac{1}{2}\frac{d^2}{d\tau^2}$ is closed, but our analysis shows more, and that is going to be important in what follows (see 7.5.1). In fact, in Subsection 7.5.1 we will need a slightly different version of this result, stated in the exercise below.

**7.3.11 Exercise** Show that for any $y \in C[-\infty, \infty]$, (7.9) has exactly one solution $x \in C[-\infty, \infty]$ and that the inverse of $\lambda I_{C[-\infty,\infty]} - \frac{d^2}{d\tau^2}$ is bounded with norm less than $\frac{1}{\lambda}$.

**7.3.12 Cores** Sometimes, it is hard to describe analytically the whole domain $\mathcal{D}(A)$ of a closed linear operator $A$. Actually, in most cases it is impossible. What is easier and possible is to find an algebraic subspace $D \subset \mathcal{D}(A)$ that characterizes $A$ in the sense that for any $x \in \mathcal{D}(A)$ there exists a sequence $x_n \in D$ such that $\lim_{n\to\infty} x_n = x$ and $\lim_{n\to\infty} A x_n = Ax$. (See e.g. 7.6.17.) Sets with this property are termed **cores** of $A$. It is clear that if $D$ is a core for $A$, then $AD$ (the image of $D$ via $A$) is dense in the range $\mathcal{R}$ of $A$. A partial converse to this remark turns out to be a useful criterion for determining whether an algebraic subspace $D \subset \mathcal{D}(A)$ is a core. Specifically, we will show that if a closed operator $A$ has the property that $\|Ax\| \geq c\|x\|$ for some positive $c$ and all $x$ in $\mathcal{D}(A)$, and if $AD$ is dense in $\mathcal{R}$, then $D$ is a core of $A$. Indeed, under these assumptions, for any $x \in \mathcal{D}(A)$, there exists a sequence $x_n$ such that $Ax_n$ converges to $Ax$. Since $\frac{1}{c}\|Ax_n - Ax\| \geq \|x_n - x\|$, the sequence $x_n$ converges to $x$, and we are done.

The role of a core for a closed operator is similar to that of a dense algebraic subspace for a bounded operator. In particular, a dense algebraic subspace is a core for any bounded linear operator. Note, however, that what is a core for one closed linear operator does not have to be a core for another closed linear operator.

**7.3.13 Exercise**     Show that $D$ is a core of a closed operator $A$, iff it is dense in $\mathcal{D}(A)$ equipped with the norm $\|\cdot\|_A$ (cf. 7.3.3).

## 7.4 Semigroups of operators

**7.4.1 Definition**     Let $\mathbb{X}$ be a Banach space, and suppose that operators $T_t, t \geq 0$, are bounded. The family $\{T_t, t \geq 0\}$ is termed a semigroup of operators (a semigroup, for short) iff

$1°$ for all $s, t \geq 0$, $T_{s+t} = T_t T_s$,

$2°$ $T_0 = I_{\mathbb{X}}$, where $I_{\mathbb{X}}$ is the identity operator in $\mathbb{X}$.

The key relation is the *semigroup property* $1°$; it establishes a homomorphism between the semigroup of positive numbers with addition as a semigroup operation, and the semigroup (the Banach algebra) $\mathcal{L}(\mathbb{X})$ of operators on $\mathbb{X}$. Families of operators that fulfill this relation enjoy surprising properties. The situation is similar to Theorem 1.6.11, which says that measurable functions that satisfy the Cauchy functional equation are continuous. Indeed, if $\{T_t, t \geq 0\}$ is a semigroup of operators in a Banach space $\mathbb{X}$, and for any $x \in \mathbb{X}$, the map $t \to T_t x$ is (Bochner) measurable, it is also continuous for $t > 0$. We will not use this theorem later on, neither shall we introduce the notion of Bochner measurability; we mention it here solely to impress the reader with the importance and far reaching consequences of the apparently simple relation $1°$. In what follows we will prove more (but simpler) results of this sort.

**7.4.2 Example**    Let $x$ be an integrable function on $\mathbb{R}^+$ and let $t$ be a non-negative number. Let $T_t x$ be a new function on $\mathbb{R}^+$ defined to be equal to 0 for $\tau < t$ and $x(\tau - t)$ for $\tau \geq t$. We will use a shorthand: $T_t x(\tau) = x(\tau - t) 1_{[t,\infty)}(\tau)$, although it is not exactly correct, for $x(\tau - t)$ is not defined for $\tau < t$. Observe that if $y$ is another integrable function such that $\int_0^\infty |x(\tau) - y(\tau)| \, d\tau = 0$, then

$$\int_0^\infty |T_t x(\tau) - T_t y(\tau)| \, d\tau = \int_t^\infty |x(\tau - t) - y(\tau - t)| \, d\tau$$

$$= \int_0^\infty |x(\tau) - y(\tau)| \, d\tau = 0.$$

Therefore, if $x$ and $y$ belong to the same equivalence class in $L^1(\mathbb{R}^+)$, then so do $T_t x$ and $T_t y$. Consequently, $T_t$ is an operator in $L^1(\mathbb{R}^+)$. To prove the semigroup property of $\{T_t, t \geq 0\}$, we take a representant $x$ of a class in $L^1(\mathbb{R})$, and calculate as follows:

$$T_s T_t x(\tau) = 1_{[s,\infty)}(\tau) T_t x(\tau - s)$$

$$= 1_{[s,\infty)}(\tau) 1_{[t,\infty)}(\tau - s - t) x(\tau - s - t)$$

$$= 1_{[s+t,\infty)}(\tau) x(\tau - s - t) = T_{t+s} x(\tau).$$

Since the choice of an $x$ from an equivalence class does not influence the result of our calculations (a.s.), the proof is complete. $\{T_t, t \geq 0\}$ is called the **semigroup of translations to the right**.

**7.4.3 Exercise**    Show that if $\{T_t, t \geq 0\}$ and $\{S_t, t \geq 0\}$ are two semigroups, and $S_t$ commutes with $T_t$ for all $t \geq 0$, i.e. $S_t T_t = T_t S_t$, then $U_t = S_t T_t$ is a semigroup.

**7.4.4 Exercise**    Let $p_r$ be the Poisson kernel defined in 1.2.29. For an integrable $x$ on the unit circle $\mathbf{C}$ and $t > 0$ let $T_t x = p_{\exp(-t)} * x$. Use 1.2.29 to check that if we let $T_0 x = x$, then $\{T_t, t \geq 0\}$ is a semigroup of operators on both the space of continuous functions on $\mathbf{C}$ and on the space of equivalence classes of integrable functions on $\mathbf{C}$. Also, show that $T_t, t \geq 0$, are contractions.

**7.4.5 Definition**    A semigroup $\{T_t, t \geq 0\}$ is said to be strongly continuous or of class $c_0$ iff

$3^\circ$ $\lim_{t \to 0} T_t x = x$, for $x \in \mathbb{X}$.

Semigroups that satisfy this condition are particularly important. In fact most modern textbooks on semigroups restrict their attention to strongly continuous semigroups [26, 43, 91, 94]; the theories of stochastic processes and PDEs supply, however, many examples of semigroups that are not strongly continuous (see [25, 54]).

**7.4.6 Exercise**   Show that if $\{T_t, t \geq 0\}$ and $\{S_t, t \geq 0\}$ are two strongly continuous semigroups, and $S_t$ commutes with $T_t$ for all $t \geq 0$ then $U_t = S_t T_t$ is a strongly continuous semigroup. In particular, if $\{T_t, t \geq 0\}$ is a semigroup, and $\omega$ is a number, then $U_t = e^{\omega t} T_t$ is a strongly continuous semigroup.

**7.4.7 Example**   Let $\mathbb{X} = BUC(\mathbb{R}^+)$ be the space of all bounded, uniformly continuous functions on $\mathbb{R}^+$, and let $T_t x(\tau) = x(\tau + t)$. Obviously, $T_t$ maps $BUC(\mathbb{R}^+)$ into itself. As in 7.4.2 we check that $\{T_t, t \geq 0\}$ is a semigroup. To show that it is of class $c_0$, take an $x \in \mathbb{X}$, and for a given $\epsilon > 0$ choose a $\delta > 0$, such that $|x(\tau) - x(\sigma)| < \epsilon$ provided $\tau, \sigma \geq 0$, and $|\tau - \sigma| < \delta$. Then, for $t < \delta$,

$$\|T_t x - x\| = \sup_{\tau \geq 0} |T_t x(\tau) - x(\tau)| = \sup_{\tau \geq 0} |x(\tau + t) - x(\tau)| \leq \epsilon.$$

This proves that $\lim_{t \to 0+} T_t x = x$. Observe, however, that $\|T_t - I\| = \sup_{\|x\|=1} \|T_t x - x\| = 2$, for all $t > 0$ (cf. 7.4.19 and 7.4.20). The reader should prove it, arguing as in 2.3.29 for example.

**7.4.8 Example**   Let $\{T_t, t \geq 0\}$ be the semigroup from 7.4.2. We will prove that it is strongly continuous. Note first that $T_t$ have norm 1; one of the ways to see this is to note that they are Markov. Therefore, it suffices to show relation $3°$ for $x$ from a dense subset $\mathbb{X}_0$ of $\mathbb{X}$. In particular, we may choose for $\mathbb{X}_0$ the set of (equivalence classes corresponding to) continuous functions with compact support in the open half-axis $(0, \infty)$. Note that $T_t$ maps $\mathbb{X}_0$ into itself, and that $\mathbb{X}_0 \subset BUC(\mathbb{R}^+)$. A modification of the reasoning from 7.4.7 shows that for $x \in \mathbb{X}_0$, $T_t x$ tends to $x$ uniformly, as $t \to 0+$. Moreover, if the support of $x$ is contained in the interval $(0, K)$ for some $K > 0$ then $\|T_t x - x\|_{L^1(\mathbb{R}^+)} \leq K \|T_t x - x\|_{BUC(\mathbb{R}^+)}$, completing the proof.

**7.4.9 Exercise**   Let $r_\alpha$ be the rotation around $(0, 0)$ of the plane $\mathbb{R}^2$ by the angle $\alpha \in \mathbb{R}^+$ and let $\mathbb{X} = C_0(\mathbb{R}^2)$. Check that $T_t x(p) = x(r_t p)$, where $p \in \mathbb{R}^2$, is a strongly continuous semigroup of operators.

**7.4.10 Example**   Prove that the semigroup from 7.4.4 is of class $c_0$.

**7.4.11** *Exponential growth* Suppose that $\{T_t, t \geq 0\}$ is a strongly continuous semigroup of operators in a Banach space $\mathbb{X}$. Then there exist constants $M \geq 1$ and $\omega \in \mathbb{R}$ such that

$$\|T_t\| \leq Me^{\omega t}. \tag{7.14}$$

*Proof* By 7.1.6, there exists a $\delta > 0$ and an $M \geq 1$, such that $\|T_t\| \leq M$ for $0 \leq t \leq \delta$ (we cannot have $M < 1$, for $\|T_0\| = 1$ – see also 7.4.13). For arbitrary $t \geq 0$, one may find an $n \in \mathbb{N} \cup \{0\}$ and $t' \in [0, \delta)$ such that $t = n\delta + t'$. By the semigroup property,

$$\|T_t\| = \|T_\delta^n T_{t'}\| \leq M^n M = Me^{n \ln M} \leq Me^{(n\delta + t')\frac{\ln M}{\delta}}.$$

Taking $\omega = \frac{\ln M}{\delta}$ completes the proof. ☐

**7.4.12 Remark** Taking $\mathbb{X} = \mathbb{R}$, and $T_t = e^{\omega t} x$, $t \geq 0$, we see that $\omega$ in (7.14) may be arbitrary. As we have remarked, $M$ must be greater than or equal to 1. It is also worth noting that for a fixed strongly continuous semigroup the minimum of the set of $\omega$ such that (7.14) holds for some $M \geq 1$ may not be attained.

**7.4.13 Exercise** Suppose that $\{T_t, t \geq 0\}$ is a semigroup, and that $\|T_t\| \leq Me^{\omega t}$, for some $\omega \in \mathbb{R}$, $M \in \mathbb{R}^+$ and all $t > 0$. Prove that $0 \leq M < 1$ implies $T_t = 0$ for all $t > 0$.

**7.4.14** *Continuity* If $\{T_t, t \geq 0\}$ is a strongly continuous semigroup, then for all $x \in \mathbb{X}$, the function $t \to T_t x$ is strongly continuous in $\mathbb{R}^+$ (right-continuous at 0).

*Proof* Right-continuity at $t = 0$ is secured by the definition of a strongly continuous semigroup. If $t > 0$ then $\lim_{h \to 0+} T_{t+h} x = \lim_{h \to 0+} T_h T_t x = T_t x$, since $T_t$ belongs to $\mathbb{X}$. Also, for suitable $\omega$ and $M$,

$$\limsup_{h \to 0+} \|T_{t-h} x - T_t x\| \leq \limsup_{h \to 0+} \|T_{t-h}\|_{\mathcal{L}(\mathbb{X})} \|x - T_h x\|$$

$$\leq \limsup_{h \to 0+} Me^{\omega(t-h)} \|x - T_h x\| = 0.$$

☐

**7.4.15** *The infinitesimal generator* Let $\{T_t, t \geq 0\}$ be a strongly continuous semigroup of operators in a Banach space $\mathbb{X}$. Let $\mathcal{D}(A)$ denote the set of all $x \in \mathbb{X}$, such that the limit

$$Ax := \lim_{h \to 0+} \frac{T_h x - x}{h} \tag{7.15}$$

exists. It is clear that if $x$ and $y$ belong to $D(A)$ then so does $\alpha x + \beta y$ for all real $\alpha$ and $\beta$, and that $A(\alpha x + \beta y) = \alpha A x + \beta A y$. Thus, $A$ is a linear operator. As we shall see, in general $A$ is not continuous and in particular defined only on an algebraic subspace of $\mathbb{X}$. Nevertheless, $\mathcal{D}(A)$ is never empty, and in fact dense in $\mathbb{X}$. To see that consider $\int_0^t T_s x \, \mathrm{d}s$ where $x \in \mathbb{X}, t > 0$. Since

$$T_h \int_0^t T_s x \, \mathrm{d}s = \int_0^t T_{s+h} x \, \mathrm{d}s = \int_h^{t+h} T_s x \, \mathrm{d}s = \int_0^{t+h} T_s x \, \mathrm{d}s - \int_0^h T_s x \, \mathrm{d}s,$$
(7.16)

we also have

$$\frac{1}{h}\left(T_h \int_0^t T_s x \, \mathrm{d}s - \int_0^t T_s x \, \mathrm{d}s\right) = \frac{1}{h}\int_t^{t+h} T_s x \, \mathrm{d}s - \frac{1}{h}\int_0^h T_s x \, \mathrm{d}s. \quad (7.17)$$

Arguing as in 7.2.3, we see that the last two expressions tend to $T_t x$ and $x$, respectively. This proves that $\int_0^t T_s x \, \mathrm{d}s$ belongs to $\mathcal{D}(A)$ and

$$A \int_0^t T_s x \, \mathrm{d}s = T_t x - x. \quad (7.18)$$

Moreover, by linearity, $\frac{1}{t}\int_0^t T_s x \, \mathrm{d}s$ also belongs to $\mathcal{D}(A)$ for all $t > 0$ and $x \in \mathbb{X}$. Since, again as in 7.2.3, $\lim_{t\to 0} \frac{1}{t}\int_0^t T_s x \, \mathrm{d}s = x$, our claim is proven.

**7.4.16 Example**    What is the generator of the semigroup from example 7.4.7? If $x \in BUC(\mathbb{R}^+)$ belongs to $\mathcal{D}(A)$, then the limit

$$\lim_{h\to 0+} \frac{T_h x(\tau) - x(\tau)}{h} = \lim_{h\to 0+} \frac{x(\tau + t) - x(\tau)}{h} = x'(\tau)$$

exists (even uniformly in $\tau \geq 0$). Therefore, $x \in \mathcal{D}(A)$ must be differentiable with $x'$ in $BUC(\mathbb{R}^+)$. Suppose, conversely, that $x$ is differentiable with $x' \in BUC(\mathbb{R}^+)$. Then, by the Lagrange Theorem, for any non-negative $\tau$ and $h > 0$, there exists a $0 \leq \theta \leq 1$ such that $x(\tau + h) - x(\tau) = h x'(\tau + \theta h)$. Since $x'$ is uniformly continuous, for any $\epsilon > 0$ there exists a $\delta > 0$ such that $|x'(\tau) - x'(\sigma)| < \epsilon$ provided $|\tau - \sigma| < \delta$ and $\tau$ and $\sigma$ are non-negative. Thus, for $0 < h < \delta$,

$$\sup_{\tau \geq 0}\left|\frac{T_h x(\tau) - x(\tau)}{h} - x'(\tau)\right| = \sup_{\tau \geq 0}|x'(\tau + \theta h) - x'(\tau)| \leq \epsilon,$$

proving that $x$ belongs to $\mathcal{D}(A)$, and $Ax = x'$. Hence, the generator is completely characterized.

**7.4.17 Exercise** What is the infinitesimal generator of the semigroup introduced in 7.4.2?

**7.4.18 Example** Let $\{T_t, t \geq 0\}$ be a strongly continuous semigroup with generator $A$. Let $\lambda \in \mathbb{R}$ be given; we define $S_t = e^{-\lambda t} T_t$, and denote its generator by $B$. Then, $\mathcal{D}(A) = \mathcal{D}(B)$ and $Bx = Ax - \lambda x$.

*Proof* For all $x \in \mathbb{X}$,

$$
\lim_{t \to 0} \left\{ \frac{1}{t}(S_t x - x) - \frac{1}{t}(T_t x - x) \right\} = \lim_{t \to 0} \frac{1}{t}(S_t x - T_t x)
$$

$$
= \lim_{t \to 0} \frac{1}{t}(e^{-\lambda t} - 1)T_t x = -\lambda x.
$$

Thus, the limit $Bx = \lim_{t \to 0} \frac{1}{t}(S_t x - x)$ exists whenever the limit $Ax = \lim_{t \to 0} \frac{1}{t}(T_t x - x)$ exists, and $Bx = Ax - \lambda x$. □

**7.4.19 Example** Let $B \in \mathcal{L}(\mathbb{X})$ be a bounded linear operator, and let $T_t = e^{tB}, t \geq 0$. $\{T_t, t \geq 0\}$ is a semigroup of operators by 2.3.13. We claim that

$$
\lim_{t \to 0} \|T_t - I\|_{\mathcal{L}(\mathbb{X})} = 0, \tag{7.19}
$$

and

$$
\lim_{t \to 0} \left\| \frac{T_t - I}{t} - B \right\|_{\mathcal{L}(\mathbb{X})} = 0. \tag{7.20}
$$

We note that relation (7.20) implies (7.19), and so we may restrict ourselves to proving this last formula. We have

$$
\left\| \frac{e^{tB} - I}{t} - B \right\| = \left\| \frac{1}{t} \sum_{n=2}^{\infty} \frac{t^n B^n}{n!} \right\| \leq \frac{1}{|t|} \sum_{n=2}^{\infty} \frac{|t|^n \|B\|^n}{n!}
$$

$$
= \frac{1}{|t|}(e^{\|B\|t} - \|B\|t - 1),
$$

and $\lim_{s \to 0^+} \frac{1}{s}(e^{as} - as - 1) = 0$, for any number $a$.

This proves both that $\{T_t, t \geq 0\}$ is a strongly continuous semigroup and that $B$ is its infinitesimal generator. Note, however, that (7.19) is much stronger than the definition of strong continuity of a semigroup (cf. 7.4.7). In 7.4.20 we show that, conversely, if a semigroup $\{T_t, t \geq 0\}$ satisfies (7.19) then there exists an operator $B \in \mathcal{L}(\mathbb{X})$ such that $T_t = e^{tB}$ – such semigroups are said to be continuous in the uniform topology. Strongly continuous semigroups form a much wider class and their generators are usually not bounded.

**7.4.20 Proposition**       Let $\mathbb{A}$ be a Banach algebra with unit $u$, and suppose that $a_t, t \geq 0$, is a family such that $a_t a_s = a_{t+s}$ and $\lim_{t \to 0} \|a_t - u\| = 0$. Then there exists an element $a$ of $\mathbb{A}$ such that $a_t = \exp ta$.

*Proof*   By assumption $t \mapsto a_t$ is continuous, hence integrable on finite intervals. Moreover, setting $b_t := \frac{1}{t} \int_0^t a_s \, ds, t > 0$, we see that $\lim_{t \to 0} b_t = u$. By 6.1.5, for small $t$, the inverse of $b_t$ exists.

On the other hand, calculating as in (7.16) and (7.17), we see that

$$\frac{a_h - u}{h} b_t = \frac{1}{t} \left[ \frac{1}{h} \int_t^{t+h} a_s \, ds - \frac{1}{h} \int_0^h a_s \, ds \right] \xrightarrow[h \to 0+]{} \frac{1}{t} (a_t - u).$$

Thus, there exists the limit $\lim_{h \to 0+} \frac{a_h - u}{h} = \lim_{h \to 0+} \frac{a_h - u}{h} b_t b_t^{-1} = \frac{1}{t}(a_t - u)b_t^{-1} =: a$. We note that by definition, $a$ commutes with all $a_t$. By the semigroup property, $\lim_{h \to 0+} \frac{a_{t+h} - a_t}{h} = a_t \lim_{h \to 0+} \frac{a_h - u}{h} = a_t a$, and $\lim_{h \to 0+} \frac{a_{t-h} - a_t}{-h} = \lim_{h \to 0+} a_{t-h} \lim_{h \to 0+} \frac{a_h - u}{h} = a_t a$, proving that the derivative of $a_t$ exists and equals $a a_t = a_t a$. A similar argument using 7.4.19 shows that $c_t = \exp(at)$ has the same property.

Finally, we take a $t_0 > 0$ and define $d_t = c_t a_{t_0 - t}$. Then, $d_t$ is differentiable with the derivative equal to $ac_t a_{t_0 - t} - c_t a a_{t_0 - t} = 0$. Therefore, $d_t$ is constant and equals identically $d_0 = a_{t_0}$. This means that $c_t a_{t_0 - t} = a_{t_0}$ for all $0 \leq t \leq t_0$. In particular, for $t = t_0$ we obtain $c_{t_0} = a_{t_0}$, completing the proof.       $\square$

· **7.4.21 Exercise**    Even though the formula $T(t) = \exp(At)$ is elegant and simple there are few interesting cases where it can be applied to give an explicit form of $T(t)$. The reason is of course that $\exp(At)$ involves all the powers of $A$. Even if $A$ is a finite matrix, calculations may be very tiresome, especially if the dimension of $A$ is large. One usual technique in such a case is diagonalizing $A$, i.e. representing it, if possible, as $UBU^{-1}$ where $U$ is an invertible matrix and $B$ has all entries zero except on the diagonal, to obtain $\exp(At) = U \exp(Bt) U^{-1}$. Here, $\exp(Bt)$ has all entries zero except on the diagonal where they are equal to $e^{\lambda_i t}$; $\lambda_i$ being eigenvalues of $A$ and entries of the diagonal of $B$. Another technique is to use the Cayley–Hamilton Theorem which says that $A$ satisfies its own characteristic equation. Hence, if $A$ is of dimension $n$, $A^n$ is a polynomial in $I, A, ..., A^{n-1}$ and so $e^{At} = \sum_{i=0}^{n-1} p_i(t) A^i$ where $p_i(t)$ are polynomials. The reader may want to check his knowledge of linear algebra in proving that

$$e^{At} = \frac{1}{5} \begin{bmatrix} 5 & 0 & 0 \\ 5 - 4e^{-t} - e^{-6t} & 2e^{-t} + 3e^{-6t} & 2e^{-t} - 2e^{-6t} \\ 5 - 6e^{-t} + e^{-6t} & 3e^{-t} - 3e^{-6t} & 3e^{-t} + 2e^{-6t} \end{bmatrix},$$

where $A = \begin{bmatrix} 0 & 0 & 0 \\ 2 & -4 & 2 \\ 0 & 3 & -3 \end{bmatrix}$.

**7.4.22** *Isomorphic semigroups* Let $\mathbb{X}$ and $\mathbb{Y}$ be two Banach spaces and let $J : \mathbb{X} \to \mathbb{Y}$ be an (isometric) isomorphism of $\mathbb{X}$ and $\mathbb{Y}$. Suppose that $\{S_t, t \geq 0\}$ is a strongly continuous semigroup of operators in $\mathbb{X}$, with the generator $B$. Then $\{U_t, t \geq 0\}$, where $U_t = JS_tJ^{-1}$, is a strongly continuous semigroup of operators in $\mathbb{Y}$ and its generator $C$ equals $C = JBJ^{-1}$. To be more specific: $y \in \mathcal{D}(C)$ iff $J^{-1}y \in \mathcal{D}(B)$, and $Cy = JBJ^{-1}y$. The semigroup $\{U_t, t \geq 0\}$ is said to be (isometrically) isomorphic to $\{S_t, t \geq 0\}$.

*Proof* Obviously, $U_0 = JS_0J^{-1} = JI_{\mathbb{X}}J^{-1} = I_{\mathbb{Y}}$, and

$$U_tU_s = JS_tJ^{-1}JS_sJ^{-1} = JS_tS_sJ^{-1} = JS_{t+s}J^{-1} = U_{t+s}.$$

Moreover, if $y \in \mathbb{Y}$, then $x = J^{-1}y \in \mathbb{X}$, so that $\lim_{t\to 0^+} S_tJ^{-1}y = J^{-1}y$. Therefore $\lim_{t\to 0^+} JS_tJ^{-1}y = JJ^{-1}y = y$. Finally, the limit $\tilde{y} = \lim_{t\to 0^+} \frac{JS_tJ^{-1}y - y}{t}$ exists iff there exists $\tilde{x} = \lim_{t\to 0^+} \frac{S_tJ^{-1}y - J^{-1}y}{t}$, and $\tilde{y} = J\tilde{x}$, proving that $J^{-1}y$ belongs to $\mathcal{D}(B)$ iff $y \in \mathcal{D}(C)$ and $Cy = JBJ^{-1}y$. $\square$

**7.4.23 Exercise** Suppose that $\mathbb{X}_1$ is a subspace of a Banach space $\mathbb{X}$. Let $\{T_t, t \geq 0\}$ be a strongly continuous semigroup of linear operators with generator $A$, such that $T_t\mathbb{X}_1 \subset \mathbb{X}_1$. Prove that $\{S_t, t \geq 0\}$ where $S_t = (T_t)_{|\mathbb{X}_1}$ is the restriction of $\{T_t, t \geq 0\}$ to $\mathbb{X}_1$ is a strongly continuous semigroup of operators in the Banach space $\mathbb{X}_1$, with the generator $B$ given by

$$\mathcal{D}(B) = \mathcal{D}(A) \cap \mathbb{X}_1 = \{x \in \mathbb{X}_1; Ax \in \mathbb{X}_1\}, \qquad Bx = Ax, x \in \mathcal{D}(B).$$

**7.4.24 Example** Let $l_r^1, r > 0$ be the space of sequences $(\xi_n)_{n\geq 1}$ such that $\sum_{n=1}^{\infty} |x_n|r^n < \infty$, considered in 5.2.5. When equipped with the norm $\|(x_n)_{n\geq 1}\| = \sum_{n=1}^{\infty} |x_n|r^n$, $l_r^1$ is a Banach space. This space is isometrically isomorphic to $l^1$ and the isomorphism is given by:

$$l^1 \ni (\xi_n)_{n \geq 1} \xrightarrow{J} \left(\frac{\xi_n}{r^n}\right)_{n \geq 1} \in l_r^1, \qquad (7.21)$$

$$l_r^1 \ni (\xi_n)_{n \geq 1} \xrightarrow{J^{-1}} (\xi_n r^n)_{n \geq 1} \in l^1. \qquad (7.22)$$

Let $L, R$ and $I$ be defined as in 2.3.9 and let $L_r, R_r$ and $I_r$ be the left translation, right translation and the identity operator in $l_r^1$, respectively. Consider the semigroup $e^{t(aL_r + bR_r + cI_r)}, t \geq 0$, in $l_r^1$ generated by the bounded operator $A = aL_r + bR_r + cI_r$. This semigroup is isometrically isomorphic to the semigroup $e^{tJ^{-1}AJ}, t \geq 0$, generated by the bounded operator $J^{-1}AJ$. Since

$$J^{-1} L_r J (\xi_n)_{n \geq 1} = J^{-1} L_r \left(\frac{\xi_n}{r^n}\right)_{n \geq 1} = J^{-1} \left(\frac{\xi_{n+1}}{r^{n+1}}\right)_{n \geq 1} = \frac{1}{r}(x_{n+1})_{n \geq 1},$$

$$(7.23)$$

$$J^{-1} R J (\xi_n)_{n \geq 1} = J^{-1} R \left(\frac{\xi_n}{r^n}\right)_{n \geq 1} = J^{-1} \left(\frac{\xi_{n-1}}{r^{n-1}}\right)_{n \geq 1} = r(\xi_{n-1})_{n \geq 1},$$

$$(7.24)$$

we have $J^{-1}AJ = \frac{a}{r}L + brR + cI$. In other words, $e^{t(aL_r + bR_r + cI_r)}, t \geq 0$, in $l_r^1$ is isometrically isomorphic to $e^{t(\frac{a}{r}L + brR + cI)}, t \geq 0$ in $l^1$. See 7.4.43 for an application.

7.4.25 **Exercise**    Arguing as in 7.4.14, prove that if $x$ belongs to $\mathcal{D}(A)$, then so does $T_t x$. Moreover, the function $t \to T_t x$ is continuously differentiable in $\mathbb{R}^+$ (has right derivative at $t = 0$), and

$$\frac{\mathrm{d}T_t x}{\mathrm{d}t} = AT_t x = T_t Ax, \quad t \geq 0.$$

More generally, define $\mathcal{D}(A^n)$ by induction as the set of all $x \in \mathcal{D}(A^{n-1})$ such that $A^{n-1}x$ belongs to $\mathcal{D}(A)$ and prove that if $x$ belongs to $\mathcal{D}(A^n)$ then so does $T_t x$. Moreover, $t \to T_t x$ is then $n$-times differentiable and

$$\frac{\mathrm{d}^n T_t x}{\mathrm{d}t^n} = A^n T_t x = T_t A^n x, \qquad t \geq 0.$$

7.4.26 **Corollary**    Using 7.4.25 and 7.2.3 we see that an element of $x \in \mathbb{X}$ belongs to $\mathcal{D}(A)$ iff there exists a $y \in \mathbb{X}$ such that $T_t x = x + \int_0^t T_s y \, \mathrm{d}s$. In such a case $Ax = y$.

**7.4.27 Exercise** A matrix $(p_{i,j})_{i,j\in\mathbb{I}}$, where $\mathbb{I}$ is a countable set of indexes, is said to be a stochastic matrix if its entries are non-negative and $\sum_{j\in\mathbb{I}} p_{i,j} = 1, i \in \mathbb{I}$. In such a case, $P(\xi_i)_{i\in\mathbb{I}} = \left(\sum_{i\in\mathbb{I}} \xi_i p_{i,j}\right)_{j\in\mathbb{I}}$ is a Markov operator in $l^1(\mathbb{I})$. A matrix $(q_{i,j})_{i,j\in\mathbb{I}}$ is said to be a $Q$-**matrix**, or an **intensity matrix** or a **Kolmogorov matrix** if $q_{i,j} \geq 0$ for $i \neq j$ and $\sum_{j\in\mathbb{I}} q_{i,j} = 0$. Show that if $\mathbb{I}$ is finite, then $Q$ is an intensity matrix iff $P(t) = e^{Qt}$ is a stochastic matrix for all $t \geq 0$.

**7.4.28** *An infinitesimal generator is closed* The infinitesimal generator of a strongly continuous semigroup is closed. Indeed, if $x_n \in \mathcal{D}(A)$ then

$$T_t x_n = x_n + \int_0^t T_s A x_n \, ds, \qquad t \geq 0, n \geq 1.$$

Moreover, if $\lim_{n\to\infty} x_n = x$, then $T_t x_n$ tends to $T_t x, t \geq 0$. Finally, if $\lim_{n\to\infty} A x_n = y$, then $\|T_s A x_n - T_s y\| \leq M e^{\omega s} \|A x_n - y\|$ for suitable constants $M$ and $\omega$. Hence,

$$\left\| \int_0^t T_s A x_n \, ds - \int_0^t T_s y \, ds \right\| \leq t \sup_{0 \leq s \leq t} \|T_s x_n - T_s y\|$$

which tends to zero, as $t \to 0$. Therefore $T_t x = x + \int_0^t T_s y \, ds$, so that $x$ belongs to $\mathcal{D}(A)$ and $Ax = y$.

**7.4.29 Exercise** *Semigroup restricted to the domain of its generator* Suppose that $\{T_t, t \geq 0\}$ is a strongly continuous semigroup with generator $A$. Consider the Banach space $(\mathcal{D}(A), \|\cdot\|_A)$ from Example 7.3.3. Since $T_t \mathcal{D}(A) \subset \mathcal{D}(A)$, $\{T_t, t \geq 0\}$ may be considered as a semigroup on $(\mathcal{D}(A), \|\cdot\|_A)$. Prove that it is strongly continuous and find its generator.

**7.4.30** *The Laplace transform of a semigroup* Let $\{T_t, t \geq 0\}$ be a strongly continuous semigroup and let $M$ and $\omega$ be constants such that (7.14) is satisfied. For $x \in \mathbb{X}$, $u > 0$, and $\lambda > \omega$ consider the integral $\int_0^u e^{-\lambda s} T_s x \, ds$ (note that the integrand is continuous). For $v > u$,

$$\left\| \int_0^v e^{-\lambda s} T_s x \, ds - \int_0^u e^{-\lambda s} T_s x \, ds \right\| \leq \int_u^v e^{-\lambda s} T_s x \, ds$$

$$\leq M \|x\| \int_u^v e^{-(\lambda-\omega)s} \, ds,$$

which tends to zero, as $u$ and $v$ tend to infinity. Since $\mathbb{X}$ is a Banach space, we may define (see 2.2.47)

$$R_\lambda x = \lim_{u\to\infty} \int_0^u e^{-\lambda s} T_s x \, ds = \int_0^\infty e^{-\lambda s} T_s x \, ds.$$

Note that $R_\lambda, \lambda > \omega$, are bounded operators and

$$\|R_\lambda\| \le M \int_0^\infty e^{-(\lambda-\omega)s} \, ds = \frac{M}{\lambda - \omega}. \tag{7.25}$$

**7.4.31 Example**   We will show that

$$\int_0^\infty e^{-\lambda t} T(t)\phi \, dt = e_\lambda * \phi, \qquad \lambda > 0, \phi \in L^1(\mathbb{R}^+)$$

where $\{T(t), t \ge 0\}$ is the semigroup of translations to the right from 7.4.2 and $e_\lambda \in L^1(\mathbb{R}^+)$ are defined by their representatives $e_\lambda(\tau) = e^{-\lambda\tau}$ (as in 2.2.49). Since $\{e_\mu, \mu > 0\}$ is linearly dense in $L^1(\mathbb{R}^+)$, it suffices to consider $\phi = e_\mu$. Direct computation of $e_\mu * 1_{[0,t)}$ shows that $T(t)e_\mu = \mu e_\mu * 1_{[0,t)} - 1_{[0,t)} + e_\mu$. Using 2.2.49 and (6.6), for $\lambda > 0$,

$$\int_0^\infty e^{-\lambda t} T(t)e_\mu = \mu \int_0^\infty e^{-\lambda t} 1_{[0,t)} * e_\mu \, dt - \int_0^\infty e^{-\lambda t} 1_{[0,t)} \, dt + \frac{1}{\lambda} e_\mu$$

$$= \frac{\mu}{\lambda} e_\mu * e_\lambda - \frac{1}{\lambda} e_\lambda + \frac{1}{\lambda} e_\mu = e_\lambda * e_\mu.$$

**7.4.32 *The resolvent of A***   Let $\{T_t, t \ge 0\}$ be a strongly continuous semigroup and let $M$ and $\omega$ be constants such that (7.14) is satisfied. Fix $\lambda > \omega$. An element $x \in \mathbb{X}$ belongs to $\mathcal{D}(A)$ iff there exists a $y \in \mathbb{X}$ such that $x = R_\lambda y$. Moreover,

$$R_\lambda(\lambda I_{\mathbb{X}} - A)x = x, \qquad x \in \mathcal{D}(A), \tag{7.26}$$

$$(\lambda I_{\mathbb{X}} - A)R_\lambda y = y, \qquad y \in \mathbb{X}. \tag{7.27}$$

In other words, *the Laplace transform of a semigroup is the resolvent of its infinitesimal generator.*

*Proof*   Instead of $\lambda I_{\mathbb{X}}$ we often write simply $\lambda$. If $x \in \mathcal{D}(A)$, then $e^{-\lambda t} T_t(A - \lambda)x = \frac{d}{dt}[e^{-\lambda t} T_t x]$. Therefore,

$$R_\lambda(\lambda - A)x = -\int_0^\infty e^{-\lambda t} T_t(A - \lambda)x \, dt = -\lim_{u\to\infty} \int_0^u \frac{d}{dt}[e^{-\lambda t} T_t x] \, dt$$

$$= \lim_{u\to\infty} [x - e^{-\lambda u} T_u x] = x,$$

since $\|e^{-\lambda u} T_u x\| \le M e^{-(\lambda-\omega)u} \|x\|$. In particular, if $x$ belongs to $\mathcal{D}(A)$ then $x = R_\lambda y$ for $y = \lambda x - Ax$.

We need to prove that $R_\lambda y$ belongs to $\mathcal{D}(A)$ for all $y \in \mathbb{X}$. As in

7.4.18, let $B$ be the generator of the semigroup $S_t = e^{-\lambda t}T_t$. By 7.4.15, the vector $\int_0^u e^{-\lambda t}T_t y \, dt = \int_0^u S_t y \, dt$ belongs to $\mathcal{D}(B) = D(A)$ and

$$A \int_0^u S_t y \, dt = B \int_0^u S_t y \, dt + \lambda \int_0^u S_t y \, dt = S_u y - y + \lambda \int_0^u S_t y \, dt.$$

Moreover, $\lim_{u \to \infty} \int_0^u S_t y \, dt = R_\lambda y$, and $\lim_{u \to \infty}[S_u y - y + \lambda \int_0^u S_t y \, dt] = \lambda R_\lambda y - y$. Since $A$ is closed, $R_\lambda y$ is a member of $\mathcal{D}(A)$ and $AR_\lambda y = \lambda R_\lambda y - y$, as desired. $\qquad\square$

**7.4.33 Remark**  Relation (7.27) shows that if $A$ is a generator of a strongly continuous semigroup in a Banach space $\mathbb{X}$, then for any $y \in \mathbb{X}$, there exists at least one solution to the equation

$$\lambda x - Ax = y, \qquad \lambda > \omega, \tag{7.28}$$

namely $x = R_\lambda y$, and (7.26) shows that there exists at most one solution to this equation. We have encountered an operator with this property in 7.3.6. In particular, since $R_\lambda$ is bounded (cf. 7.25), we have proven once again that the infinitesimal generator of a strongly continuous semigroup must be closed.

**7.4.34 Remark**  Relations given in 7.4.32 show that the generator determines the semigroup or, in other words, different semigroups have different generators. Indeed, 7.4.32 shows that the generator, or, to be more specific, the resolvent of the generator determines the Laplace transform of the semigroup it generates. Moreover, using Exercise 2.3.31 and the argument from 7.2.1 (i.e. employing functionals to reduce the problem to real-valued functions) we show that the Laplace transform of a continuous (!) Banach space valued function determines this function.

**7.4.35 Example**  Let $v > 0$ be given. Throughout this subsection $\mathbb{Y} = BUC(\mathbb{R})$ and $\mathbb{X} = BUC_1(\mathbb{R})$ is the space of differentiable $x \in \mathbb{Y}$ with $x' \in BUC(\mathbb{R})$. The norm in $\mathbb{X}$ is given by $\|x\|_\mathbb{X} = \|x\|_\mathbb{Y} + \|x'\|_\mathbb{Y}$. The space $\mathbb{X} \times \mathbb{Y}$ is equipped with the norm $\|(x,y)\|_{\mathbb{X} \times \mathbb{Y}} = \|x\|_\mathbb{X} + \|y\|_\mathbb{Y}$ (in what follows the subscripts in the norms will be omitted). Let the family $\{G(t), t \in \mathbb{R}\}$ of operators in $\mathbb{X}$ be given by

$$G(t)\begin{pmatrix} x \\ y \end{pmatrix} = \begin{pmatrix} C(t)x + \int_0^t C(u)y \, du \\ \frac{dC(t)x}{dt} + C(t)y \end{pmatrix} \tag{7.29}$$

where $C(t)x(\tau) = \frac{1}{2}x(\tau + vt) + \frac{1}{2}x(\tau - vt)$, so that $(\int_0^t C(u)y \, du)(\tau) = \frac{1}{2}\int_{-t}^t y(\tau + vu) \, du$ and $\frac{dC(t)x}{dt} = \frac{v}{2}x'(\tau + vt) - \frac{v}{2}x'(\tau - vt)$. The family

$\{C(t), t \in \mathbb{R}\}$ is an example of **cosine operator function** in that it satisfies the **cosine functional equation** (Exercise 7.4.36):

$$2C(t)C(s) = C(t+s) + C(t-s). \tag{7.30}$$

It is clear that $C(t)$ maps $\mathbb{Y}$ into $\mathbb{Y}$ and is a contraction operator. Moreover, $C(t)$ leaves $\mathbb{X}$ invariant and $\int_0^t C(u)y\,\mathrm{d}u$ maps $\mathbb{Y}$ into $\mathbb{X}$ with

$$(C(t)x)'(\tau) = C(t)x'(\tau), \quad \left(\int_0^t C(s)y\,\mathrm{d}s\right)'(\tau) = \frac{1}{2v}(y(\tau+vt) - y(\tau-vt)). \tag{7.31}$$

Hence, $C(t) : \mathbb{X} \to \mathbb{X}$ and $\int_0^t C(u)\,\mathrm{d}u : \mathbb{Y} \to \mathbb{X}$ are bounded. Similarly, $\frac{\mathrm{d}C(t)}{\mathrm{d}t}$ is a bounded linear operator from $\mathbb{X}$ to $\mathbb{Y}$.

A direct computation based on (7.30) shows that $\{G(t), t \in \mathbb{R}\}$ is a group of operators, i.e. $G(t)G(s) = G(t+s), t, s \in \mathbb{R}$.

Clearly, $\lim_{t\to 0} C(t)y = y$ strongly in $\mathbb{Y}$, and by (7.31), $\lim_{t\to 0} C(t)x = x$ strongly in $\mathbb{Y}_1$. As a result, $\lim_{t\to 0} \int_0^t C(s)y\,\mathrm{d}s = 0$ strongly in $\mathbb{Y}$. Using the other relation in (7.31), $\lim_{t\to 0} \int_0^t C(s)y\,\mathrm{d}s = 0$ strongly in $\mathbb{X}$, as well. Finally, $\lim_{t\to 0} \frac{\mathrm{d}C(t)x}{\mathrm{d}t} = 0$ strongly in $\mathbb{Y}$ for $x \in \mathbb{X}$. This shows that $\{G(t), t \in \mathbb{R}\}$ is strongly continuous.

To find the generator $A$ of *the semigroup* $\{G(t), t \geq 0\}$, we reason as follows. Let $\mathcal{D}(\frac{\mathrm{d}^2}{\mathrm{d}\tau^2})$ be the set of twice differentiable $y$ in $\mathbb{Y}$ with $y'' \in \mathbb{Y}$. A direct calculation shows that for $\lambda > 0, x \in \mathbb{Y}$ there exists the Laplace transform

$$L_\lambda x(\tau) := \int_0^\infty e^{-\lambda t} C(t)x(\tau)\,\mathrm{d}t = \frac{1}{2v}e^{\frac{\lambda}{v}\tau}\int_\tau^\infty e^{-\frac{\lambda}{v}\sigma}x(\sigma)\,\mathrm{d}\sigma$$
$$+ \frac{1}{2v}e^{-\frac{\lambda}{v}\tau}\int_{-\infty}^\tau e^{\frac{\lambda}{v}\sigma}x(\sigma)\,\mathrm{d}\sigma.$$

This implies that $L_\lambda x$ is twice differentiable with $(L_\lambda x)'' = \frac{\lambda^2}{v^2}L_\lambda x - \frac{\lambda}{v^2}x \in \mathbb{Y}$ (if $x \in \mathbb{X}$, then $(L_\lambda x)'' \in \mathbb{X}$). Since for $y \in \mathbb{Y}$, $L_\lambda^\sharp y := \int_0^\infty e^{-\lambda t}\int_0^t C(s)x(\tau)\,\mathrm{d}s\,\mathrm{d}t = \lambda^{-1}L_\lambda y$, $L_\lambda^\sharp y \in \mathcal{D}(\frac{\mathrm{d}^2}{\mathrm{d}\tau^2})$. Hence, the Laplace transform of the first coordinate in (7.29) belongs to $\mathcal{D}(\frac{\mathrm{d}^2}{\mathrm{d}\tau^2})$, too. Similarly, we check that the Laplace transform of the other coordinate in (7.29) belongs to $\mathbb{X}$. This shows that $\mathcal{D}(A) \subset D(\frac{\mathrm{d}^2}{\mathrm{d}\tau^2}) \times \mathbb{X}$. On the other hand, for $(x, y) \in D(\frac{\mathrm{d}^2}{\mathrm{d}\tau^2}) \times \mathbb{X}$,

$$\lim_{t\to 0+}\frac{1}{t}\left\{G(t)\begin{pmatrix} x \\ y \end{pmatrix} - \begin{pmatrix} x \\ y \end{pmatrix}\right\} = \begin{pmatrix} y \\ v^2 x'' \end{pmatrix} \quad \text{strongly in } \mathbb{X}. \tag{7.32}$$

Hence $\mathcal{D}(A) = \mathcal{D}(\frac{\mathrm{d}^2}{\mathrm{d}\tau^2}) \times \mathbb{X}$ and $A(x, y) = (y, v^2 x'')$.

**7.4.36 Exercise** Show that the family $\{C(t), t \in \mathbb{R}\}$ defined in 7.4.35 is a cosine operator function.

**7.4.37 Exercise** Prove (7.32).

**7.4.38 Exercise** Let $\mathbb{X} = c_0$ and $T_t (\xi_n)_{n \geq 1} = (e^{-nt}\xi_n)_{n \geq 1}$. Show that $\{T_t, t \geq 0\}$ is a strongly continuous semigroup and that the operator

$$A (\xi_n)_{n \geq 1} = - (n\xi_n)_{n \geq 1}, \quad \mathcal{D}(A) = \{(\xi_n)_{n \geq 1} \in c_0; (n\xi_n)_{n \geq 1} \in c_0\}$$

is its infinitesimal generator.

**7.4.39 Exercise** Use (7.26) and (7.27) to show that the resolvent $R_\lambda$ satisfies the Hilbert equation (cf. (6.6))

$$(\lambda - \mu)R_\lambda R_\mu = R_\mu - R_\lambda, \quad \lambda, \mu > \omega. \tag{7.33}$$

Then, show that, by (7.25), $\mathbb{R}^+ \ni \lambda \to R_\lambda$ is continuous (see (8.38), if needed). Argue by induction that it is also infinitely differentiable with $\frac{\mathrm{d}^n}{\mathrm{d}\lambda^n} R_\lambda = (-1)^n n! R_\lambda^{n+1}$.

**7.4.40 Semigroups and the Cauchy problem** Let $A$ be the infinitesimal generator of a strongly continuous semigroup $\{T_t, t \geq 0\}$. The Cauchy problem

$$\frac{\mathrm{d}x_t}{\mathrm{d}t} = Ax_t, t \geq 0, \quad x_0 = x \in D(A) \tag{7.34}$$

where $x_t$ is a sought-for differentiable function with values in $D(A)$, has the unique solution $x_t = T_t x$.

*Proof* By 7.4.25 we merely need to prove uniqueness of solutions. To this end suppose that $x_t$ is a solution to (7.34), fix $t > 0$ and consider $y_s = T_{t-s}x_s, 0 < s < t$. Since $\frac{y_{s+h}-y_s}{h} = T_{t-s-h}\frac{x_{s+h}-x_s}{h} + \frac{T_{t-s-h}-T_{t-s}}{h}x_s$ for suitably small $|h|$, and the operators $T_t$ are bounded in any compact subinterval of $\mathbb{R}^+$, $y_s$ is differentiable and

$$\frac{\mathrm{d}}{\mathrm{d}s}y_s = T_{t-s}\frac{\mathrm{d}}{\mathrm{d}s}x_s - T_{t-s}Ax_s = T_{t-s}(Ax_s - Ax_s) = 0.$$

(See Example 7.1.9.) Hence, $T_{t-s}x_s$ is constant, and since $\lim_{s \to t} y_s = T_t x$, $T_{t-s}x_s = T_t x$, for all $0 < s < t$. Letting $s \to t$, we obtain $x_t = T_t x$. $\qquad \square$

**7.4.41 Corollary** If $\{T_t, t \geq 0\}$ and $\{S_t, t \geq 0\}$ are two strongly continuous semigroups with generators $A$ and $B$ respectively, and if $A = B$, then $\{T_t, t \geq 0\} = \{S_t, t \geq 0\}$.

*Proof* By 7.4.40, $T_t x = S_t x$ for all $t \geq 0$ and $x \in \mathcal{D}(A) = \mathcal{D}(B)$. Since $\mathcal{D}(A)$ is a dense set, we are done. $\qquad\square$

**7.4.42 Example** By 7.4.35, the equation

$$\begin{cases} \frac{dx(t)}{dt} = y(t), & x(0) = x \in \mathcal{D}(\frac{d^2}{d\tau^2}), \\ \frac{dy(t)}{dt} = v^2 \frac{d^2}{d\tau^2} x(t), & y(0) = y \in BUC_1(\mathbb{R}), \end{cases} \tag{7.35}$$

in $BUC_1(\mathbb{R}) \times BUC(\mathbb{R})$ has the unique solution $\begin{pmatrix} x(t) \\ y(t) \end{pmatrix} = G(t) \begin{pmatrix} x \\ y \end{pmatrix}, t \geq 0$ with $G(t)$ given by (7.29). A function $x(t, \tau)$ satisfying (7.35) satisfies also the **wave equation:**

$$\frac{\partial^2 x(t, \tau)}{\partial t^2} = v^2 \frac{\partial^2 x(t, \tau)}{\partial \tau^2}, \qquad x(0, \tau) = x(\tau), \frac{\partial}{\partial t} x(0, \tau) = y(\tau); \tag{7.36}$$

the first coordinate in (7.29) is the solution to (7.36) while the second is its derivative with respect to $t$.

**7.4.43 Example** The infinite system of equations

$$x_1'(t) = \lambda x_1(t) - (b + d)x_1(t) + dx_2(t),$$
$$x_i'(t) = \lambda x_i(t) - (b + d)x_i(t) + dx_{i+1}(t) + bx_{i-1}(t), \qquad i \geq 2$$

where $d > b > 0$ and $\lambda \in \mathbb{R}$ are parameters, was introduced and studied in [108] as a model of behavior of a population of cells that are resistant to a cancer drug. In this model, there are infinitely many types of resistant cells, and $x_i(t)$ is the number of resistant cells of type $i$ at time $t$. It was natural to ask for criteria for the decay, as $t \to \infty$, of the population $\sum_{i=1}^{\infty} x_i(t)$ of resistant cells. Also the weighted sums $\sum_{i=1}^{\infty} x_i(t)r^i$ where $r > 0$ were of interest and it led the authors to considering the problem in the spaces $l_r^1$ introduced in 5.2.5.

Our system may be written as a differential equation in $l_r^1$ in the form

$$\frac{dx(t)}{dt} = (\lambda - b - d)x(t) + [dL + bR]x(t)$$

where $L$ and $R$ stand for left and right translation in $l_r^1$ (see 2.3.9). In particular, the solution to this system is given by the exponential function of the operator appearing on the right-hand side above. By 7.4.24, this exponential function is isometrically isomorphic to the exponential function of the operator $(\lambda - b - d)I + \frac{d}{r}L + brR$ in $l^1$. By 2.3.16, the norm of the exponential function of $(\lambda - b - d)I + \frac{d}{r}L + brR$ equals $\exp\{\frac{t}{r}(br^2 + (\lambda - b - d)r + d)\}$. This converges to zero as $t \to 0$, iff $br^2 + (\lambda - b - d)r + d$ is negative. Treating this binomial as a function of $r$ with a parameter $\lambda$ and fixed $d$ and $b$, we consider $\Delta(\lambda) = [\lambda - (b + d)]^2 - 4bd = (\lambda - \lambda_1)(\lambda - \lambda_2)$, where $\lambda_1 = (\sqrt{d} - \sqrt{b})^2$, $\lambda_2 = (\sqrt{d} + \sqrt{b})^2$. If the binomial is to be negative, we must assume that either $\lambda > \lambda_2$ or $\lambda < \lambda_1$, and that $r$ belongs to the open interval $(r_1, r_2)$ where $r_i = \frac{b+d-\lambda+(-1)^i\sqrt{\Delta(\lambda)}}{2b}, i = 1, 2$, are the corresponding roots of the binomial. Note that in the case $\lambda > \lambda_2$ these roots are negative, and thus we must choose $\lambda < \lambda_1$, as it was proven in [108] in a different way.

**7.4.44** *Cores of generators* Suppose that $\{T_t, t \geq 0\}$ is a strongly continuous semigroup with generator $A$ and that an algebraic subspace $D \subset \mathcal{D}(A)$ is dense in $\mathcal{D}(A)$. If $T_t D \subset D, t \geq 0$ then $D$ is a core of $A$.

*Proof* Let $\omega$ and $M$ be constants such that (7.14) holds. By (7.26) and (7.27), the relation $y = (\lambda - A)x$ is equivalent to $x = R_\lambda y$, for $\lambda > \omega$. Thus, using (7.25), $\|(\lambda - A)x\| \geq \frac{\lambda - \omega}{M}\|x\|$, for all $x \in \mathcal{D}(A)$. By 7.3.12, it suffices therefore to show that $(\lambda - A)D$ is dense in $\mathbb{X}$. Since $D$ is dense in $\mathbb{X}$, it is enough to show that the closure of $(\lambda - A)D$ contains $D$.

The reason why this relation is true is that for any $x$, $R_\lambda x$ belongs to the closure of the linear span of elements of the form $T_t x, t \geq 0$. Here is the complete argument: note first that for any $x \in \mathbb{X}$, the sequence

$$x_n = \frac{1}{n}\sum_{k=1}^{n^2} e^{-\lambda\frac{k}{n}} T_{\frac{k}{n}} x \tag{7.37}$$

converges to $R_\lambda x$. Indeed, for $t \geq s$,

$$\|e^{-\lambda t}T_t x - e^{-\lambda s}T_s x\| \leq \|T_s e^{-\lambda s}\| \, \|e^{-\lambda(t-s)}T_{t-s}x - x\|,$$

and our claim follows since the norm of $x_n - R_\lambda x$ may be estimated by

$$\|x_n - \int_0^n e^{-\lambda t}T_t x \, \mathrm{d}t\| + \|\int_n^\infty e^{-\lambda t}T_t x \, \mathrm{d}t\|$$

$$\leq \left\|\sum_{k=1}^{n^2} \int_{\frac{k-1}{n}}^{\frac{k}{n}} \left[e^{-\lambda\frac{k}{n}}T_{\frac{k}{n}}x - e^{-\lambda t}T_t x\right] \mathrm{d}t\right\| + \frac{M}{\lambda - \omega}e^{-(\lambda-\omega)n}$$

$$\leq \sup_{0\leq h\leq \frac{1}{n}} \|e^{-\lambda h}T_h x - x\| \sum_{k=1}^{n^2} \int_{\frac{k-1}{n}}^{\frac{k}{n}} \|e^{-\lambda t}T_t\| \, dt + \frac{M}{\lambda - \omega}e^{-(\lambda-\omega)n}$$

$$\leq \sup_{0\leq h\leq \frac{1}{n}} \|e^{-\lambda h}T_h x - x\| \int_0^\infty \|e^{-\lambda t}T_t\| \, dt + \frac{M}{\lambda - \omega}e^{-(\lambda-\omega)n}$$

$$\leq \left\{ \sup_{0\leq h\leq \frac{1}{n}} \|e^{-\lambda h}T_h x - x\| + e^{-(\lambda-\omega)n} \right\} \frac{M}{\lambda - \omega}.$$

Now, take $x \in D$ and consider $x_n$ defined in (7.37). By assumption, $x_n$ belongs to $D$, and the sequence $\lambda x_n - A x_n = \frac{1}{n}\sum_{k=1}^{n^2} e^{-\lambda \frac{k}{n}}T_{\frac{k}{n}}(\lambda x - Ax)$ converges to $R_\lambda(\lambda x - Ax) = x$. It implies that $x$ belongs to the closure of $(\lambda - A)D$ and completes the proof.    □

**7.4.45 Exercise**   Show that the set of all functions that are infinitely many times differentiable with derivatives in $BUC(\mathbb{R}^+)$ is a core of the generator of the semigroup from 7.4.7.

**7.4.46 *The representation of $L^1(\mathbb{R}^+)$ related to a bounded semigroup***
Let $\{T(t), t \geq 0\}$ be a strongly continuous semigroup of equibounded operators, i.e. let (7.14) be satisfied with $\omega = 0$. Moreover, let $\phi$ be a continuous function that is a member of $L^1(\mathbb{R}^+)$. For any $x \in \mathbb{X}$, the map $\mathbb{R}^+ \ni t \to \phi(t)T(t)x$ is continuous and one may consider its Riemann integral on an interval, say $[0, u]$. Arguing as in 7.4.30, one proves that the improper integral

$$H(\phi)x = \int_0^\infty \phi(t)T(t)x \, dt$$

exists also, and we obtain

$$\|H(\phi)x\| \leq M\|\phi\|_{L^1(\mathbb{R}^+)}\|x\|_{\mathbb{X}}.$$

This implies that, for $\phi$ fixed, $H(\phi)$ is a bounded linear operator in $\mathbb{X}$. On the other hand, by fixing $x$ and varying $\phi$ we see that such a bounded linear operator may be defined for any $\phi \in L^1(\mathbb{R}^+)$, because any such $\phi$ may be approximated by a sequences of continuous elements of $L^1(\mathbb{R}^+)$, say $\phi_n$, and then $H(\phi)$ may be defined as the limit of $H(\phi_n)$. In other words we use 2.3.33 to extend a bounded linear map $L^1(\mathbb{R}^+) \ni \phi \to H(\phi) \in \mathcal{L}(\mathbb{X})$ from a dense subset of $L^1(\mathbb{R}^+)$, where we have $\|H(\phi)\|_{\mathcal{L}(\mathbb{X})} \leq M\|\phi\|_{L^1(\mathbb{R}^+)}$ to the whole of $L^1(\mathbb{R}^+)$. †

In particular if $\phi(\tau) = e_\lambda(\tau) = e^{-\lambda\tau}$, where $\lambda > 0$, then $H(\phi) = R_\lambda$ is

† The operator $H(\phi)$ is in fact the strong Bochner integral of $\phi(t)T(t)$.

the resolvent of the semigroup $\{T(t), t \geq 0\}$. Moreover, approximating the indicator function of an interval $[0, t]$ by a sequence of continuous functions, we check that $H(1_{[0,t]})x = \int_0^t T(s)x \, ds$.

Using (6.6) and Exercise 7.4.39, for $\lambda, \mu > 0, \lambda \neq \mu$ we have

$$(\lambda - \mu)H(e_\lambda * e_\mu) = H(e_\mu - e_\lambda) = H(e_\lambda) - H(e_\mu) = R_\lambda - R_\mu$$
$$= (\lambda - \mu)R_\lambda R_\mu = (\lambda - \mu)H(e_\lambda)H(e_\mu)$$

Since $\lambda \to e_\lambda \in L^1(\mathbb{R}^+)$ is continuous, this implies that

$$H(e_\lambda * e_\mu) = H(e_\lambda)H(e_\mu)$$

for all positive $\lambda$ and $\mu$, and since the set $\{e_\lambda, \lambda > 0\}$ is linearly dense in $L^1(\mathbb{R}^+)$, the map $\phi \to H(\phi)$ is proven to be a homomorphism of the Banach algebra $L^1(\mathbb{R}^+)$.

**7.4.47 Exercise** Let $\{T_t, t \geq 0\}$ be a strongly continuous semigroup in a Banach space $\mathbb{X}$ and let $A$ be its generator. Show that $\bigcap_{n=1}^\infty \mathcal{D}(A^n)$ is dense in $\mathbb{X}$.

**7.4.48 Exercise** This exercise prepares the reader for the Trotter–Kato Theorem and for the proof of the Hille–Yosida Theorem (both to be presented in Chapter 8). Let $\{T_n(t), t \geq 0\}$ be a sequence of strongly continuous equibounded semigroups, i.e. semigroups such that (7.14) is satisfied with $\omega = 0$ and $M > 0$ that does not depend on $n$. Let $A_n$ be the generators of these semigroups and let $H_n$ be the corresponding homomorphisms of $L^1(\mathbb{R}^+)$. Prove that the following conditions are equivalent.

(a) For all $\lambda > 0$, $(\lambda - A_n)^{-1}$ converges strongly (as $n \to \infty$).
(b) For all $t > 0$, $\int_0^t T_n(s) \, ds$ converges strongly.
(c) For all $\phi \in L^1(\mathbb{R}^+)$, $H_n(\phi)$ converges strongly.

If one of these conditions holds, then $\phi \mapsto H(\phi) = \lim_{n \to \infty} H_n(\phi)$ is a homomorphism of the algebra $L^1(\mathbb{R}^+)$.

**7.4.49 Exercise** Show the implication (a)$\Rightarrow$(b) in the previous exercise, arguing as follows. (a) Let $b(\mathbb{X})$ be the space of bounded sequences $(x_n)_{n \geq 1}$, $x_n \in \mathbb{X}$ with the norm $\| (x_n)_{n \geq 1} \|_{b(\mathbb{X})} = \sup_{n \geq 1} \|x_n\|$, and let $\mathcal{U}(t) (x_n)_{n \geq 1} = \left( \int_0^t T_n(s)x_n \, ds \right)_{n \geq 1} \in b(\mathbb{X})$. Show that $\mathbb{R}^+ \ni t \mapsto \mathcal{U}(t) (x_n)_{n \geq 1}$ is continuous in $b(\mathbb{X})$ and $\|\mathcal{U}(t)\|_{\mathcal{L}(b(\mathbb{X}))} \leq Mt$. Conclude that the improper integral $\int_0^\infty e^{-\lambda t}\mathcal{U}(t)\,(x_n)_{n\geq 1}\,dt$ exists and check that it equals

$$\frac{1}{\lambda}\left(\int_0^\infty e^{-\lambda t}\int_0^t T_n(s)x_n\,ds\right)_{n\geq 1} = \frac{1}{\lambda}\left((\lambda - A_n)^{-1}x_n\right)_{n\geq 1}.$$

(b) Let $c(\mathbb{X})$ be the subspace of $b(\mathbb{X})$ composed of convergent sequences. Show that the operators $\mathcal{U}(t), t \geq 0$ leave $c(\mathbb{X})$ invariant. To this end, note that otherwise, by 5.1.13, there would exist a functional $F \in b(\mathbb{X})^*$ such that $F\,(x_n)_{n\geq 1} = 0$ for $(x_n)_{n\geq 1} \in c(\mathbb{X})$ and $F(\mathcal{U}(t)\,(x_n)_{n\geq 1}) \neq 0$ for some $t > 0$ and $(x_n)_{n\geq 1} \in c(\mathbb{X})$. This would contradict the form of the Laplace transform of $\overline{U}(t)$ found in (a).

7.4.50 *A preparation for semigroup-theoretical proof of the CLT* Quite often we are interested in a situation where $\{T_n(t), t \geq 0\}, n \geq 1$, considered in the two previous subsections are not semigroups, and yet approximate a semigroup in a certain sense. We may assume in particular that $\{T_n(t), t \geq 0\}, n \geq 1$, are families of equibounded operators in a Banach space $\mathbb{X}$, i.e. that $\|T_n(t)\| \leq M$ for all $t \geq 0$ and some constant $M > 0$. Under some measurability conditions (notably, if functions $t \to T_n(t)x$ are Bochner measurable), it makes sense to define $H_n(\phi)x = \int_0^\infty \phi(t)T_n(t)x\,dt$, for $\phi \in L^1(\mathbb{R}^+)$. The case we need for the proof of the CLT is that of piecewise continuous functions $t \mapsto T_n(t)x$ with countable number of points of discontinuity. In such a case, we may first define operators $H_n(\phi)$ for continuous $\phi$ with compact support as a Riemann integral, and then extend the definition to the whole of $L^1(\mathbb{R}^+)$. Of course, now the $H_n$ are not homomorphisms of $L^1(\mathbb{R}^+)$ but merely operators from $L^1(\mathbb{R}^+)$ to $\mathcal{L}(\mathbb{X})$. Nevertheless, the following are equivalent.

(a)  For all $\lambda > 0$, $H_n(e_\lambda)$ converges strongly (as $n \to \infty$).

(b)  For all $t > 0$, $\int_0^t T_n(s)x\,ds$ converges strongly.

(c)  For all $\phi \in L^1(\mathbb{R}^+)$, $H_n(\phi)$ converges strongly.

As an example let us consider a sequence $T_n, n \geq 1$, of contractions, and a sequence of positive numbers $(h_n)_{n\geq 1}$ such that $\lim_{n\to\infty} h_n = 0$. We define $T_n(t) = T_n^{[t/h_n]}$ and assume that $\lim_{n\to\infty}(\lambda - A_n)^{-1} =: R_\lambda$ exists for all $\lambda > 0$, where $A_n = \frac{1}{h_n}(T_n - I)$. This will imply that all three conditions (a)–(c) above hold. To this end we note that $\|T_n^{[t/h_n]}\| \leq 1$, and for $\lambda > 0$ write

$$
\int_0^\infty e^{-\lambda t} T_n^{[t/h_n]} x \, dt = \sum_{k=0}^\infty \int_{kh_n}^{(k+1)h_n} e^{-\lambda t} \, dt \, T_n^k x
$$

$$
= \frac{h_n \lambda_n}{\lambda} e^{-\lambda h_n} \sum_{k=0}^\infty (e^{-\lambda h_n} T_n)^k x = \frac{h_n \lambda_n}{\lambda} e^{-\lambda h_n} \left[ I - e^{-\lambda h_n} T_n \right]^{-1} x
$$

$$
= \frac{h_n \lambda_n}{\lambda} \left( e^{\lambda h_n} - T_n \right)^{-1} x = \frac{\lambda_n}{\lambda} \left[ \frac{e^{\lambda h_n} - 1}{h_n} - \frac{T_n - I}{h_n} \right]^{-1} x
$$

$$
= \frac{\lambda_n}{\lambda} (\lambda_n - A_n)^{-1} x \tag{7.38}
$$

where $\lambda_n = \frac{e^{\lambda h_n} - 1}{h_n}$. Hence, $\|\mu(\mu - A_n)^{-1}\| \leq \nu_n(\mu) \int_0^\infty e^{-\nu_n(\mu)t} \, dt = 1$ where $\nu_n(\mu) = \frac{1}{h_n} \ln(h_n \mu + 1)$. Consequently, $\|\mu R_\mu\| \leq 1$. Moreover, $\lambda \mapsto (\lambda - A_n)^{-1}$ satisfies the Hilbert equation. Thus,

$$
\|(\lambda - A_n)^{-1} - (\mu - A_n)^{-1}\| \leq \|(\mu - \lambda)(\lambda - A_n)^{-1}(\mu - A_n)^{-1}\| \leq \frac{|\lambda - \mu|}{\lambda \mu},
$$

for all $\lambda, \mu > 0$. By $\lim_{n \to \infty} \lambda_n = \lambda$, this enables us to prove that the limit of (7.38) exists and equals $R_\lambda x$, as desired. Note that $\lambda \mapsto R_\lambda$ satisfies the Hilbert equation, as a limit of $\lambda \mapsto (\lambda - A_n)^{-1}$. Hence $\phi \mapsto H(\phi) = \lim_{n \to \infty} H_n(\phi)$ *is* a representation of $L^1(\mathbb{R}^+)$. We will continue these considerations in 8.4.18.

## 7.5 Brownian motion and Poisson process semigroups

7.5.1 *A semigroup of operators related to Brownian motion*    Let $w(t)$, $t \geq 0$, be a Brownian motion, and let $\mu_t$ be the distribution of $w(t)$. Define the family $\{T_t, t \geq 0\}$ in $BM(\mathbb{R})$ by the formula $T_t = T_{\mu_t}$. In other words,

$$
(T_t x)(\tau) = E \, x(\tau + w(t)). \tag{7.39}
$$

I claim that $\{T_t, t \geq 0\}$ is a semigroup of operators. Before proving this note that $\|T_t\| \leq 1$, so that $\{T_t, t \geq 0\}$ is the family of contractions, i.e. (7.14) holds with $M = 1$ and $\omega = 0$. This will be the case for all semigroups discussed in Sections 7.5 and 7.6.

Properties of the operators $T_t, t \geq 0$, reflect properties of the measures $\mu_t, t \geq 0$, and thus of the process $w(t), t \geq 0$. The property $2°$ of the definition of a semigroup of operators is an immediate consequence of the fact that $\mu_0$ is a Dirac measure (point mass) at $0$, $\int x \, d\mu_0 = x(0)$, which in turn follows from $w(0) = 0$ a.s. Property $1°$ will be proven once we show that

$$\mu_t * \mu_s = \mu_{s+t}. \tag{7.40}$$

Brownian motion has independent increments, so that $w(t + s)$ is a sum of two independent random variables $w(t+s) = [w(t+s) - w(s)] + w(s)$. Consequently,

$$\mu_{t+s} = \mathbb{P}_{w(t+s)} = \mathbb{P}_{w(t+s)-w(s)} * \mathbb{P}_{w(s)} = \mu_t * \mu_s;$$

the last equality following from the fact that $w(t + s) - w(s)$ has the same distribution as $w(t)$.

Moreover, we claim that $\{T_t : t \geq 0\}$ when restricted to $C[-\infty, \infty]$† is a strongly continuous semigroup. Indeed, for any $\epsilon > 0$, by Chebyshev's inequality,

$$\mu_t(-\epsilon, \epsilon)^{\complement} = \mathbb{P}\{|w(t)| \geq \epsilon\} \leq \frac{E\, w^2(t)}{\epsilon^2} = \frac{t}{\epsilon^2};$$

hence $\lim_{t\to 0} \mu_t(-\epsilon, \epsilon)^{\complement} = 0$. Therefore, by 5.4.12, $\mu_t$ converge weakly to $\delta_0$, the Dirac measure at 0, proving our claim by 5.4.18.

Finally, we claim that the domain $\mathcal{D}(A)$ of the infinitesimal generator of $\{T_t, t \geq 0\}$ equals $\mathcal{D}(\frac{d^2}{d\tau^2})$, the set of twice differentiable functions with $x'' \in C[-\infty, \infty]$, and $Ax = \frac{1}{2}x''$. To prove that $\lim_{t\to 0+} \frac{1}{t}\{T_t x - x\} = \frac{1}{2}x''$, for $x \in \mathcal{D}(\frac{d^2}{d\tau^2})$, we take a sequence $t_n$ of positive numbers with $\lim_{n\to\infty} t_n = 0$, and assume without loss of generality that $t_n \leq 1$. Then, $T(t_n) = T_{w(t_n)} = T_{\sqrt{t_n}w(1)}$ so that $\lim_{n\to\infty} \frac{1}{t_n}(T(t_n)x - x) = \lim_{n\to\infty} \frac{1}{t_n}(T_{\sqrt{t_n}w(1)} - x) = \frac{1}{2}x''$ by Lemma 5.5.1 with $a_n = \sqrt{t_n}$ and $X = w(1)$. (This lemma was used to prove the Central Limit Theorem; the Central Limit Theorem and the fact that $\frac{1}{2}x''$ is the generator of the Brownian motion semigroup are very much related.)

We still need to prove that $\mathcal{D}(A) \subset \mathcal{D}(\frac{d^2}{d\tau^2})$. There are several ways to do that; the first, most direct one is to note that, by 7.4.32, $\mathcal{D}(A)$ equals the range of $R_\lambda$. Moreover,

$$\begin{aligned} R_\lambda x(\tau) &= \int_0^\infty e^{-\lambda t} \frac{1}{\sqrt{2\pi t}} \int_{-\infty}^\infty e^{-\frac{(\tau-\sigma)^2}{2t}} x(\sigma)\, d\sigma\, dt \\ &= \int_{-\infty}^\infty k(\lambda, \tau - \sigma)x(\sigma)\, d\sigma \end{aligned} \tag{7.41}$$

† The choice of space is more or less arbitrary here; we could have taken $C_0(\mathbb{R})$ or $L^1(\mathbb{R})$ instead. See Subsection 8.1.15.

where $k(\lambda, \tau - \sigma) = \int_0^\infty e^{-\lambda t} \frac{1}{\sqrt{2\pi t}} e^{-\frac{(\tau-\sigma)^2}{2t}} dt$. This last integral was calculated in 6.4.7 to be $\frac{1}{\sqrt{2\lambda}} e^{-\sqrt{2\lambda}|\tau-\sigma|}$. Thus, $R_\lambda x(\tau)$ equals

$$\frac{1}{\sqrt{2\lambda}} e^{-\tau\sqrt{2\lambda}} \int_{-\infty}^\tau e^{\sigma\sqrt{2\lambda}} x(\sigma) \, d\sigma + \frac{1}{\sqrt{2\lambda}} e^{\tau\sqrt{2\lambda}} \int_\tau^\infty e^{-\sigma\sqrt{2\lambda}} x(\sigma) \, d\sigma$$

which implies that $R_\lambda x$ is twice differentiable. Performing the necessary differentiation we see that $(R_\lambda x)'' = 2\lambda R_\lambda x - 2x$, so that $(R_\lambda x)''$ belongs to $C[-\infty, \infty]$ and we are done (cf. (7.12)!).

The second method relies on 7.3.10 (or, rather on 7.3.11) and 7.4.32. Suppose that an $x$ belongs to $\mathcal{D}(A) \setminus \mathcal{D}(\frac{d^2}{d\tau^2})$, and consider $\lambda x - Ax \in C[-\infty, \infty]$. By 7.3.11, there exists a $y \in \mathcal{D}(\frac{d^2}{d\tau^2})$ such that $\lambda y - \frac{1}{2}y'' = \lambda x - Ax$. On the other hand, $\lambda y - \frac{1}{2}y'' = \lambda y - Ay$, and this contradicts the fact that $\lambda - A$ is one-to-one.

The third and final method uses (7.4.44) and (7.3.6). If $x$ is differentiable with $x' \in C[-\infty, \infty]$, then, by (1.2.12) $T_t x(\tau) = E\, x(\tau + w(t))$ is differentiable and its derivative $E\, x'(\tau + w(t))$ belongs to $C[-\infty, \infty]$. Analogously, if $x$ belongs to $\mathcal{D}(\frac{d^2}{d\tau^2})$, then so does $T_t x$. Therefore, $\mathcal{D}(\frac{d^2}{d\tau^2})$, is a core for $A$. Hence, for any $x \in \mathcal{D}(A)$ there exists a sequence $x_n \in \mathcal{D}(\frac{d^2}{d\tau^2})$, such that $x_n$ converges to $x$, and $Ax_n$ converges to $Ax$. But $Ax_n = \frac{1}{2}x_n''$, and the operator $\frac{1}{2}\frac{d^2}{d\tau^2}$, defined on $\mathcal{D}(\frac{d^2}{d\tau^2})$, is closed, and so $x$ belongs to $\mathcal{D}(\frac{d^2}{d\tau^2})$ and $Ax = \frac{1}{2}x''$.

**7.5.2 Remark**    As a by-product we obtain that $u(t, \tau) = E\, x(\tau + w(t))$ is the only bounded (in $t \geq 0$) solution to the Cauchy problem for the heat equation:

$$\frac{\partial}{\partial t} u(t, \tau) = \frac{1}{2}\frac{\partial^2}{\partial \tau^2} u(t, \tau), \qquad u(0, \tau) = x(\tau), \qquad x \in \mathcal{D}\left(\frac{d^2}{d\tau^2}\right).$$

This important relation is indeed only the peak of an iceberg. There is a large class of important PDE and integro-differential equations of second order for which one may construct probabilistic solutions. Moreover, stochastic processes are closely related to integro-differential operators in the way the operator $\frac{d^2}{d\tau^2}$ is related to Brownian motion. In fact as all properties of the Brownian motion are hidden in the operator $\frac{d^2}{d\tau^2}$, so the properties of some other processes are hidden in their "generating" operators.

**7.5.3 Exercise**    Show that $\lambda R_\lambda$, where $R_\lambda$ is the resolvent of the Brownian motion semigroup, is the operator related to the bilateral exponential distribution with parameter $a = \sqrt{2\lambda}$.

**7.5.4 Exercise** Consider the process $w(t) + at$, where $a$ is a constant and $w(t)$ is the Brownian motion. Introduce the corresponding family of operators, show that this is a strongly continuous semigroup on $C[-\infty, \infty]$, and find its generator.

**7.5.5** *The semigroup of operators related to the Poisson process* As in 2.3.45, let $X_n$ be independent random variables with the same exponential distribution of parameter $a > 0$, defined on a probability space $(\Omega, \mathcal{F}, \mathbb{P})$. Also, let $S_0 = 0$, and $S_n = \sum_{i=1}^n X_i$, for $n \geq 1$. For any $\omega \in \Omega$, let $N(t) = N_t(\omega)$ be the largest integer $k$ such that $S_k(\omega) \leq t$. Certainly $N_0 = 0$ (almost surely). Moreover, for any $t \geq 0$, $N(t)$ is a random variable, for if $\tau$ is a real number then

$$\{\omega \in \Omega; N_t(\omega) \leq \tau\} = \{\omega \in \Omega; N_t(\omega) \leq [\tau]\} = \{\omega \in \Omega; S_{[\tau]+1}(\omega) > t\}$$

and this last set is measurable. It is also quite easy to see what is the distribution $\mu_t$ of $N(t)$. Certainly, this is a measure that is concentrated on the set of integers. Moreover,

$$
\begin{aligned}
\mu_t(\{k\}) &= \mathbb{P}\{N(t) = k\} = \mathbb{P}\{S_k \leq t\} - \mathbb{P}\{S_{k+1} \leq t\} \\
&= \int_0^t a^k \frac{\tau^{k-1}}{(k-1)!} \mathrm{e}^{-a\tau}\, \mathrm{d}\tau - \int_0^t a^{k+1} \frac{\tau^k}{k!} \mathrm{e}^{-a\tau}\, \mathrm{d}\tau \\
&= \int_0^t \frac{\mathrm{d}}{\mathrm{d}\tau}\left[ a^k \frac{\tau^k}{k!} \mathrm{e}^{-a\tau} \right]\, \mathrm{d}\tau = \frac{a^k t^k}{k!} \mathrm{e}^{-at}.
\end{aligned}
$$

The family $N(t), t \geq 0$ is called the Poisson process (on $\mathbb{R}^+$). Define the related operators in $BM(\mathbb{R})$† by $T_t = T_{\mu_t}$. In other words,

$$T_t x(\tau) = E\, x(\tau + N(t)) = \int x(\tau + \varsigma)\, \mu_t(\varsigma) = \sum_{n=0}^\infty x(\tau + n) \frac{\lambda^n t^n}{n!} \mathrm{e}^{-\lambda t}. \tag{7.42}$$

Since $\mu_t$ is concentrated on natural numbers, it may be viewed as a member of $l^1(\mathbb{N}_0)$. In the notation of 5.2.3 and 6.3.1,

$$\mu_t = \mathrm{e}^{-at} \sum_{k=0}^\infty \frac{a^k t^k}{k!} e_k = \mathrm{e}^{-at} \sum_{k=0}^\infty \frac{a^k t^k}{k!} e_1^{*k} = \mathrm{e}^{-at} \exp(ate_1)$$

$$= \exp(-at(e_1 - e_0)). \tag{7.43}$$

It means that $\{\mu_t, t \geq 0\}$, being the exponent function, is a convolution semigroup:

$$\mu_t * \mu_s = \mu_{t+s}.$$

† See footnote to 7.5.1.

On the other hand $l^1(\mathbb{N}_0)$ is a subalgebra of (signed) measures and from 2.3.20 we remember that a map $\mu \to T_\mu$ is a representation of the algebra of signed measures in the spaces $C_0(\mathbb{R})$ and $BM(\mathbb{R})$. Since $T_{e_1}$ is a translation operator $Ax(\tau) = x(\tau + 1)$, (7.43) gives

$$T_t = T_{\mu_t} = \mathrm{e}^{-at}\mathrm{e}^{atA} = \mathrm{e}^{-at} \sum_{n=0}^{\infty} \frac{a^n t^n A^n}{n!}; \qquad (7.44)$$

where $A^0 = I$; the semigroup related to the Poisson process is the exponent function of $a(I - A)$.

Formally, (7.42) and (7.44) are identical, there is a difference though in the sense they are understood. Formula (7.42) establishes merely the fact that for any $\tau$ the numerical series on its the right-hand side converges to $T_t x(\tau)$. Formula (7.44) proves not only that the convergence is actually uniform in $\tau$; which means that the series converges in the sense of the norm in $BM(\mathbb{R})$, but also that the convergence is uniform in $x$, when $x$ varies in a (say: unit) ball in $BM(\mathbb{R})$. In other words, the series converges in the sense of the norm in $\mathcal{L}(BM(\mathbb{R}))$.

As a corollary, $u(t, \tau) = E\, u(\tau + N(t))$, where $N(t)$ is the Poisson process with parameter $a$, and $u$ is any member of $BM(\mathbb{R})$, is a solution to the following Cauchy problem:

$$\frac{\partial u(t, \tau)}{\partial t} = au(t, \tau + 1) - au(t, \tau), \qquad u(0, \tau) = u(\tau), t \in \mathbb{R} \qquad (7.45)$$

**7.5.6 Remark** The semigroup $\{T_t = \mathrm{e}^{t(aA-aI)}, t \geq 0\}$ related to the Poisson process has a natural extension to a *group* $T_t = \mathrm{e}^{t(aA-aI)}, t \in \mathbb{R}$. Note, however, that the operators $T_{-t}$, $t > 0$ (except for being inverses of $T_t$) do not have clear probabilistic interpretation.

**7.5.7 Exercise** Prove directly that the Poisson process has stationary and independent increments, i.e. that for all $0 \leq t_1 < t_2 < ... < t_n$, the random variables $N(t_2) - N(t_1), ..., N(t_n) - N(t_{n-1})$ are independent and $N(t_i) - N(t_{i-1})$ has the Poisson distribution with parameter $a(t_i - t_{i-1})$.

**7.5.8 Exercise** Let $Y_n$, $n \geq 1$, be a sequence of independent identically distributed random variables. Suppose that $N(t)$ is a Poisson process that is independent from these variables. Consider the random process $p_t = \sum_{n=1}^{N(t)} Y_n$, where we agree that $\sum_{n=1}^{0} Y_n = 0$. Processes of this form are termed compound Poisson processes. If $Y_n = 1$, for all $n \geq 1$, $p_t$ is a (simple) Poisson process. If $Y_n$ admit only two values: 1 and $-1$, both with probability $\frac{1}{2}$, $p_t$ is a symmetric random walk on integers.

Find an explicit expression for the semigroup of operators related to $p_t$ and repeat the analysis made for the Poisson process to find the Cauchy problem of which $u(t, \tau) = E\, u(\tau + p_t)$ is a solution. Check directly that this equation is satisfied. Write this equation for the random walk.

**7.5.9 Exercise**     Find the probability generating function of a compound Poisson process if $Y_n$ takes values in $\mathbb{N}$ and the probability generating function of $Y_n$ is given. Find the characteristic function of this process if $Y_n$ takes values in $\mathbb{R}$ and the characteristic function of $Y_n$ is given.

**7.5.10 Exercise**     Let $R_\lambda, \lambda > 0$, be the resolvent of the Poisson semigroup. Show that $\lambda R_\lambda$ is the operator related to the geometric distribution with parameter $p = \frac{\lambda}{\lambda + a}$.

## 7.6 More convolution semigroups

**7.6.1 Definition**     A family $\{\mu_t, t \geq 0\}$ of Borel measures on $\mathbb{R}$ is said to be a convolution semigroup of measures iff (a) $\mu_0 = \delta_0$, (b) $\mu_t$ converges weakly to $\delta_0$, as $t \to 0+$, and (c)

$$\mu_t * \mu_s = \mu_{t+s}, \qquad t, s \geq 0. \tag{7.46}$$

**7.6.2 Example**     Let $b > 0$. The measures $\{\mu_t, t \geq 0\}$ with gamma densities $x_t(\tau) = \frac{b^t}{\Gamma(t)} \tau^{t-1} e^{-b\tau} 1_{\mathbb{R}^+}(\tau)$, for $t > 0$ and $\mu_0 = \delta_0$, form a convolution semigroup. The semigroup property should have been proven by the reader in 1.2.33. Moreover, if $X_t$ is a random variable with distribution $\mu_t$, then $E\, X_t = tb^{-1}$ and $D^2 X_t = tb^{-2}$. Hence, for any $\epsilon > 0$, $\mathbb{P}\{|X_t - tb^{-1}| > \epsilon\}$ tends to zero, as $t \to 0+$, being dominated by $tb^{-2}\epsilon^{-2}$ (by Chebyshev's inequality). This implies that $\mu_t$ converges to $\delta_0$, as $t \to 0+$.

**7.6.3 Definition**     A stochastic process $X_t, t \geq 0$, is said to be a **Lévy process** iff (a) $X_0 = 0$ a.s., (b) almost all its paths are right-continuous and have left limits, (c) for all $t \geq s \geq 0$, the variable $X_t - X_s$ is independent of $\sigma(X_u, 0 \leq u \leq s)$ and has the same distribution as $X_{t-s}$.

As we have already seen, Brownian motion and the Poisson process are examples of Lévy processes. In fact they constitute the most prominent and most important examples of Lévy processes [66], [100].

**7.6.4** *Lévy processes and convolution semigroups* Lévy processes are in a natural way related to convolution semigroups of measures. In fact, the distributions $\mu_t, t \geq 0$ of a Lévy process $X_t, t \geq 0$ form a convolution semigroup. Indeed, $\mu_0 = \delta_0$ by point (a) of the definition of Lévy process. Moreover, by (b) of this definition, $X_t$ converges a.s. to $X_0$, as $t \to 0+$. Hence, by 5.8.1, $X_t$ converges to $X_0$ weakly, as well. Finally, by (c) in 7.6.3, $X_{t+s} = (X_{t+s} - X_s) + X_s$ is a sum of two independent random variables, and $\mathbb{P}_{X_{t+s}-X_s} = \mathbb{P}_{X_t} = \mu_t$. Hence, $\mu_{t+s} = \mathbb{P}_{X_{t+s}} = \mathbb{P}_{X_{t+s}-X_s} * \mathbb{P}_{X_s} = \mu_t * \mu_s$.

**7.6.5** *Cauchy flights* Not all processes with distributions forming convolution semigroups are Lévy processes. For example, a Cauchy process is by definition a process that satisfies conditions (a) and (c) of definition 7.6.3 and $X_t, t > 0$ has a Cauchy measure with parameter $t$, i.e. if $\mu_t$ has the density $x_t(\tau) = \frac{1}{\pi} \frac{t}{t^2+\tau^2}$, yet its sample paths are discontinuous (see [36]).

Moreover, the distributions $\mu_t, t \geq 0$, of Cauchy random variables $X_t$ with parameter $t$ form a convolution semigroup *regardless of whether the increment $X_t - X_s$ is independent of $X_s$* or not. Indeed, since the Fourier transform $\hat{x}(\varsigma) = \int_{-\infty}^{\infty} e^{i\varsigma\tau} x_t(\tau) \, d\tau$ of $x_t$ equals $e^{-t|\varsigma|}$, by 6.4.12, we have $\hat{x}_t \hat{x}_s = \hat{x}_{s+t}$. By 6.4.11, this proves that $\mu_{t+s} = \mu_t * \mu_s$. Moreover, for any $\epsilon > 0$,

$$\mu_t(-\epsilon, \epsilon) = \frac{1}{\pi} \int_{-\epsilon}^{\epsilon} \frac{t}{t^2 + \tau^2} \, d\tau = \frac{2}{\pi} \arctan \frac{\epsilon}{t} \xrightarrow[t \to 0+]{} 1,$$

proving that $\mu_t$ converges to $\delta_0$.

In other words, forming a convolution semigroup of measures is a property of distributions and not of random variables.

**7.6.6** *Examples of generators of convolution semigroups* Let $\{\mu_t, t \geq 0\}$ be a convolution semigroup on $\mathbb{R}$. Define the semigroup $\{T_t, t \geq 0\}$ in $\mathbb{X} = BUC(\mathbb{R})$ by $T_t = T_{\mu_t}$. By 7.1.12, this is a $c_0$ semigroup in $\mathbb{X}$; its generator $A$ will be called the generator of the convolution semigroup $\{\mu_t, t \geq 0\}$.

In general, finding an explicit form of $A$ is difficult, if possible at all. As we shall see, however, the domain of $A$ contains the space $\mathbb{X}_2$ of all twice differentiable functions in $\mathbb{X}$ with both derivatives in $\mathbb{X}$, and $A$ restricted to $\mathbb{X}_2$ can be described in more detail. Here are two examples.

(a) *The Cauchy semigroup* We have

$$\frac{1}{t}(T_t x - x)(\sigma) = \frac{1}{\pi} \int_{\mathbb{R}} \frac{x(\tau + \sigma) - x(\sigma)}{\tau^2 + t^2} \, d\tau.$$

Since $y(\tau) = \frac{\tau}{\tau^2+1}$ is odd, this equals

$$\frac{1}{\pi} \int_{\mathbb{R}} \left[ \frac{x(\tau+\sigma) - x(\sigma) - x'(\sigma)y(\tau)}{\tau^2}(\tau^2+1) \right] \frac{\tau^2}{\tau^2+1} \frac{1}{\tau^2+t^2} \, d\tau.$$

Since $y(0) = 0$, $y'(0) = 1$, and $y''(0) = 0$, by de l'Hospital's rule, the expression in brackets tends, for $x \in \mathbb{X}_2$, to $\frac{1}{2}x''(\sigma)$, as $\tau \to 0$. Also, it is continuous as a function of $\tau$, and bounded. Hence, it belongs to $BC(\mathbb{R})$. Moreover, the measures $\nu_t$ with densities $\frac{\tau^2}{\tau^2+1}\frac{1}{\tau^2+t^2}$ converge weakly, as $t \to 0$, to the measure $m$ with density $\frac{1}{\tau^2+1}$. Hence, for $x \in \mathbb{X}_2$, and any $\sigma$, the above expression tends to

$$\frac{1}{\pi} \int_{\mathbb{R}} \left[ \frac{x(\tau+\sigma) - x(\sigma) - x'(\sigma)y(\tau)}{\tau^2} \right] d\tau.$$

Using 7.1.12, this convergence may be proven to be uniform in $\sigma$. We skip the details now, as we will prove this result in a more general setting later – see 7.6.14.

(b) *The gamma semigroup* In this case,

$$\frac{1}{t}(T_t x - x)(\sigma) = \frac{b^t}{t\Gamma(t)} \int_0^\infty \left[ \frac{x(\tau+\sigma) - x(\sigma) - x'(\sigma)y(\tau)}{\tau^2} \right] e^{-b\tau}\tau^{t+1} \, d\tau$$

$$+ x'(\sigma)\frac{b^t}{t\Gamma(t)} \int_0^\infty \frac{1}{\tau^2+1} e^{-b\tau}\tau^t \, d\tau.$$

For any $\lambda \geq 0$, changing variables $\tau' = (\lambda+b)\tau$,

$$\frac{b^t}{t\Gamma(t)} \int_0^\infty e^{-\lambda\tau}e^{-b\tau}\tau^{t+1} \, d\tau = \frac{b^t}{t\Gamma(t)} \frac{1}{(\lambda+b)^{t+2}}\Gamma(t+2)$$

$$= b^t \frac{1}{(\lambda+b)^{t+2}}(t+1).$$

This converges to $\frac{1}{(\lambda+b)^2}$, as $t \to 0$. Thus, the measures $\nu_t$ with densities $\frac{b^t}{t\Gamma(t)}e^{-b\tau}\tau^{t+1}$ converge weakly to the measure $m$ with density $\frac{\tau}{b^2}e^{-b\tau}$. Similarly, the second integral converges to $x'(\sigma)\frac{1}{b} \int_0^\infty \frac{1}{\tau^2+1}e^{-b\tau} \, d\tau$. Thus, for any $\sigma$, $\frac{1}{t}(T_t x - x)(\sigma)$ converges to

$$\int_0^\infty \left[ \frac{x(\tau+\sigma) - x(\sigma) - x'(\sigma)y(\tau)}{\tau^2} \right] e^{-\tau}\frac{\tau}{b^2} \, d\tau + \frac{x'(\sigma)}{b} \int_0^\infty \frac{e^{-b\tau}}{\tau^2+1} \, d\tau.$$

Again, the convergence may be shown to be uniform in $\sigma$.

**7.6.7 Generating functional** Let $\{\mu_t, t \geq 0\}$ be a convolution family of measures on $\mathbb{R}$, $\{T_t, t \geq 0\}$ be the corresponding semigroup of operators in $\mathbb{X} = BUC(\mathbb{R})$, and $A$ be its generator. Let $F : C \supset \mathcal{D}(F) \to \mathbb{R}$ be the **generating functional** of $\{\mu_t, t \geq 0\}$, defined as

$$Fx = \lim_{t \to 0+} F_t x$$

on the domain $\mathcal{D}(F) = \{x \in \mathbb{X} | \lim_{t \to 0+} F_t x \text{ exists}\}$, where

$$F_t x = \frac{1}{t}(T_t x(0) - x(0)).$$

It is clear that $\mathcal{D}(A) \subset \mathcal{D}(F)$. Moreover, if $x \in \mathcal{D}(A)$ then $Ax(\sigma) = \lim_{t \to 0+} \frac{1}{t}(T_t x(\sigma) - x(\sigma)) = \lim_{t \to 0+} F_t x_\sigma$, where $x_\sigma(\tau) = x(\sigma + \tau), \sigma \in \mathbb{R}$. In particular, $x_\sigma$ belongs to $\mathcal{D}(F)$ and $Ax(\sigma) = Fx_\sigma$. Hence, the values of $A$ may be recovered from the values of $F$.

The key result concerning $F$ is that:

$$\mathbb{X}_2 \subset \mathcal{D}(F), \tag{7.47}$$

where $\mathbb{X}_2$ is the set of all twice differentiable functions in $\mathbb{X}$ with both derivatives in $\mathbb{X}$.

We note that $\mathbb{X}_2$ when equipped with the norm $\|x\|_2 = \|x\|_{\mathbb{X}} + \|x'\|_{\mathbb{X}} + \|x''\|_{\mathbb{X}}$ (or with the equivalent norm, $\|x\|_2^* = \|x\|_{\mathbb{X}} + \|x''\|_{\mathbb{X}}$), is a Banach space – see 7.3.3 and 7.3.6. Moreover, if $x \in \mathbb{X}$ is differentiable with $x' \in \mathbb{X}$, then so is $T_t x$ and $(T_t x)' = T_t x' \in \mathbb{X}$. Similarly, if $x \in \mathbb{X}_2$, then $(T_t x)'' = T_t x''$. Hence, $T_t, t \geq 0$, leave $\mathbb{X}_2$ invariant, $\|T_t x\|_2 \leq \|x\|_2$, and $(T_t)_{|\mathbb{X}_2}, t \geq 0$, is a strongly continuous semigroup of operators. In particular, the domain $\mathcal{D}_2$ of its infinitesimal generator is dense in $\mathbb{X}_2$ (in the sense of the norm $\| \cdot \|_2$); certainly, $\mathcal{D}_2 \subset \mathcal{D}(A)$. Besides this remark, for the proof of (7.47) we will need the following set of lemmas.

**7.6.8 Lemma** For every $\delta > 0$ we have

$$\sup_{t > 0} \frac{1}{t} \mu_t(-\delta, \delta)^{\complement} < \infty.$$

*Proof* Let $x(\tau) = |\tau| \wedge \delta$. There exists a $y \in \mathcal{D}_2$, lying within $\delta/6$ distance from $x$. Let $\tau_0 \in [-\delta/6, \delta/6]$ be such that $\min_{\tau \in [-\delta/6, \delta/6]} y(\tau) = y(\tau_0)$. Note that $|y(\tau_0)| \leq \delta/6$. Let $z(\tau) = y(\tau - \tau_0) - y(\tau_0), \tau \in \mathbb{R}$. Then, $z \in \mathcal{D}_2, z \geq 0$, and $z(0) = 0$. Moreover, for $|\tau| > \delta$, we have $z(\tau) \geq y(\tau - \tau_0) - \delta/6 - \delta/6 \geq \frac{5}{6}\delta - \frac{2}{6}\delta = \frac{\delta}{2}$, since $|\tau - \tau_0| > \frac{5}{6}\delta$.

The function $f(t) = F_t z = \frac{1}{t} \int_{\mathbb{R}} z \, d\mu_t$ is continuous in $t \in \mathbb{R}_*^+$ with $\lim_{t \to 0} f(t) = Fz$ and $\lim_{t \to \infty} f(t) = 0$, hence bounded. Therefore, $\sup_{t > 0} \frac{1}{t} \mu_t(-\delta, \delta)^{\complement} \leq \frac{2}{\delta} \sup_{t > 0} \frac{1}{t} \int_{\mathbb{R}} z \, d\mu_t$ is finite. $\qquad\square$

**7.6.9 Lemma**      Let $\delta > 0$. Then, there exist $y^\flat$ and $z^\flat$ in the set $C^\infty$ of infinitely differentiable functions with all derivatives in $\mathbb{X}$, such that $y^\flat(\tau) = \tau$ and $z^\flat(\tau) = \tau^2$ whenever $|\tau| \le \delta$.

We leave the proof of this lemma as an exercise.

**7.6.10 Lemma**      (a) There exist $y^\sharp$ and $z^\sharp$ in $\mathcal{D}_2$ such that $y^\sharp = 0$ and $(y^\sharp)'(0) = 1$, and $z^\sharp(0) = (z^\sharp)'(0) = 0$ and $(z^\sharp)''(0) = 2$. (b) There exists a $\delta > 0$ such that $\sup_{t>0} \frac{1}{t} \int_{(-\delta,\delta)} \tau^2 \, \mu_t(\,\mathrm{d}\tau)$ is finite.

*Proof*  (a) We may not have $x'(0) = 0$ for all $x \in \mathcal{D}_2$, for then we would have $\inf_{x \in \mathcal{D}_2} \|x - y^\flat\|_2 \ge |(y^\flat)'(0)| = 1$, contrary to the fact that $\mathcal{D}_2$ is dense in $\mathbb{X}_2$. Hence, there exists an $x \in \mathcal{D}_2$ with $x'(0) \ne 0$ and, consequently, $y^\sharp = \frac{1}{x'(0)}(x - x(0)1_{\mathbb{R}})$ possesses the required properties.

To prove the other relation, we note that the operator

$$Px = x - x(0)1_{\mathbb{R}} - x'(0)y^\sharp \tag{7.48}$$

mapping $\mathbb{X}_2$ into $\mathbb{Y} = \{x \in \mathbb{X}_2 | x(0) = x'(0) = 0\}$ is linear and bounded with the norm not exceeding $2 + \|y^\sharp\|_2$. Moreover, it is onto (since $Px = x$ for $x \in \mathbb{Y}$) and leaves $\mathcal{D}_2$ invariant. Hence, $\mathcal{D}_2 \cap \mathbb{Y}$ is dense in $\mathbb{Y}$, $\mathcal{D}_2$ being dense in $\mathbb{X}_2$. Therefore, there is an $x \in \mathcal{D}_2 \cap \mathbb{Y}$ such that $x''(0) \ne 0$, for otherwise we would have $\inf_{x \in \mathcal{D}_2 \cap \mathbb{Y}} \|x - z^\flat\|_2 \ge |(z^\flat)''(0)| = 2$. Finally, $z^\sharp = \frac{2}{x''(0)}x$ possesses the required properties.

(b) By the Taylor formula, $z^\sharp(\tau) = \frac{\tau^2}{2}(z^\sharp)''(\theta\tau)$, for some $\theta = \theta(\tau), 0 \le \theta \le 1$. By continuity of $(z^\sharp)''$, there exists a $\delta > 0$ such that $(z^\sharp)''(\tau) \ge 1$ whenever $|\tau| \le \delta$. For such a $\tau$, $z^\sharp(\tau) \ge \frac{1}{2}\tau^2$.

Since $z^\sharp(0) = 0$, $F_t z^\sharp = \frac{1}{t} \int_{\mathbb{R}} z^\sharp \, \mathrm{d}\mu_t$. Moreover, arguing as in the proof of Lemma 7.6.8, we see that $\sup_{t>0} |F_t z^\sharp|$ is finite. Therefore,

$$\sup_{t>0} \frac{1}{t} \int_{(-\delta,\delta)} \tau^2 \, \mu_t(\,\mathrm{d}\tau) \le 2 \sup_{t>0} \frac{1}{t} \int_{(-\delta,\delta)} z^\sharp(\tau) \, \mu_t(\,\mathrm{d}\tau)$$

$$\le 2 \sup_{t>0} \frac{1}{t} \int_{\mathbb{R}} z^\sharp(\tau) \, \mu_t(\,\mathrm{d}\tau) + 2 \sup_{t>0} \frac{1}{t} \int_{(-\delta,\delta)^{\complement}} z^\sharp(\tau) \, \mu_t(\,\mathrm{d}\tau)$$

$$\le 2 \sup_{t>0} |F_t z^\sharp| + 2\|z^\sharp\|_2 \sup_{t>0} \frac{1}{t} \mu_t(-\delta,\delta)^{\complement}$$

which is finite by Lemma 7.6.10.      $\square$

*Proof  (of relation (7.47))*  We need to show that the limit $\lim_{t\to 0} F_t x$ exists for all $x \in \mathbb{X}_2$. By definition this limit exists for all $x$ in $\mathcal{D}_2$ and this set is dense in $\mathbb{X}_2$. Therefore, it suffices to show that there exists a constant $c$ such that $\sup_{t>0} |F_t x| \leq c\|x\|_2, x \in \mathbb{X}_2$. Since both $1_\mathbb{R}$ and $y^\sharp$ in (7.48) belong to $\mathcal{D}(F)$, our task reduces to showing that there exists a $c$ such that

$$\sup_{t>0} |F_t P x| \leq c\|x\|_2, \qquad x \in \mathbb{X}_2. \tag{7.49}$$

By the Taylor theorem, $Px(\tau) = \frac{\tau^2}{2}(Px)''(\theta\tau)$ where $\theta = \theta(\tau)$ and $0 \leq \theta \leq 1$. Thus, $|Px(\tau)| \leq \|Px\|_2 \frac{\tau^2}{2}$. Now the estimate

$$|Px(\tau)| \leq \|Px\|_2 \frac{\tau^2}{2} 1_{(-\delta,\delta)} + \|Px\|_2 1_{(-\delta,\delta)^\complement}$$

implies (7.49) with

$$c = c(\delta) = \|P\| \left[ \sup_{t>0} \frac{1}{t} \int_{-(\delta,\delta)} \frac{\tau^2}{2} \mu_t(\,\mathrm{d}\tau) + \sup_{t>0} \frac{1}{t} \mu_t(-\delta,\delta)^\complement \right]$$

which may be made finite by Lemmas 7.6.8 and 7.6.10. $\qquad\square$

Before we establish the form of the generator of a convolution semigroup we need two more lemmas and the following definition.

**7.6.11 Definition**    A distribution $\mu$ on $\mathbb{R}$ is said to be **symmetric** if its transport $\mu^\mathrm{s}$ via the map $\tau \mapsto -\tau$ equals $\mu$. In other words, if $X$ is a random variable with distribution $\mu$, then $-X$ has the distribution $\mu$, too.

**7.6.12 Lemma**    Let $\mu$ be a symmetric distribution on $\mathbb{R}$. Then, for any $\delta > 0$ and $k \in \mathbb{N}$,

$$\mu^{*k}[-\delta,\delta]^\complement \geq \frac{1}{2}(1 - \mathrm{e}^{k\mu[-\delta,\delta]^\complement}).$$

*Proof* Let $S_k = \sum_{i=1}^k X_i$ be the sum of $k$ independent random variables with distribution $\mu$. We claim that

$$\mathbb{P}[|S_k| > \delta] \geq \frac{1}{2}\mathbb{P}\left\{ \max_{1 \leq i \leq k} |X_i| > \delta \right\}. \tag{7.50}$$

To prove this we note first that by assumption for any Borel sets $B_i \in \mathcal{B}(\mathbb{R}), i = 1, ..., k$, the probability

$$\mathbb{P}\{X_i \in B_i, i = 1, ..., k\} = \prod_{i=1}^k \mathbb{P}\{X_i \in B_i\}$$

does not change when one or more of the $X_i$ is replaced by $-X_i$. By 1.2.7, this implies that the same is true for $\mathbb{P}\{(X_1,...,X_k) \in B\}$ where $B \in \mathcal{B}(R^k)$, and, consequently, for the probability $\mathbb{P}\{f(X_1,...,X_k) \in B\}$ where $B \in \mathcal{B}(R^l), l \in \mathbb{N}$ and $f : R^k \to \mathbb{R}^l$ is Borel measurable. In particular, for any $i = 1,...,k$, $p_{i,1} = p_{i,2}$ where

$$
p_{i,\alpha} = \mathbb{P}\left\{|X_j| \leq \delta, j = 1,...,i-1, X_i > \delta, (-1)^\alpha \sum_{j=1,...,k, j\neq i} X_j \geq 0\right\},
$$

$\alpha = 1, 2$. On the other hand, $p_{i,1} + p_{i,2}$ is no less than $q_i := \mathbb{P}\{|X_j| \leq \delta, j = 1,...,i-1, X_i > \delta\}$. Hence, $p_{i,2} \geq \frac{1}{2}q_i$. Similarly, $p'_{i,1} \geq \frac{1}{2}q'_i$ where $p'_{i,\alpha}$ and $q'_i$ are defined in the same way as $p_{i,\alpha}$ and $q_i$, respectively, but with $X_i > \delta$ replaced by $X_i < -\delta$.

Let $\tau = \tau(\omega) = \min\{i = 1,...,k \,|\, |X_i(\omega)| > \delta\}$, where $\min \emptyset = \infty$. Then $q_i + q'_i = \mathbb{P}\{\tau = i\}$. Moreover,

$$
\mathbb{P}\{|S_k| > \delta\} \geq \mathbb{P}\{|S_k| > \delta, \tau < \infty\} = \sum_{i=1}^{k} \mathbb{P}\{|S_k| > \delta, \tau = i\}
$$

$$
\geq \sum_{i=1}^{n}(p_{i,2} + p'_{i,1}) \geq \frac{1}{2}\sum_{i=1}^{n}(q_i + q'_i)
$$

$$
= \frac{1}{2}\sum_{i=1}^{n}\mathbb{P}\{\tau = i\} = \frac{1}{2}\mathbb{P}(\tau < \infty) = \frac{1}{2}\mathbb{P}\left\{\max_{1\leq i\leq k}|X_i| > \delta\right\},
$$

as claimed. Next, by independence, $\mathbb{P}\{\max_{1\leq i\leq k}|X_i| \leq \delta\} \leq (\mu[-\delta,\delta])^k \leq e^{-k\mu[-\delta,\delta]^{\complement}}$ since for $x \leq 1$, $x \leq e^{-(1-x)}$. Combining this with (7.50), $\mathbb{P}\{|S_k| > \delta\} \geq \frac{1}{2}\left(1 - e^{k\mu[-\delta,\delta]^{\complement}}\right)$. This is the same as our thesis, since $S_k$ has distribution $\mu^{*k}$. □

**7.6.13 Lemma**    Let $t_n, n \geq 0$ be a sequence of positive numbers such that $\lim_{n\to\infty} t_n = 0$. For any $\epsilon > 0$ we may choose a $\delta > 0$ so that

$$
\sup_{n\geq 1}[t_n^{-1}]\mu_{t_n}[-\delta,\delta]^{\complement} < \epsilon.
$$

*Proof 1*    Let us recall that an $m \in \mathbb{R}$ is said to be a **median** of a distribution $\mu$ on $\mathbb{R}$ if both $(-\infty, m]$ and $[m, \infty)$ have $\mu$ measure at least $\frac{1}{2}$. Note that $\mu$ has at least one median.

For $n \geq 1$, let $m_n$ be a median for $\mu_{t_n}$. Since the measures $\mu_{t_n}$ tend to Dirac measure at 0, we have $\lim_{n\to\infty} m_n = 0$. Indeed, without loss of generality we may assume $m_n \geq 0$, because the measures $\mu_{t_n}$ with $m_n <$

0 may be replaced by $\mu_{t_n}^s$ (see Definition 7.6.11). Moreover, we may not have $m_n \geq m$ for some $m > 0$ and infinitely many $n$, because otherwise for infinitely many $n$ we would have $\mu_{t_n}(-\infty, m) \leq \mu_{t_n}(-\infty, m_n) \leq \frac{1}{2}$ contrary to the fact that $\mu_{t_n}(-\infty, m)$ converges to 1, as $n \to \infty$.

**2** Fix $n \geq 1$. Let $X_n$ and $X'_n$ be two independent random variables with distribution $\mu_{t_n}$. The events $\{X_n > m_n + \delta, X'_n \leq m_n\}$ and $\{X_n < m_n - \delta, X'_n \geq m_n\}$ are disjoint and their union is contained in $\{|X_n - X'_n| > \delta\}$. On the other hand, their probabilities equal $\mathbb{P}\{X_n > m_n + \delta\}\mathbb{P}\{X'_n \leq m_n\} \geq \frac{1}{2}\mathbb{P}\{X_n > m_n + \delta\}$ and $\mathbb{P}\{X_n < m_n - \delta\}\mathbb{P}\{X'_n \geq m_n\} \geq \frac{1}{2}\mathbb{P}\{X_n < m_n - \delta\}$. Hence, $2\mathbb{P}\{|X_n - X'_n| > \delta\} \geq \mathbb{P}\{|X_n - m_n| > \delta\}$, i.e.

$$[t_n^{-1}]\mu_{t_n}[m_n - \delta, m_n + \delta]^{\complement} \leq 2[t_n^{-1}]\mu_{t_n} * \mu_{t_n}^s[-\delta, \delta].$$

Thus, for $\delta > \sup_{n \geq 1} |m_n|$,

$$[t_n^{-1}]\mu_{t_n}[-2\delta, 2\delta]^{\complement} \leq 2[t_n^{-1}]\mu_{t_n} * \mu_{t_n}^s[-\delta, \delta]^{\complement}. \tag{7.51}$$

**3** For $\epsilon > 0$ let $\eta = \frac{1}{2}(1 - e^{-\epsilon/2})$ and let $\delta > \sup_{n \geq 1} |m_n|$ be a point of continuity for $\mu_1 * \mu_1^s$ large enough so that $\mu_1 * \mu_1^s[-\delta, \delta]^{\complement} < \eta$. By 5.4.18, 5.4.20 and 7.4.14, the measures $\mu_{[t_n^{-1}]t_n} * \mu_{[t_n^{-1}]t_n}^s$ converge weakly to $\mu_1 * \mu_1^s$. Therefore, for $n$ larger than some $n_0$, $\mu_{[t_n^{-1}]t_n} * \mu_{[t_n^{-1}]t_n}^s[-\delta, \delta]^{\complement} < \eta$. By (7.50) and (7.51),

$$[t_n^{-1}]\mu_{t_n}[-2\delta, 2\delta]^{\complement} < \epsilon, \tag{7.52}$$

for $n \geq n_0$. On the other hand, for each $1 \leq n \leq n_0$, we may choose a $\delta$ such that (7.52) holds for this $n$. Therefore, for sufficiently large $\delta$, (7.52) holds for all $n$. This implies our thesis. $\square$

**7.6.14 *The form of the generator*** Let $\{\mu_t, t \geq 0\}$ be a convolution semigroup and let $A$ be the generator of the corresponding semigroup $\{T_t, t \geq 0\}$ of operators in $\mathbb{X} = BUC(\mathbb{R})$. Then, $\mathbb{X}_2 \subset \mathcal{D}(A)$. Moreover, there exists an $a \in \mathbb{R}$ and a finite Borel measure $m$ on $\mathbb{R}$ such that

$$Ax(\sigma) = ax'(\sigma) \tag{7.53}$$

$$+ \int_{\mathbb{R}} [x(\tau + \sigma) - x(\sigma) - x'(\sigma)y(\tau)]\frac{\tau^2 + 1}{\tau^2} m(d\tau), \qquad x \in \mathbb{X}_2,$$

where $y(\tau) = \frac{\tau}{\tau^2 + 1}$.

*Proof* **1** Define the measures $\nu_t, t > 0$, by $\nu_t(d\tau) = \frac{1}{t}\frac{\tau^2}{\tau^2 + 1}\mu_t(d\tau)$. In other words, $\nu_t$ is absolutely continuous with respect to $\mu_t$ and has a density $\frac{1}{t}\frac{\tau^2}{\tau^2+1}$ For appropriate choice of $\delta > 0$ (Lemmas 7.6.8 and 7.6.10),

$$\sup_{t>0} \nu_t(\mathbb{R}) \leq \sup_{t>0} \frac{1}{t} \int_{(-\delta,\delta)} \tau^2 \, \mu_t(\,\mathrm{d}\tau) + \sup_{t>0} \frac{1}{t}\mu_t(-\delta,\delta)^{\mathbb{C}} < \infty.$$

Let $t_n, n \geq 1$, be a numerical sequence such that $\lim_{n\to\infty} t_n = 0$. The measures $\frac{1}{\nu_t(\mathbb{R})}\nu_t$ are probability measures. Hence, there exists a sequence $n_k, k \geq 1$, with $\lim_{k\to\infty} n_k = \infty$, such that the measures $\frac{1}{\nu_{t_{n_k}}(\mathbb{R})}\nu_{t_{n_k}}$ converge weakly to a probability measure on $[-\infty,\infty]$, and the numerical sequence $\nu_{t_{n_k}}(\mathbb{R})$ converges, too. Hence, $\nu_{t_{n_k}}$ converge to a finite measure $m$ on $[-\infty,\infty]$. Furthermore, $m$ is concentrated on $\mathbb{R}$. Indeed, it is clearly so when $\lim_{n\to\infty} \nu_{t_{n_k}}(\mathbb{R}) = 0$, for then $\nu_{t_{n_k}}$ converge weakly to zero measure. In the other case, $\nu_{t_{n_k}}(\mathbb{R}), k \geq 1$, is bounded away from 0, and Lemma 7.6.13 shows that the sequence $\frac{1}{\nu_{t_{n_k}}(\mathbb{R})}\nu_{t_{n_k}}, k \geq 1$, is tight, so that its limit is concentrated on $\mathbb{R}$, and so is $m$.

For $x \in \mathbb{X}_2$, let $(Wx)(\tau) = \frac{\tau^2}{\tau^2+1}x(\tau)$. Clearly, $Wx \in \mathbb{X}_2$, and $(Wx)(0) = 0$. Hence,

$$FWx = \lim_{t\to 0} \frac{1}{t} \int_{\mathbb{R}} Wx \, \mathrm{d}\mu_t = \lim_{t\to 0} \int_{\mathbb{R}} x \, \mathrm{d}\nu_t = \lim_{k\to\infty} \int_{\mathbb{R}} x \, \mathrm{d}\nu_{t_{n_k}} = \int_{\mathbb{R}} x \, \mathrm{d}m.$$

Since $\mathbb{X}_2$ is dense in $\mathbb{X}$, this determines the measure $m$.

We have proved that any sequence $v_{t_n}, n \geq 1$, with $\lim_{n\to\infty} t_n = 0$, has a subsequence converging weakly to the unique Borel measure $m$ on $\mathbb{R}$. Therefore, $\lim_{t\to 0+} v_t = m$ (weakly).

**2** For $x \in \mathbb{X}_2$, let $z_\sigma(\tau) = [x(\sigma+\tau) - x(\sigma) - x'(\sigma)y(\tau)] \frac{\tau^2+1}{\tau^2}, \tau, \sigma \in \mathbb{R}$. Since $y(0) = 0$, $y'(0) = 1$, and $y''(0) = 0$, by de l'Hospital's rule, $\lim_{\tau\to 0} z_\sigma(\tau) = \frac{1}{2}x''(\sigma)$. Also, for any $\delta > 0$, $z_\sigma$ is seen to be uniformly continuous in $\tau \in (-\delta,\delta)^{\mathbb{C}}$ and bounded. This implies $z_\sigma \in \mathbb{X}$. Moreover,

$$\frac{1}{t}\left[(T_t x)(\sigma) - x(\sigma)\right] = \frac{1}{t} \int_{\mathbb{R}} \left[x(\tau+\sigma) - x(\sigma)\right] \mu_t(\,\mathrm{d}\tau)$$

$$= x'(\sigma)F_t y + \int_{\mathbb{R}} z_\sigma \, \mathrm{d}\nu_t.$$

For any $\sigma \in \mathbb{R}$, this converges to $ax'(\sigma) + \int_{\mathbb{R}} z_\sigma \, \mathrm{d}m$, where $a = Fy$. Hence, our task reduces to showing that $\lim_{t\to 0} \int_{\mathbb{R}} z_\sigma \, \mathrm{d}\nu_t = \int_{\mathbb{R}} z_\sigma \, \mathrm{d}m$ uniformly in $\sigma \in \mathbb{R}$.

To this end, we check directly that the functions $\tilde{z}_\sigma(\tau) = z_\sigma(\tau)\frac{\tau^2}{\tau^2+1}$ are equicontinuous at any $\tau \in \mathbb{R}$. Hence, $z_\sigma$ are equicontinuous at $\tau \neq 0$. Also, writing $\frac{1}{2}x''(\sigma) - \frac{x(\sigma+\tau)-x(\sigma)-x'(\sigma)y(\tau)}{\tau^2}$ as $\frac{1}{2}x''(\sigma) - \frac{1}{2}x''(\sigma+\theta\tau) + \frac{x'(\sigma)(y(\tau)-\tau)}{\tau^2}$ where $\theta = \theta(\tau), 0 \leq \theta \leq 1$ (by the Taylor expansion), and using uniform continuity of $x''$ we see that $z_\sigma$ are equicontinuous at $\tau = 0$, too. Finally, by the Taylor formula, $|z_\sigma(\tau)| \le \|x\|_2(1 + \|y\|_2)$ whenever $|\tau| \le 1$; when $|\tau| > 1$ it does not exceed $2\|x\|_2(2 + \|y\|)$. This implies uniform convergence, by 7.1.11. $\qquad\square$

**7.6.15 Exercise**   If $x \in \mathbb{X}_2$, then so does $x_\sigma$ where $x_\sigma(\tau) = x(\sigma + \tau)$, and we have $\|x\|_{\mathbb{X}_2} = \|x_\sigma\|_{\mathbb{X}_2}$ (cf. (7.6.7)). Moreover, $t^{-1}[T_t x(\sigma) - x(\sigma)] = t^{-1}[T_t x_\sigma(0) - x_\sigma(0)]$. Use this and (7.49) to show directly that $\mathbb{X}_2 \subset \mathcal{D}(A)$.

**7.6.16 Examples**   In the case of the Brownian motion semigroup, $m$ is the Dirac measure at 0 and $a = 0$. In the case of the Cauchy semigroup, $m$ is the measure with density $\frac{1}{\tau^2+1}$ (with respect to Lebesgue measure) and $a = 0$. In the case of the gamma semigroup, $m$ has the density $\tau e^{-b\tau}$, and $a = \frac{1}{b} \int_0^\infty \frac{1}{\tau^2+1} e^{-b\tau} \, d\tau$.

**7.6.17 Corollary**   *The set $\mathbb{X}_2$ is a core for $A$. In particular, $A$ is fully determined by 7.6.14.*

*Proof*   The operators $T_t$ leave $\mathbb{X}_2$ invariant and $\mathbb{X}_2 \subset \mathcal{D}(A)$. Moreover, $\mathbb{X}_2$ is dense in $\mathbb{X}$, hence dense in $\mathcal{D}(A)$, as well. The result follows now by 7.4.44. $\qquad\square$

**7.6.18** *The Lévy–Khintchine formula*   Let $\{\mu_t, t \ge 0\}$ be a convolution semigroup on $\mathbb{R}$. There exists a finite Borel measure $m$ and a constant $a$ such that

$$\int_{\mathbb{R}} e^{\mathrm{i}\tau\xi} \mu_t(\,d\tau) = \exp\left\{ \mathrm{i}t\xi a + t \int_{\mathbb{R}} \left( e^{\mathrm{i}\xi\tau} - 1 - \frac{\mathrm{i}\xi\tau}{\tau^2 + 1} \right) \frac{\tau^2 + 1}{\tau^2} \, m(\,d\tau) \right\}. \tag{7.54}$$

*Proof*   Fix $\xi \in \mathbb{R}$. Let $x_1(\tau) = \cos(\xi\tau)$ and $x_2(\tau) = \sin(\xi\tau)$. We have $x_j \in \mathbb{X}_2 \subset \mathcal{D}(A), j = 1, 2$. Hence, $\frac{d}{dt} T_t x_j = A T_t x_j, j = 1, 2$. In particular, $\frac{d}{dt}[T_t x_j(0)] = \left[\frac{d}{dt} T_t x_j\right](0) = F T_t x_j, j = 1, 2$. Thus, $t \mapsto \phi(t) = \int_{\mathbb{R}} e^{\mathrm{i}\tau\xi} \mu_t(\,d\tau) = T_t x_1(0) + \mathrm{i} T_t x_2(0)$ is differentiable, too, and $\frac{d}{dt}\phi(t) = F T_t x_1 + \mathrm{i} F T_t x_2$. Furthermore, by (7.53), $F T_t x_j$ is the sum of $a \int_{\mathbb{R}} x'_j \, d\mu_t$ and

$$\int_{\mathbb{R}} \left[ \int_{\mathbb{R}} x_j(\tau + \sigma) \, \mu_t(\,d\sigma) - \int_{\mathbb{R}} x_j \, d\mu_t - \int_{\mathbb{R}} x'_j \, d\mu_t \, \frac{\tau}{\tau^2 + 1} \right] \frac{\tau^2 + 1}{\tau^2} \, m(\,d\tau).$$

Also, $x_1'(\tau) + ix_2'(\tau) = i\xi e^{i\xi\tau}$ and $x_1(\tau + \sigma) + ix_2(\tau + \sigma) = e^{i\xi\tau}e^{i\xi\sigma}$. Therefore, $FT_t x_1 + iFT_t x_2$ equals

$$a i\xi\phi(t) + \phi(t) \int_{\mathbb{R}} \left( e^{i\xi\tau} - 1 - \frac{i\xi\tau}{\tau^2 + 1} \right) \frac{\tau^2 + 1}{\tau^2} \, m(\, d\tau) =: b\phi(t).$$

In other words, $\phi(t)$ is the solution of the equation $\frac{d}{dt}\phi(t) = b\phi(t)$ satisfying $\phi(0) = \int_{\mathbb{R}} e^{i\tau\xi} \mu_0(\, d\tau) = 1$. Hence, $\phi(t) = e^{bt}$. $\qquad\square$

## 7.7 The telegraph process semigroup

7.7.1 *Orientation*    The example we are going to present now is somewhat unusual. Since the time of publication of the pioneering paper by Kolomogorov [73], it has been well known that there is a close connection between stochastic processes and partial differential equations of second order. Partial differential equations (PDEs) of second order form three distinct classes: the elliptic, the parabolic and the hyperbolic equations. One of the reasons for such a classification is the fact that properties of PDEs differ greatly depending on which class they belong to. Now, the second order PDEs that are known to be related to stochastic processes are of elliptic or parabolic type (see e.g. [35], [33], [38], [42], [113]). The process that we are going to describe now, however, is related to a *hyperbolic* PDE known as the telegraph equation. A probabilistic formula for the solutions to this equation was introduced by S. Goldstein [44] and M. Kac [60]. Years later, J. Kisyński [71] has recognized the fact that a modified process introduced by Kac is a process with stationary increments, provided increments are considered in the sense of the group that we now call by his name. Let us also note that the discovery of Kac, followed by papers by R. J. Griego and R. Hersh, marked the beginning of interest in so-called random evolutions (see e.g. [96] where also an abundant bibliography is given).

Let us describe the result obtained by Kac. Let $a > 0$ and $v$ be two real numbers. The equation

$$\frac{\partial^2 y(t,\tau)}{\partial t^2} + 2a\frac{\partial y(t,\tau)}{\partial t} = v^2 \frac{\partial^2 y(t,\tau)}{\partial \tau^2} \qquad (7.55)$$

is called the **telegraph equation**. From the theory of PDEs it is known that it has exactly one solution if we require additionally that

$$y(0,\tau) = y(\tau), \qquad \text{and} \qquad \frac{\partial y(0,\tau)}{\partial t} = 0, \qquad (7.56)$$

where $y$ is a sufficiently regular function.

M. Kac has shown that the solution is given by:

$$y(t, \tau) = \frac{1}{2} E \left[ y(\tau + v\xi(t)) + y(\tau - v\xi(t)) \right], \qquad (7.57)$$

where

$$\xi(t) = \xi_a(t) = \int_0^t (-1)^{N(u)} \, du, \quad t \geq 0, \qquad (7.58)$$

and $N(t), t \geq 0$, is a Poisson process with parameter $a$. It is worth noting that this formula generalizes the well-known solution to the **wave equation**. The wave equation is the equation

$$\frac{\partial^2 y(t, \tau)}{\partial t^2} = v^2 \frac{\partial^2 y(t, \tau)}{\partial \tau^2}; \qquad (7.59)$$

and it is just a question of a change of variables to see that its solution satisfying (7.56) is (compare 7.4.42)

$$y(t, \tau) = \frac{1}{2} [y(\tau + vt) + y(\tau - vt)]. \qquad (7.60)$$

Certainly, the only difference between the telegraph equation and the wave equation is the second term on the left-hand side of (7.55); and if $a = 0$ the telegraph equation becomes the wave equation. In such a case, however, the Poisson process degenerates to a family of random variables that are all equal to 0 and $\int_0^t (-1)^{N(u)} \, du = t$ (a.s.), so that (7.57) reduces to (7.60).

**7.7.2 Exercise** From the theory of PDEs it is well-known that the unique solution to (7.55)–(7.56) is given by (see e.g. [23], [97])

$$y(t, \tau) = \left( a + \frac{1}{2} \frac{\partial}{\partial t} \right) \left( e^{-at} \int_{-t}^t I_0 \left( a\sqrt{t^2 - \sigma^2} \right) y(\tau + v\sigma) \, d\sigma \right)$$

$$= \frac{1}{2} e^{-at} [y(\tau + vt) + y(\tau - vt)]$$

$$+ \frac{a}{2} e^{-at} \int_{-t}^t I_0 \left( a\sqrt{t^2 - \sigma^2} \right) y(\tau + v\sigma) \, d\sigma$$

$$+ \frac{a}{2} e^{-at} \int_{-t}^t \frac{t}{\sqrt{t^2 - \sigma^2}} I_1 \left( a\sqrt{t^2 - \sigma^2} \right) y(\tau + v\sigma) \, d\sigma,$$

where $I_0(z) = \sum_{k=0}^\infty \frac{1}{k!k!} \left( \frac{z}{2} \right)^{2k}$ and $I_1(z) = \sum_{k=0}^\infty \frac{1}{k!(k+1)!} \left( \frac{z}{2} \right)^{2k+1}$ $(= \frac{d}{dz} I_0(z))$ are modified Bessel functions of order zero and one, respectively.

Show that

$$e^{-at} + \frac{a}{2} e^{-at} \int_{-t}^{t} I_0 \left(a\sqrt{t^2 - \sigma^2}\right) d\sigma$$

$$+ \frac{a}{2} e^{-at} \int_{-t}^{t} \frac{t}{\sqrt{t^2 - \sigma^2}} I_1 \left(a\sqrt{t^2 - \sigma^2}\right) d\sigma = 1, \qquad (7.61)$$

and conclude that, for each $t \geq 0$, there exists a random variable $X(t)$ such that $y(\tau, t) = E\, y(\tau + X(t))$. Equation (7.57) fully describes $X(t)$.

### 7.7.3 *The telegraph process*    Let us study in more detail the process

$$\xi(t) = \int_0^t (-1)^{N(u)} du, \qquad t \geq 0,$$

often called the **telegraph process** or **Poisson–Kac process**. It describes the position of a particle that at time 0 starts its movement to the right with velocity 1 and then changes the direction of the movement (but not the absolute value of its speed) at times of jumps of the Poisson process. From the physical point of view, to describe the movement completely we should know not only its position at time $t$ but also its velocity. In our case all we need is the position of the point and the direction in which it is moving. Comparing the process $\int_0^t (-1)^{N(u)} du$ with the Brownian motion also suggests the need for another coordinate. If we know that at time $t$ the Brownian motion was at a point $\tau$, we know also the distribution of the Brownian motion at $s \geq t$, specifically, we have $w(s) \sim N(\tau, t - s)$. However, this is not the case with the process $\int_0^t (-1)^{N(u)} du$; knowing that $\int_0^t (-1)^{N(u)} du = \tau$ does not determine the distribution of this process in the future $s > t$. If $(-1)^{N(t)} = 1$ this distribution is going to have a bigger mass on $[\tau, \infty)$ than in the case $(-1)^{N(t)} = -1$. Motivated by such or similar reasons, Kisyński has introduced the process

$$\mathbf{g}_t = \left(v \int_0^t (-1)^{N(u)} du, (-1)^{N(t)}\right) \qquad (7.62)$$

and the related group $\mathbb{G} = \mathbb{R} \times \{-1, 1\}$, defined in 1.2.24. Then he proved that, for any non-negative $t$ and $s$,

(i) the random vectors $\mathbf{g}_{t+s}\mathbf{g}_t^{-1}$ and $\mathbf{g}_t$ are independent, and
(ii) the random vectors $\mathbf{g}_{t+s}\mathbf{g}_t^{-1}$ and $\mathbf{g}_s$ have the same distribution,

and derived (7.57) as a consequence of this fact.

To prove (i) and (ii) we may argue as follows. First, using $(\tau, k)^{-1} = (-k\tau, k)$ we calculate $\mathbf{g}_{t+s}\mathbf{g}_t^{-1}$ to be equal to

$$(v \int_0^{t+s} (-1)^{N(u)} \, du, (-1)^{N(t+s)})((-1)^{N(t)}v \int_0^t (-1)^{N(u)} \, du, (-1)^{N(t)})$$

$$= (v \int_t^{t+s} (-1)^{N(u)-N(t)} \, du, (-1)^{N(t+s)-N(t)}).$$

Next, for $n \geq 1$ we consider $s_k = \frac{k}{2^n}s, t_k = \frac{k}{2^n}t, 0 \leq k \leq 2^n - 1$ and random variables

$$Y_n = v \sum_{k=0}^n \frac{t}{2^n}(-1)^{N(t_k)}, \quad Z_n = v \sum_{k=0}^n \frac{s}{2^n}(-1)^{N(t+s_{k+1})-N(t)}.$$

For $n, m \geq 1$, let $\mathcal{G}_n$ and $\mathcal{F}_m$ be $\sigma$-algebras generated by the random variables $N(t_{k+1})-N(t_k), 0 \leq k < 2^n$, and $N(t+s_{i+1})-N(t+s_i), 0 \leq i < 2^m$, respectively. Since these random variables are mutually independent, by 1.4.12, $\mathcal{G}_n$ is independent of $\mathcal{F}_m$. Thus, the $\pi$-system $\bigcup_{n\geq 1} \mathcal{G}_n$ is a subset of the $\lambda$-system $\mathcal{F}_m^\perp$ and the $\pi$–$\lambda$ theorem implies $\sigma\left(\bigcup_{n\geq 1} \mathcal{G}_n\right) \subset \mathcal{F}_m$, i.e. the $\sigma$-algebra $\sigma\left(\bigcup_{n\geq 1} \mathcal{G}_n\right)$ is independent of $\mathcal{F}_m$. A similar argument shows that $\sigma\left(\bigcup_{m\geq 1} \mathcal{F}_m\right)$ is independent of $\sigma\left(\bigcup_{n\geq 1} \mathcal{G}_n\right)$. On the other hand,

$$\mathbf{g}_t = \lim_{n\to\infty} \left(Y_n, (-1)^{N(t_{2^n-1})}\right)$$

$$\mathbf{g}_{t+s}\mathbf{g}_t^{-1} = \lim_{n\to\infty} \left(Z_n, (-1)^{N(t+s_{2^n-1})-N(t)}\right),$$

and the random variables $N(t_k) = \sum_{l=1}^k [N(t_l)-N(t_{l-1})]$ are $\mathcal{G}_n$ measurable and $N(t+s_i)-N(t) = \sum_{l=1}^i [N(t+s_l)-N(t+s_{l-1})]$ are $\mathcal{F}_m$ measurable. Hence, $\mathbf{g}_t$ is independent of $\mathbf{g}_{t+s}\mathbf{g}_t^{-1}$. Condition (b) is proven similarly; the main idea of the proof is that by 7.5.7, $N(t+s_{k+1}) - N(t+s_k)$, $0 \leq k < 2^{n-1}$, have the same distribution as $N(s_{k+1}) - N(s_k)$, $0 \leq k < 2^{n-1}$.

7.7.4 *The telegraph process semigroup*  Let $\mu_t$ be the distribution of $\mathbf{g}_t$; $\mu_t$ is a probability measure on $\mathbb{G}$. Writing $\mathbf{g}_{t+s} = (\mathbf{g}_{t+s}\mathbf{g}_t^{-1})\mathbf{g}_t$ and using (i) and (ii) we see that $\mu_{t+s} = \mu_t * \mu_s$ (convolution in the sense of 1.2.22, compare the analysis following (7.40)). Therefore, by (2.16), the operators

$$T_t x(h) = T_{\mu_t} x(h) = \int x(h'h)\mu_t(\,dh') = E\, x(\mathbf{g}_t h), \quad h \in \mathbb{G}, x \in BUC(\mathbb{G}),$$

form a semigroup: $T_t T_s = T_{\mu_t} T_{\mu_s} = T_{\mu_t * \mu_s} = T_{s+t}$. To see that $\{T_t, t \geq 0\}$ is strongly continuous it suffices to show that $\mu_t$ converges weakly to $\delta_{(0,1)}$, as $t \to 0+$. (Compare 7.1.12.) To this end, consider $\mu_t$ as a measure on $\mathbb{R}^2$. Then, $\mu_t(Ball^{\complement})$, where $Ball^{\complement}$ is the complement of the ball with center at $(0,1)$ and with radius $r < 1$, is the probability that $N(t)$ is odd, or $|v\xi(t)| > r$. If $t < r/v$ the latter inequality may not be satisfied, and we see that our probability equals $e^{-at} \sinh at$ which tends to 0 as $t \to 0$, proving our claim.

**7.7.5 *The generator of the telegraph semigroup***   One way to find the generator of $\{T_t, t \geq 0\}$ is to compare this semigroup with the semigroup $\{S_t, t \geq 0\}$ where

$$S_t x(h) = x((vt, 1)h) \qquad t \geq 0, h \in \mathbb{G}$$

whose generator is easy to find. Specifically, the reader should check that the domain $\mathcal{D}(B)$ of the infinitesimal generator $B$ of $\{S_t, t \geq 0\}$ is given by

$$\mathcal{D}(B) = \{x \in BUC(\mathbb{G}), x \text{ is differentiable and } x' \in BUC(\mathbb{G})\} \quad (7.63)$$

where we say that $x \in BUC(\mathbb{G})$ is differentiable iff $x_i(\tau) = x(\tau, i), i = 1, -1$, are differentiable with $x_i' \in BUC(\mathbb{R})$, and put $x'(\tau, i) = ix_i'(\tau)$. Moreover, $B$ is given by

$$Bx = vx'. \tag{7.64}$$

Now, let $C = T_{\delta_{(0,-1)}}$. Taking $h = (\tau, k) \in \mathbb{G}$, we have

$$(e^{at} T_t x)(h) = e^{at} E\, x(\mathbf{g}_t h) 1_{[N(t)=0]} + e^{at} E\, x(\mathbf{g}_t h) 1_{[N(t)=1]}$$
$$+ e^{at} E\, x(\mathbf{g}_t h) 1_{[N(t) \geq 2]}.$$

If $N(t) = 0$, which happens with probability $e^{-at}$, then $\mathbf{g}_t = (vt, 1)$. Hence, $(e^{at} T_t x)(h) = (S_t x)(h) + e^{at} E\, x(\mathbf{g}_t h) 1_{[N(t)=1]} + o(t)$ where $o(t) \leq e^{at} \|x\| \mathbb{P}\{N(t) \geq 2\} = \|x\|(e^{at} - 1 - at)$ so that $o(t)/t \to 0$ as $t \to 0$. Therefore, for any $x \in BUC(\mathbb{G})$

$$\limsup_{t \to 0} \frac{1}{t} \|e^{at} T_t x - S_t x - tCx\|$$

is the $\limsup$, as $t \to 0$, of

$$\frac{e^{at}}{t} \sup_{\tau \in \mathbb{R}, k=1,-1} \left| E\left[ x(v \int_0^t (-1)^{N(u)}\, du + \tau, -k) - x(\tau, -k) \right] 1_{[N(t)=1]} \right|$$

which equals zero by uniform continuity of $x$ and the fact that $\mathbb{P}[N(t) = 1] = te^{-at}$.

This shows that the difference between $\frac{e^{at}T_t x - x}{t}$ and $\frac{S_t x - x}{t}$ converges to $Cx$. In particular, the domains of infinitesimal generators of semigroups $\{S_t, t \geq 0\}$ and $\{e^{at}T_t, t \geq 0\}$ coincide and the generator of the latter equals $B + C$. Finally, by 7.4.18, $Ax = Bx + Cx - ax$ on the domain $\mathcal{D}(A)$ equal to $\mathcal{D}(B)$ defined in (7.63).

**7.7.6 Exercise** Prove (7.63) and (7.64).

**7.7.7 Proof of Eq. (7.57)** Take an $x \in \mathcal{D}(A)$ and define $x_k(\tau, t) = T_t x(\tau, k)$. By 7.7.5, the function $t \to x_k(\tau, t)$ is the unique solution of the system of equations:

$$\frac{\partial x_k(\tau, t)}{\partial t} = kv \frac{\partial x_k(\tau, t)}{\partial \tau} + ax(\tau, -k) - ax(\tau, k), k = 1, -1.$$

Sometimes it is convenient to write this system in a matrix form:

$$\begin{pmatrix} \frac{\partial x_1(\tau,t)}{\partial t} \\ \frac{\partial x_{-1}(\tau,t)}{\partial t} \end{pmatrix} = \begin{pmatrix} v\frac{\partial}{\partial \tau} - a & a \\ a & -v\frac{\partial}{\partial \tau} - a \end{pmatrix} \begin{pmatrix} x_1(\tau,t) \\ x_{-1}(\tau,t) \end{pmatrix}. \tag{7.65}$$

By 7.4.25, if $x$ belongs to $\mathcal{D}(A^2)$, i.e. if the components of $x$ are twice differentiable with second derivatives in $BUC(\mathbb{R})$, then the $x_k(\tau, t)$ are also twice differentiable and $t \to x_k(\tau, t)$ satisfies

$$\begin{pmatrix} \frac{\partial^2 x_1(\tau,t)}{\partial t^2} \\ \frac{\partial^2 x_{-1}(\tau,t)}{\partial t^2} \end{pmatrix} = \begin{pmatrix} v\frac{\partial}{\partial \tau} - a & a \\ a & -v\frac{\partial}{\partial \tau} - a \end{pmatrix}^2 \begin{pmatrix} x_1(\tau,t) \\ x_{-1}(\tau,t) \end{pmatrix}. \tag{7.66}$$

In particular we may take a twice differentiable $y \in BUC(\mathbb{R})$ with $y'' \in BUC(\mathbb{R})$, define $x \in BUC(\mathbb{G})$ to be (isomorphic with) the pair $(y, y)$ (compare 2.2.38) and consider $y(t, \tau) = \frac{1}{2}[x_1(\tau, t) + x_{-1}(\tau, t)]$ and $z(t, \tau) = \frac{1}{2}[x_1(\tau, t) - x_{-1}(\tau, t)]$. Then, by (7.65),

$$\frac{\partial y}{\partial t} = \frac{1}{2}(1,1) \begin{pmatrix} v\frac{\partial}{\partial \tau} - a & a \\ a & -v\frac{\partial}{\partial \tau} - a \end{pmatrix} \begin{pmatrix} x_1(\tau,t) \\ x_{-1}(\tau,t) \end{pmatrix} = v\frac{\partial z}{\partial \tau};$$

in particular $\frac{\partial}{\partial t} y(0, \tau) = \frac{\partial z}{\partial \tau} z(0, \tau) = 0$. Furthermore, by (7.66),

$$\frac{\partial^2 y}{\partial t^2} = \frac{1}{2}(1,1) \begin{pmatrix} (v\frac{\partial}{\partial \tau} - a)^2 + a^2 & -2a^2 \\ -2a^2 & (v\frac{\partial}{\partial \tau} + a)^2 + a^2 \end{pmatrix} \begin{pmatrix} x_1(\tau,t) \\ x_{-1}(\tau,t) \end{pmatrix}$$

$$= \frac{1}{2} \left( v^2 \frac{\partial^2}{\partial \tau^2} - 2av\frac{\partial}{\partial \tau}, v^2 \frac{\partial^2}{\partial \tau^2} + 2av\frac{\partial}{\partial \tau} \right) \begin{pmatrix} x_1(\tau,t) \\ x_{-1}(\tau,t) \end{pmatrix}$$

$$= v^2 \frac{\partial^2 y}{\partial \tau^2} - 2av\frac{\partial z}{\partial \tau}$$

$$= v^2 \frac{\partial^2 y}{\partial \tau^2} - 2a\frac{\partial y}{\partial t},$$

i.e. $y$ satisfies the telegraph equation with $y(0,\tau) = y(\tau)$, and $\frac{\partial y}{\partial t}(0,\tau) = 0$. Finally, to obtain (7.57) it suffices to note that

$$x_1(\tau,t) = E\, y(\tau + v\xi(t))1_{\{N(t) \text{ is even}\}} + E\, y(\tau - v\xi(t))1_{\{N(t) \text{ is odd}\}},$$
$$x_{-1}(\tau,t) = E\, y(\tau + v\xi(t))1_{\{N(t) \text{ is odd}\}} + E\, y(\tau - v\xi(t))1_{\{N(t) \text{ is even}\}}.$$
$$(7.67)$$

**7.7.8 Exercise** From the theory of PDEs it is known that the unique solution to (7.55) with initial conditions $y(0,\tau) = 0$ and $\frac{\partial y}{\partial t}(0,\tau) = z(\tau)$ is given by (see e.g. [23], [97])

$$y(t,\tau) = \frac{1}{2}e^{-at}\int_{-t}^{t} I_0\left(a\sqrt{t^2 - \sigma^2}\right)z(\tau + v\sigma)\,d\sigma.$$

Conclude that, for each $t \geq 0$, there exists a random variable $Y(t)$ such that $y(t,\tau) = \frac{1}{2}E\, z(\tau + vY(t))$ – compare (7.68), below.

**7.7.9 Exercise** (a) Repeat the analysis from subsection 7.7.7 with $x \in BUC(\mathbb{G})$ equal to the pair $(y,-y)$ where $y \in BUC(\mathbb{R})$ is twice differentiable with $y'' \in BUC(\mathbb{R})$, to see that $y(t,\tau)$ defined there is a solution of the telegraph equation with initial conditions $y(0,\tau) = 0$ and $\frac{\partial y}{\partial t}(0,\tau) = vy'(\tau)$. (b) Use (a) to show that

$$y(t,\tau) = \frac{1}{2}E\, y(\tau + v\xi(t)) + \frac{1}{2}E\, y(\tau - v\xi(t)) + \frac{1}{2}E\int_{-\xi(t)}^{\xi(t)} z(\tau + v\sigma)\,d\sigma$$
$$(7.68)$$

solves the telegraph equation with initial conditions $y(0,\tau) = y(\tau)$ and $\frac{\partial y}{\partial t}(0,\tau) = z(\tau)$ where $y \in BUC(\mathbb{R})$ is twice differentiable with $y'' \in BUC(\mathbb{R})$ and $z \in BUC(\mathbb{R})$ is differentiable with $z' \in BUC(\mathbb{R})$ (see also (8.64))

Relation (7.68), due to Kisyński [71], is a counterpart of the classic d'Alembert's formula for the solution of the wave equation:

$$u(t,\tau) = \frac{1}{2}[y(\tau + vt) + y(\tau - vt)] + \frac{1}{2}\int_{-t}^{t} z(\tau + v\sigma)\,d\sigma,$$

where $y$ and $z$ are the initial conditions stated above (compare Exercise 7.4.42).

## 7.8 Convolution semigroups of measures on semigroups

In this subsection we present a short discussion of convolution semigroups on two important classes of topological semigroups: the (one-dimensional) Lie groups and the discrete semigroups. We start by presenting an example.

### 7.8.1 Convolution semigroup of measures on $\mathbb{R}_*^+$

What is the general form of a convolution semigroup on the group $\mathbb{R}_*^+$ with multiplication as a group product? It is quite easy to answer this question once we note that this group is isomorphic to $\mathbb{R}$ with addition (via the map $\mathbb{R}_*^+ \ni u \mapsto \ln u \in \mathbb{R}$). This implies that $X_t, t \geq 0$, is a Lévy process on $\mathbb{R}_*^+$ iff $\ln X_t, t \geq 0$, is a Lévy process on $\mathbb{R}$.

In what follows we will find the form of the infinitesimal generator of a convolution semigroup on $\mathbb{R}_*^+$. Before we do this, let us note that the isomorphism described above preserves "differential structure" of $\mathbb{R}_*^+$. To be more specific, a function $x$ on $\mathbb{R}_*^+$ is differentiable iff $z = x \circ \exp$ is differentiable on $\mathbb{R}$ and we have $z'(\tau) = x'(e^\tau)e^\tau$ or, which is the same $x'(u) = z'(\ln u)\frac{1}{u}$. A similar statement is true for higher derivatives. This will be important in what follows.

Let $\{\mu_t, t \geq 0\}$ be a convolution semigroup on $\mathbb{R}_*^+$. In other words, we assume that $\mu_t * \mu_s = \mu_{t+s}$ where $*$ denotes convolution on $\mathbb{R}_*^+$ as defined in 1.2.22. Also, $\mu_0 = \delta_1$ (the Dirac delta at 1), and $\lim_{t \to 0+} \mu_t = \mu_0$ weakly – note that 1 is the neutral element of $\mathbb{R}_*^+$ and as such it plays the same rôle as 0 does in $\mathbb{R}$.

For $t \geq 0$, let $\nu_t$ be the transport of $\mu_t$ via exp, i.e. $\int_{\mathbb{R}} z \, d\nu_t = \int_{\mathbb{R}_*^+} z \circ \ln d\mu_t$, $x \in BM(\mathbb{R})$. A direct calculation shows that

$$\int_{\mathbb{R}} z \, \mathrm{d}(\nu_t * \nu_s) = \int_{\mathbb{R}_*^+} z \circ \ln \mathrm{d}(\mu_t * \mu_s) = \int_{\mathbb{R}_*^+} z \circ \ln \mathrm{d}\mu_{t+s} = \int_{\mathbb{R}} z \, \mathrm{d}\nu_{t+s}$$

(the first convolution in $\mathbb{BM}(\mathbb{R}_*^+)$, the second in $\mathbb{BM}(\mathbb{R})$), i.e. that $\nu_t * \nu_s = \nu_{t+s}$. Also, $\nu_0 = \delta_0$ and, by 5.6.2, $\lim_{t \to 0+} \nu_t = \nu_0$. Hence, $\{\nu_t, t \geq 0\}$ is a convolution semigroup on $\mathbb{R}$. Moreover, the corresponding semigroup $\{S_t, t \geq 0\}$ of operators $S_t = T_{\nu_t}$ in $BUC(\mathbb{R})$ is (isometrically) isomorphic to the semigroup $\{T_t, t \geq 0\}$, $T_t = T_{\mu_t}$ on $BUC(\mathbb{R}_*^+)$. The isomorphism $J : BUC(\mathbb{R}_*^+) \to BUC(\mathbb{R})$ is given by $Jx = x \circ \ln$. But $J$ is more than an isomorphism of two Banach spaces – it preserves the differential structure of $\mathbb{R}_*^+$.

Let $A$ be the infinitesimal generator of $\{S_t, t \geq 0\}$ and $B$ be the infinitesimal generator of $\{T_t, t \geq 0\}$. If $x \in BUC(\mathbb{R}_*^+)$ is twice differentiable with $x'' \in BUC(\mathbb{R}_*^+)$, then $Jx$ is twice differentiable and $(Jx)'' \in BUC(\mathbb{R})$. By 7.6.14, $Jx$ belongs to $\mathcal{D}(A)$ and $AJx$ is given by (7.53) with $x$ replaced by $Jx$. By 7.4.22, $x$ belongs to $\mathcal{D}(B)$ and

$Bx = J^{-1}AJx$. Using $J^{-1}(Jx)'(u) = ux'(u)$ we obtain

$$Bx(u) = aux'(u) + \int_{\mathbb{R}} [x(e^\tau u) - x(u) - ux'(u)y(\tau)] \frac{\tau^2 + 1}{\tau^2} \, m(\mathrm{d}\tau)$$

$$= aux'(u)$$
$$+ \int_{\mathbb{R}_*^+} [x(vu) - x(u) - ux'(u)y_\diamond(v)] \frac{(\ln v)^2 + 1}{(\ln v)^2} \, m_\diamond(\mathrm{d}v),$$

where $y_\diamond = y \circ \ln$ and $m_\diamond$ is the transport of $m$ via $\exp$.

There are striking similarities between (7.53) and this formula, especially when written as

$$Bx(u) = aDx(u) \tag{7.69}$$
$$+ \int_{\mathbb{R}_*^+} [x(vu) - x(u) - Dx(u)y_\diamond(v)] \frac{(\ln v)^2 + 1}{(\ln v)^2} \, m_\diamond(\mathrm{d}v),$$

where $Dx = J^{-1}(Jx)'$ so that $Dx(u) = ux'(u)$. Note that the function $\frac{(\ln v)^2 + 1}{(\ln v)^2}$ has singularity at 1, but as $v \to 1$, the expression in the brackets tends to $D^2x(1)$. Also, $y_\diamond$ inherits properties of $y$: $Dy_\diamond = J^{-1}y'$ so that $Dy_\diamond(u) = y'_{|\tau=\ln u}$; similarly $D^2y_\diamond(u) = y''_{|\tau=\ln u}$. In particular, $y_\diamond(1) = 0$ and $Dy_\diamond(1) = 1$.

Relations (7.53) and (7.69) are particular cases of Hunt's Theorem, as explained in the next subsection.

### 7.8.2 Convolution semigroups of measures on a one-dimensional Lie group

The Kisyński group and the group from 7.8.1 are examples of a one-dimensional **Lie group**. Roughly speaking, a one-dimensional Lie group is a topological group that is locally isomorphic to $\mathbb{R}$ and the involved isomorphism preserves "a (local) differential structure".

On any one-dimensional Lie group $\mathbb{G}$ there exists exactly one (up to a constant) natural way to define a derivative $Dx$ of a "smooth" function $x$. In the case where $\mathbb{G} = \mathbb{R}$, $Dx = x'$, if $\mathbb{G} = \mathbb{R}_*^+$, $Dx(u) = ux'(u)$, and if $\mathbb{G}$ is the Kisyński group $D(x_1, x_2) = (x_1', -x_2')$ – see 2.2.38, compare 7.7.5. Moreover, there exists a function $z$ on $\mathbb{G}$, called the **Hunt function**, that near the neutral element $e$ of $\mathbb{G}$ mimics the properties of $z(\tau) = \tau^2$ defined on $\mathbb{R}$. In particular, $z(e) = Dz(e) = 0$ and $D^2z(e) = 2$. Also, there is a $y$ on $\mathbb{G}$ that plays the rôle of $y$ appearing in (7.53) in that $y(e) = 0$ and $Dy(e) = 1$. Finally, for every convolution semigroup $\{\mu_t, t \geq 0\}$ the set of bounded uniformly continuous twice differentiable functions with bounded uniformly continuous second derivative is contained in the domain of the generating functional $Fx = \lim_{t \to 0+} \frac{1}{t}(\int_{\mathbb{G}} x \, \mathrm{d}\mu_t - x(e))$, and there exist a constant $a \in \mathbb{R}$ and a measure $m$ on $\mathbb{G}$, possibly not finite but such that $\int z \, dm$ is finite, such that

$$Fx = aDx(e) + \int_{\mathbb{G}} [x(g) - x(e) - Dx(e)y(g)] \, m(\,dg).$$

A similar, but more complicated, formula is true for $n$-dimensional Lie groups. The proof of this theorem, due to **Hunt**, may be found in [55].

**7.8.3 Example** *Two-dimensional Dirac equation* Let $\mathbb{H} = \mathbb{R} \times \mathbb{Z}$ be the non-commutative group with multiplication rule $(\tau, k)(\sigma, l) = (\tau(-1)^l + \sigma, k + l)$. ($\mathbb{H}$ is isomorphic to the subgroup $\mathbb{H}_1 = \{(\tau, k, l)| k = (-1)^l\}$ of the direct product of the Kisyński group $\mathbb{G}$ and the group $\mathbb{Z}$ of integers.) We note that the natural derivative $D$ on $\mathbb{H}$ is given by $Dx(\tau, k) = (-1)^k \frac{\partial}{\partial \tau} x(\tau, k)$; $D$ is defined on the subset $BUC_1(\mathbb{H})$ of $BUC(\mathbb{H})$ of functions $x$ such that $\tau \mapsto x(\tau, k)$ is differentiable for all $k \in \mathbb{Z}$, and $y(\tau, k) = Dx(\tau, k) = (-1)^k \frac{\partial}{\partial \tau} x(\tau, k)$ belongs to $BUC(\mathbb{H})$.

The process $\mathbf{g}_t = (\xi(t), N(t))$, where $N(t), t \geq 0$ is a Poisson process and $\xi(t) = \int_0^t (-1)^{N(s)} \, ds$, has independent, identically distributed increments in the group $\mathbb{H}$ in the sense that relations (i) and (ii) of 7.7.3 hold.

We want to find the generator $A$ of the corresponding semigroup $\{T_t, t \geq 0\}$,

$$T_t x(h) = E \, x(\mathbf{g}_t h), \qquad h \in \mathbb{H}, x \in BUC(\mathbb{H}). \tag{7.70}$$

To this end we compare it with the semigroup $\{S_t, t \geq 0\}$, $S_t x(\tau, k) = e^{-t} x((t, 0)(\tau, k)) = e^{-t} x(\tau + (-1)^k t, k)$. We note that the domain of the infinitesimal generator $B$ of $\{S_t, t \geq 0\}$ equals $BUC_1(\mathbb{H})$, and $Bx = Dx - x$ where $D$ is the natural derivative on $\mathbb{H}$. Moreover, when $N(t) = 0$, which happens with probability $e^{-t}$, $\mathbf{g}_t = (t, 0)$ and so $E \, 1_{\{N(t)=0\}} x(\mathbf{g}_t h) = S_t x(h)$. Similarly, $N(t) = 1$ with probability $te^{-t}$ and then $\mathbf{g}_t = (\xi(t), 1)$. Since $|\xi(t)| \leq t$, we have

$$\sup_{h \in \mathbb{H}} \left| \frac{1}{t} E \, 1_{\{N(t)=1\}} x(\mathbf{g}_t h) - e^{-t} x((0, 1)h) \right|$$
$$\leq e^{-t} \sup_{\sigma, \tau \in \mathbb{R}, k \in \mathbb{Z}, |\sigma - \tau| \leq t} |x(\tau, k) - x(\sigma, k)| \xrightarrow[t \to 0]{} 0.$$

Therefore,

$$\frac{1}{t}(T_t - S_t)x(h) = \frac{1}{t} E 1_{\{N(t)=1\}} x(\mathbf{g}_t h) + \frac{1}{t} E 1_{\{N(t) \geq 2\}} x(\mathbf{g}_t h)$$

tends to $x\left((0,1)h\right)$ (in the norm of $BUC(\mathbb{H})$), because

$$\sup_{h\in\mathbb{H}}\left|\frac{1}{t}E1_{\{N(t)\geq 2\}}x(\mathbf{g}_t h)\right| \leq \|x\|_{BUC(\mathbb{H})}\frac{1}{t}\mathbb{P}\{N(t)\geq 2\} \xrightarrow[t\to 0]{} 0.$$

This shows that $\mathbb{D}(A) = \mathbb{D}(B)$ and $Ax = Bx + Cx = Dx + Cx - x$ where $Cx(h) = x\left((0,1)h\right)$, i.e. $Cx(\tau,k) = x(\tau,k+1)$ ($C$ is the operator related to the Dirac measure at the point $(0,1)$). Defining $U_t = e^t T_t$ we obtain that, for $x \in BUC(\mathbb{H})$,

$$\frac{dU_t x}{dt} = (D+C)U_t x \quad \text{and} \quad U_0 x = x. \tag{7.71}$$

Our findings may be applied to give a probabilistic formula for the solution of the **two-dimensional Dirac equation**:

$$\frac{\partial u(t,\tau)}{\partial t} = \frac{\partial u(t,\tau)}{\partial \tau} + zv, \qquad u(0,\tau) = u(\tau),$$

$$\frac{\partial v(t,\tau)}{\partial t} = -\frac{\partial v(t,\tau)}{\partial \tau} + zu, \qquad v(0,\tau) = v(\tau), \tag{7.72}$$

where $z$ is a complex number with modulus 1, and $u$ and $v$ are differentiable members of $BUC(\mathbb{H})$ with derivative in this space. Indeed, we may set $x(\tau,k) = z^k u(\tau)$ for even $k$ and $x(\tau,k) = z^k v(\tau)$ otherwise. Then $x$ belongs to (complex) $BUC_1(\mathbb{H})$ and (7.71) is satisfied. In particular

$$\frac{dU_t x(\tau,0)}{dt} = (D+C)U_t x(\tau,0),$$

$$\frac{dU_t x(\tau,1)}{dt} = (D+C)U_t x(\tau,1).$$

Defining $u(t,\tau) = U_t x(\tau,0)$ and $v(t,\tau) = \frac{1}{z}U_t x(\tau,1)$, since

$$(DU_t x)(\tau,0) = \frac{\partial}{\partial \tau}U_t x(\tau,0),$$

$$(CU_t x)(\tau,0) = U_t x\left((0,1)(\tau,0)\right) = U_t x(\tau,1) = zv(t,\tau),$$

$$(DU_t x)(\tau,1) = -\frac{\partial}{\partial \tau}U_t x(\tau,1),$$

$$(CU_t x)(\tau,1) = U_t x\left((0,1)(\tau,1)\right) = U_t x(\tau,2) = z^2 U_t x(\tau,0),$$

we see that $u(t,\tau)$ and $v(t,\tau)$ solve the Dirac equation (7.72).

By (7.70), using matrix notation we may write the solution to the Dirac equation as

$$\begin{pmatrix} u(t,\tau) \\ v(t,\tau) \end{pmatrix} = Ee^t z^{N(t)} \begin{pmatrix} 0,1 \\ 1,0 \end{pmatrix}^{N(t)} \begin{pmatrix} u(\tau+\xi(t)) \\ v(\tau-\xi(t)) \end{pmatrix}.$$

### 7.8.4 Continuous convolution semigroups of measures on the Klein group

What is the form of a convolution semigroup $\{\mu_t, t \geq 0\}$ of probability measures on the Klein group $\mathbb{G}$? It may be difficult to answer this question directly. However, we may be helped by the fact that the Gelfand transform (6.23) establishes an isometric isomorphism between the algebra $L^1(\mathbb{G})$ and the algebra $\mathbb{R}^4$ with coordinatewise multiplication. Indeed, it is easy to find all convolution semigroups, say $\{\nu_t, t \geq 0\}$, on $\mathbb{R}^4$. First of all, coordinates of $\nu_t$ must be non-negative, and we check that logarithms of coordinates satisfy the Cauchy equation. Hence, $\nu_t$ must be of the form $(r_1^t, r_2^t, r_3^t, r_4^t)$ for some non-negative $r_i$. Now, if $\mu_t$ is a convolution semigroup of probability measures on $\mathbb{G}$, then the Gelfand transform $G\mu_t$ is a convolution semigroup on $\mathbb{R}^4$, and by the definition of $G$, the first coordinate of $G\mu_t$ is 1 for all $t \geq 0$. Hence $G\mu_t = (1, r_2^t, r_3^t, r_4^t)$ for some non-negative $r_i, i = 2, 3, 4$, and $\mu_t = \frac{1}{4}G(1, r_2^t, r_3^t, r_4^t)^\mathrm{T}$; the $r_i$ must be chosen in such a way that $\mu_t$ has positive coordinates (we will see how to do that shortly).

To continue our analysis, note that by (1.12) (see (2.17)), the measure $\mu_t = (a_1(t), a_2(t), a_3(t), a_4(t)) \in L^1(\mathbb{G})$ may be represented as the matrix

$$
A(t) = \begin{pmatrix} a_1(t) & a_2(t) & a_3(t) & a_4(t) \\ a_2(t) & a_1(t) & a_4(t) & a_3(t) \\ a_3(t) & a_4(t) & a_1(t) & a_2(t) \\ a_4(t) & a_3(t) & a_2(t) & a_1(t) \end{pmatrix},
$$

i.e. as an operator in $L^1(\mathbb{G})$. Since $\{\mu_t, t \geq 0\}$ is a convolution semigroup, the matrices $A(t)$ satisfy $A(t)A(s) = A(t+s)$ (the left-hand side is the product of matrices). Moreover, as $t \to 0$, matrices $A(t)$ converge to the identity matrix coordinatewise, and thus uniformly with respect to all coordinates. By 7.4.20, $A(t) = e^{tB}$ where $B = \lim_{t\to 0} \frac{A(t)-I}{t} = A'(0)$. Now, the derivative of $\frac{1}{4}G(1, r_2^t, r_3^t, r_4^t)^\mathrm{T}$ at $t = 0$ equals

$$
\frac{1}{4}G(0, \ln r_2, \ln r_3, \ln r_4)^\mathrm{T} = (\alpha_1, \alpha_2, \alpha_3, \alpha_4)^\mathrm{T},
$$

where $\alpha_1 = \ln r_2 r_3 r_4$, and

$$
\alpha_2 = \ln r_4 (r_2 r_3)^{-1}, \alpha_3 = \ln r_3 (r_2 r_4)^{-1}, \alpha_4 = \ln r_2 (r_3 r_4)^{-1},
$$

or

$$
r_2 = e^{-\frac{1}{2}(\alpha_2 + \alpha_3)}, r_3 = e^{-\frac{1}{2}(\alpha_2 + \alpha_4)}, r_4 = e^{-\frac{1}{2}(\alpha_3 + \alpha_4)}.
$$

Therefore,

$$B = \frac{1}{4} \begin{pmatrix} \alpha_1 & \alpha_2 & \alpha_3 & \alpha_4 \\ \alpha_2 & \alpha_1 & \alpha_4 & \alpha_3 \\ \alpha_3 & \alpha_4 & \alpha_1 & \alpha_2 \\ \alpha_4 & \alpha_3 & \alpha_2 & \alpha_1 \end{pmatrix}.$$

We have $\sum_{i=1}^{4} \alpha_i = 0$. Writing $e^{Bt} = e^{\alpha_1 t} e^{(B-\alpha_1)t}$, we see that the vector $\frac{1}{4}G(1, r_2^t, r_3^t, r_4^t)$ has non-negative coordinates if $\alpha_i, i \geq 2$, are non-negative.

There are striking similarities between the form of $B$ and the matrix in (2.17). Writing

$$B = a \begin{pmatrix} 0 & \beta_2 & \beta_3 & \beta_4 \\ \beta_2 & 0 & \beta_4 & \beta_3 \\ \beta_3 & \beta_4 & 0 & \beta_2 \\ \beta_4 & \beta_3 & \beta_2 & 0 \end{pmatrix} - a \begin{pmatrix} 1 & 0 & 0 & 0 \\ 0 & 1 & 0 & 0 \\ 0 & 0 & 1 & 0 \\ 0 & 0 & 0 & 1 \end{pmatrix},$$

where $a = -\alpha_1, \beta_i = \frac{1}{4a}\alpha_i, i \geq 2$, we see that the value of the operator $B$ on a measure, say $(a_1, a_2, a_3, a_4) \in \mathbb{BM}(\mathbb{G})$ is $a$ times the convolution of probability measures $(0, \beta_2, \beta_3, \beta_4)$ and $(a_1, a_2, a_3, a_4)$ minus $a(a_1, a_2, a_3, a_4)$. Hence $\mu(t)$ is the exponent of $a(0, \beta_2, \beta_3, \beta_4) - a(1, 0, 0, 0)$ in our convolution algebra. This is surprisingly "clean" and elegant result, especially when obtained after such a long calculation. It suggests that these were special properties of the Klein group that blurred our analysis and that we should look for more general principles to obtain a more general result. This suggestion is strengthen by Theorem 3, p. 290 in [40]. Before continuing, the reader may wish to try to guess and prove the general form of a convolution semigroup on the group of congruences modulo $p$.

### 7.8.5 *Continuous convolution semigroups of measures on a discrete semigroup*

Let $\mathbb{G}$ be a discrete semigroup with identity element. All measures on $\mathbb{G}$ are members of the algebra $l^1(\mathbb{G})$ of measures that are absolutely continuous with respect to the counting measure. If $\{\mu(t), t \geq 0\}$ is a convolution semigroup of probability measures in $\mathbb{G}$, then there exists a probability measure $x \in l^1(\mathbb{G})$, and a positive number $a$ such that $\mu(t) = \exp at(x - \delta)$ where $\delta$ is the Dirac measure at the identity of the semigroup.

*Proof* For simplicity of notation we assume that $\mathbb{G}$ has infinitely many elements; the reader will easily make minor modifications needed for the case where the number of elements in $\mathbb{G}$ is finite. By 5.8.6, the assumption that $\mu(t)$ converges in weak* topology to $\delta$ implies that it converges in the strong topology. By 7.4.20, there exists an element $y$ of $l^1(\mathbb{G})$ such that $\mu(t) = \exp\{yt\}$. The Banach space $l^1(\mathbb{G})$ is isometrically isomorphic to $l^1(\mathbb{N}_0)$ and a measure $\mu(t)$ may be identified with a sequence $(m_i(t))_{i \geq 0}$ where we may arrange $m_0(t)$ to be the mass of $\mu(t)$ at the identity element of the semigroup. Since $y = \lim_{t \to 0} \frac{\mu(t) - \delta}{t}$, the coordinates of $y$ are given by $\eta_0 = \lim_{t \to 0} \frac{m_0(t) - 1}{t}$, and $\eta_i = \lim_{t \to 0} \frac{m_i(t)}{t}, i \geq 1$, and all except $\eta_0$ are non-negative. Since $y \in l^1(\mathbb{G})$, $\sum_{i=1}^{\infty} \eta_i < \infty$ and except for the trivial case this sum is not equal to zero. Moreover, $\sum_{i=0}^{\infty} \eta_i = 0$, because $\mu(t)$ are probability measures. Hence one may take $a = \sum_{i=1}^{\infty} \eta_i = -\eta_0$ and $x = \frac{1}{a}y + \delta$. $\qquad\square$

**7.8.6 Remark** Calculations related to the Klein group presented in subsection 7.8.4 have this advantage that they lead to an explicit formula for transition probability matrices, while in 7.8.5 only the form of the generator is given. From 7.4.21 we know that one is sometimes able to obtain the transition matrices from the generating matrix using diagonalization procedure (see also 8.4.31). A closer look at our calculations reveals that in the case of Klein group, the Gelfand transform is exactly this procedure. This is a simple case of a more general result – as explained in [37] Section 2.2.1, the abstract Fourier transform is a way of decomposing a Banach space, like $BUC(\mathbb{G})$, into a possibly infinite number of subspaces that are left invariant by translation operators.

# 8

# Markov processes and semigroups of operators

This chapter provides more advanced theory of semigroup of operators needed for the treatment of Markov processes. The main theorem is the Hille–Yosida theorem. Also, we establish perturbation and approximation results; the Central Limit Theorem, theorems on approximation of Brownian motion and the Ornstein–Uhlenbeck process by random walks, the Feynman–Kac formula and Kingman's coalescence are obtained as corollaries of these results.

## 8.1 Semigroups of operators related to Markov processes

We start with an example of a semigroup related to a process that is not a Lévy process.

8.1.1 *A semigroup of operators related to reflected Brownian motion*
Consider again the semigroup $\{T_t, t \geq 0\}$ from 7.5.1, acting in $BUC(\mathbb{R})$. Let $BUC_e(\mathbb{R})$ be the subspace of $BUC(\mathbb{R})$ composed of even functions. Note that $T_t$ leaves this space invariant. Indeed, the distribution of $w(t)$ is the same as that of $-w(t)$ so that if $x \in BUC_e(\mathbb{R})$, then

$$T_t x(-\tau) = E\, x(-\tau + w(t)) = E\, x(-\tau - w(t)) = E\, x(\tau + w(t)) = T_t x(\tau).$$

By Exercise 7.4.23, the domain $\mathcal{D}(B)$ of the generator of the semigroup $\{S_t, t \geq 0\}$ of restrictions of $T_t$ to $BUC_e(\mathbb{R})$ is given by

$$\mathcal{D}(B) = \{x \in BUC_e(\mathbb{R}); x \text{ is twice differentiable with } x'' \in BUC_e(\mathbb{R})\}.$$

In particular, $x'(0) = 0$ for all $x \in \mathcal{D}(B)$.

Next, $BUC_e(\mathbb{R})$ is isometrically isomorphic to the space $BUC(\mathbb{R}^+)$ of bounded uniformly continuous functions on $\mathbb{R}^+$. The isomorphism is given by $Jx(\tau) = x(\tau), \tau \geq 0, x \in BUC_e(\mathbb{R})$. Note that $J^{-1}y(\tau) =$

$y(|\tau|), \tau \in \mathbb{R}, y \in BUC(\mathbb{R}^+)$. We claim that the domain $\mathcal{D}(C)$ of the infinitesimal generator of $T_t^{\mathrm{r}} = JS_t J^{-1}$ equals

$$\{y \in BUC(\mathbb{R}^+); y'' \text{ exists and belongs to } BUC(\mathbb{R}^+) \text{ and } y'(0) = 0\}. \tag{8.1}$$

(At $\tau = 0$ the derivative is the right-hand derivative.) To this end note first that, by 7.4.22, $\mathcal{D}(C)$ is the image of $\mathcal{D}(B)$ via $J$, so that if $y$ belongs to $\mathcal{D}(C)$, then it must be twice differentiable with $y'' \in BUC(\mathbb{R}^+)$ and $y'(t) = 0$, being the restriction of an $x \in BUC_{\mathrm{e}}(\mathbb{R})$. On the other hand, if $y$ is twice differentiable with $y'' \in BUC_{\mathrm{e}}(\mathbb{R})$, and $y'(0) = 0$, then the function $x$ defined as $x(\tau) = y(|\tau|)$ is certainly twice differentiable in $\mathbb{R} \setminus \{0\}$, and

$$x'(\tau) = \frac{\tau}{|\tau|} y'(|\tau|), \quad x''(\tau) = y(|\tau|), \quad \tau \neq 0.$$

Also, the right-hand derivative of $x$ at zero equals $\lim_{\tau \to 0^+} \frac{x(\tau) - x(0)}{\tau} = \lim_{\tau \to 0^+} \frac{y(\tau) - y(0)}{\tau} = y'(0) = 0$; and, analogously, the left-hand derivative equals $-y'(0) = 0$. Thus, $x'(0)$ exists and equals 0. Finally,

$$\lim_{\tau \to 0} \frac{x'(\tau) - x'(0)}{\tau} = \lim_{\tau \to 0} \frac{y'(|\tau|)}{|\tau|} = \lim_{\sigma \to 0^+} \frac{y(\sigma)}{\sigma} = y''(0).$$

So, $x$ belongs to $\mathcal{D}(B)$ and $y = Jx$, which proves that $\mathcal{D}(C)$ is the set described in (8.1).

As a result, for any $y \in \mathcal{D}(C)$, the function $y(t, \tau) = T_t^{\mathrm{r}} y(\tau)$ is the unique solution to the following Cauchy problem:

$$\frac{\partial y(\tau, t)}{\partial t} = \frac{1}{2} \frac{\partial^2 y(\tau, t)}{\partial \tau^2}, \quad \tau > 0, \quad y'(0, t) = 0, \quad y(\tau, 0) = y(\tau). \tag{8.2}$$

We may write an explicit formula for $T_t^{\mathrm{r}}$ :

$$\begin{aligned} T_t^{\mathrm{r}} y(\tau) &= JS_t J^{-1} y(\tau) = S_t J^{-1} y(\tau) = E\, J^{-1} y(\tau + w(t)) \\ &= E\, y\left(|\tau + w(t)|\right). \end{aligned} \tag{8.3}$$

We may also write:

$$\begin{aligned} T_t^{\mathrm{r}} y(\tau) &= E\, 1_{\tau + w(t) \geq 0} y[\tau + w(t)] + E\, 1_{\tau + w(t) < 0} y[-(\tau + w(t))] \\ &= \frac{1}{\sqrt{2\pi t}} \int_0^\infty y(\sigma) e^{-\frac{(\sigma - \tau)^2}{2t}} \, d\sigma + \frac{1}{\sqrt{2\pi t}} \int_{-\infty}^0 y(-\sigma) e^{-\frac{(-(\sigma - \tau))^2}{2t}} \, d\sigma \\ &= \frac{1}{\sqrt{2\pi t}} \int_0^\infty y(\sigma) e^{-\frac{(\sigma - \tau)^2}{2t}} \, d\sigma + \frac{1}{\sqrt{2\pi t}} \int_0^\infty y(\sigma) e^{-\frac{(\sigma + \tau)^2}{2t}} \, d\sigma \\ &= \int_0^\infty y(\sigma) k(t, \tau, \sigma) \, d\sigma, \end{aligned} \tag{8.4}$$

where

$$k(t, \tau, \sigma) = \frac{1}{\sqrt{2\pi t}} \left[ e^{-\frac{(\sigma - \tau)^2}{2t}} + e^{-\frac{(\sigma + \tau)^2}{2t}} \right]. \qquad (8.5)$$

Equivalently,

$$T_t^{\mathrm{r}} y(\tau) = \int_0^\infty y(\sigma) K(t, \tau, \mathrm{d}\sigma) \qquad (8.6)$$

where for fixed $t$ and $\tau$, $K(t, \tau, \cdot)$ is a measure on $\mathbb{R}^+$ with density $k(t, \tau, \sigma)$. In other words, $k(t, \sigma, \tau)$ is a density of the random variable $|\tau + w(t)|$, and $K(t, \tau, \cdot)$ is its distribution.

We note that

$$(T_t^{\mathrm{r}} T_s^{\mathrm{r}} x)(\tau) = \int_{\mathbb{R}^+} (T_s^{\mathrm{r}} x)(\sigma) K(t, \tau, \mathrm{d}\sigma)$$

$$= \int_{\mathbb{R}^+} \int_{\mathbb{R}^+} x(\varsigma) K(s, \sigma, \mathrm{d}\varsigma) K(t, \tau, \mathrm{d}\sigma). \qquad (8.7)$$

On the other hand, by the semigroup property this equals

$$(T_{t+s}^{\mathrm{r}} x)(\tau) = \int_{\mathbb{R}^+} x(\varsigma) K(t + s, \tau, \mathrm{d}\varsigma).$$

Applying 1.2.20 to the measures $\int_{\mathbb{R}^+} K(s, \sigma, \cdot) K(t, \tau, \mathrm{d}\sigma)$ and $K(t + s, \tau, \cdot)$ we obtain the Chapman–Kolmogorov equation:

$$\int_{\mathbb{R}^+} K(s, \sigma, B) K(t, \tau, \mathrm{d}\sigma) = K(t + s, \tau, B). \qquad (8.8)$$

**8.1.2 Exercise**  Let $t > 0$. Show that there is no measure $\mu$ such that $T_t^{\mathrm{r}} = T_\mu$ in the sense of 2.3.17. (Look for the hint there!)

**8.1.3 Exercise**  Let $BUC_{\mathrm{n}}(\mathbb{R})$ be the space of odd functions in $BUC(\mathbb{R})$ ('n' stands for "not even"). Show that the Brownian motion semigroup leaves this subspace invariant.

The process $w_{\mathrm{r}}(t) = |\tau + w(t)|$ is termed **reflected Brownian motion** starting at $\tau \geq 0$. More generally, if $w(t), t \geq 0$, is a Brownian motion and $X$ is a non-negative independent random variable, then $w_{\mathrm{r}}(t) = |X + w(t)|$ is a reflected Brownian motion with initial distribution $\mathbb{P}_X$. Though it has continuous paths, reflected Brownian motion is not a Lévy process, as it is not space homogeneous (see Exercise 8.1.2, above). It belongs to a more general class of *Markov processes*. Similarly, the Chapman–Kolmogorov equation (8.8) is a more general principle than (7.46) – see 8.1.12.

**8.1.4 Definition**    A process $X_t, t \geq 0$ is said to be a **Markov process** if for every $t \geq 0$, the $\sigma$-algebra $\sigma\{X_s, s \geq t\}$ depends on $\mathcal{F}_t = \sigma\{X_s, s \leq t\}$ only through $\sigma(X_t)$ – see 3.3.20. This is often expressed shortly by saying that the future (behavior of the process) depends on the past only through the present. Equivalently, by 3.3.20 we say that the past depends on the future only through the present, or that given the present, the future and the past are independent.

**8.1.5 Remark**    The definition of a Markov process remains the same when we treat the case where $X_t, t \geq 0$, are random vectors or, even more generally, random elements, i.e. measurable functions with values in an abstract measurable space $(S, \mathcal{F})$; $S$ is quite often a topological space and $\mathcal{F}$ is the $\sigma$-algebra of Borel subsets of $S$.

**8.1.6 *Equivalent conditions***    In checking whether a process is Markov, it is convenient to have a condition that is equivalent to the above definition but may turn out to be easier to check. To this end, we note that if $X(t), t \geq 0$, is a Markov process then, for every $n$ and $t \leq t_1 \leq ... \leq t_n$ and Borel sets $B_i, i = 1, ..., n$,

$$\mathbb{P}(X(t_i) \in B_i, i = 1, ..., n \,|\mathcal{F}_t) = \mathbb{P}(X(t_i) \in B_i, i = 1, ..., n \,|X(t)). \quad (8.9)$$

On the other hand, this condition implies that $X(t), t \geq 0$, is Markov. Indeed, for any $A \in \mathcal{F}_t$, both sides of the equation

$$\mathbb{P}(A \cap B) = \int_A \mathbb{E}(1_B | X(t)) \, \mathrm{d}\mathbb{P}$$

are finite measures as functions of $B$, and (8.9) shows that these measures are equal on the $\pi$-system of sets of the form $B = \{X(t_i) \in B_i, i = 1, ..., n\}$, $B_i \in \mathcal{B}(\mathbb{R})$, that generates $\mathcal{F}_t$. Hence, our claim follows by the the $\pi$–$\lambda$ theorem.

Yet another, still simpler condition for the process to be Markov is

$$\mathbb{P}(X(s) \in B \,|\mathcal{F}_t) = \mathbb{P}(X(s) \in B \,|X(t)), \qquad s \geq t, B \in \mathcal{B}(\mathbb{R}). \quad (8.10)$$

Clearly, (8.9) implies (8.10). To see the other implication, we note first that by a standard argument, (8.10) implies

$$\mathbb{E}(f(X(s))|\mathcal{F}_t) = \mathbb{E}(f(X(s))|X(t)) \quad (8.11)$$

for any bounded, Borel measurable function $f$. Next, taking $t_1 \leq t_2$ larger than $t$, using the tower property and the fact that the $1_{X(t_1)\in B_1}$

is $\mathcal{F}_{t_1}$ measurable, we calculate $\mathbb{P}(X(t_1) \in B_1, X(t_2) \in B_2 | \mathcal{F}_t)$ as

$$
\begin{aligned}
\mathbb{E}\left(1_{X(t_1) \in B_1} 1_{X(t_2) \in B_2} | \mathcal{F}_t\right) &= \mathbb{E}\left(\mathbb{E}\left(1_{X(t_1) \in B_1} 1_{X(t_2) \in B_2} | \mathcal{F}_{t_1}\right) | \mathcal{F}_t\right) \\
&= \mathbb{E}\left(1_{X(t_1) \in B_1} \mathbb{E}\left(1_{X(t_2) \in B_2} | \mathcal{F}_{t_1}\right) | \mathcal{F}_t\right) \\
&= \mathbb{E}\left(1_{X(t_1) \in B_1} \mathbb{E}\left(1_{X(t_2) \in B_2} | X(t_1)\right) | \mathcal{F}_t\right)
\end{aligned}
$$

with the last equality following by (8.9). Now, since $\mathbb{E}\left(1_{X(t_2) \in B_2} | X(t_1)\right)$ is $\sigma(X(t_1))$ measurable, it equals $g(X(t_1))$ for some Borel measurable $g$. Hence, by (8.10), the last conditional expectation equals

$$
\begin{aligned}
&\mathbb{E}\left(1_{X(t_1) \in B_1} \mathbb{E}\left(1_{X(t_2) \in B_2} | X(t_1)\right) | X(t)\right) \\
&= \mathbb{E}\left(\mathbb{E}\left(1_{X(t_1) \in B_1} 1_{X(t_2) \in B_2} | X_{t_1}\right) | X(t)\right) \\
&= \mathbb{E}\left(1_{X(t_1) \in B_1} 1_{X(t_2) \in B_2} | X(t)\right),
\end{aligned}
$$

establishing (8.9) for $n = 2$. The reader should find it an easy exercise now to give the details of an induction argument leading from (8.10) to (8.9).

**8.1.7 Easy exercise**   Give these details.

**8.1.8 Exercise**   Show that a process $X(t), t \geq 0$, is Markov iff (8.11) holds for all $f \in C(\overline{\mathbb{R}})$ (see 6.5.6).

**8.1.9** *Lévy processes are Markov*   The class of Lévy processes is a subclass of the class of Markov processes. To see that, let $X(t), t \geq 0$, be a Lévy process. We need to show that (8.11) holds for all $f \in C(\overline{\mathbb{R}})$ (see Exercise 8.1.8 above). Since $X(s)$ may be written as the sum of two independent random variables $X(s) - X(t)$ and $X(t)$, it suffices to show that if $X$ and $Y$ are random variables, $X$ is independent from a $\sigma$-algebra $\mathcal{F}$ and $Y$ is $\mathcal{F}$ measurable, then

$$
\mathbb{E}(f(X+Y)|\mathcal{F}) = \mathbb{E}(f(X+Y)|Y), \qquad f \in C(\overline{\mathbb{R}}). \tag{8.12}
$$

Now, if $f \in C(\overline{\mathbb{R}})$, then $(\tau, \sigma) \mapsto f(\tau + \sigma)$ is a member of $C(\overline{\mathbb{R}^2})$. Hence, by 6.5.7, $\mathbb{E}(f(X+Y)|\mathcal{F}) = \int_{\mathbb{R}} f(\tau+Y) \mathbb{P}_X(\mathrm{d}\tau)$. By the Fubini Theorem, $\sigma \mapsto \int_{\mathbb{R}} f(\tau + \sigma) \mathbb{P}_X(\mathrm{d}\tau)$ is Borel measurable. Hence, $\omega \mapsto \int_{\mathbb{R}} f(\tau + Y(\omega)) \mathbb{P}_X(\mathrm{d}\tau)$ is a well-defined, $\sigma(Y)$ measurable random variable. This implies (8.12).

**8.1.10 Exercise**   Show that the reflected Brownian motion is a Markov process.

**8.1.11** *Transition functions of Markov processes*    The distribution of the position of a Markov process at time $t$ given that at time $s$ it was at $\tau$ is in general a function of both $s$ and $t$. Markov processes for which it does depend on $s$ and $t$ only through the difference $t - s$ are termed **time-homogeneous Markov processes**. In what follows, we consider only such processes.

The kernel $K$ related to the reflected Brownian motion is an example of a **transition function**. The transition function of a (time-homogeneous) Markov process is a function $K(t, \tau, B)$ (‘$K$’ for “kernel”) of three variables $t \geq 0, p \in S, B \in \mathcal{F}$, where $(S, \mathcal{F})$ is a measurable space; $S$ is the set of possible values of the process; $K$ satisfies the following properties.

(a)   $K(t, p, \cdot)$ is a probability measure on $(S, \mathcal{F})$, for all $t \geq 0, p \in S$.
(b)   $K(0, p, \cdot) = \delta_p$ (delta measure at $p$).
(c)   $K(t, \cdot, B)$ is measurable for all $t \geq 0$ and $B \in \mathcal{F}$.
(d)   The Chapman–Kolmogorov equation is satisfied:

$$\int_S K(s, q, B) K(t, p, \mathrm{d}q) = K(t + s, p, B). \tag{8.13}$$

We say that a family $\{X(t), t \geq 0\}$ of random variables on a probability space $(\Omega, \mathcal{F}, \mathbb{P})$ with values in $S$ is a Markov process with transition function $K$ if for $t > s$:

$$\mathbb{P}(X(t) \in B | X(s)) = K(t - s, X(s), B), \qquad B \in \mathcal{B}(S). \tag{8.14}$$

Hence, the measure $K(t, p, \cdot)$ is the distribution of the position of the process at time $t$ given that at time zero it started at $p$.

**8.1.12 Exercise**    Let $\{\mu_t, t \geq 0\}$ be a convolution semigroup of measures, and let $K(t, \tau, B) = \mu_t(B - \tau)$. Prove that $K$ is a transition function, and in particular that the Chapman–Kolmogorov equation is satisfied.

**8.1.13 Example**    If $S = \mathbb{N}$, a measure $\mu$ on $S$ may be identified with a sequence $x = (\xi_n)_{n \geq 1}$ where $\xi_n = \mu(\{n\})$. Similarly, given a transition function $K$ on $\mathbb{N}$ we may define $p_{n,m}(t) = K(t, n, \{m\}), n, m \geq 1, t \geq 0$. Then, $p_{n,n}(0) = 1, p_{n,m}(0) = 0, n \neq m, p_{n,m} \geq 0$ and $\sum_{m \geq 1} p_{n,m} = 1$. Moreover, by the Chapman–Kolmogorov equation $p_{n,m}(s + t) = K(s + t, n, \{m\}) = \int_{\mathbb{N}} K(s, m, \{n\}) K(t, n, \mathrm{d}m) = \sum_{m \geq 1} p_{m,n}(s) p_{n,m}(t)$. In other words, $P(s)P(t) = P(s + t)$ where $P(t) = (p_{n,m}(t))_{n,m \geq 1}$ in the sense of multiplication of matrices. Such a family of matrices is called a **semigroup of transition matrices**.

Conversely, given a semigroup of transition matrices $\{P(t), t \geq 0\}$, $P(t) = (p_{n,m}(t))_{n,m \geq 1}, t \geq 0$, we may define a transition function by $K(s, n, B) = \sum_{m \in B} p_{n,m}(s)$, $B \subset \mathbb{N}$. The details are left to the reader.

**8.1.14 Example**    *The Ornstein–Uhlenbeck process*    Given $\alpha, \gamma > 0$ and a Wiener process $w(t), t \geq 0$, we define the **Ornstein–Uhlenbeck process** starting at 0 as

$$X(t) = \gamma e^{-\alpha t} \int_0^t e^{\alpha s} \, dw(s).$$

We follow Breiman [19] to show that it is a Gaussian process and a time-homogeneous Markov process, and find its transition function. Let $k \in \mathbb{N}$ and $0 = t_0 < t_1 < ... < t_k$ be given. Define $\Delta w_{i,j,n} = w(s_{i+1,j,n}) - w(s_{i,j,n})$, $s_{i,j,n} = t_{j-1} + \frac{i}{n}(t_j - t_{j-1}), i = 0, ..., n-1, j = 1, ..., k, n \in \mathbb{N}$. Then,

$$X(t_j) = \gamma e^{-\alpha t_j} \sum_{l=1}^{j} \int_{t_{l-1}}^{t_l} e^{\alpha s} \, dw(s) \qquad (8.15)$$

$$= \gamma e^{-\alpha t_j} \lim_{n \to \infty} \sum_{l=1}^{j} \sum_{i=0}^{n-1} e^{\alpha s_{i,l,n}} \Delta w_{i,l,n}$$

with the limit in $L^2(\Omega)$. Therefore, for any coefficients $a_i, i = 1, ..., k$, $\sum_{j=1}^{k} a_j X(t_j)$ is the limit of $\gamma \sum_{l=1}^{k} b_l \sum_{i=0}^{n-1} e^{\alpha s_{i,l,n}} \Delta w_{i,l,n}$ where $b_l = \sum_{j=l}^{k} a_j e^{-\alpha t_j}$. By Theorem 4.1.5, the approximating sum being normal as the sum of independent normal random variables, $\sum_{j=1}^{k} a_j X(t_j)$ is normal, too. Using (8.15) again, by independence of increments of the Brownian motion and $E(\Delta_{i,1,n})^2 = \frac{t_1}{n}$,

$$E\, X(t_2) X(t_1) = \lim_{n \to \infty} \gamma^2 e^{-\alpha(t_1 + t_2)} \sum_{i=0}^{n-1} \sum_{l=0}^{n-1} e^{\alpha(s_{i,1,n} + s_{l,1,n})} E \Delta_{i,1,n} \Delta_{l,1,n}$$

$$+ \lim_{n \to \infty} \gamma^2 e^{-\alpha(t_1 + t_2)} \sum_{i=0}^{n-1} \sum_{l=0}^{n-1} e^{\alpha(s_{i,1,n} + s_{l,2,n})} E \Delta_{i,1,n} \Delta_{l,2,n}$$

$$= \lim_{n \to \infty} \gamma^2 e^{-\alpha(t_1 + t_2)} \sum_{i=0}^{n-1} e^{2\alpha s_{i,1,n}} \frac{t_1}{n}$$

$$= \gamma^2 e^{-\alpha(t_1 + t_2)} \int_0^{t_1} e^{2\alpha s} \, ds.$$

For $s > t$, let $Z = Z(s,t) = X(s) - e^{-\alpha(s-t)}X(t)$. $Z$ is a normal variable with mean 0. Also, for $r \leq t$,

$$EZX(r) = \gamma^2 e^{-\alpha(r+s)} \int_0^r e^{2\alpha u}\,du - \gamma^2 e^{-\alpha(s-t)}e^{-\alpha(t+r)} \int_0^r e^{2\alpha u}\,du = 0,$$

proving that $Z$ is independent of $X(r)$. Therefore, $Z$ is independent of $\mathcal{F}_t = \sigma(X(r), r \leq t)$. Furthermore, this allows us to compute the variance of $Z$ to be

$$EZ^2 = EZX(s) = \gamma^2 e^{-2\alpha s} \int_0^s e^{2\alpha u}\,du - \gamma^2 e^{-\alpha(s-t)}e^{-\alpha(s+t)} \int_0^t e^{2\alpha u}\,du$$

$$= \gamma^2 e^{-2\alpha s} \int_t^s e^{2\alpha u}\,du = \frac{\gamma^2}{2\alpha}(1 - e^{-2\alpha(s-t)}).$$

Consequently, using 6.5.7, for any $f \in C(\overline{\mathbb{R}})$, $\mathbb{E}(f(X(s))|\mathcal{F}_t)$ equals $\mathbb{E}(f(Z+e^{-\alpha(s-t)}X(t))|\mathcal{F}_t) = \int_{\mathbb{R}} f(\tau+e^{-\alpha(s-t)}X(t))\,\mathbb{P}_Z(\,d\tau)$. As in 8.1.9, we argue that this is $\mathbb{E}(f(X(s))|X(t))$, and, as in Exercise 8.1.8, extend this relation to all $f \in BM(\mathbb{R})$. This shows that the Ornstein–Uhlenbeck process is a Markov process. Moreover, taking $f = 1_B, B \in \mathcal{B}(\mathbb{R})$, we have $\mathbb{P}(X(s) \in B|X(t)) = \int_{\mathbb{R}} 1_B(\tau + e^{-\alpha(s-t)}X(t))\,\mathbb{P}_Z(\,d\tau) = \mathbb{P}(Z \in B - e^{-\alpha(s-t)}X(t))$. In other words, since $Z \sim N(0, \frac{\gamma^2}{2\alpha}(1 - e^{-2\alpha(s-t)}))$, $K$, defined to be, for any $t > 0$ and $\tau \in \mathbb{R}$, the distribution of a normal variable with mean $e^{-\alpha t}\tau$ and variance $\frac{\gamma^2}{2\alpha}(1 - e^{-2\alpha t})$, is a transition function of this process. Clearly, the Ornstein–Uhlenbeck process is a time-homogeneous Markov process.

**8.1.15** *Semigroups of operators related to transition functions of Markov processes*   With a transition function one may associate a family of operators in $\mathbb{BM}(S)$ by

$$(U_t\mu)(B) = \int_S K(t,p,B)\,\mu(\,dp); \qquad (8.16)$$

it is clear that $U_t\mu$ is a measure. In particular, if $\mu$ is a probability measure, then $U_t\mu$ is a probability measure. Moreover, by the Chapman–Kolmogorov equation, $\{U_t, t \geq 0\}$ is a semigroup of operators. To check that $\|U_t\| = 1, t \geq 0$, we may use the minimal representation of a charge, as described in 1.3.6, and the fact that $U_t$ maps non-negative measures into non-negative measures.

Formula (8.16) has a clear interpretation: if $X(t), t \geq 0$, is a Markov process with transition function $K$ and initial distribution $\mu$, then $U_t\mu$ is the distribution of the process at time $t$. However, at least historically, instead of dealing with this semigroup directly, it has been easier to consider the semigroup

$$T_t x(p) = \int_S x(q) K(t, p, dq), \qquad t \geq 0, \qquad (8.17)$$

defined in $BM(S)$. This semigroup is dual to $\{U_t, t \geq 0\}$ in the sense that we may treat a member of $BM(S)$ as a functional on $\mathbb{BM}(S)$ given by $\mu \mapsto \int_S x \, d\mu =: (x, \mu)$, and we have $(x, U_t) = (T_t x, \mu)$. To be more specific: the dual to $U_t$ coincides with $T_t$ on $BM(S)$. Clearly, $\|T_t\| = 1$, and $T_t 1_S = 1_S, t \geq 0$.

Note that in contradistinction to the case of Lévy processes, when $S = \mathbb{R}$ we may not claim that this semigroup maps $C_0(\mathbb{R})$ or $BUC(\mathbb{R})$ into itself. In general, all we may claim is that it maps $BM(S)$ into itself. Also, the operators $T_t$ are non-negative in that they leave the cone of non-negative functions invariant. If $S$ is locally compact and the semigroup leaves $C_0(S)$ invariant and is a strongly continuous semigroup as restricted to this subspace, we say that $\{T_t, t \geq 0\}$ is a Feller semigroup, that the related process is a **Feller process**, and/or that the kernel $K$ is a Feller kernel. We note that Lévy processes are Feller processes.

**8.1.16 Example** Let $S = \mathbb{N}$. The space of measures on $\mathbb{N}$ is isometrically isomorphic with $l^1(\mathbb{N})$, the space of absolutely summable sequences $x = (\xi_n)_{n \geq 1}$. Moreover, $BM(S)$ is isometrically isomorphic with $l^\infty$, the space of bounded sequences $y = (\eta_n)_{n \geq 1}$. For a transition family given by a semigroup of transition matrices $P(t) = (p_{n,m}(t))_{n,m \geq 1}, t \geq 0$, the semigroup $\{U_t, t \geq 0\}$ in $l^1$ is given by $U_t x = (\sigma_n(t))_{n \geq 1}$ where $\sigma_n(t) = (U_t x)(\{n\}) = \sum_{m=1}^\infty p_{m,n}(t)\xi_m$. In other words, $U_t x$ is the matrix product $x P(t)$ where $x$ is treated as a horizontal vector. Similarly, the semigroup $\{T_t, t \geq 0\}$ in $l^1$ is given by $T_t y = (\varsigma_n(t))_{n \geq 1}$, where $\varsigma_n(t) = \sum_{m=1}^\infty \eta_m p_{n,m}(t)$. In other words, $T_t x$ is the matrix product $P(t)y$ where $y$ is treated as a vertical vector.

**8.1.17 *Semigroups in $C(S)$ and transition kernels*** Let $S$ be a compact space and $\{T_t, t \geq 0\}$ be a semigroup of non-negative contraction operators in $C(S)$ such that $T_t 1_S = 1_S$. Then, there exists the unique transition function $K$ such that (8.17) holds for all $x \in C(S)$.

*Proof* For $p \in S$ and $t \geq 0$, the map $x \mapsto T_t x(p)$ is a non-negative linear functional in that $x \geq 0$ implies $T_t x(p) \geq 0$. Also it is bounded with the norm not exceeding $\|T_t\| = 1$. By the Riesz Theorem, there exists a Borel measure $K(t, p, \cdot)$ such that (8.17) holds. Since $T_t 1_S = 1_S$, $K(t, p, \cdot)$ is a probability measure.

We will show that $K$ is a transition function. Clearly, conditions (a) and (b) of Definition 8.1.11 are satisfied, the latter following by $T_0 x = x$. We need to show (c) and (d).

We provide the proof in the case of a metric space. The argument in the general case is similar, but more technical. If $S$ is a metric space and $B$ is a closed set, then $1_B$ may be approximated by a pointwise limit of continuous functions, say $x_k, k \geq 1$, as in the proof of 5.6.3. Then, $\int_S x_k(q) K(t, p, dq)$ converges to $K(t, p, B)$. On the other hand, by (8.17), $K(t, p, B)$ is a pointwise limit of continuous, hence measurable, functions $T_t x_k(p)$. This shows (c) for closed $B$. Now, the family of sets $B$ for which $K(t, \cdot, B)$ is measurable is a $\lambda$-system. Since the Borel $\sigma$-algebra is generated by the $\pi$-system of closed sets, by the $\pi$–$\lambda$ theorem, $K(t, \cdot, B)$ is measurable for all Borel $B$.

Note that by the semigroup property we have (compare the final part of the argument in 8.1.1):

$$\int_S \int_S x(r) K(s, q, dr) K(t, p, dq) = \int_S x(r) K(t + s, p, dr), \quad x \in C(S).$$

Approximating $1_B$ by continuous functions $x_k$ as above, we obtain the Chapman–Kolmogorov equation with closed $B$, and then, extend the result to all Borel sets.

Uniqueness follows directly from (8.17) and the Riesz Theorem. $\square$

**8.1.18 Exercise** Show that an operator $T$ in $BM(S)$ (or $BC(S)$) is a non-negative contraction iff $0 \leq Tx \leq 1$ provided $0 \leq x \leq 1$.

**8.1.19 Exercise** Show that if $K$ is a transition function of a Markov process $\{X(t), t \geq 0\}$, then $T_s x(X(t)) = \mathbb{E}(x(X(t + s))|\mathcal{F}_t)$, where $\mathcal{F}_t = \sigma(X(u), u \leq t)$.

**8.1.20 Exercise** Let $\{X(t), t \geq 0\}$ be a Lévy process. Show that the corresponding semigroup is given by $T_t x(\tau) = E\, x(\tau + X(t))$.

**8.1.21 Exercise** Let $\{X(t), t \geq 0\}$ be the Ornstein–Uhlenbeck process constructed by means of a Brownian motion $\{w(t), t \geq 0\}$. Show that the corresponding semigroup is given by $T_t x(\tau) = E\, x(e^{-\alpha t}\tau + w(\beta(t)))$ where $\beta(t) = \frac{\gamma^2}{2\alpha}(1 - e^{-2\alpha t})$.

8.1.22  *The minimal Brownian motion*     A slight change in the reasoning from 8.1.1 leads to another interesting process known as the **minimal Brownian motion**. By Exercise 8.1.3, the Brownian motion semigroup leaves the space $BUC_n(\mathbb{R})$ invariant. On the other hand, this space is isomorphic to $BUC_0(\mathbb{R}^+)$ of uniformly continuous functions $x$ on $\mathbb{R}^+$ that vanish at $\tau = 0$. The isomorphism $I : BUC_0(\mathbb{R}^+) \to BUC_n(\mathbb{R})$ is given by $(Ix)(\tau) = \mathrm{sgn}(\tau)y(|\tau|)$, $x \in BUC_0(\mathbb{R}^+)$. The inverse $I^{-1}$ is given by $I^{-1}y(\tau) = y(\tau), \tau \geq 0$, $y \in BUC_n(\mathbb{R})$. We define the semigroup $\{T_t^m, t \geq 0\}$ in $BUC_0(\mathbb{R}^+)$ as the semigroup isomorphic to the restriction of $T_t$ to $BUC_n(\mathbb{R})$. Calculating as in (8.3) one obtains:

$$(T_t^m x)(\tau) = E\,\mathrm{sgn}(\tau + w(t))x(|\tau + w(t)|). \qquad (8.18)$$

Moreover,

$$T_t^m x(\tau) = \int_0^\infty x(\sigma)K(t,\tau,\,\mathrm{d}\sigma) \qquad (8.19)$$

where $K(t,\tau,\cdot)$ is the measure with density

$$k(t,\tau,\sigma) = \mathrm{e}^{-\frac{(\tau-\sigma)^2}{2t}} - \mathrm{e}^{-\frac{(\tau+\sigma)^2}{2t}}. \qquad (8.20)$$

The generator $A_m$ of $T_t^m$ is given by: $A_m x = \frac{1}{2}x''$, on the domain $\mathcal{D}(A_m)$ composed of $x \in BUC_0(\mathbb{R}^+)$ that are twice differentiable with $x'' \in BUC_0(\mathbb{R}^+)$. For any $x \in \mathcal{D}(A_m)$, the function $x(t) = T_t^m x$ is a unique solution to the Cauchy problem

$$\frac{\partial x(\tau,t)}{\partial t} = \frac{1}{2}\frac{\partial^2 x(\tau,t)}{\partial \tau^2}, \quad \tau > 0, x(0,t) = 0, x(\tau,0) = x(\tau). \qquad (8.21)$$

The process introduced in this way may be described as follows: after starting at $\tau > 0$, it evolves as a free Brownian motion until it touches the barrier $\tau = 0$. At this moment, it disappears from the space. This process is termed the **minimal Brownian motion**.

As a result, the distribution $K(t,\tau,\cdot)$ with density (8.20) is not a probability measure, for there is a positive probability that at time $t$ the process will no longer be in the space. To accommodate such situations, in Definition 8.1.11 (a) the requirement that $K(t,\tau,\cdot)$ is the probability measure is relaxed and it is assumed that $K(t,\tau,S) \leq 1$, instead. Another way of dealing with such phenomena is to introduce an additional point $\Delta$, referred to as the **cemetery** (or **coffin state**), where all the disappearing trajectories of the process are sent and from where none of them ever returns. More details about this in the next example and in 8.1.26.

Note, finally, that the state space $S$ of the minimal Brownian motion is the open half-axis $\mathbb{R}_*^+$, while the state of the reflected Brownian motion is $\mathbb{R}^+$.

**8.1.23 Exercise**    Using the procedure from 8.1.1 construct the semigroup related to "reflected Cauchy process" and a "minimal Cauchy process". Construct also "reflected" and "minimal" compound Poisson process from the compound Poisson process defined in 7.5.8 with symmetric distribution of $Y_n$.

**8.1.24 Absorbed Brownian motion**    Let $K(t, \tau, B)$, $t \geq 0$, $\tau > 0$, be the transition function on $\mathbb{R}_*^+$ related to the minimal Brownian motion. Define $\tilde{K}(t, \tau, B)$ on $\mathbb{R}^+$ by $\tilde{K}(t, 0, B) = 1_B(0)$ and

$$\tilde{K}(t, \tau, B) = K(t, \tau, B)1_{B^0}(0) + [1 - K(t, \tau, \mathbb{R}_*^+)]1_B(0), \quad \tau > 0. \quad (8.22)$$

Note that $[1 - K(t, \tau, \mathbb{R}_*^+)]$ is the probability that, at time $t$, the minimal Brownian motion that started at $\tau$ is no longer in $\mathbb{R}_*^+$. Therefore, the above formula says that we modify the minimal Brownian motion by requiring that after the process touches the barrier $\tau = 0$ it stays there for ever. This new process is termed the **absorbed Brownian motion**. The procedure just described is of course a particular case of the procedure from the previous subsection with $\Delta = 0$. The reader will check that $\tilde{K}$ is a transition function in $\mathbb{R}^+$.

Define the operators $T_t^a$ in $BUC(\mathbb{R}^+)$ by

$$T_t^a x(\tau) = \int_{\mathbb{R}^+} x(\sigma)\tilde{K}(t, \tau, d\sigma), \quad t \geq 0. \quad (8.23)$$

If $\{T_t^m, t \geq 0\}$ denotes the semigroup of the minimal Brownian motion, then by (8.22),

$$T_t^a x = x(0)1_{\mathbb{R}^+} + T_t^m(x - x(0)1_{\mathbb{R}^+});$$

note that $x - x(0)1_{\mathbb{R}^+}$ belongs to $BUC_0(\mathbb{R}^+)$. Using this and the semigroup property and strong continuity of $\{T_t^m, t \geq 0\}$ we see that $\{T_t^a, t \geq 0\}$ is a strongly continuous semigroup. Furthermore,

$$\frac{T_t^a x - x}{t} = \frac{T_t^m[x - x(0)1_{\mathbb{R}^+}] - [x - x(0)1_{\mathbb{R}^+}]}{t},$$

proving that $x$ belongs to the domain $\mathcal{D}(A_a)$ of the infinitesimal generator of $\{T_t^a, t \geq 0\}$ iff $x - x(0)1_{\mathbb{R}^+}$ belongs to the domain $\mathcal{D}(A_m)$ of the infinitesimal generator of $\{T_t^m, t \geq 0\}$. Thus, $A_a x = \frac{1}{2}x''$, on the domain $\mathcal{D}(A_a)$ composed of $x \in BUC(\mathbb{R}^+)$ that are twice differentiable with

$x'' \in BUC(\mathbb{R}^+)$. Moreover, for any $x \in \mathcal{D}(A_a)$, the function $x(t) = T_t^a x$ is the unique solution to the Cauchy problem

$$\frac{\partial x(\tau, t)}{\partial t} = \frac{1}{2} \frac{\partial^2 x(\tau, t)}{\partial \tau^2}, \quad \tau > 0, x''(0, t) = 0, x(\tau, 0) = x(\tau). \quad (8.24)$$

**8.1.25 Remark**　Subsections 8.1.1, 8.1.22 and 8.1.24 illustrate the fact that the behavior of a Markov process on a boundary (in our case at the point $\tau = 0$) is reflected in the form of the domain of the infinitesimal generator of the process, and, consequently in the boundary condition of the related Cauchy problem. The boundary condition $x'(0) = 0$ corresponds to reflection, $x''(0) = 0$ corresponds to absorption, and $x(0) = 0$ describes the fact that the process disappears from the space upon touching the boundary (sometimes called a non-accessible boundary). Elastic Brownian motion, a process with still another type of behavior at the boundary, sometimes called the sticky barrier phenomenon, will be introduced in 8.2.18 below. See [58], [88], [100], [109] for more about this fascinating subject.

**8.1.26 *Semigroups in $C_0(S)$ and transition functions***　Let $S$ be a locally compact space (but not a compact space) and let $S_\Delta$ be the one-point compactification of $S$. Let $\{T_t, t \geq 0\}$ be a semigroup of non-negative contraction operators in $C_0(S)$. Then, there exists the unique transition function $K$ on $S_\Delta$ such that (8.17) holds for all $x \in C_0(S)$, and $K(t, \Delta, \cdot) = \delta_\Delta$.

*Proof*　We will show that the operators

$$T_t^\Delta x = x(\Delta) 1_{S_\Delta} + T_t(x - x(\Delta) 1_{S_\Delta})$$

form a semigroup of non-negative contraction operators in $C(S_\Delta)$. The semigroup property of $\{T_t^\Delta, t \geq 0\}$ follows directly from the semigroup property of $\{T_t, t \geq 0\}$.

For $y \in C_0(S)$ let $y^+ = \max(x, 0)$ and $y^- = \max(0, -x)$. Both $y^+$ and $y^-$ belong to $C_0(S)$ and we have $y = y^+ - y^-$. Since $T_t y = T_t y^+ - T_t y^-$ and the elements $T_t y^+$ and $T_t y^-$ are non-negative, $(T_t y)^+ \leq T_t y^+$. For $x \in C(S_\Delta)$ let $y = x - a 1_{S_\Delta}$, where $a = x(\Delta)$. To prove that the operators $T_t$ are non-negative, we need to show that $a 1_{S_\Delta} + T_t y \geq 0$ provided $a 1_{S_\Delta} + y \geq 0$. The inequality $a 1_{S_\Delta} + y \geq 0$, however, implies $y^- \leq a$ and, hence, $\|y^-\| \leq a$. Since $T_t, t \geq 0$ are contraction operators, $\|T_t y^-\| \leq a$, i.e. $T_t y^- \leq a 1_{S_\Delta}$. Hence, $a 1_{S_\Delta} + T_t y = a 1_{S_\Delta} + T_t y^+ - T_t y^- \geq 0$, as desired.

Using non-negativity and $T_t^\Delta 1_{S_\Delta} = 1_{S_\Delta}$, since $|x| \le \|x\| 1_{S_\Delta}$, we have $|T_t x| \le T_t \|x\| 1_{S_\Delta} = \|x\| 1_{S_\Delta}$. This implies $\|T_t\| \le 1$, as desired. Therefore, there exists a transition function on $S_\Delta$ such that

$$T_t^\Delta x(p) = \int_{S_\Delta} x(q) K(t, p, dq), \qquad p \in S_\Delta.$$

This implies (8.17) with $x \in C_0(S)$; for such $x$ integrals over $S_\Delta$ and $S$ are the same. The rest follows by $T_t^\Delta x(\Delta) = x(\Delta)$. $\square$

**8.1.27 Remark** In general $K(t, p, S) \le 1, p \in S$ as $K(t, p, \{\Delta\}) \ge 0$.

**8.1.28 *Pseudo-Poisson process*** Let $K(p, B)$ be a probability measure on a measurable space $(S, \mathcal{F})$ for each $p \in S$, and a measurable function if $B \in \mathcal{F}$ is fixed. Let us define $K^n$, $n \ge 0$, inductively by $K^0(p, B) = 1_B(p)$ and $K^{n+1}(p, B) = \int_S K^n(q, B) K(p, dq)$. In particular, $K^1 = K$. By induction one shows that $K^n$, $n \ge 0$, are probability measures for each fixed $p \in S$, and measurable functions if $B \in \mathcal{F}$ is fixed.

Now, for a given $a > 0$ define $K(t, p, B)$ as $\sum_{n=0}^\infty e^{-at} \frac{a^n t^n}{n!} K^n(p, B)$. Then, conditions (a)–(c) of Definition 8.1.11 hold. Probably the easiest way to prove the Chapman–Kolmogorov equation is to note that the semigroup $(S_t x)(p) = \int_\mathbb{R} x(q) K(t, p, dq)$ related to these measures is given by $S_t = e^{-at} e^{atK}$. Here $K$ is the operator $(Kx)(p) = \int_S x(q) K(p, dq)$.

The operator $K$ and the semigroup $\{S_t, t \ge 0\}$ act in the space $BM(S)$ of bounded measurable functions on $S$. The infinitesimal generator of $\{S_t, t \ge 0\}$ is $a(K - I)$. The reader should write the Cauchy problem of which $x_t = S_t x$ is a solution for $x \in BM(S)$.

The realization of the process with transition function defined above looks as follows. Suppose that the process starts at $p \in S$. Then, it stays there for a random, exponential time with parameter $a$. Next, it jumps to a random point $q$ and the distribution of its position after this jump is $K(p, \cdot)$. The process continues in the same way later. This is the so-called **pseudo-Poisson process**, compare 7.5.8. It is interesting that all Markov processes are limits of pseudo-Poisson processes: this is a by-product of Yosida's proof of the Hille–Yosida theorem. See Section 8.2, below.

**8.1.29 Example** *Markov property and the Poisson formula* The Markov property of a process may exhibit itself in interesting analytic ways. The example we want to present now involves the classical **Poisson formula** for the solution of the **Dirichlet problem** in a disc. The

Dirichlet problem in a disc may be posed in the following way: given a continuous function $\phi_0$ on a unit circle $C$ on the complex plane find a function $\phi$ continuous in the unit disc, where $|z| \leq 1$, and harmonic in the open unit ball, where $|z| < 1$. In other words, we want to have a continuous function such that

$$\Delta\phi(x,y) = \frac{\partial^2\phi}{\partial x^2}(x,y) + \frac{\partial^2\phi}{\partial y^2}(x,y) = 0, \ x^2 + y^2 < 1,$$

$$\phi(x,y) = \phi_0(x,y), \qquad\qquad x^2 + y^2 = 1.$$

It turns out that this problem is well-posed and the unique solution is given by (see e.g. [95], [97], [103])

$$\phi(x,y) = \int_{-\pi}^{\pi} p_r(e^{i(\alpha-\theta)})\phi_0(e^{i\theta})\,d\theta = \phi_0 * p_r(e^{i\alpha}) \qquad (8.25)$$

where $x = r\cos\alpha, y = r\sin\alpha$ ($0 \leq r < 1$), and $p_r$ is the Poisson kernel defined in Exercise 1.2.29. In this exercise the reader has checked (I hope he did) that $p_r \geq 0$, $\int_{-\pi}^{\pi} r_r(\theta)\,d\theta = 1$, and $p_r * p_s = p_{rs}$, $0 \leq r, s < 1$.

The first two relations have a nice interpretation once we know that (see e.g. [103]) the space of harmonic functions in the unit disc (harmonic in the open disc, continuous in the disc) is isometrically isomorphic to the space of continuous function on the unit circle, both spaces with supremum norm. Hence, the map $\phi_0 \mapsto \phi \mapsto \phi(x,y)$ is a bounded, non-negative linear functional. Since this functional may be proven to have norm one (by the so-called maximum principle), there must exist a probability measure $\mathbb{P}_{x,y}$ on $C$ such that $\phi(x,y) = \int_C \phi(e^{i\theta})\,\mathbb{P}_{x,y}(d\theta)$. By (8.25), $p_r(e^{i(\alpha-\cdot)})$ is a density of this measure.

A still deeper insight is given by the following probabilistic solution to the Dirichlet problem (see [34], [93], [113] etc., or the classic [30]):

$$\phi(x,y) = E\,\phi_0((x,y) + w(\tau)), \qquad (8.26)$$

where $w(t), t \geq 0$, is a two-dimensional Brownian motion (i.e. a pair of independent one-dimensional Brownian motions) and $\tau$ is the random time when $(x,y) + w(t)$ touches the unit circle for the first time.

A word of explanation is needed here. As we have seen, at any time $t$, a Markov process starts afresh, forgetting the whole past. Some processes, the two-dimensional Brownian motion among them, possess the stronger property, the **strong Markov property** and start afresh at some random times. These times are Markov times $\tau$, i.e., similarly to 3.7.11, at time $t$ we must know whether $\tau$ has happened or not. In other words, $\{\tau \le t\} \in \mathcal{F}_t$ where $\mathcal{F}_t$ is an appropriate filtration. The time of touching the unit circle for the first time is a Markov time. After that, the Brownian motion starts afresh.

Armed with this knowledge, let us rewrite the last of the three properties of the Poisson kernel as

$$p_{\frac{r}{s}} * p_s = p_r, \tag{8.27}$$

where it is assumed that $0 \le r < s < 1$. Comparing (8.25) and (8.26) we see that $p_s(e^{i(\alpha-\cdot)})$ is the density of distribution of the position of the process $se^{i\alpha} + w(t)$ at the time when it touches the unit circle for the first time. Hence, using the scaling property of the Brownian motion we may show that $p_{\frac{r}{s}}(e^{i(\alpha-\cdot)})$ is the distribution of the process $re^{i\alpha} + w(t)$ when it touches the circle with radius $s$ for the first time. After starting at $r^{i\alpha}$ where $r < s$, and before touching the unit circle for the first time, the process must touch the circle with radius $s$. Formula (8.27) expresses thus the fact that after touching the circle with radius $s$ the process starts afresh. Conditional on reaching the circle with radius $s$ at a point $se^{i\theta}$, the distribution of the position at the time of touching the unit circle is given by the kernel $p_s(e^{i(\theta-\cdot)})$, and the unconditional distribution is obtained by integration over positions $se^{i\theta}$. In this sense, (8.27) is very similar to the Chapman–Kolmogorov equation.

## 8.2 The Hille–Yosida Theorem

Given a transition family $K$ one may construct a Markov process such that (8.14) holds (see e.g. [38], [113]). However, transition functions are rarely given explicitly, and the same is true about the semigroups of operators. Therefore, instead of giving an explicit formula for the semigroup or specifying the appropriate transition function, we often restrict ourselves to describing the generator of the semigroup. The main theorem of this section, the **Hille–Yosida–Feller–Phillips–Miyadera Theorem**,† characterizes operators that are generators of strongly continuous semigroups. Characterization of generators of semigroups defined by means of transition families will be given in the next section.

8.2.1 *The Hille–Yosida Theorem*    Let $\mathbb{X}$ be a Banach space. An operator $A : \mathbb{X} \supset \mathcal{D}(A) \to \mathbb{X}$ is the generator of a strongly continuous

---

† Although the original formulation of this theorem was given independently by Hille and Yosida in the case $\omega = 0$ and $M = 1$, and the general case was discovered later, independently by Feller, Phillips and Miyadera, in what follows for simplicity we will often call this theorem the **Hille–Yosida Theorem**.

semigroup $\{T_t, t \geq 0\}$ in $\mathbb{X}$ such that

$$\|T_t\| \leq Me^{\omega t}, \qquad t \geq 0 \tag{8.28}$$

for some $M \geq 1$ and $\omega \in \mathbb{R}$ (see (7.14)) iff the following three conditions hold.

(a)  $A$ is closed and densely defined.
(b)  All $\lambda > \omega$ belong to the resolvent set of $A$: this means that for all $\lambda > \omega$ there exists a bounded linear operator $R_\lambda = (\lambda - A)^{-1} \in \mathcal{L}(\mathbb{X})$, i.e. the unique operator such that $(\lambda - A)R_\lambda x = x, x \in \mathbb{X}$ and $R_\lambda(\lambda - A)x = x, x \in \mathcal{D}(A)$.
(c)  For all $\lambda > \omega$ and all $n \geq 1$,

$$\|R_\lambda^n\| \leq \frac{M}{(\lambda - \omega)^n}. \tag{8.29}$$

*The proof* of this crucial theorem is the main subject of this section. To be more specific, this section is devoted to the proof of sufficiency of conditions (a)–(c). This is because necessity of (a)–(c) is proven in a straightforward manner. In fact, necessity of (a) and (b) was shown in 7.4.15 and 7.4.32, respectively. To prove (c) we note that according to 7.4.32, $R_\lambda$ is the Laplace transform of the semigroup. Hence, by the semigroup property $R_\lambda^n x = \int_0^\infty \cdots \int_0^\infty e^{-\lambda \sum_{i=1}^n t_i} T(\sum_{i=1}^n t_i) x \, dt_1 \ldots dt_n$, which implies $\|R_\lambda^n x\| \leq M\|x\| \left( \int_0^\infty e^{-(\lambda - \omega)t} \, dt \right)^n$, as desired.

We note that, by 7.4.34, $A$ may generate only one semigroup; the point is to prove its existence.

8.2.2  *Reduction to the case* $\omega = 0$    We note that it is enough to consider the case $\omega = 0$, i.e. the case where

$$\|\lambda^n R_\lambda^n\| \leq M. \tag{8.30}$$

Indeed, if $A$ satisfies (a)–(c) then $B = A - \omega$ is a closed, densely defined operator satisfying (a)–(c) with $\omega = 0$ for we have $(\lambda - B)^{-1} = (\lambda + \omega - A)^{-1}, \lambda > 0$. Let $\{S_t, t \geq 0\}$ be the semigroup of operators generated by $B$; we have $\|S_t\| \leq M$. Define $T_t = e^{\omega t} S_t$. Then, by 7.4.18, the infinitesimal generator of $\{T_t, t \geq 0\}$ is $B + \omega I = A$. Finally, $\|T_t\| = e^{\omega t} \|S_t\| \leq e^{\omega t} M$.

The sufficiency of conditions (a)–(c) in the Hille–Yosida theorem follows from the following result.

**8.2.3 Theorem**    Let $A$ be a closed, not necessarily densely defined, operator in a Banach space $\mathbb{X}$, and let $\mathbb{X}' = cl\mathcal{D}(A)$. Suppose that all $\lambda > 0$ belong to the resolvent set of $A$. Then, (8.30) holds iff there exists a family $\{U(t), t \geq 0\}$ of bounded linear operators such that $\|U(t) - U(s)\| \leq M(t-s), 0 \leq s \leq t$, and

$$\lambda \int_0^\infty e^{-\lambda t} U(t) x \, dt = R_\lambda x, \qquad x \in \mathbb{X}. \tag{8.31}$$

In such a case, there exists a strongly continuous semigroup $\{T_t, t \geq 0\}$ on $\mathbb{X}'$ such that $\|T_t\| \leq M$, and $\int_0^\infty e^{-\lambda t} T_t x \, dt = R_\lambda x, x \in \mathbb{X}'$. In particular, the infinitesimal generator of $\{T_t, t \geq 0\}$ is the operator $A_\mathrm{p}$, termed the **part of** $A$ **in** $\mathbb{X}'$, defined by

$$\mathcal{D}(A_\mathrm{p}) = \{x \in \mathcal{D}(A) | Ax \in \mathbb{X}'\}, \quad A_\mathrm{p} x = Ax.$$

**8.2.4 Definition**    The family $\{U(t), t \geq 0\}$ described above is termed the **integrated semigroup** related to $A$. Clearly, $A$ determines the Laplace transform of $U(t)$ and hence $U(t)$ itself.

**8.2.5** *The Yosida approximation*    The key role in the proof of Theorem 8.2.3 is played by the operators

$$A_\lambda = \lambda^2 R_\lambda - \lambda I = \lambda(\lambda R_\lambda - I), \qquad \lambda > 0,$$

called the **Yosida approximation**, and more specifically by their exponents $e^{A_\lambda t}$ which will be shown to approximate $T_t$, as $\lambda \to \infty$. Let us therefore look at some properties of $A_\lambda$ and exhibit examples.

**8.2.6 Lemma**    Let $A$ be as in 8.2.3 and assume that (8.30) holds. Then

$$\|e^{A_\lambda t}\| \leq M, \tag{8.32}$$

$$\lim_{\lambda \to \infty} \lambda R_\lambda x = x, \qquad x \in \mathbb{X}', \tag{8.33}$$

and

$$\lim_{\lambda \to \infty} A_\lambda x = Ax, \qquad x \in \mathcal{D}(A). \tag{8.34}$$

*Proof*  Since $e^{A_\lambda t} = e^{-\lambda t} e^{\lambda^2 R_\lambda t}$, to show (8.32) it suffices to prove that $\|e^{\lambda^2 R_\lambda t}\| \leq e^{\lambda t} M$, and we write

$$\|e^{\lambda^2 R_\lambda t}\| \leq \sum_{n=0}^\infty \frac{\|\lambda^{2n} R_\lambda^n\|}{n!} \leq M \sum_{n=0}^\infty \frac{\lambda^n}{n!} = e^{\lambda t} M.$$

Since $\mathcal{D}(A)$ is dense in $\mathbb{X}'$, and $\|\lambda R_\lambda\| \leq M$, it is enough to check (8.33) for $x \in \mathcal{D}(A)$. For such an $x$, however, by (7.26), $\lambda R_\lambda x - x = R_\lambda Ax$, and $\|R_\lambda Ax\| \leq \frac{M}{\lambda}\|Ax\|$ converges to 0 as $\lambda \to \infty$.

Referring to (7.26) again we see that $A_\lambda x = \lambda R_\lambda Ax, x \in \mathcal{D}(A)$, so that (8.34) follows from (8.33). $\qquad\square$

Let us note that conditions (8.33) and (8.34) are counterparts of $3°$ in 7.4.5, and (7.15), respectively.

**8.2.7 Example** Let us consider the operator $A$ in $C_0(\mathbb{R})$ given by $Ax = x''$ for all twice differentiable functions $x \in C_0(\mathbb{R})$ with $x'' \in C_0(\mathbb{R})$. An analysis similar to the one given in 7.3.10 shows that $R_\lambda$ exists for all $\lambda > 0$ and that $\lambda R_\lambda$ is the operator related to the bilateral exponential distribution, say $\mu_\lambda$, with parameter $a = \sqrt{2\lambda}$ (see 7.5.3). Thus, (8.33) expresses the fact that, as $\lambda \to \infty$, the $\mu_\lambda$ converge weakly to the Dirac measure at 0. Moreover, the exponential of $A_\lambda$ is the operator related to the probability measure $e^{-\lambda t}e^{t\lambda\mu_\lambda}$ with exponential function taken in the sense of the algebra $\mathbb{BM}(\mathbb{R})$. By 6.2.9, the Fourier transform of this measure, a function of $\xi \in \mathbb{R}$, equals $e^{-\lambda t}e^{\lambda\hat{\mu}_\lambda(\xi)t}$, which by 6.4.11 equals $e^{-\lambda t}e^{\frac{2\lambda^2}{2\lambda+\xi^2}t}$, and tends to $e^{-t\frac{\xi^2}{2}}$. By the Continuity Theorem, $e^{-\lambda t}e^{\lambda\mu_\lambda t}$ converges to $N(0,t)$ and so by 5.4.18, $e^{A_\lambda t}$ converges to the operator related to the normal distribution. In other words, $e^{A_\lambda t}$ approximates the Brownian motion semigroup in $C_0(\mathbb{R})$.

**8.2.8 Example** Let $A$ be the operator of first derivative in $C_0(\mathbb{R})$ with suitably chosen (how?) domain. Then $\lambda R_\lambda$ exists for all $\lambda > 0$ and is the operator related to the exponential distribution $\mu_\lambda$ with parameter $\lambda$. Relation (8.33) has thus a familiar interpretation. Moreover, the Laplace transform of the measure related to $e^{A_\lambda t}$, a function of $\xi > 0$, equals $e^{-\lambda t}e^{\lambda\frac{\lambda}{\xi+\lambda}t}$ and converges to $e^{-\xi t}$, the Laplace transform of the Dirac measure at $t$. Hence $e^{A_\lambda t}$ converges to the semigroup of translations to the left, $T_t x(\tau) = x(\tau + t)$.

**8.2.9 Example** Let $Ax(\tau) = ax(\tau + 1) - ax(\tau), a > 0, x \in C_0(\mathbb{R})$ (or, say, $x \in BM(\mathbb{R})$). One checks that $(\lambda - A)^{-1}$ exists for all $\lambda > 0$, and $\lambda R_\lambda$ is the operator related to the geometric distribution $\mu_\lambda$ with parameter $p = \frac{\lambda}{\lambda+a}$ (see 6.1.5 and 7.5.10). Therefore, the probability generating function of the probability measure related to $e^{A_\lambda t}$, a function of $s \in [0,1]$, equals $e^{-\lambda t}e^{\frac{\lambda^2}{\lambda+(1-s)a}t}$. As $\lambda \to \infty$, this converges to $e^{-at}e^{-ast}$, which is the probability generating function of the Poisson distribution with parameter $at$. In other words, $e^{A_\lambda t}$ converges to the semigroup related to the Poisson process.

8.2.10 **Example**    Let $\mathbb{X} = C([0, \infty])$ be the space of continuous functions on $\mathbb{R}^+$ with a limit at infinity. Define the domain $\mathcal{D}(B)$ of an operator $B$ to be the set all differentiable functions $x \in \mathbb{X}$ such that $x(0) = 0$
and $x' \in \mathbb{X}$, and set $Bx = -x'$. The differential equation $\lambda x + x' = 0$,
where $\lambda > 0$ has infinitely many solutions $x(\tau) = ce^{-\lambda \tau}$ indexed by the
constant $c$. However, in $\mathcal{D}(B)$ there is only one solution to this equation,
namely the zero function. On the other hand, for any $y \in \mathbb{X}$ the function
$x(\tau) = \int_0^\tau e^{-\lambda(\tau-\varsigma)} y(\varsigma) \, d\varsigma$ satisfies $\lambda x + x' = y$ and belongs to $\mathcal{D}(B)$ (we
use de l'Hospital's rule to show that, if additionally $y$ is non-negative, $x$
has a limit at infinity, and then by linearity extend this claim to all $y$).
This shows that $(\lambda - B)^{-1}$ exists for $\lambda > 0$, and we have

$$\|(\lambda - B)^{-1}y\| \leq \sup_{\tau \geq 0} \int_0^\tau e^{-\lambda(\tau-\varsigma)} \, d\varsigma \, \|y\| = \int_0^\infty e^{-\lambda\varsigma} \, d\varsigma \, \|y\| = \frac{1}{\lambda}\|y\|,$$

so that the estimate (8.30) is satisfied with $M = 1$. We will show, however, that the Yosida approximation does not converge. To this end note
first that we have $(\lambda - B)^{-1}y = e_\lambda * y$ where $e_\lambda(\tau) = e^{-\lambda\tau}$. Next, if we
take any continuous function $z$ defined on $\mathbb{R}$ with compact support in
$\mathbb{R}_*^+$ then

$$\int_0^\infty [(\lambda - B)^{-1}y](\tau) \, z(\tau) \, d\tau = \int_0^\infty y(\tau) \, R_\lambda z(\tau) \, d\tau$$

where $R_\lambda$ is the operator considered in 8.2.8; this is a particular case of
(5.17). Thus,

$$\int_0^\infty (e^{B_\lambda t}y)(\tau)z(\tau) \, d\tau = \int_0^\infty y(\tau)(e^{A_\lambda t}z)(\tau) \, d\tau$$

where $A_\lambda$ is the Yosida approximation of the operator $A$ from 8.2.8 and
$B_\lambda$ is the Yosida approximation of $B$. We know that $e^{A_\lambda t}z$ converges
uniformly to the translation of $z$ to the left. For a $y$ with compact support
this implies that the last interval converges to $\int_0^\infty y(\tau)z(\tau+t) \, d\tau$. Hence
if $e^{B_\lambda t}y$ converges uniformly to some $T_t y$ then $\int_0^\infty T_t y(\tau)z(\tau) \, d\tau$ must be
equal to this last integral. Since $z$ is arbitrary, $T_t y(\tau) = 0$ for $\tau < t$ and
$T_t y(\tau) = y(\tau-t)$ for $\tau \geq 0$. This is a contradiction as long as we consider
$y$ with $y(0) \neq 0$, because then $T_t y$ defined above is not continuous and on
the other hand is supposed to be a uniform limit of continuous functions.

The reason why the Yosida approximation fails to converge here is the
fact that $\mathcal{D}(B)$ is not dense in $\mathbb{X}$; the closure $\mathbb{X}'$ of $\mathcal{D}(B)$ is the space of
all $x \in \mathbb{X}$ with $x(0) = 0$. The reader has probably noticed that for such
functions the argument presented above does not lead to contradiction
since the function $T_t y$ *is* continuous. In fact one may prove that for

$y \in \mathbb{Y}$, $e^{B_\lambda t}y$ converges strongly to the semigroup $\{T_t, t \geq 0\}$ in $\mathbb{Y}$, of translations to the right defined above.

**8.2.11 Example**     Let $l^\infty$ denote the space of bounded sequences $x = (\xi_n)_{n\geq 1}$ equipped with the supremum norm, and $A$ be the operator given by $Ax = (-n\xi_n)_{n\geq 1}$ defined on the domain $\mathcal{D}(A) \subset l^\infty$ composed of all $x$ such that $(-n\xi_n)_{n\geq 1}$ belongs to $l^\infty$. For any $\lambda > 0$ and $y = (\eta_n)_{n\geq 1} \in l^\infty$, the resolvent equation $\lambda x - Ax = y$ is equivalent to the system of infinitely many equations $\lambda \xi_n + n\xi_n = \eta_n, n \geq 1$, and thus has the unique solution in $\mathcal{D}(A)$ given by $\xi_n = \frac{1}{\lambda+n}\eta_n$. In other words, $\lambda R_\lambda y = \left(\frac{\lambda}{\lambda+n}\eta_n\right)_{n\geq 1}$, and so $\|\lambda R_\lambda\| \leq 1$, and the estimate (8.30) is satisfied with $M = 1$. Moreover, any $x \in \mathcal{D}(A)$ is also a member of $c_0$ because $|\xi_n| \leq \frac{1}{n}\|(n\xi_n)_{n\geq 1}\|$, and so $\mathbb{X}'$ is contained in $c_0$. Considering sequences $(\xi_n)_{n\geq 1}$ that are eventually zero, we prove that $\mathbb{X}'$ actually equals $c_0$. Despite this fact the Yosida approximation converges for all $x \in l^\infty$. Indeed, note first that

$$e^{A_\lambda t}x = \left(e^{-\lambda t}e^{\frac{\lambda^2}{\lambda+n}t}\xi_n\right)_{n\geq 1} = \left(e^{-\frac{\lambda n}{\lambda+n}t}\xi_n\right)_{n\geq 1}$$

and that $\lim_{\lambda\to\infty} e^{-\frac{\lambda n}{\lambda+n}t} = e^{-nt}, t \geq 0$. Defining

$$T_t x = \left(e^{-nt}\xi_n\right)_{n\geq 1}, x \in l^\infty,$$

we have, for any $k \in \mathbb{N}$,

$$\|T_t x - e^{A_\lambda t}x\| \leq \sup_{n\geq 1} |e^{-\frac{\lambda n}{\lambda+n}t} - e^{-nt}| \|x\|$$

$$\leq \sup_{1\leq n<k} |e^{-\frac{\lambda n}{\lambda+n}t} - e^{-nt}| \|x\| + 2e^{-\frac{\lambda k}{\lambda+k}t}\|x\| \qquad (8.35)$$

since the sequence $e^{-\frac{\lambda n}{\lambda+n}t}, n \geq 1$, decreases and, for $n \geq k$, $e^{-nt} \leq e^{-kt} \leq e^{-\frac{\lambda k}{\lambda+k}t}$. Now, for arbitrary $t > 0$ and $\epsilon > 0$ we choose a $k$ so that $2e^{-kt} < \frac{\epsilon}{2}$; in fact one such $k$ may be chosen for all $t$ larger than a given $t_0$. With this $k$ fixed we may choose a $\lambda$ large enough so that the supremum in (8.35) is less than $\frac{\epsilon}{2}$ and so is $2e^{-\frac{k\lambda}{\lambda+k}t}$, uniformly in any interval $[t_0, t_1]$ where $0 < t_0 < t_1$.

We have thus proven that the Yosida approximation converges for any $x \in l^\infty$ uniformly in any interval $[t_0, t_1]$. If $x$ belongs to $c_0$, the term $2e^{-\frac{\lambda k}{\lambda+k}t}\|x\|$ in (8.35) may be replaced by $2\sup_{n\geq k}|\xi_n|$ and consequently one may prove that for $x \in c_0$ the Yosida approximation converges uniformly in any interval $[0, t_1]$. The situation is typical in that the Yosida approximation of a densely defined operator always converges uniformly on compact subsets of $\mathbb{R}^+$ and, for non-densely defined operators, if it converges at all, it converges uniformly on compact subintervals of $\mathbb{R}_*^+$.

To the best of my knowledge, the reason why the Yosida approximation converges sometimes despite the fact that $\mathcal{D}(A)$ is not dense in $\mathbb{X}$ (and such situations happen quite often) is not fully known yet; only particular cases were studied. In particular there is no useful criterion for such convergence that would cover all important cases.

**8.2.12 Lemma** Suppose that the assumptions of 8.2.6 are satisfied. Then, for all $\mu, \lambda > 0$,

$$(\mu - A_\lambda)^{-1} = \frac{1}{\lambda + \mu} + \left(\frac{\lambda}{\lambda + \mu}\right)^2 R_{\frac{\lambda\mu}{\lambda+\mu}}. \tag{8.36}$$

Furthermore, $\lim_{\lambda \to \infty} (\mu - A_\lambda)^{-1} = R_\mu$, and the representations $H_\lambda$ of the algebra $L^1(\mathbb{R}^+)$ related to the exponential functions of the Yosida approximation converge strongly to a representation $H$ such that $H(e_\lambda) = R_\lambda$. In particular, $\|H\| \leq \limsup_{\lambda \to \infty} \|H_\lambda\| \leq M$.

*Proof* Observe that, by the Hilbert equation,

$$(I - \nu R_\lambda)(I + \nu R_{\lambda-\nu}) = (I + \nu R_{\lambda-\nu})(I - \nu R_\lambda) = I, \tag{8.37}$$

provided $\lambda - \nu > 0, \lambda, \nu > 0$. Fix $\mu, \lambda > 0$. For $\nu = \frac{\lambda^2}{\lambda+\mu}$ we have $\lambda - \nu = \frac{\lambda\mu}{\lambda+\mu} > 0$ and, by (8.37),

$$(I - \nu R_\lambda)^{-1} = \left(I - \frac{\lambda^2}{\lambda + \mu} R_\lambda\right)^{-1} = I + \frac{\lambda^2}{\lambda + \mu} R_{\frac{\lambda\mu}{\lambda+\mu}}.$$

Thus, for all $\mu, \lambda > 0$,

$$(\mu - A_\lambda)^{-1} = \left(\mu + \lambda - \lambda^2 R_\lambda\right)^{-1} = \frac{1}{\lambda + \mu}\left(I - \frac{\lambda^2}{\lambda + \mu} R_\lambda\right)^{-1}$$

$$= \frac{1}{\lambda + \mu} + \left(\frac{\lambda}{\lambda + \mu}\right)^2 R_{\frac{\lambda\mu}{\lambda+\mu}},$$

proving (8.36).

Observe now that the map $\mathbb{R}_*^+ \ni \lambda \to R_\lambda$ is continuous in the operator norm; indeed, by the Hilbert equation and (8.30),

$$\|R_\lambda - R_\mu\| \leq |\lambda - \mu| \frac{M^2}{\lambda\mu}. \tag{8.38}$$

Hence $\lim_{\lambda \to \infty} (\mu - A_\lambda)^{-1} = R_\mu$ and the rest follows by 7.4.48. $\qquad\square$

**8.2.13 Exercise**    Suppose that (8.30) holds. By (8.32), $(\mu - A_\lambda)^{-1} = \int_0^\infty e^{-\mu t} e^{tA_\lambda} dt$ satisfies $\|(\mu - A)^{-n}\| \le \frac{M}{\mu^n}$. Prove this estimate directly using (8.36).

*Proof (of Theorem 8.2.3)*

(a)  If (8.30) holds, then by 7.4.48 (or 7.4.49) and 8.2.12, the limit $U(t) = \lim_{\lambda \to \infty} U_\lambda(t)$, where $U_\lambda(t) = \int_0^t e^{A_\lambda s} ds$, exists in the strong topology (it exists in the operator topology as well, but we will not need this result here). Hence, by (8.32), $\|U(t) - U(s)\| \le M(t - s), t \ge s$.

Next, for $\mu > 0$,

$$\mu \int_0^\infty e^{-\mu t} U(t) \, dt = \mu \lim_{\lambda \to \infty} \int_0^\infty e^{-\mu t} \int_0^t e^{A_\lambda s} \, ds \, dt$$

$$= \lim_{\lambda \to \infty} \int_0^\infty e^{-\mu t} e^{A_\lambda t} \, dt = \lim_{\lambda \to \infty} (\mu - A_\lambda)^{-1} = R_\mu.$$

Here, the second equality follows by simple calculation (integration by parts) and the third by Lemma 8.2.12. The first equality is a direct consequence of the Dominated Convergence Theorem for Banach space valued functions; since we do not have it at our disposal, however, we need to establish it in another way. To this end we note that, since both $\| \int_T^\infty e^{-\lambda t} U(t) \, dt\|$ and $\| \int_T^\infty e^{-\lambda t} U_\lambda(t) \, dt\|$, where $T > 0$, are bounded by $M \int_T^\infty e^{-\lambda t} t \, dt$, given $\epsilon > 0$ we may choose a $T$ such that the norms of both integrals are less than $\epsilon$. Hence, it suffices to show that, for $x \in \mathbb{X}$ with $x \ne 0$, $U_\lambda(t)x$ converges to $U(t)x$ uniformly on any interval $[0, T]$. Given $\epsilon > 0$, we choose a $k \in \mathbb{N}$ with $k > 3TM\|x\|\epsilon^{-1}$. Then, we choose a $\lambda_0$ such that for $\lambda > \lambda_0$, $\|U_\lambda(t_i)x - U(t_i)x\| < \frac{\epsilon}{3}$, for $i = 0, ..., k$, where $t_i = \frac{iT}{k}$. Then, for any $t \in [0, T]$ there exists an $i$ with $|t - t_i| \le \frac{T}{k}$. Therefore, by the triangle inequality applied to $U(t)x - U_\lambda(t)x = [U(t)x - U(t_i)x] + [U(t_i)x - U_\lambda(t_i)x] + [U_\lambda(t_i)x - U_\lambda(t)x]$, for $\lambda > \lambda_0$, $\|U(t)x - U_\lambda(t)x\| < \epsilon$, as desired (comp. 5.7.19).

Hence, the family $\{U(t), t \ge 0\}$ has the required properties. Conversely, suppose that (8.31) is satisfied and $\|U(t) - U(s)\| \le M(t - s)$. Let $F \in \mathbb{X}^*$ and $x \in \mathbb{X}$. The real-valued function $t \to F(U(t)x)$ is Lipschitz continuous with $|F(U(t)) - F(U(s))| \le M\|F\| \|x\|(t - s)$. Hence, it is differentiable almost everywhere and $|\frac{d}{dt}F(U(t)x)| \le M\|F\| \|x\|$. Integrating by parts, $F(R_\lambda x) = \int_0^\infty e^{-\lambda t} \frac{d}{dt} F(U(t)x) \, dt$. Hence,

$$\|R_\lambda\| = \sup_{\|x\|=1} \sup_{\|F\|_{\mathbb{X}^*}=1} |F(R_\lambda x)| \le \int_0^\infty e^{-\lambda t} M \, dt = \frac{M}{\lambda}. \tag{8.39}$$

As in (8.38), this implies that $\lambda \to R_\lambda$ is continuous. Consequently, using an induction argument, by the Hilbert equation (7.33), it is differentiable with

$$\frac{d^n}{d\lambda^n} R_\lambda = (-1)^n n! R_\lambda^{n+1}. \tag{8.40}$$

On the other hand, for any $F$ and $x$ as above, a direct calculation shows that

$$\frac{d^n}{d\lambda^n} F(R_\lambda x) = \int_0^\infty e^{-\lambda t}(-t)^n \frac{d}{dt} F(U(t)x)\, dt.$$

Arguing as in (8.39), we obtain $\|\frac{d^n}{d\lambda^n} R_\lambda\| \le M \int_0^\infty e^{-\lambda t} t^n\, dt = \frac{Mn!}{\lambda^{n+1}}$ and this gives (8.30) by (8.40).

(b) Let $x \in \mathcal{D}(A)$. Observe that $\lambda R_\lambda x \in \mathcal{D}(A)$, and that $A(\lambda R_\lambda x) = \lambda R_\lambda A x \in D(A) \subset \mathbb{X}'$. This means that if $x \in \mathcal{D}(A)$, then $\lambda R_\lambda x \in \mathcal{D}(A_p)$. Moreover, by (8.33), $\lim_{\lambda \to \infty} \lambda R_\lambda x = x$, proving that $\mathcal{D}(A_p)$ is dense in $\mathcal{D}(A)$. Thus, $\mathcal{D}(A_p)$ is dense in $\mathbb{X}'$, too.

Let $x \in \mathcal{D}(A_p)$. We write

$$e^{A_\lambda t} x - x = \int_0^t \frac{d}{ds} e^{A_\lambda s} x\, ds = \int_0^t e^{A_\lambda s} A_\lambda x\, ds = U_\lambda(t) A_\lambda x.$$

By (8.34), $A_\lambda x$ converges, as $\lambda \to \infty$, to $Ax$. Moreover, $\|U_\lambda(t)\| \le Mt$ and the operators $U_\lambda(t)$ converge strongly to $U(t)$. Hence, the limit

$$T(t)x = \lim_{\lambda \to \infty} e^{A_\lambda t} x = x + U(t)Ax \tag{8.41}$$

exists for all $x \in \mathcal{D}(A_p)$. Since $\mathcal{D}(A_p)$ is dense in $\mathbb{X}'$ and we have (8.32), this limit exists for all $x \in \mathbb{X}'$, and $\|T(t)\|_{\mathcal{L}(\mathbb{X}')} \le M$. Clearly $e^{A_\lambda t} x$ belongs to $\mathbb{X}'$ and so does $T(t)x$. The semigroup property of $\{T(t), t \ge 0\}$ follows from the semigroup property of $\{e^{A_\lambda t}, t \ge 0\}$. Furthermore, by $\|U(t)Ax\| \le Mt\|Ax\|$, we have $\lim_{t \to 0+} T(t)x = x$, $x \in \mathcal{D}(A_p)$. Since $\|T(t)\| \le M$ and $D(A_p)$ is dense in $\mathbb{X}'$, $\lim_{t \to 0+} T(t)x = x$, $x \in \mathbb{X}'$. Finally, by (8.31), for $x \in \mathcal{D}(A_p)$ and $\lambda > 0$,

$$\int_0^\infty e^{-\lambda t} T(t)x\, dt = \frac{1}{\lambda}(x + R_\lambda Ax) = R_\lambda x; \tag{8.42}$$

since the operators on both sides of this equality are bounded and $D(A_p)$ is dense in $\mathbb{X}'$, this equality is true for all $x \in \mathbb{X}'$.

(c) By (8.42) and integration by parts,

$$\lambda \int_0^\infty e^{-\lambda t} \int_0^t T(s)x\, ds\, dt = R_\lambda x, \qquad x \in \mathbb{X}'.$$

Comparing this with (already established) equality (8.31), by 7.4.34, $U(t)x = \int_0^t T(s)x\, ds\, dt$, $x \in \mathbb{X}'$. In particular, (8.41) takes on the form

$T(t)x = x + \int_0^t T(s)Ax\,ds$, $x \in \mathcal{D}(A_p)$. By 7.4.26 this implies that $\mathcal{D}(A_p)$ is a subset of the domain $\mathcal{D}(G)$ of the infinitesimal generator $G$ of $\{T(t), t \geq 0\}$, and $Ax = Gx$ for $x \in \mathcal{D}(A_p)$. Finally, if $x \in \mathcal{D}(G)$ then there exists a $y \in \mathbb{X}'$ and a $\lambda > 0$ such that $x = (\lambda - G)^{-1}y$. On the other hand, $(\lambda - G)^{-1}y$ is the Laplace transform of $t \mapsto T(t)y$ and hence, by (8.42), equals $R_\lambda y$. This shows $x \in \mathcal{D}(A)$. Since $y$ belongs to $\mathbb{X}'$, $Ax = AR_\lambda y = \lambda R_\lambda y - y \in \mathbb{X}'$, showing that $x \in \mathcal{D}(A_p)$, i.e. $\mathcal{D}(G) \subset \mathcal{D}(A_p)$.                                               □

**8.2.14 Corollary**    One of the by-products of the proof is that the semigroup generated by $A$ is the limit, as $\lambda \to \infty$, of exponential functions $e^{A_\lambda t}$. This important result has many applications, one of them we will need later is that if the operators $R_\lambda$ are non-negative, then so are $T_t$. The proof of the converse statement is elementary.

**8.2.15 Exercise**    Show that, for all $x \in \mathbb{X}$, $\lambda > 0$ and $t \geq 0$,

$$T(t)R_\lambda x = U(t)\lambda R_\lambda - U(t)x + R_\lambda x.$$

**8.2.16 *The algebraic version of the Hille–Yosida theorem***    Under assumptions of 8.2.3, (8.30) is satisfied iff there exists a representation $H$ of $L^1(\mathbb{R}^+)$ in $\mathcal{L}(\mathbb{X})$ such that $H(e_\lambda) = R_\lambda, \lambda > 0$, where $e_\lambda \in L^1(\mathbb{R}^+)$ are defined by their representatives $e_\lambda(\tau) = e^{-\lambda\tau}$ (as in 2.2.49). In such a case $U(t) = H(1_{[0,t)})$,

$$\mathbb{X}' = \{x \in \mathbb{X} \mid x = H(\phi)y, \phi \in L^1(\mathbb{R}^+), y \in \mathbb{X}\}, \qquad (8.43)$$

and

$$T(t)x = T(t)H(\phi)y = H(S(t)\phi)y \qquad (8.44)$$

where $\{S(t), t \geq 0\}$ is the semigroup of translations to the right in $L^1(\mathbb{R}^+)$ defined in 7.4.2.

*Proof*    A direct calculation shows that $e_\lambda * e_\lambda(\tau) = \tau e^{-\lambda\tau}$. More generally, by induction we show that $e_\lambda^{*n}(\tau) = \frac{\tau^{n-1}}{(n-1)!}e^{-\lambda\tau}$. In particular, $\|e_\lambda^{*n}\|_{L^1(\mathbb{R}^+)} = \frac{1}{\lambda^n}$. Therefore, if $H$ is a representation of $L^1(\mathbb{R}^+)$ such that $H(e_\lambda) = R_\lambda$ and $\|H\| \leq M$, then $\|R_\lambda^n\|_{\mathcal{L}(\mathbb{X})} = \|H(e_\lambda^{*n})\|_{\mathcal{L}(\mathbb{X})} \leq M\|e_\lambda^{*n}\|_{L^1(\mathbb{R}^+)} = \frac{M}{\lambda^n}$, i.e. $R_\lambda$ satisfies (8.30). The other implication was shown in Lemma 8.2.12.

Next, $\|H(1_{[0,t)}) - H(1_{[0,s)})\| = \|H(1_{[s,t)})\| \leq M(t - s), t \geq s$, so that $t \mapsto H(1_{[0,t)})$ is continuous and there exists the (improper, Riemann) integral $\int_0^\infty e^{-\lambda t} H(1_{[0,t)})\,dt$ which, by Exercise 2.2.49, equals

$H\left(\int_0^\infty e^{-\lambda t} 1_{[0,t)}\,\mathrm{d}t\right) = \frac{1}{\lambda}H(e_\lambda) = \frac{1}{\lambda}R_\lambda$. By (8.31) and 7.4.34, this implies $U(t) = H(1_{[0,t)})$.

Furthermore, by $H(e_\lambda) = (\lambda - A)^{-1}$, we have $D(A) = \{x \in \mathbb{X} |\, x = H(e_\lambda)y, y \in \mathbb{X}\} \subset \mathbb{X}''$ where $\mathbb{X}''$ is the right-hand side of (8.43). Hence $\mathbb{X}' \subset cl\mathbb{X}''$. Moreover, since $e_\lambda, \lambda > 0$, are linearly dense in $L^1(\mathbb{R}^+)$, $\mathbb{X}'' \subset \mathbb{X}'$. Therefore, $\mathbb{X} = cl\mathbb{X}''$. Now, Cohen's Factorization Theorem shows that $\mathbb{X}''$ is closed, and (8.43) follows.

Finally, $t \mapsto H(S(t)\phi)y$ is continuous, and bounded by $M\|\phi\|\,\|y\|$; hence its Laplace transform exists. By 7.4.31, it equals

$$H(\int_0^\infty e^{-\lambda t}S(t)\phi\,\mathrm{d}t)y = H(e_\lambda * \phi)y = H(e_\lambda)H(\phi)y = R_\lambda H(\phi)y.$$

This shows both (8.44) and the fact that the definition does not depend on the choice of $\phi$ and $y$ but solely on $x = H(\phi)y$. □

**8.2.17 Remarks** (a) Without Factorization Theorem 6.7.1, we could merely prove that $\mathbb{X}'$ is the closure of the right-hand side of (8.43), and that (8.44) holds for $x$ of the form $x = H(\phi)y$.

(b) With cosmetic changes, our argument shows *existence* of both the integrated semigroup related to $A$ and the semigroup generated by $A_\mathrm{p}$. In other words, we have an independent proof of (the sufficiency part of) the Hille–Yosida theorem.

**8.2.18 Example** *Elastic Brownian motions* Given a number $\lambda > 0$, and a continuous function $y : \mathbb{R}^+ \to \mathbb{R}$, with a finite limit at infinity, let us consider the differential equation

$$\lambda x - \frac{1}{2}x'' = y \tag{8.45}$$

where $x$ is supposed to be twice continuously differentiable. Fix $\epsilon > 0$. We claim that there exists a unique twice continuously differentiable function $x$ satisfying (8.45), such that the limit $\lim_{\tau \to \infty} x(\tau)$ exists, and

$$x(0) = \epsilon x'(0). \tag{8.46}$$

To see that, note first that looking for solutions to Equation (8.45) of the form $x(\tau) = e^{-\sqrt{2\lambda}\tau}z(\tau)$, using the Laplace transform, or another standard method, one can easily find out that the general solution of the equation is

$$x(\tau) = C_1 e^{\sqrt{2\lambda}\tau} + C_2 e^{-\sqrt{2\lambda}\tau} + \sqrt{\frac{2}{\lambda}} \int_0^\tau \sinh[\sqrt{2\lambda}(\sigma - \tau)]y(\sigma)\,d\sigma$$

$$= \left[ C_1 - \frac{1}{\sqrt{2\lambda}} \int_0^\tau e^{-\sqrt{2\lambda}\tau} y(\sigma)\,d\sigma \right] e^{\sqrt{2\lambda}\tau}$$

$$+ \left[ C_2 + \frac{1}{\sqrt{2\lambda}} \int_0^\tau e^{\sqrt{2\lambda}\sigma} y(\sigma)\,d\sigma \right] e^{-\sqrt{2\lambda}\tau}. \tag{8.47}$$

Observe that, by de l'Hospital's rule,

$$\lim_{\tau \to \infty} \left[ C_2 + \frac{1}{\sqrt{2\lambda}} \int_0^\tau e^{\sqrt{2\lambda}\sigma} y(\sigma)\,d\sigma \right] e^{-\sqrt{2\lambda}\tau} = \lim_{\tau \to \infty} y(\tau).$$

Thus $\lim_{\tau \to \infty} x(\tau)$ exists iff the limit

$$\lim_{\tau \to \infty} \left[ C_1 - \frac{1}{\sqrt{2\lambda}} \int_0^\tau e^{-\sqrt{2\lambda}\sigma} y(\sigma)\,d\sigma \right] e^{\sqrt{2\lambda}\tau}$$

exists. This is the case when

$$C_1 = C_1(\lambda, y) = \frac{1}{\sqrt{2\lambda}} \int_0^\infty e^{-\sqrt{2\lambda}\sigma} y(\sigma)\,d\sigma. \tag{8.48}$$

Now, we demand additionally that (8.46) holds, to obtain

$$C_1 + C_2 = \epsilon\sqrt{2\lambda}(C_1 - C_2),$$

i.e.

$$C_2 = C_2(\lambda, \epsilon, y) = \frac{\epsilon\sqrt{2\lambda} - 1}{\epsilon\sqrt{2\lambda} + 1} C_1, \tag{8.49}$$

and this completes the proof of our claim. We observe also that, by (8.45), $\lim_{\tau \to \infty} x''(\tau)$ exists.

Let us consider the space $\mathbb{X} = C[0, \infty]$ of all continuous functions $x : \mathbb{R}^+ \to \mathbb{R}$ with a finite limit at infinity, equipped with the norm $\|x\| = \sup_{\tau \geq 0} |x(\tau)|$. Given $\epsilon > 0$, define the domain of an operator $A_\epsilon$ as the set of all twice continuously differentiable functions $x$ with $x'' \in \mathbb{X}$, which satisfy (8.46), and set $A_\epsilon x = \frac{1}{2}x''$.

We claim now that for every $\epsilon > 0$, the operator $A_\epsilon$ is the infinitesimal generator of a positive contraction semigroup $\{T_t^\epsilon, t \geq 0\}$, acting in $\mathbb{X}$. To see that note first that the domain $D(A_\epsilon)$ of $A_\epsilon$ is dense in $\mathbb{X}$. Indeed, the set of all twice continuously differentiable functions is dense in $\mathbb{X}$, and, for every $\delta > 0$ and every twice continuously differentiable function $x \in \mathbb{X}$ with $x'' \in \mathbb{X}$ such that $x \notin D(A_\epsilon)$, there exists a function $x_\delta$ enjoying $\|x_\delta - x\| < \delta$ and $x_\delta \in D(A_\epsilon)$. For example, we may put $x_\delta(\tau) =$

$x(\tau) + ae^{-b\tau}$ where $a = \frac{1}{2}\min\left(\delta, |\epsilon x'(0) - x(0)|\right)\operatorname{sgn}(\epsilon x'(0) - x(0)), b =$
$\frac{1}{\epsilon}\left[\frac{\epsilon x'(0) - x(0)}{a} - 1\right]$ and observe that $b \geq \frac{1}{\epsilon}\left(2\frac{|\epsilon x'(0) - x(0)|}{|\epsilon x'(0) - x(0)|} - 1\right) = \frac{1}{\epsilon} > 0$
and $\epsilon(x_\delta)'(0) = \epsilon x'(0) - [\epsilon x'(0) - x(0) - a] = x(0) + a = x_\delta(0)$ and
$\|x - x_\delta\| = |a| < \delta$.

Furthermore, by (8.47)–(8.49), we have, for $\lambda > 0$, $\tau \geq 0$,

$$R_\lambda x(\tau) := (\lambda - A_\epsilon)^{-1} x(\tau)$$

$$= \frac{1}{\sqrt{2\lambda}}\int_0^\infty e^{-\sqrt{2\lambda}|\tau - \sigma|} y(\sigma)\,d\sigma$$

$$+ H(\epsilon\sqrt{2\lambda})\frac{1}{\sqrt{2\lambda}}\int_0^\infty e^{-\sqrt{2\lambda}(\sigma + \tau)} y(\sigma)\,d\sigma$$

$$= \frac{1}{\sqrt{2\lambda}}\int_{-\infty}^\infty e^{-\sqrt{2\lambda}|\tau - \sigma|} y^*(\sigma)\,d\sigma \qquad (8.50)$$

where $H(u) = \frac{u-1}{u+1}$ and

$$y^*(\tau) = \begin{cases} y(\tau), & \tau \geq 0, \\ H(\epsilon\sqrt{2\lambda})y(-\tau), & \tau < 0. \end{cases}$$

The values of the linear fractional function $H$ lie in $[-1, 1]$ for non-negative $u$. Thus $|H(\epsilon\sqrt{2\lambda})| < 1$ and $\sup_{\tau \in \mathbb{R}}|y^*(\tau)| = \sup_{\tau \geq 0}|y(\tau)|$, so that, for $\lambda > 0$,

$$\|R_\lambda(A_\epsilon)x\| \leq \frac{1}{\sqrt{2\lambda}}\int_{-\infty}^\infty e^{-\sqrt{2\lambda}|\sigma|}\,d\sigma\|y^*\| = \frac{\|y\|}{\lambda}, \qquad (8.51)$$

which, by the Hille–Yosida theorem, proves our claim.

The semigroups $T_t^\epsilon, t \geq 0$ are related to so-called **elastic Brownian motion** and the parameter $\epsilon$ measures the degree in which the barrier $\tau = 0$ is "sticky". Some insight into the way the particle behaves at the boundary may be gained from the following analysis.

Let us rewrite (8.50) in the form

$$R_\lambda(A_\epsilon)x(\tau) = \frac{1}{\sqrt{2\lambda}}\int_0^\infty \left[e^{-\sqrt{2\lambda}|\sigma - \tau|} + e^{-\sqrt{2\lambda}(\sigma + \tau)}\right] y(\sigma)\,d\sigma$$

$$- \frac{2}{\epsilon\sqrt{2\lambda} + 1}\frac{1}{\sqrt{2\lambda}}\int_0^\infty e^{-\sqrt{2\lambda}(\sigma + \tau)} y(\sigma)\,d\sigma. \qquad (8.52)$$

Notice that the function $\mathbb{R}_*^+ \ni \tau \to \frac{1}{\epsilon\tau + 1} \in \mathbb{R}$ as well as the derivative of the positive function $\mathbb{R}_*^+ \ni \tau \to \sqrt{2\tau} \in \mathbb{R}$ are completely monotone (see [41] p. 415 for the definition) and, thus, by Criterion 2, p. 417 and Theorem 1a, p. 416 of the above mentioned monograph, for any $\epsilon \geq 0$

there exists a measure $\mu_\epsilon$ on $\mathbb{R}^+$ such that

$$\frac{1}{\epsilon\sqrt{2\lambda+1}} = \int_0^\infty e^{-\lambda t}\mu_\epsilon\,dt.$$

By setting $\lambda = 0$ we see that $\mu_\epsilon$ is a probability measure. By 6.4.7 the transition probability of the stochastic process governed by the semigroup $T_t^\epsilon, t \geq 0$, is given by

$$K_\epsilon(t,\tau,\Gamma) = \frac{1}{\sqrt{2\pi t}}\int_\Gamma [e^{-\frac{(\tau-\sigma)^2}{2t}} + e^{-\frac{(\tau+\sigma)^2}{2t}}]\,d\sigma \qquad (8.53)$$

$$-2\int_\Gamma\int_{[0,t)} \frac{1}{\sqrt{2\pi(t-s)}}e^{-\frac{(\tau+\sigma)^2}{2(t-s)}}\mu_\epsilon(ds)\,d\sigma, t \geq 0, \tau \geq 0.$$

Comparing this formula with (8.5), (8.20) and (8.22) we see that the measure $\mu_\epsilon$ governs the probability of annihilation of the Brownian traveller touching the screen $\tau = 0$. Here we are not able to dive more into this difficult subject. In [58] pp. 45–47 a more detailed probabilistic treatment based on the employment of P. Levy's local time is presented. A more modern presentation may be found in [100].

8.2.19 *The Phillips Perturbation Theorem*    Suppose that $A$ is the generator of a strongly continuous semigroup $\{T(t), t \geq 0\}$ satisfying (7.14), and $B$ is a bounded linear operator. Then, $A + B$ with domain $\mathcal{D}(A)$ is the generator of a strongly continuous semigroup $\{S(t), t \geq 0\}$ such that

$$\|S(t)\| \leq Me^{(\omega+M\|B\|)t}. \qquad (8.54)$$

*Proof* Let $\mathbb{X}$ be the space where $\{T(t), t \geq 0\}$ is defined. Define inductively bounded linear operators $S_n(t) \in \mathcal{L}(\mathbb{X})$, by $S_0(t) = T(t), t \geq 0$, and

$$S_{n+1}(t)x = \int_0^t T(t-s)BS_n(s)x\,ds, \qquad t \geq 0, x \in \mathbb{X}, n \geq 0. \qquad (8.55)$$

By induction, $\|S_n(t)\| \leq \frac{M^{n+1}\|B\|^n t^n}{n!}e^{\omega t}$, and the series $\sum_{n=0}^\infty S_n(t)$ converges in the operator norm, uniformly in $t$ in compact intervals. Its sum, $S(t)$, is a strongly continuous family of operators, being a limit of strongly continuous families, and (8.54) holds. Moreover, a straightforward calculation shows that, for $\lambda > \omega, n \geq 0$, $\int_0^\infty e^{-\lambda s}S_{n+1}(s)x\,ds = R_\lambda B\int_0^\infty e^{-\lambda s}S_n(s)x\,ds$, where $R_\lambda x = \int_0^\infty e^{-\lambda t}S_0(t)x\,dt$ is the resolvent of $A$ (note that $S_{n+1}$ is a *convolution* of $S_n$ and $T$ and use linear functionals as in 7.2.1 and 7.2.4 to justify the change of the order of integration). Hence, by induction $\int_0^\infty e^{-\lambda t}S_n(t)x\,dt = (R_\lambda B)^n R_\lambda x$. Using (8.54) and uniform convergence of the series on compact intervals we argue that for $\lambda > \omega + M\|B\|$,

$$\int_0^\infty e^{-\lambda t} S(t)x\, dt = \sum_{n=0}^\infty \int_0^\infty e^{-\lambda t} S_n(t)x\, dt = \sum_{n=0}^\infty (R_\lambda B)^n R_\lambda x.$$

For such a $\lambda$ and $x \in \mathcal{D}(A)$, we have $\sum_{n=0}^\infty (R_\lambda B)^n R_\lambda (\lambda - A - B)x = \sum_{n=0}^\infty (R_\lambda B)^n x - \sum_{n=0}^\infty (R_\lambda B)^{n+1} x = x$. Similarly, for $x \in \mathbb{X}$ and $n \in \mathbb{N}$, the sum $\sum_{k=0}^n (R_\lambda B)^k R_\lambda x = R_\lambda \sum_{k=0}^n (BR_\lambda)^k x$ belongs to $\mathcal{D}(A)$ and $(\lambda - A - B)\sum_{k=0}^n (R_\lambda B)^k R_\lambda x = x - (BR_\lambda)^{n+1} x$. Since $\|BR_\lambda\| \le \frac{M\|B\|}{\lambda - \omega} < 1$, we have $\lim_{n\to\infty} (BR_\lambda)^{n+1} x = 0$. $\lambda - A - B$ being closed, $\sum_{n=0}^\infty (R_\lambda B)^n R_\lambda x \in \mathcal{D}(A)$ and $(\lambda - A - B)\sum_{n=0}^\infty (R_\lambda B)^n R_\lambda x = x$, proving that

$$\int_0^\infty e^{-\lambda t} S(t)x\, dt = (\lambda - A - B)^{-1}x, \qquad x \in \mathbb{X}, \lambda > \omega + M\|B\|. \quad (8.56)$$

This implies $\frac{d^n}{d\lambda^n}(\lambda - A - B)^{-1}x = \int_0^\infty e^{-\lambda t}(-t)^n S(t)x\, dt, n \ge 0$. By the Hilbert equation, $\frac{d^n}{d\lambda^n}(\lambda - A - B)^{-1} = (-1)^n n!(\lambda - A - B)^{-(n+1)}$ (see Exercise 7.4.39). Hence, by the Hille–Yosida theorem, (8.54) shows that $A + B$ is the generator of a strongly continuous semigroup. (8.56) proves now that the semigroup generated by $A + B$ equals $\{S(t), t \ge 0\}$ – see 7.4.34. □

**8.2.20 Corollary** The semigroup generated by $A + B$ is given by $S(t) = \sum_{n=0}^\infty S_n(t)$, with the limit in operator topology being uniform in $t$ in compact intervals.

**8.2.21 Example** If $B$ commutes with all $T(t), t \ge 0$, then $S(t) = T(t)\exp(tB)$. Offer two proofs of this result: a proof based on 8.2.20 and a direct one.

**8.2.22 Example** *A semigroup generated by an integro-differential equation* Let us consider the semigroup $\{U_t, t \ge 0\}$ defined by (8.16) with the transition kernel $K$ related to the Brownian motion. To be more specific, $K(t, \tau, \cdot)$ is the measure with density $k(t, \tau, \sigma) = \frac{1}{\sqrt{2\pi t}} e^{-\frac{(\sigma - \tau)^2}{2t}}$. If a probability measure $\mu$ is absolutely continuous with respect to Lebesgue measure, and has a density $x$, then also the measure $U_t\mu$ is absolutely continuous, and has a density $y(\sigma) = \int_{-\infty}^\infty \frac{1}{\sqrt{2\pi t}} e^{-\frac{(\sigma - \varsigma)^2}{2t}} x(\varsigma)\, d\varsigma$. In other words, $\{U_t, t \ge 0\}$ leaves the subspace $L^1(\mathbb{R})$ of $\mathbb{BM}(\mathbb{R})$ invariant. A straightforward calculation shows that this semigroup, as restricted to

$L^1(\mathbb{R})$, is a strongly continuous semigroup of Markov operators. The infinitesimal generator of the restricted semigroup is $Ax = \frac{1}{2}\frac{\mathrm{d}^2 x}{\mathrm{d}\tau^2}$ defined for $x \in \mathcal{D}(A) \subset L^1(\mathbb{R})$, composed of $x$ such that $x''$ exists almost everywhere, is absolutely integrable and $x(\tau) = x'(\tau) + \int_0^\tau x''(\sigma)\,\mathrm{d}\sigma$.

Moreover, let us consider a stochastic kernel $b$, i.e. a measurable function of two real variables such that $b \geq 0$, $\int_\mathbb{R} b(\tau,\sigma)\,\mathrm{d}\sigma = 1, \tau \in \mathbb{R}$. In other words, for each $\tau \in \mathbb{R}, b(\tau,\cdot)$ is a density of a measure, say $K(\tau,\cdot)$, on $\mathbb{R}$. Given a non-negative number $a$ and such a kernel, we may construct the transition family related to a pseudo-Poisson process, as described in 8.1.28. The special form of $K(\tau,\cdot)$ forces the related semigroup $\{U_t, t \geq 0\}$ in $\mathbb{BM}(\mathbb{R})$, as was the case with the Brownian motion semigroup discussed above, to leave the space $L^1(\mathbb{R})$ invariant. Indeed, for $\tau \in \mathbb{R}$, the kernels $K^n(\tau,\cdot), n \geq 1$, have densities $b_n(\tau,\sigma)$ given inductively by $b_1 = b$, $b_{n+1}(\tau,\sigma) = \int_\mathbb{R} b(\tau,\varsigma)b_n(\varsigma,\sigma)\,\mathrm{d}\varsigma$. Therefore, $K(t,\tau,\cdot)$ has a mass of $\mathrm{e}^{-at}$ at $\tau = 0$ and, apart from this point, a density

$$b(t,\tau,\cdot) = \sum_{n=1}^{\infty} \mathrm{e}^{-at}\frac{a^n t^n}{n!}b_n(\tau,\cdot), \quad \text{(convergence in } L^1(\mathbb{R})\text{)}.$$

We note that $\|b_n(\tau,\cdot)\|_{L^1(\mathbb{R})} = 1$ and $\|b(t,\tau,\cdot)\|_{L^1(\mathbb{R})} = 1 - \mathrm{e}^{-at}$, $\tau \in \mathbb{R}, t \geq 0$. Consequently, if $\mu$ is absolutely continuous with respect to the Lebesgue measure and has a density $x$, then $U_t\mu$ is absolutely continuous, too, and has a density $y(\sigma) = \mathrm{e}^{-at}x(\sigma) + \int_\mathbb{R} b(t,\tau,\sigma)x(\tau)\,\mathrm{d}\tau$.

The last formula implies, furthermore, that $\{U_t, t \geq 0\}$ restricted to $L^1(\mathbb{R})$ is continuous in the operator norm and that its infinitesimal generator is $aBx - ax$ where $Bx(\tau) = \int_\mathbb{R} b(\tau,\sigma)x(\sigma)\,\mathrm{d}\sigma$, and is a Markov, hence bounded, operator.

By the Phillips Perturbation Theorem, the operator $\frac{1}{2}\frac{\mathrm{d}^2 x}{\mathrm{d}\tau^2} + aBx$ defined on $\mathcal{D}(A)$ is the generator of a strongly continuous semigroup in $L^1(\mathbb{R})$, say $\{S(t), t \geq 0\}$. Using 8.2.20, we see that $S(t)$ are non-negative operators. To be more specific, an inductive argument shows that for a density $x \in L^1$, $S_n(t)x$ is non-negative and $\int_\mathbb{R} S_n(t)x(\tau)\,\mathrm{d}\tau = \frac{a^n t^n}{n!}$; in particular $\|S_n(t)\| = \frac{a^n t^n}{n!}$. Hence, in agreement with (8.54), $\|S(t)\| = \mathrm{e}^{at}$. Now, the semigroup $\{\mathrm{e}^{-at}S(t), t \geq 0\}$ generated by $\frac{1}{2}\frac{\mathrm{d}^2 x}{\mathrm{d}\tau^2} + aBx - ax$ with domain $\mathcal{D}(A)$ is a strongly continuous semigroup, and operators $\mathrm{e}^{-at}S(t)$ are Markov operators.

Consequently, the integro-differential equation

$$\frac{\partial x(t,\tau)}{\partial t} = \frac{1}{2}\frac{\partial^2 x(t,\tau)}{\partial \tau^2} + a\int_\mathbb{R} b(\tau,\sigma)x(\sigma)\,\mathrm{d}\sigma - ax(t,\tau), \quad x(0,\cdot) \in \mathcal{D}(A),$$

$$(8.57)$$

describes the evolution of densities of a Brownian motion that at the epochs of the Poisson process with intensity $a$ is perturbed by jumps of the pseudo-Poisson process discussed above.

**8.2.23 Example** Let $\mathbb{X}$ and $\mathbb{Y}$ be two Banach spaces, and let $K : \mathbb{X} \to \mathbb{Y}$ and $\Theta : \mathbb{Y} \to \mathbb{X}$ be two bounded linear operators such that $K\Theta y = y, y \in \mathbb{Y}$; in the non-trivial case, though, we *do not* have $\Theta K x = x, x \in \mathbb{X}$. Also, let $\{T(t), t \geq 0\}$ and $\{P(t), t \geq 0\}$ be two strongly continuous semigroups in $\mathbb{X}$ and $\mathbb{Y}$, respectively, and let $A$ be the generator of $\{T(t), t \geq 0\}$. We assume that

$$P(t)K = KT(t), \qquad t \geq 0. \tag{8.58}$$

To persuade yourself that such semigroups and operators exist, see [16] pp. 60–62. We note here that letting $P^\sharp(t) = \Theta P(t)K, t \geq 0$, we obtain $P^\sharp(t)P^\sharp(s) = P^\sharp(t+s)$ and that $[0, \infty) \ni t \mapsto P^\sharp(t)x$ is strongly continuous for all $x \in \mathbb{X}$, yet in general $\{P^\sharp(t), t \geq 0\}$ is *not* a strongly continuous semigroup, since $P^\sharp(0) = \Theta K \neq I_\mathbb{X}$.

We claim that the semigroup $\{S(t), t \geq 0\}$ generated by $A + \Theta K$ (with domain $\mathcal{D}(A)$) is given by

$$S(t) = T(t) + \int_0^t e^s T(t-s)\Theta P(s)K \, \mathrm{d}s = T(t) + \int_0^t e^s T(t-s)P^\sharp(s) \, \mathrm{d}s. \tag{8.59}$$

To prove it, by the Phillips Perturbation Theorem, it suffices to show that

$$S_n(t) = \int_0^t \frac{s^{n-1}}{(n-1)!} T(t-s)P^\sharp(s) \, \mathrm{d}s, \qquad n \geq 1, t \geq 0. \tag{8.60}$$

By (8.58), this relation holds for $n = 1$. To prove the induction step we note that $KT(s-u)\Theta P(u) = P(s)$. Hence,

$$\begin{aligned}
S_{n+1}(t) &= \int_0^t T(t-s)\Theta \int_0^s \frac{u^{n-1}}{(n-1)!} KT(s-u)\Theta P(u)K \, \mathrm{d}u \, \mathrm{d}s \\
&= \int_0^t \int_0^s \frac{u^{n-1}}{(n-1)!} \, \mathrm{d}u \, T(t-s)\Theta P(s)K \, \mathrm{d}s \\
&= \int_0^t \frac{s^n}{n!} T(t-s)\Theta P(s)K \, \mathrm{d}s,
\end{aligned}$$

as desired.

**8.2.24 Example**  (See [70]) Let $a > 0$ and $v > 0$ be given. Consider the operator $A_{a,v}(x, y) = (y, v^2 x'' - 2ay)$ in $\mathbb{X} \times \mathbb{Y} = BUC_1(\mathbb{R}) \times BUC(\mathbb{R})$ with domain $\mathcal{D}(A_{a,v}) = \mathcal{D}(\frac{d^2}{d\tau^2}) \times \mathbb{X}$ where $\mathcal{D}(\frac{d^2}{d\tau^2})$ is defined in 7.4.35. In 7.4.35, we have proved that $A_{0,v}$ is a generator of a semigroup in $\mathbb{X} \times \mathbb{Y}$. Since, $A_{a,v} = A_{0,v} + B_a$ where $B_a(x, y) = (0, -2ay)$ is a bounded linear operator, $A_{a,v}$ is also a generator of a semigroup $\{T_{a,v}(t), t \geq 0\}$. Since $T_{a,v}(t) \in \mathcal{L}(\mathbb{X} \times \mathbb{Y})$, there exist operators $S_{ij}(t) = S_{ij}(t, a, v), i, j = 0, 1$ such that $S_{00}(t) \in \mathcal{L}(\mathbb{X}), S_{01}(t) \in \mathcal{L}(\mathbb{Y}, \mathbb{X}), S_{10}(t) \in \mathcal{L}(\mathbb{X}, \mathbb{Y}), S_{11}(t) \in \mathcal{L}(\mathbb{Y})$ and

$$T_{a,v}(t) = \begin{pmatrix} S_{00}(t, a, v) & S_{01}(t, a, v) \\ S_{10}(t, a, v) & S_{11}(t, a, v) \end{pmatrix}, \tag{8.61}$$

i.e. for $x \in \mathbb{X}$ and $y \in \mathbb{Y}$,

$$T_{a,v}(t)\begin{pmatrix} x \\ y \end{pmatrix} = \begin{pmatrix} S_{00}(t, a, v)x & S_{01}(t, a, v)y \\ S_{10}(t, a, v)x & S_{11}(t, a, v)y \end{pmatrix}.$$

By 7.4.40, for $x \in \mathcal{D}(\frac{d^2}{d\tau^2})$ and $y \in \mathbb{X}$ the system

$$\begin{cases} \frac{dx(t)}{dt} = y(t), & x(0) = x, \\ \frac{dy(t)}{dt} = v^2 \frac{d^2}{d\tau^2}x(t) - 2ay(t), & y(0) = y, \end{cases} \tag{8.62}$$

has a unique solution. On the other hand, for such $x$ and $y$, by 7.7.9,

$$x(t, \tau) = \frac{1}{2}E\, x(\tau + v\xi(t)) + \frac{1}{2}E\, x(\tau - v\xi(t)) + \frac{1}{2}E \int_{-\xi(t)}^{\xi(t)} y(\tau + v\sigma)\, d\sigma$$

and $y(t, \tau) = \frac{\partial}{\partial t}x(t, \tau)$ solve (8.62). By the density argument, it shows that for $x \in \mathbb{X}$ and $y \in \mathbb{Y}$,

$$S_{00}(t)x(\tau) = \frac{1}{2}E\, x(\tau + v\xi(t)) + \frac{1}{2}E\, x(\tau - v\xi(t)),$$

$$S_{01}(t)y(\tau) = \frac{1}{2}E \int_{-\xi(t)}^{\xi(t)} y(\tau + v\sigma)\, d\sigma,$$

$$S_{10}(t)x(\tau) = \frac{v}{2}E\, x'(\tau + v\xi(t))1_{E(t)} - \frac{v}{2}E\, x'(\tau + v\xi(t))1_{E(t)^c}$$

$$+ \frac{v}{2}E\, x'(\tau - v\xi(t))1_{E(t)^c} - \frac{v}{2}E\, x'(\tau - v\xi(t))1_{E(t)},$$

$$S_{11}(t)y(\tau) = \frac{1}{2}E\, [y(\tau + v\xi(t)) + y(\tau - v\xi(t))]1_{E(t)}$$

$$- \frac{1}{2}E\, [y(\tau + v\xi(t)) + y(\tau - v\xi(t))]1_{E(t)^c}, \tag{8.63}$$

where $E(t)$ is the event that $N(t)$ is even and $E(t)^{\complement}$ is its complement (note that, for a fixed $t$, the probability that $\xi(t)$ is not differentiable at $t$ equals 0).

Another interesting formula for $S_{01}$ may be derived from the fact that for $(x, y) \in \mathcal{D}(A_{a,v})$

$$A_{a,v} T_{a,v}(t) \begin{pmatrix} x \\ y \end{pmatrix} = T_{a,v}(t) A_{a,v} \begin{pmatrix} x \\ y \end{pmatrix}.$$

Specifically, taking $x = 0$ and $y \in \mathbb{X}$, we obtain $S_{11}(t)y = S_{00}y - 2aS_{01}y$. Hence, by (8.63),

$$S_{01}(t)y(\tau) = \frac{1}{2a} E[y(\tau + v\xi(t)) + y(\tau - v\xi(t))]1_{E(t)^{\complement}}. \tag{8.64}$$

The density argument extends this formula to all $y \in \mathbb{Y}$.

**8.2.25 Remark**    In view of (7.68), (8.64) shows that

$$x(t, \tau) = \frac{1}{2} E\, x(\tau + v\xi(t)) + \frac{1}{2} E\, x(\tau - v\xi(t))$$

$$+ \frac{1}{2a} E[y(\tau + v\xi(t)) + y(\tau - v\xi(t))]1_{E(t)^{\complement}} \tag{8.65}$$

solves the telegraph equation with initial conditions $x(0, \tau) = x(\tau)$ and $\frac{\partial x}{\partial t}(0, \tau) = y(\tau)$ where $x \in \mathcal{D}(\frac{d^2}{d\tau^2})$ and $y \in \mathbb{X}$.

**8.2.26 Corollary**    Formulae (8.63) and (8.64) allow us to estimate the norms of $S_{ij}$. Specifically, it is clear that $\|S_{00}(t)x\|_{\mathbb{Y}} \le \|x\|_{\mathbb{Y}}$ and that for $x \in \mathbb{X}$, $(S_{00}(t)x)' = S_{00}(t)x'$. Hence, $\|S_{00}(t)\|_{\mathcal{L}(\mathbb{X},\mathbb{X})} \le 1$. Moreover, for $y \in \mathbb{Y}$, by (8.63), $(S_{01}(t)y)'(\tau) = \frac{1}{2v} E\,[y(\tau + v\xi(t)) - y(\tau - v\xi(t))]$. This implies $\|(S_{01}(t)y)'\|_{\mathbb{Y}} \le \frac{1}{v}\|y\|_{\mathbb{Y}}$. Since, by (8.64), $\|S_{01}(t)y\|_{\mathbb{Y}} \le \frac{1}{a}\mathbb{P}(E(t)^{\complement})\|y\|_{\mathbb{Y}}$ and $\mathbb{P}(E(t)^{\complement}) = \frac{1-e^{-at}}{2} \le \frac{1}{2}$, we obtain $\|S_{01}(t)\|_{\mathcal{L}(\mathbb{X},\mathbb{Y})} \le (2a)^{-1} + v^{-1}$. Moreover, using (8.63) again, $\|S_{10}(t)\|_{\mathcal{L}(\mathbb{Y},\mathbb{X})} \le v[\mathbb{P}(E(t)) + \mathbb{P}(E(t)^{\complement})] = v$. Similarly, $\|S_{11}(t)\|_{\mathcal{L}(\mathbb{Y},\mathbb{Y})} \le 1$.

It may be shown that the order of the estimate for the norm of $S_{10}(t)$ is the finest in that $\|S_{10}(t)\| \ge \frac{1}{2}v$.

## 8.3 Generators of stochastic processes

In this section we give a characterization of generators of Feller processes and Markov chains.

**8.3.1 Definition**    Let $S$ be a locally compact space. An operator $A$ : $C_0(S) \supset D(A) \to C_0(S)$ is said to satisfy the **positive maximum principle** if for any $x \in \mathcal{D}(A)$ and $p \in S$, $x(p) = \sup_{q \in S} x(q) \geq 0$ implies $Ax(p) \leq 0$.

**8.3.2 Example**    Let $S = \mathbb{R}$, and $A = a\frac{d^2}{d\tau^2}$, where $a \geq 0$, on the domain $\mathcal{D}(A)$ composed of all twice differentiable functions in $C_0(\mathbb{R})$ such that $x'' \in C_0(\mathbb{R})$. If the global, or even local, maximum of $x \in \mathcal{D}(A)$ is attained at $\tau \in \mathbb{R}$, then $x'(\tau) = 0$ and, by the Taylor formula, $x''(\tau) \leq 0$. Hence, $A$ satisfies the positive maximum principle.

**8.3.3 Exercise**    Show that $Ax = ax'$, where $a \in \mathbb{R}$, defined on a suitable domain $\mathcal{D}(A) \subset C_0(\mathbb{R})$ satisfies the positive maximum principle.

**8.3.4 *Generators of Feller processes I***    Let $S$ be a locally compact space. An operator $A$ in $C_0(S)$ is the generator of a semigroup related to a Feller kernel iff

(a) $\mathcal{D}(A)$ is dense in $C_0(S)$,
(b) $A$ satisfies the positive maximum principle,
(c) for some $\lambda_0 > 0$, the range of the operator $\lambda_0 - A$ equals $C_0(S)$.

*Proof*

*(Necessity)* The necessity of (a) and (c) follows directly from the Hille–Yosida theorem. To show (b), suppose that $\{T_t, t \geq 0\}$ given by (8.17) is a Feller semigroup with generator $A$, $x \in \mathcal{D}(A)$, and the global maximum of $x$ is attained at $p$ with $x(p) \geq 0$. Then $\|x^+\| = x(p)$, where as before, $x^+ = \max(x, 0)$. Since $Ax = \lim_{t \to 0} \frac{1}{t}(T_t x - x)$ in $C_0(S)$, we have $Ax(p) = \lim_{t \to 0} \frac{1}{t}(T_t x(p) - x(p))$. But, the operators $T_t$ are non-negative. Hence, $T_t x(p) \leq T_t x^+(p) = \int_S x^+(q) K(t, p, \mathrm{d}q) = \int_{S_\Delta} x^+(q) K(t, p, \mathrm{d}q) \leq x(p) K(t, p, S_\Delta) = x(p)$. This implies $Ax(p) \leq 0$.

*(Sufficiency)* Suppose that $A$ satisfies (a) through (c). Suppose also that for some $\lambda > 0$, $y \in C_0(S)$ and $x \in \mathcal{D}(A)$, $\lambda x - Ax = y$. Also, let $p$ be such that $|x(p)| = \|x\|$. If $x(p) \geq 0$ then by the positive maximum principle, $y(p) \geq \lambda x(p)$. Hence, $\|y\| \geq \lambda x(p) = \lambda \|x\|$. If $x(p) < 0$, the same argument applied to $-x$ gives $\| - y\| \geq \lambda\| - x\|$. So, in both cases $\|y\| \geq \lambda \|x\|$. In other words, if the range of $\lambda - A$ equals $C_0(S)$, $(\lambda - A)^{-1}$ exists and its norm does not exceed $\frac{1}{\lambda}$. In particular, since this is true for $\lambda_0$, $A$ is closed.

Let $\rho(A)$ denote the set of $\mu > 0$ such that the range of $\mu - A$ equals

$C_0(S)$. We need to show that $\rho(A) = (0, \infty)$. Let $\lambda \in \rho(A)$ and $\mu \in (0, 2\lambda)$. Then $|\mu - \lambda| \|(\lambda - A)^{-1}\| < 1$, and the series

$$R_\mu = \sum_{n=0}^{\infty} (\lambda - \mu)^n (\lambda - A)^{-(n+1)}$$

converges in the operator norm. Moreover, writing $\mu - A$ as $\mu - \lambda + \lambda - A$, for any $y \in C_0(S)$, $(\mu - A) \sum_{n=0}^{N} (\lambda - \mu)^n (\lambda - A)^{-(n+1)} y$ equals

$$\sum_{n=0}^{N} (\lambda - \mu)^n (\lambda - A)^{-n} y - \sum_{n=0}^{N} (\lambda - \mu)^{n+1} (\lambda - A)^{-(n+1)} y$$

$$= y - (\lambda - \mu)^{N+1} (\lambda - A)^{-(N+1)} y \xrightarrow[N \to \infty]{} y.$$

This implies that $\lim_{n \to \infty} A \sum_{n=0}^{N} (\lambda - \mu)^n (\lambda - A)^{-(n+1)} y$ exists and equals $\mu R_\mu y - y$. Since $A$ is closed, $R_\mu y \in \mathcal{D}(A)$, and $A R_\mu y = \mu R_\mu y - y$. In other words, for every $y \in C_0(S)$, there exists an $x \in \mathcal{D}(A)$, namely $x = R_\mu y$, such that $\mu x - A x = y$. This proves that $\lambda \in \rho(A)$ implies $(0, 2\lambda) \subset \rho(A)$. Now, if $\mu \in (0, 4\lambda)$, then $\frac{1}{4}\mu + \lambda \in (0, 2\lambda)$ and so $(0, \frac{1}{2}\mu + 2\lambda) \subset \rho(A)$, implying $(0, 4\lambda) \subset \rho(A)$. By the induction argument $\lambda \in \rho(A)$ implies $(0, 2^n\lambda) \subset \rho(A)$. Since, by (b), $\rho(A)$ is non-empty, our claim is proved.

By the Hille–Yosida theorem, $A$ generates a strongly continuous semigroup $\{T_t, t \geq 0\}$ in $C_0(S)$. What is left is to show that the operators $T_t$ are non-negative, and this will be proved once we prove that $(\lambda - A)^{-1}$ are non-negative. To this end we take $x \geq 0$. Then, $y = (\lambda - A)^{-1} x$ belongs to $\mathcal{D}(A)$ and satisfies the equation $\lambda y - A y = x$. If $y(q) < 0$ for some $q \in S$, then there exists a $p$ such that $-y(p) = \sup_{q \in S}[-y(q)] > 0$. By the positive maximum principle $-Ay(p) \leq 0$. Hence, $x(p) \leq \lambda y(p) < 0$, a contradiction. $\square$

**8.3.5 Remarks** (i) Condition (c) may be replaced by:

(c$'$) for all $\lambda > 0$, the range of the operator $\lambda - A$ equals $C_0(S)$.

(ii) If $S$ is compact, $A$ satisfies the **maximum principle**: if $x \in \mathcal{D}(A)$ and the global maximum is attained at $p$, then $Ax(p) \leq 0$, regardless of whether $x(p) \geq 0$ or not. The difference lies in the fact that in such a case $1_S$ belongs to $C_0(S)$; hence $x(p)1_S - x \in C_0(S)$ is non-negative, and we obtain $T_t x(p) \leq x(p)$. Note that if $S$ is locally compact, but not compact, and $x(p) = \sup_{q \in S} x(q)$, where $x \in C_0(S)$, then automatically $x(p) \geq 0$.

(iii) Proving condition (c), called the **range condition**, is usually the hardest part of the proof of the fact that an operator $A$ is a generator.

(iv) As a by-product of the proof, we obtain that operators satisfying the positive maximum principle are dissipative. A linear operator $A :$ $\mathbb{X} \supset \mathcal{D}(A) \to \mathbb{X}$ is said to be **dissipative** if for all $x \in \mathcal{D}(A)$ and $\lambda > 0$, $\|\lambda x - Ax\| \geq \lambda \|x\|$.

**8.3.6 Exercise** Use 8.3.4 to show that the operators $A_\epsilon$ defined in 8.2.18 are generators of Feller semigroups.

**8.3.7 Exercise** Let $C^2[0,1]$ be the subspace of $C[0,1]$ composed of twice differentiable functions with second derivative in $C[0,1]$. Use 8.3.4 to show that the operators $A_{\epsilon,\delta}$ defined by $A_{\epsilon,\delta}x = \frac{1}{2}x''$ on

$$\mathcal{D}(A_{\epsilon,\delta}) = \{x \in C^2[0,1]; x(0) - \epsilon x'(0) = 0, x(1) + \delta x'(1) = 0\},$$

are generators of Feller processes with values in $[0,1]$. These processes are elastic Brownian motions with two barriers: at 0 and at 1.

**8.3.8 Proposition** The reasoning used in proving that (c) implies (c′) is worth closer attention (compare [38] p. 12, [43] p. 46, etc.). Here we give another example of a situation where it applies. Suppose $\mathbb{A}$ is a Banach algebra and $H_n : L^1(\mathbb{R}^+) \to \mathbb{A}, n \geq 1$, are homomorphisms of the algebra $L^1(\mathbb{R}^+)$, such that $\|H_n\| \leq M$ for some $M > 0$. Suppose that the limit $\lim_{n\to\infty} H_n(e_\lambda)$ exists for some $\lambda > 0$. Then, it exists for all $\lambda > 0$.

*Proof* Let $\rho$ be the set of $\lambda > 0$ such that the above limit exists. We have seen that it suffices to show that $\lambda \in \rho$ implies $(0, 2\lambda) \subset \rho$. A minor modification of the argument shows that all that needs to be showed is that $\lambda \in \rho$ implies $(0, 2\lambda) \cap (\lambda - 1, \lambda + 1) \subset \rho$. Let $\lambda, \mu > 0$. By the Hilbert equation (6.6), $e_\mu = e_\lambda + (\mu - \lambda)e_\lambda e_\mu$. An induction argument shows that $e_\mu = \sum_{k=1}^n e_\lambda^{*k}(\mu - \lambda)^{k-1} + (\mu - \lambda)^n e_\lambda^{*n} e_\mu, n \geq 0$. Suppose $|\mu - \lambda| < \lambda$. Then, letting $n \to \infty$, $e_\mu = \sum_{k=1}^\infty e_\lambda^{*k}(\mu - \lambda)^{k-1}$, since $\|e_\lambda^{*n}\|_{L^1(\mathbb{R}^+)} = \frac{1}{\lambda^n}$. Therefore, $H_n(e_\lambda) = \sum_{k=1}^\infty H_n(e_\lambda^{*k})(\mu - \lambda)^{k-1} = \sum_{k=1}^\infty [H(e_\lambda)]^k(\mu - \lambda)^{k-1}$. If $R_\lambda = \lim_{n\to\infty} H_n(e_\lambda)$ exists, and $|\mu - \lambda| < 1 \wedge \lambda$, this series converges, as $n \to \infty$, to $R_\mu := \sum_{k=1}^\infty R_\lambda^k(\mu - \lambda)^{k-1}$. To see that we take an $\epsilon > 0$, choose an $l \in \mathbb{N}$ such that $M \sum_{k=l}^\infty |\mu - \lambda|^{k-1} \lambda^{-k} \leq \epsilon/3$, and then choose an $n \geq 0$ such that $\|R_\lambda^k - H_n(e_\lambda^k)\|_{\mathbb{A}} \leq \epsilon/(3d)$ for all $1 \leq k \leq l - 1$, where $d = \frac{1}{1 - |\lambda - \mu|}$. Clearly $\lim_{n\to\infty} H_n(e_\lambda^{*k}) = \lim_{n\to\infty} [H_n(e_\lambda)]^k = R_\lambda^k$ for all $k \geq 1$; hence $\|R_\lambda^k\| \leq M\lambda^{-k}$. Therefore,

$\|R_\mu - H_n(\mu)\|_A$ does not exceed $\epsilon(3d)^{-1} \sum_{k=1}^{l-1} |\lambda - \mu|^{k-1} + 2M \sum_{k=l}^{\infty} |\mu - \lambda|^{k-1} \lambda^{-k} \leq \epsilon/3 + 2\epsilon/3 = \epsilon$. $\qquad\square$

Sometimes it is convenient to have the following version of theorem 8.3.4.

**8.3.9 Generators of Feller processes II** Let $S$ be a locally compact space. An operator $A$ in $C_0(S)$ is the generator of a semigroup related to a Feller kernel iff

(a) $\mathcal{D}(A)$ is dense in $C_0(S)$,

(b) if $x \in \mathcal{D}(A)$, $\lambda > 0$ and $y = \lambda x - Ax$, then $\lambda \inf_{p \in S} x(p) \geq \inf_{p \in S} y(p)$,

(c) for some $\lambda_0 > 0$, the range of the operator $\lambda_0 - A$ equals $C_0(S)$.

*Proof* *(Necessity)* Let $\{T_t, t \geq 0\}$ be the semigroup generated by $A$. By (8.17), we have $T_t y(p) \geq \inf_{p \in S} y(p)$ (note that $\inf_{p \in S} y(p) = \inf_{p \in S_\Delta} y(p)$). Moreover, if assumptions of (b) are satisfied, then $\lambda x(p) = \lambda \int_0^\infty e^{-\lambda t} T_t y(p) \, dt \geq \inf_{t \geq 0} T_t y(p) \geq \inf_{p \in S} y(p)$. This proves (b). The rest is clear.

*(Sufficiency)* Let $x, y$ and $\lambda$ be as in (b). Taking $-y$ and $-x$ instead of $y$ and $x$, respectively, we obtain $\lambda \sup_{p \in S} x(p) \leq \sup_{p \in S} y(p)$. Together with (b) this gives $\lambda \|x\| \leq \|y\| = \|\lambda x - Ax\|$, i.e. dissipativity of $A$. Hence, as in the proof of 8.3.4 we argue that $A$ generates a semigroup of contractions. Moreover, (b) implies that $R_\lambda y \geq 0$ provided $y \geq 0$. Therefore, the semigroup generated by $A$ is a semigroup of non-negative operators and 8.1.26 applies. $\qquad\square$

The problem with applying theorems 8.3.4 and 8.3.9 is that the whole domain of an operator is rarely known explicitly, and we must be satisfied with knowing its core. Hence, we need to characterize operators which may be *extended* to a generator of a Feller semigroup. In particular, such operators must be closable. A linear operator $A : \mathbb{X} \supset \mathcal{D}(A) \to \mathbb{X}$ is said to be **closable** if there exists a closed linear operator $B$ such that $Bx = Ax$ for $x \in \mathcal{D}(A)$.

Let us recall that a graph $G_A$ of an operator $A$ is defined as $G_A = \{(x, y) \in \mathbb{X} \times \mathbb{X}; x \in \mathcal{D}(A), y = Ax\}$.

**8.3.10 Lemma** Let $A$ be a linear operator in a Banach space $\mathbb{X}$. The following conditions are equivalent:

(a) $A$ is closable,

(b) the closure of the graph $G_A$ of $A$ in the space $\mathbb{X} \times \mathbb{X}$ equipped with the norm $\|(x,y)\| = \|x\| + \|y\|$ is a graph of a closed operator,

(c) if $x_n \in \mathcal{D}(A), n \geq 1$, $\lim_{n \to \infty} x_n = 0$ and $\lim_{n \to \infty} Ax_n$ exists, then $\lim_{n \to \infty} Ax_n = 0$.

*Proof* We will show that $(b) \Longrightarrow (a) \Longrightarrow (c) \Longrightarrow (b)$.

Clearly, $(b)$ implies (a). Also, if $A$ is closable and $B$ is its closed extension, then $x_n, n \geq 1$, described in (c) belong to $\mathcal{D}(B)$ and $Bx_n = Ax_n$. Since $B$ is closed, $\lim_{n \to \infty} Bx_n = B0 = 0$, proving (c).

Since the closure $cl\, G_A$ of $G_A$ is a subspace of $\mathbb{X} \times \mathbb{X}$ (in particular, it is closed), we are left with proving that (c) implies that $cl\, G_A$ is a graph of an operator (see 7.3.8). To this end we need to show that $(x, y), (x, y') \in cl\, G_A$ implies $y = y'$. By linearity, it suffices to show that $(0, y) \in cl\, G_A$ implies $y = 0$. Now, $(0, y) \in cl\, G_A$ iff there exists a sequence $x_n \in \mathcal{D}(A), n \geq 1$, such that $\lim_{n \to \infty} x_n = 0$ and $\lim_{n \to \infty} Ax_n = y$. By (c), this implies $y = 0$. □

The following example shows that there are operators that are not closable.

**8.3.11 Example**    Let $\mathbb{X} = C[0,1]$ and define $Ax(\tau) = x'(0)$ on $\mathcal{D}(A)$ composed of all $x \in \mathbb{X}$ such that $x'(0)$ (the right-hand derivative) exists. Let $x_n(\tau) = \max(2^{-n} - |\tau - 2^{-n}|, 0)$. Then $x_n \in \mathcal{D}(A)$ with $x'_n(0) = 1$. Moreover, $\|x_n\| = 2^{-n} \to 0$ as $n \to \infty$ and $\lim_{n \to \infty} Ax_n = 1_{[0,1]} \neq 0$.

**8.3.12 Definition**    The **closure** $\overline{A}$ of a closable operator $A$ is the unique closed operator such that $G_{\overline{A}} = cl\, G_A$.

**8.3.13** *(Pre)-generators of Feller processes I*    Let $S$ be a locally compact space and $A$ be a linear operator $A : C_0(S) \supset \mathcal{D}(A) \to C_0(S)$. $A$ is closable and its closure $\overline{A}$ generates a Feller semigroup iff:

(a) $\mathcal{D}(A)$ is dense in $C_0(S)$,

(b) if $x \in \mathcal{D}(A)$, $\lambda > 0$ and $y = \lambda x - Ax$, then $\lambda \inf_{p \in S} x(p) \geq \inf_{p \in S} y(p)$,

(c) the range of $\lambda - A$ is dense in $C_0(S)$ for some $\lambda > 0$.

*Proof* Necessity is clear in view of 8.3.9. For sufficiency we note first that by (b) $A$ is dissipative; we will show that $A$ is closable. Let $x_n \in \mathcal{D}(A)$ be such that $\lim_{n \to \infty} x_n = 0$ and $\lim_{n \to \infty} Ax_n$ exists and equals, say, $y$. Let $z$ belong to $\mathcal{D}(A)$. Then $\|(\lambda - A)(\lambda x_n + z)\| \geq \lambda \|\lambda x_n + z\|$, for all $\lambda > 0$. Letting $n \to \infty$ we obtain $\|\lambda z - \lambda y - Az\| \geq \lambda \|z\|$. Dividing by

$\lambda$ and letting $\lambda \to \infty$, $\|z - y\| \geq \|z\|$. Now, by (a), we may choose a sequence $z_n \in \mathcal{D}(A)$ such that $\lim_{n\to\infty} z_n = y$. Hence, $0 \geq \|y\|$, proving that $y = 0$, as desired.

Let $\overline{A}$ be the closure of $A$. We need to show that $\overline{A}$ generates a Feller semigroup, and to this end we may apply theorem 8.3.9. Clearly, $\overline{A}$ is densely defined. To prove condition (c) of the theorem, we consider the $\lambda > 0$ from assumption (c), and suppose $y \in C_0(S)$ is such that there exists a sequence $x_n \in \mathcal{D}(\overline{A})$ with $\lim_{n\to\infty}(\lambda - \overline{A})x_n = y$. Then, by $\|x_n - x_m\| \leq \frac{1}{\lambda}\|(\lambda - \overline{A})(x_n - x_m)\|$, $x_n, n \geq 1$, is a Cauchy sequence, hence, it converges to some $x \in C_0(S)$. Since $\overline{A}$ is closed, $x$ belongs to $\mathcal{D}(\overline{A})$ and $(\lambda - \overline{A})x = y$. This shows that the range of $\lambda - \overline{A}$ is closed. Therefore, by assumption (c), it equals $C_0(S)$.

Finally, we need to show that $\overline{A}$ satisfies condition (b). This, however, is easy because for any $x \in \mathcal{D}(\overline{A})$ there exist $x_n \in \mathcal{D}(A)$ such that $\lim_{n\to\infty} x_n = x$ and $\lim_{n\to\infty} Ax_n = \overline{A}x$. By assumption (b) we have $\lambda \inf_{p\in S} x_n(p) \geq \inf_{p\in S}(\lambda x_n(p) - Ax_n(p))$. Letting $n \to \infty$, we obtain $\lambda \inf_{p\in S} x(p) \geq \inf_{p\in S}(\lambda x(p) - Ax(p))$, as desired. $\quad\square$

**8.3.14** *(Pre)-generators of Feller processes II*   Let $S$ be a locally compact space and $A$ be a linear operator $A : C_0(S) \supset \mathcal{D}(A) \to C_0(S)$. $A$ is closable and its closure $\overline{A}$ generates a Feller semigroup iff:

(a) $\mathcal{D}(A)$ is dense in $C_0(S)$,
(b) $A$ satisfies the positive maximum principle,
(c) the range of $\lambda - A$ is dense in $C_0(S)$ for some $\lambda > 0$.

*Proof*  Necessity is clear in view of 8.3.4. As for sufficiency, by 8.3.13, all we need to show is that $A$ satisfies condition (b) of this theorem. Consider $x, y$ and $\lambda$ described there. There are two possible cases: either there exists a $p$ such that $x(p) = \inf_{q\in S} x(q)$ or $\inf_{q\in S} x(q) = x(\Delta) = 0$. In the former case, $-x$ attains its maximum at $p$ and so $Ax(p) \geq 0$. Therefore, $\lambda \inf_{q\in S} x(q) = \lambda x(p) \geq \lambda x(p) - Ax(p) \geq \inf_{q\in S}\{\lambda x(q) - Ax(q)\}$, as desired. To treat the latter case, we recall that at the end of the proof of 8.3.4 we showed that the positive maximum principle implies that $x \geq 0$ provided $y \geq 0$. This means that $\lambda \inf_{q\in S} x(q) \geq 0 = \inf_{q\in S} y(q)$. $\quad\square$

**8.3.15 Example**   Let $S = [0,1]$ and consider the operator $Ax(s) = s(1-s)x''(s)$ defined for all polynomials on $[0,1]$. It is clear that $A$ is densely defined and satisfies the maximum principle. Moreover, if $x(s) = \sum_{i=0}^{n} a_i s^i$ then $Ax(s) = \sum_{i=1}^{n-1}(i+1)ia_{i+1}s^i - \sum_{i=2}^{n} i(i-1)a_i s^i$. Hence, for a polynomial $y(s) = \sum_{i=0}^{n} b_i s^i$ and a number $\lambda > 0$, the equation $\lambda x - Ax = y$ is satisfied iff the coefficients satisfy the system

$$(\lambda+i(i-1))a_i - i(i+1)a_{i+1} = b_i, \quad 0 \le i \le n-1, \quad (\lambda+n(n-1))a_n = b_n.$$
$$(8.66)$$

Since this system has a solution ($a_n$ is calculated from the last equation and substituted to the previous-to-last, which allows the calculation of $a_{n-1}$, and so on), the range of $A$ is dense in $C[0,1]$. Therefore, theorem 8.3.14 applies and the closure of $A$ is the generator of a Feller semigroup in $C[0,1]$. (See also 8.4.20.)

**8.3.16 Exercise**   It is possible to characterize $\overline{A}$ introduced above in more detail: show that $\mathcal{D}(\overline{A})$ is composed of functions that are twice differentiable in $(0,1)$ with $\lim_{s\to 0+} s(1-s)x''(s) = \lim_{s\to 1-} s(1-s)x''(s) = 0$ and we have $\overline{A}x(s) = s(1-s)x''(s), s \in (0,1), \overline{A}x(1) = \overline{A}x(0) = 0$.

We now turn to generators of continuous-time Markov chains; we will characterize generators of strongly continuous semigroups $\{U(t), t \ge 0\}$ in $l^1 = l^1(\mathbb{N})$ of the form $U(t)x = xP(t)$ where $\{P(t), t \ge 0\}$ is a semigroup of transition matrices and $xP(t)$ is the matrix product. Note that, in contradistinction to the case of Feller semigroups, we thus study the evolution of distributions of a Markov chain and not the related evolution given by the dual semigroup – see 8.1.15 and 8.1.16. As we have seen in 7.4.27, there is a one-to-one correspondence between finite intensity matrices and semigroups of finite transition matrices. If the matrices fail to be finite the situation is more complicated in that different semigroups of transition matrices may be related to the same intensity matrix. In the remainder of this section in the set of propositions we present a result due to T. Kato [54, 65] which explains this situation in more detail.

We recall that $(\xi_n)_{n\ge 1} \in l^1$ is said to be a distribution iff $\xi_n \ge 0, n \ge 1$, and $F(\xi_n)_{n\ge 1} = 1$ where the functional $F \in (l^1)^*$ is given by $F(\xi_n)_{n\ge 1} = \sum_{n=1}^{\infty} \xi_n$; the set of densities is denoted by **D**. A linear, not necessarily bounded operator $A$ in $l^1$ is said to be non-negative if it maps $\mathcal{D}(A) \cap (l^1)^+$ into $(l^1)^+$, where $(l^1)^+$ is the non-negative cone, i.e. the set of non-negative $(\xi_n)_{n\ge 1} \in l^1$. For $x$ and $y$ in $l^1$ we write $x \le y$ or $y \le x$ if $y - x \in (l^1)^+$. For two operators, $A$ and $B$, in $l^1$ we write $A \le B$ or $B \ge A$ iff $B - A$ is non-negative. An operator $A$ (defined on the whole of $l^1$) is said to be Markov if it leaves **D** invariant; it is said to be sub-Markov iff it is non-negative and $FAx \le Fx$ for $x \in \mathbf{D}$. Markov and sub-Markov operators are contractions. As in 5.2.1, we write $e_i = (\delta_{i,n})_{n\ge 1}, i \ge 1$.

**8.3.17 Exercise** Let $x_n, n \geq 1$, be a sequence of elements of $l^1$ such that $0 \leq x_n \leq x_{n+1}, n \geq 1$, and $\|x_n\| \leq M, n \geq 1$, for some $M > 0$. Show that $x_n$ converges.

**8.3.18 Definition** Let $Q = (q_{i,j})_{i,j\in\mathbb{N}}$ be an intensity matrix. We define the domain of an operator $A_0$ to be the linear span of $e_i, i \geq 0$, and put $A_0 e_i = (q_{i,n})_{n\geq 1}$. Furthermore, the operator $D$ ("$D$" for "diagonal") with domain $\mathcal{D}(D) = \{(\xi_n)_{n\geq 1} \in l^1 | (q_{n,n}\xi_n)_{n\geq 1} \in l^1\}$ is defined by $D(x_n)_{n\geq 1} = (q_{n,n}\xi_n)_{n\geq 1}$; note that $-D$ is non-negative.

**8.3.19 Proposition** The operator $O$ ($O$ for "off diagonal") given by $O(\xi_n)_{n\geq 1} = \left(\sum_{i\geq 1, i\neq n} \xi_i q_{i,n}\right)_{n\geq 1}$ is well-defined on $\mathcal{D}(D)$ and $\|Ox\| \leq \|Dx\|$ for $x \in \mathcal{D}(D)$ and $\|Ox\| = \|Dx\|$ for $x \in \mathbf{D} \cap \mathcal{D}(D)$. Moreover, for any $0 \leq r < 1$, the operator $D + rO$ with domain $\mathcal{D}(D)$ is the generator of a strongly continuous semigroup of sub-Markov operators in $l^1$.

*Proof* For $x \in \mathcal{D}(D)$, $\sum_{n=1}^{\infty} \left|\sum_{i\geq 1, i\neq n} \xi_i q_{i,n}\right|$ does not exceed

$$\sum_{n=1}^{\infty} \sum_{i\geq 1, i\neq n} |\xi_i q_{i,n}| = \sum_{i=1}^{\infty} |\xi_i| \sum_{n\geq 1, n\neq i}^{\infty} q_{i,n} = \sum_{i=1}^{\infty} |\xi_i| (-q_{n,n}) = \|Dx\|,$$

with equality iff $(\xi_n)_{n\geq 1}$ is non-negative. This proves the first claim.

For $r = 0$ the second claim is immediate: the semigroup generated by $D$, say $\{S(t), t \geq 0\}$, is given by $S(t)(\xi_n)_{n\geq 1} = (e^{q_{n,n}t}\xi_n)_{n\geq 1}$. To treat the general case we note first that for $\lambda > 0$ we have $(\lambda - D)^{-1}(\xi_n)_{n\geq 1} = \left(\frac{1}{\lambda - q_{n,n}}\xi_n\right)_{n\geq 1}$ and

$$B_\lambda := O(\lambda - D)^{-1} \qquad (8.67)$$

is well-defined. Moreover, we have $\|B_\lambda x\| \leq \sum_{n\geq 1} \sum_{i\geq 1, i\neq n} \frac{q_{i,n}}{\lambda - q_{i,i}} |\xi_i| = \sum_{i\geq 1} \sum_{n\geq 1, n\neq i} \frac{q_{i,n}}{\lambda - q_{i,i}} |\xi_i| = \sum_{i\geq 1} \frac{-q_{i,i}}{\lambda - q_{i,i}} |\xi_i| \leq \sum_{i\geq 1} |\xi_i| = \|x\|$. Hence, $B_\lambda$ is a contraction and for any $0 \leq r < 1$ the series $\sum_{n=0}^{\infty} r^n B_\lambda^n (= (I + rB_\lambda)^{-1})$ converges in the operator norm. Let

$$R_{\lambda,r} = (\lambda - D)^{-1} \sum_{n=0}^{\infty} r^n B_\lambda^n.$$

By definition $R_{\lambda,r}x$ belongs to $\mathcal{D}(D)$ and $(\lambda - D)R_{\lambda,r}x = \sum_{n=0}^{\infty} r^n B_\lambda^n x$, and $rOR_{\lambda,r}x = rB_\lambda \sum_{n=0}^{\infty} r^n B_\lambda^n x = \sum_{n=1}^{\infty} r^n B_\lambda^n x$. Hence, $(\lambda - D - rO)R_{\lambda,r}x = x$. Similarly, $R_{\lambda,r}(\lambda - D - rO)x = x, x \in \mathcal{D}(D)$. This shows that $R_{\lambda,r} = (\lambda - D - rO)^{-1}$ and in particular that $D + rO$ is closed. Moreover, if $x \geq 0$ then $y = R_{\lambda,r}x$ is non-negative, too; indeed, $(\lambda - D)^{-1} \geq 0$ and $O \geq 0$ and so $B_\lambda \geq 0$. Hence, using $-D \geq 0$ and $O \geq 0$, $\|\lambda y\| \leq \|\lambda y\| + (1 - r)\|Dy\| = \|\lambda y - Dy\| - r\|Dy\| = \|\lambda y - Dy\| - r\|Oy\| \leq \|\lambda y - Dy - rOy\| = \|x\|$ with the previous-to-last step following by the triangle inequality. This shows that $\lambda R_{\lambda,r}$ is sub-Markov and in particular $\|\lambda R_{\lambda,r}\| \leq 1$. Since $D + rO$ is densely defined, by the Hille–Yosida theorem it generates a strongly continuous semigroup of operators. This is a semigroup of sub-Markov operators since the approximating exponential functions of the Yosida approximation are formed by such operators – see 8.2.14. □

**8.3.20 Proposition**     As $r \uparrow 1$, the semigroups $\{S_r(t), t \geq 0\}$ converge strongly to a strongly continuous semigroup $\{S(t), t \geq 0\}$ of sub-Markov operators generated by an extension of $D + O$ (hence, an extension of $A_0$, as well).

We postpone the *proof* to the next section (Subsection 8.4.16) where we will have approximation theorems for operator semigroups at our disposal.

**8.3.21 Proposition**     Let $\{S(t), t \geq 0\}$ be the semigroup defined in 8.3.20 and suppose that the generator $A$ of a strongly continuous semigroup $\{T(t), t \geq 0\}$ is an extension of the operator $A_0$. Then $A$ is also an extension of $D + O$ and, if $T(t) \geq 0, t \geq 0$, then $S(t) \leq T(t), t \geq 0$. *We say that* $\{S(t), t \geq 0\}$ *is the minimal semigroup related to* $Q$.

*Proof*  Suppose that $x = \sum_{n=1}^{\infty} \xi_n e_n$ belongs to $\mathcal{D}(D)$. By definition of $\mathcal{D}(D)$, so do $x_N := \sum_{n=1}^{N} \xi_n e_n, N \geq 1$, and since $Ae_n = A_0 e_n = (D + O)e_n$, we have $Ax_N = (D + O)x_N$. Moreover, $\lim_{n \to \infty} Dx_N = Dx$ and so, by $\|O(x_N - x)\| \leq \|D(x_N - x)\|$, $\lim_{N \to \infty} Ox_N = Ox$. Therefore $\lim_{N \to \infty} Ax_N$ exists and equals $(D + O)x$ and, obviously, $\lim_{N \to \infty} x_N = x$. Since $A$ is closed, being the generator of a semigroup, $x$ belongs to $\mathcal{D}(A)$ and $Ax = (D + O)x$, proving the first claim.

Next, we note that $(\lambda - A)^{-1}$ exists for sufficiently large $\lambda > 0$. For $y \in \mathcal{D}(D)$, we may write

$$(1 - r)Oy = Ay - Dy - rOy = (\lambda - D - rO)y - (\lambda - A)y.$$

Taking $y = (\lambda - D - rO)^{-1}x, x \in l^1$, and applying $(\lambda - A)^{-1}$ to the left-most and right-most sides of the above equality,

$$(1 - r)(\lambda - A)^{-1}O(\lambda - D - rO)^{-1}x = (\lambda - A)^{-1}x - (\lambda - D - rO)^{-1}x.$$

Since all operators on the left-hand side are non-negative, for large $\lambda >$ 0 we have $(\lambda - D - rO)^{-1} \leq (\lambda - A)^{-1}$. As we shall see in 8.4.16, $(\lambda - D - rO)^{-1}$ converges to $(\lambda - G)^{-1}$ where $G$ is the generator of $\{S(t), t \geq 0\}$. Hence, $(\lambda - G)^{-1} \leq (\lambda - A)^{-1}$. This implies the second claim – see 8.2.14. $\qquad \Box$

**8.3.22 Proposition** Let $\{S(t), t \geq 0\}$ be the semigroup defined in Subsection 8.3.20. The following are equivalent:

(a) $\{S(t), t \geq 0\}$ is a semigroup of Markov operators;
(b) for any $\lambda > 0$, $\lim_{n\to\infty} B_\lambda^n = 0$ strongly;
(c) for any $\lambda > 0$, $Range(\lambda - D - O)$ is dense in $l^1$;
(d) for any $\lambda > 0$, $Range(\lambda - A_0)$ is dense in $l^1$;
(e) for any $\lambda > 0$, $Range(I - B_\lambda)$ is dense in $l^1$;
(f) if for some $\lambda > 0$ and $a = (\alpha_n)_{n\geq 1} \in l^\infty$ we have $Qa = \lambda a$ (where $Qa$ is the product of the matrix $Q$ and the column-vector $a$), then $a = 0$.

*If one and hence all of these conditions hold, the matrix $Q$ is said to be* **non-explosive.**

*Proof* Condition (a) holds iff $\lambda R_\lambda$ is a Markov operator for all $\lambda > 0$ (cf. 8.2.14). On the other hand,

$$I + O \sum_{k=0}^{n} (\lambda - D)^{-1} B_\lambda^k = (\lambda - D) \sum_{k=0}^{n} (\lambda - D)^{-1} B_\lambda^k + B_\lambda^{n+1}. \quad (8.68)$$

Therefore, for $x \geq 0$, $\|x\| + \|O \sum_{k=0}^{n} (\lambda - D)^{-1} B_\lambda^k\| = \|\lambda \sum_{k=0}^{n} (\lambda - D)^{-1} B_\lambda^k x\| + \|D \sum_{k=0}^{n} (\lambda - D)^{-1} B_\lambda^k x\| + \|B_\lambda^{n+1} x\|$, since $-D \geq 0$. By 8.3.19 this gives, $\|x\| = \|\lambda \sum_{k=0}^{n} (\lambda - D)^{-1} B_\lambda^k x\| + \|B_\lambda^{n+1} x\|$. Letting $n \to \infty$ we see that $\|x\| = \|\lambda R_\lambda\|$ iff $\lim_{n\to\infty} \|B_\lambda^n x\| = 0$ (the fact that $\lim_{n\to\infty} \sum_{k=0}^{n} (\lambda - D)^{-1} B_\lambda^k x = R_\lambda x$ is proved in 8.4.16). This shows (a) $\Leftrightarrow$ (b).

Next we show (b) $\Rightarrow$ (c) $\Rightarrow$ (d) $\Rightarrow$ (e) $\Rightarrow$ (b). To prove the first implication we rewrite (8.68) as $(\lambda - D - O) \sum_{k=0}^{n} (\lambda - D)^{-1} B_\lambda^k x = x + B_\lambda^{n+1} x, x \in l^1$; this relation shows that if (b) holds any $x$ may be approximated by elements of $Range(\lambda - D - O)$, as desired. To see that (c) implies (d) we note that for any $x \in \mathcal{D}(D)$ there exist $x_n \in \mathcal{D}(A_0)$ such that $\lim_{n\to\infty} x_n = x$ and $\lim_{n\to\infty} A_0 x_n = \lim_{n\to\infty} (D + O) x_n = (D + O)x$ (see the beginning of the proof of 8.3.21); hence the range of $\lambda - D - O$ is contained in the closure of the range of $\lambda - A_0$. The fact that (d) implies (e) becomes clear once we write $I - B_\lambda = (\lambda - D)(\lambda - D)^{-1} - O(\lambda - D)^{-1} = (\lambda - D - O)(\lambda - D)^{-1}$ and note that all elements $x$ of $\mathcal{D}(D)$

are of the form $x = (\lambda - D)^{-1}y$ for some $y \in l^1$. Indeed, this shows that the range of $\lambda - D - O$ is equal to the range of $I - B_\lambda$ and we know that $\lambda - D - O$ is an extension of $A_0$. To show the last implication we note that, since $B_\lambda$ is sub-Markov, $\|B_\lambda^n x\| \leq \|B_\lambda^k x\|$ for $x \geq 0$ and $k \leq n$. Therefore, for such an $x$, $\|B_\lambda^n x\| \leq \|C_n x\|$ where $C_n = \frac{1}{n+1}\sum_{k=0}^n B_\lambda^k$. Hence, it suffices to show that $C_n$ converges strongly to 0. If $x = y - B_\lambda y$ for some $y \in l^1$, we have $C_n x = \frac{1}{n+1}\sum_{k=0}^n B_\lambda^k (I - B_\lambda)y = \frac{1}{n+1}(x - B_\lambda^{n+1}x)$. Therefore, for $x \in Range(I - B_\lambda)$, $\lim_{n\to\infty} C_n x = 0$. If (e) holds, the same is true for all $x \in l^1$ since $\|C_n\| \leq 1$.

Finally, we show (d) $\Leftrightarrow$ (f). To this end we note that (d) holds iff, for any functional $F$ on $l^1$, the relation $F(\lambda x - A_0 x) = 0$ for all $x \in \mathcal{D}(A_0)$ implies $F = 0$. By definition of $\mathcal{D}(A_0)$, $F(\lambda x - A_0 x) = 0$ for all $x \in \mathcal{D}(A_0)$ iff $F(\lambda e_i - A_0 e_i) = $ for all $i \geq 1$. On the other hand, any $F$ may be identified with an $a = (\alpha_n)_{n\geq 1} \in l^\infty$ and we have $F(\lambda e_i - A_0 e_i) = \lambda \alpha_i - \sum_{j=1}^\infty q_{i,j}\alpha_j$. $\qquad\square$

**8.3.23 Example**    Let $r_n, n \geq 1$, be a sequence of non-negative numbers. A Markov chain with intensity matrix $Q = (q_{i,j})_{i,j\geq 1}$ where

$$q_{i,j} = \begin{cases} -r_i, & j = i, \\ r_i, & j = i+1, \\ 0, & \text{otherwise}, \end{cases}$$

is said to be a **pure birth process** with rates $r_n, n \geq 1$. (In particular, the Poisson process is a pure birth process with a constant rate.) For such a $Q$, we have $(\lambda - D)^{-1}(\xi_n)_{n\geq 1} = \left(\frac{\xi_n}{\lambda + r_n}\right)_{n\geq 1}$ and in particular $(\lambda - D)^{-1}e_k = \frac{1}{\lambda + r_k}e_k$. Also $Oe_k = r_k e_{k+1}$. Hence, $B_\lambda e_k = \frac{r_k}{\lambda + r_k}e_{k+1}$ and so $B_\lambda^n e_k = \prod_{i=k}^{k+n-1}\frac{r_i}{\lambda + r_i}e_{n+k}$ and $\|B_\lambda^n e_k\| = \prod_{i=k}^{k+n-1}\frac{r_i}{\lambda + r_i}$. Since $e_k, k \geq 1$, are linearly dense in $l^1$, $B_\lambda^n$ converges strongly to 0 as $n \to \infty$ iff $\prod_{i=1}^\infty \frac{r_i}{\lambda + r_i} = 0$. This last condition is equivalent to $\sum_{i=1}^\infty (-\ln \frac{r_i}{\lambda + r_i}) = \infty$. Since $\lim_{x\to 0+}\frac{-\ln(1-x)}{x} = 1$, $Q$ is non-explosive iff $\sum_{n=1}^\infty \frac{1}{\lambda + r_n}$ diverges for all $\lambda > 0$.

**8.3.24 Example**    (See [110].) Let $d, r > 0$. Consider the Kolmogorov matrix $Q = (q_{i,j})_{i,j\geq 1}$ given by

$$q_{i,j} = \begin{cases} (i-1)r, & j = i-1, i \geq 2, \\ -(i-1)r - (i+1)d, & j = i, i \geq 1, \\ (i+1)r, & j = i+1, i \geq 1, \end{cases}$$

and 0 otherwise. To show that $Q$ is non-explosive we check that condition (f) of Proposition 8.3.22 is satisfied. The equation $Qa = \lambda a$ considered there may be rewritten as

$$(i-1)r\alpha_{i-1} - [(i-1)r + (i+1)d]\alpha_i + (i+1)d\alpha_{i+1} = \lambda\alpha_i, \qquad i \geq 1,$$

where we put $\alpha_0 = 0$. Then $\alpha_{i+1} = \left[1 + \frac{(i-1)r+\lambda}{(i+1)d}\right]\alpha_i - \frac{i-1}{i+1}\frac{r}{d}\alpha_{i-1}$, or

$$\alpha_{i+1} - \alpha_i = \frac{(i-1)r+\lambda}{(i+1)d}\alpha_i - \frac{i-1}{i+1}\frac{r}{d}\alpha_{i-1}. \tag{8.69}$$

Note that, if $\alpha_i > 0$ for some $i \geq 1$, then $\alpha_{i+1} - \alpha_i > \frac{i-1}{i+1}\frac{r}{d}(\alpha_i - \alpha_{i-1})$. Hence, by the induction argument, if $\alpha_1 > 0$ then $\alpha_{i+1} - \alpha_i > 0$ and $\alpha_i > 0$ for all $i \geq 1$. Therefore, by (8.69) again, $\alpha_{i+1} - \alpha_i \geq \frac{\lambda}{(i+1)d}\alpha_i$ or $\alpha_{i+1} \geq \left[1 + \frac{\lambda}{(i+1)d}\right]\alpha_i$ resulting in $\alpha_n \geq \alpha_1 \prod_{i=2}^{n}\left(1 + \frac{\lambda}{id}\right), n \geq 1$.

Hence, if $\alpha_1 > 0$, the limit $\lim_{n\to\infty}\alpha_n$ exists and is no less than $\prod_{i=2}^{\infty}\left(1 + \frac{\lambda}{id}\right)\alpha_1 = \infty$. Since this contradicts $a \in l^\infty$, we must have $\alpha_1 \leq 0$. But, we may not have $\alpha_1 < 0$ for then $b := -a \in l^\infty$ would satisfy $Qb = \lambda b$ while having its first coordinate positive, which we know is impossible. Thus, $\alpha_1 = 0$ and an induction argument based on (8.69) shows that $\alpha_i = 0$ for all $i \geq 1$.

**8.3.25 Remark**     Probabilistically, the reason why there are in general many semigroups related to a given $Q$ matrix may be explained as follows. Let us recall that if $X(t), t \geq 0$ is a Markov chain related to $Q$, then given that $X(t) = n$, the chain waits in this state for an exponential time with parameter $-q_{n,n}$ and then jumps to one of the other states, the probability of jumping to $k \neq n$ being $-q_{n,k}/q_{n,n}$ (if $q_{n,n} = 0$ the process stays at $n$ for ever). It is important to note that in general such a procedure defines the process only up to a certain random time $\tau$, called **explosion**. This is well illustrated by the pure birth process of Subsection 8.3.23. If the process starts at 1, then after exponential time $T_1$ with parameter $r_1$ it will be at 2, and after exponential time $T_2$ with parameter $r_2$ it will be at 3, and so on. Let us put $\tau = \sum_{n=1}^{\infty} T_n$. Is $\tau$ finite or infinite? If $\sum_{n=1}^{\infty}\frac{1}{r_n} = \infty$, then $\mathbb{P}\{\tau = \infty\} = 1$ and in the other case $\mathbb{P}\{\tau < \infty\} = 1$. Indeed, if the series converges, we may not have $\mathbb{P}\{\tau = \infty\} > 0$, as this would imply $E\tau = \infty$ while we have $E\tau = \sum_{n=1}^{\infty} E T_n = \sum_{n=1}^{\infty}\frac{1}{r_n} < \infty$. Conversely, if the series diverges, then, as we have seen in 8.3.23 we have for any $\lambda > 0$, $\prod_{n=1}^{\infty}\frac{r_n}{\lambda+r_n} = 0$. Hence, $E e^{-\tau} = \prod_{n=1}^{\infty} E e^{-T_n} = \prod_{n=1}^{\infty}\frac{r_n}{\lambda+r_n} = 0$ showing that $\mathbb{P}\{\tau = \infty\} = 1$.

This means that after the (random) time $\tau$, the process is left undefined. In other words, at *any* time $t > 0$ some paths of the process may no longer be defined (namely the paths $X(t, \omega)$ such that $\tau(\omega) < t$), and we observe only some of them – hence the probability that the process is somewhere in $\mathbb{N}$ may be (and is) strictly less than 1. The transition probabilities of the process described above form the minimal semigroup defined in 8.3.20. Now, we may introduce an additional rule for the behavior of the process after $\tau$; for example we may require that at $\tau$ it jumps back to 1 and does the same for all subsequent explosions. However, instead of the above rule, we could require that at $\tau$ it jumps to one of the even numbers, the probability of jumping to $2k$ being some $p_k$ such that $\sum_{k=1}^{\infty} p_k = 1$, and the reader will be able to find more such possibilities. All these choices lead to different processes and different semigroups – all of them, however, have transition semigroups dominating the minimal transition semigroup.

In this context it is worth mentioning that condition (f) of 8.3.22 has a nice probabilistic interpretation. It turns out, specifically, that $a = (\alpha_n)_{n \geq 1}, \alpha_n = E\{e^{-\lambda \tau} | X(0) = n\}$ solves the equation $Qa = \lambda a$ and is maximal in the sense that if $Q(\alpha'_n)_{n \geq 1} = \lambda(\alpha'_n)_{n \geq 1}$ for some $(\alpha'_n)_{n \geq 1} \in l^{\infty}$ with $\|(\alpha'_n)_{n \geq 1}\|_{l^{\infty}} \leq 1$, then $\alpha'_n \leq \alpha_n$ – see e.g. [92]. Certainly $a \neq 0$ iff $\tau \neq \infty$.

## 8.4 Approximation theorems

The Trotter–Kato Approximation Theorem establishes a connection between convergence of semigroups and convergence of their resolvents. As we have already seen in the examples of Yosida approximation, convergence of resolvents of semigroups alone does not imply convergence of semigroups on the whole of the space; in general the semigroups converge only on a subspace, perhaps on the subspace $\{0\}$. In fact, convergence of resolvents is equivalent to convergence of integrated semigroups, and to convergence of related homomorphisms – see 7.4.48. Before presenting the theorem, we illustrate this situation further by the following two examples. In the first example we need to consider a complex Banach space, but the reader should not find this a difficulty after remarks made in 6.2.7.

8.4.1 **Example**    Let $\{T(t), t \geq 0\}$, $\|T(t)\| \leq M$, be a semigroup acting in a complex Banach space $\mathbb{X}_0$ and let $R_\lambda, \lambda > 0$, be the resolvent of this semigroup. Define $\mathbb{X}$ as the Cartesian product $\mathbb{X}_0 \times \mathbb{C}$ where $\mathbb{C}$ is the field of complex numbers and set, for $x \in \mathbb{X}_0$, $z \in \mathbb{C}$, $t \geq 0$, $n \geq 1$,

$$T_n(t) \begin{pmatrix} x \\ z \end{pmatrix} = \begin{pmatrix} T(t)x \\ e^{itn}z \end{pmatrix}.$$

Let $R_{\lambda,n}$ be the resolvent of $\{T_n(t), t \geq 0\}$. We have, for $\lambda > 0$,

$$R_{\lambda,n} \begin{pmatrix} x \\ z \end{pmatrix} = \int_0^\infty e^{-\lambda t} T_n(t) \begin{pmatrix} x \\ z \end{pmatrix} \, dt = \begin{pmatrix} R_\lambda x \\ \frac{1}{\lambda - ni} z \end{pmatrix}$$

which converges to $\begin{pmatrix} R_\lambda x \\ 0 \end{pmatrix}$ in the uniform operator topology. However,

$T_n(t) \begin{pmatrix} x \\ z \end{pmatrix}$ does not have any limit, as $n \to \infty$, either in the strong or in the weak topology as long as $z \neq 0$. On the other hand, integrals $\int_0^t e^{nis} \, ds = \frac{1}{in}(e^{int} - 1)$ tend to 0, as $n \to \infty$, in agreement with 7.4.48.

**8.4.2 Example**    Let $\mathbb{X} = C_0(\mathbb{R}_*^+)$ be the space of continuous functions $x$ that satisfy $x(0) = \lim_{\tau \to \infty} x(\tau) = 0$. For $n \geq 1$, let

$$T_n(t)x(\tau) = 1_{\mathbb{R}^+}(\tau - nt)x(\tau - nt), \qquad t \geq 0,$$

and let $A_n$ be the generators of these semigroups. The set $\mathbb{X}_0 = \{x \in \mathbb{X} | \exists K(x) > 0 \text{ such that } \tau > K(x) \Rightarrow x(\tau) = 0\}$ is dense in $\mathbb{X}$. If $x \in \mathbb{X}_0$ then

$$\begin{aligned}
(\lambda - A_n)^{-1}x(\tau) &= \int_0^\infty e^{-\lambda t} T_n(t)x(\tau) \, dt \\
&= \int_0^\infty e^{-\lambda t} 1_{\mathbb{R}^+}(\tau - nt)x(\tau - nt) \, dt \\
&= \int_0^{\frac{\tau}{n}} e^{-\lambda t} x(\tau - nt) \, dt = \frac{1}{n} \int_0^\tau e^{-\frac{\lambda}{n}(\tau - \sigma)} x(\sigma) \, d\sigma \\
&\leq \frac{1}{n} K(x) \|x\|
\end{aligned}$$

which tends to 0 as $n \to \infty$. Since the operators $R_{\lambda,n}$ are equibounded in $n$, i.e. $\|R_{\lambda,n}\| \leq \frac{1}{\lambda}$, we also have $\lim_{n \to \infty} R_{\lambda,n}x = 0$, for all $x \in \mathbb{X}$. On the other hand, it is obvious that, for $t > 0$, $T_n(t)$ tends weakly, as $n \to \infty$, to 0. Thus, if the strong limit of it exists it is equal to 0, too. But, for all $n \geq 1$, $t > 0$ and $x \in \mathbb{X}$,

$$\|T_n(t)x - 0\| = \sup_{\tau \in \mathbb{R}^+} |1_{\mathbb{R}^+}(\tau - nt)x(\tau - nt)| = \|x\|.$$

This contradiction proves that although the weak convergence takes place, the strong one does not, and, once again, convergence of semigroups is not implied by convergence of their resolvents.

**8.4.3** *The Trotter–Kato Theorem*     Let $\{T_n(t), t \geq 0\}, n \geq 1$, be a sequence of strongly continuous semigroups with generators $A_n$. Suppose, furthermore, that there exists an $M > 0$ such that $\|T_n(t)\| \leq M$ and let $R_{\lambda,n} = (\lambda - A_n)^{-1}, \lambda > 0, n \geq 1$, denote the resolvents of $A_n$. If the limit

$$R_\lambda = \lim_{n \to \infty} R_{\lambda,n} \tag{8.70}$$

exists in the strong topology for some $\lambda > 0$, then it exists for all $\lambda > 0$. Moreover, in such a case, there exists the strongly continuous semigroup $\{T(t), t \geq 0\}$

$$T(t)x := \lim_{n \to \infty} T_n(t)x, \qquad x \in \mathbb{X}' \tag{8.71}$$

of operators in $\mathbb{X}' = cl(Range R_\lambda)$. The definition of $\mathbb{X}'$ does not depend on the choice of $\lambda > 0$, convergence in (8.71) is uniform in compact subintervals of $\mathbb{R}^+$ and we have $\int_0^\infty e^{-\lambda t} T(t)x \, dt = R_\lambda x, \lambda > 0, x \in \mathbb{X}'$ and $\|T(t)\|_{\mathcal{L}(\mathbb{X}')} \leq M$.

*Proof* To prove the first assertion we argue as in 8.3.8, where $\mathbb{A} = \mathcal{L}(\mathbb{X})$, replacing convergence in the operator topology by strong convergence. Moreover, the definition of $\mathbb{X}'$ does not depend on the choice of $\lambda > 0$ because $R_\lambda$ satisfies the Hilbert equation $R_\mu - R_\lambda = (\lambda - \mu)R_\lambda R_\mu$ which, written as $R_\mu x = R_\lambda(x + (\lambda - \mu)R_\mu x), x \in \mathbb{X}$, implies first $Range(R_\lambda) \supset Range(R_\mu)$ for all $\lambda, \mu > 0$ and then $Range(R_\lambda) = Range(R_\mu)$ by symmetry.

Next,

$$\mathbb{X}' = \{x \in \mathbb{X} | \lim_{\lambda \to \infty} \lambda R_\lambda x \text{ exists and equals } x\}.$$

Indeed, if we denote the right-hand side above by $\mathbb{X}''$, then by definition $\mathbb{X}' \supset \mathbb{X}''$. Also, $\mathbb{X}''$ is closed, by $\|\lambda R_\lambda\| \leq M, \lambda > 0$. Hence, to show the opposite inclusion it suffices to show that $Range R_\mu \subset \mathbb{X}''$, for some $\mu > 0$. But, if $x = R_\mu y$, then $\lambda R_\lambda x = \lambda R_\lambda R_\mu y = \mu R_\lambda R_\mu y + R_\mu y - R_\lambda y$, and since $\|R_\lambda\| \leq M\lambda^{-1}$, $\lim_{\lambda \to \infty} \lambda R_\lambda x = R_\mu y = x$, as claimed.

Hence, operators $T_n(t)$ being equibounded, to prove (8.71), it suffices to show that it holds for $x$ of the form $x = R_\lambda y$ where $\lambda > 0$ and $y \in \mathbb{X}$. This, however, will be shown once we prove that $T_n(t)R_{\lambda,n}y$ converges. By Exercise 8.2.15,

$$T_n(t)R_{\lambda,n}y = \lambda \int_0^t T_n(s)R_{\lambda,n}y \, ds - \int_0^t T_n(s)y \, ds + R_{\lambda,n}x.$$

On the other hand, by 7.4.48, there exists the strong limit $U(t) =$

$\lim_{n \to \infty} \int_0^t T_n(s) \, ds$ with $\|U(t)\|_{\mathcal{L}(\mathbb{X})} \le Mt$. This implies that the limit $\lim_{n \to \infty} T_n(t) R_{\lambda,n} y$ exists and equals $U(t) \lambda R_\lambda y - U(t) y + R_\lambda y$. In particular,

$$T(t) R_\lambda y = U(t) \lambda R_\lambda y - U(t) y + R_\lambda y, \qquad \lambda > 0, t \ge 0, y \in \mathbb{X}. \quad (8.72)$$

Clearly, $\|T(t)x\| \le M\|x\|, x \in \mathbb{X}'$. Since $R_{\lambda,n}$ commutes with $T_n(t)$, $R_\lambda$ commutes with $U(t)$. Therefore, by (8.72), $\|\lambda R_\lambda U(t) y - U(t) y\| = \|T(t) R_\lambda y - R_\lambda y\| \le (M+1)\lambda^{-1}\|y\|$ which implies $\lim_{\lambda \to \infty} \lambda R_\lambda U(t) y = U(t) y$. Hence, (8.72) shows that $T(t)x$ is a member of $\mathbb{X}'$. By the density argument the same is true for all $x \in \mathbb{X}'$. Also, in view of (8.71), it is clear that $\{T(t), t \ge 0\}$ is a semigroup. Using (8.72) and $\|U(t)\|_{\mathcal{L}(\mathbb{X})} \le Mt$ we see that $\lim_{t \to 0} T(t)x = x$ for $x$ in a dense subspace of $\mathbb{X}'$, hence, for all $x \in \mathbb{X}'$; this means that $\{T(t), t \ge 0\}$ is strongly continuous. Finally, continuity of $t \to T(t)x$ implies that convergence in (8.71) is uniform on compact subintervals of $\mathbb{R}^+$. The rest is clear. $\qquad \square$

**8.4.4 Remark** *The generator of the limit semigroup*    As Examples 8.4.1 and 8.4.2 make it clear, in general there is no closed linear operator $A$ such that $(\lambda - A)^{-1}$ equals $R_\lambda, \lambda > 0$, the limit pseudo-resolvent in the Trotter–Kato Theorem. (A family $R_\lambda, \lambda > 0$, of operators in a Banach space is said to be a **pseudo-resolvent** if it satisfies the Hilbert equation.) The point is that $R_\lambda, \lambda > 0$, are, in general, not injective. However, $\mathbb{X}' \cap Ker R_\lambda = \{0\}$ so that $R_\lambda, \lambda > 0$, restricted to $\mathbb{X}'$ are injective. Indeed, by the Hilbert equation, $R_\lambda x = 0$ implies $R_\mu x = 0$, $\lambda, \mu > 0$; in other words, $Ker R_\lambda$ does not depend on $\lambda > 0$. Hence, if $x \in \mathbb{X}' \cap Ker R_\lambda$ then $x = \lim_{\lambda \to \infty} \lambda R_\lambda x = 0$.

Therefore, for any $\lambda > 0$ we may define $Ax = \lambda x - \left((R_\lambda)_{|\mathbb{X}'}\right)^{-1} x$ on $\mathcal{D}(A) = Range R_\lambda$. By the Hilbert equation, this definition does not depend on $\lambda > 0$. A straightforward argument shows that $A$ thus defined satisfies the assumptions of the Hille–Yosida theorem (in $\mathbb{X}'$, with $\omega = 0$). Therefore it generates a strongly continuous semigroup in $\mathbb{X}'$. Since $(\lambda - A)^{-1} x = R_\lambda x = \int_0^\infty e^{-\lambda t} T(t) x \, dt, x \in \mathbb{X}'$, this semigroup is the semigroup $\{T(t), t \ge 0\}$ from the Trotter–Kato Theorem.

**8.4.5 Example** *Convergence of elastic Brownian motions*    Let $\mathbb{X} = C[0, \infty]$ and let $A_\epsilon, \epsilon > 0$, be the generators of semigroups related to elastic Brownian motions defined in 8.2.18. Observe that the formula (8.50) may be rewritten as

$$(\lambda - A_\epsilon)^{-1} x(\tau) = R_\lambda x(\tau) + \frac{2\epsilon}{\epsilon\sqrt{2\lambda} + 1} \int_0^\infty e^{-\sqrt{2\lambda}\sigma} x(\sigma) \, d\tau e^{-\sqrt{2\lambda}\tau},$$

where

$$R_\lambda x(\tau) = \frac{1}{\sqrt{2\lambda}} \int_0^\infty \left[ e^{-\sqrt{2\lambda}|\sigma - \tau|} - e^{-\sqrt{2\lambda}(\sigma + \tau)} \right] x(\sigma) \, d\sigma.$$

Therefore, for every $x \in \mathbb{X}$ and $\lambda > 0$,

$$\left\| (\lambda - A_\epsilon)^{-1} x - R_\lambda x \right\| = \left| \frac{2\epsilon}{\epsilon\sqrt{2\lambda} + 1} \right| \sup_{\tau \geq 0} \left| \int_0^\infty e^{-\sqrt{2\lambda}\sigma} x(\sigma) \, d\sigma e^{-\sqrt{2\lambda}\tau} \right|$$

$$\leq \left| \frac{2\epsilon}{\epsilon\sqrt{2\lambda} + 1} \right| \frac{1}{\sqrt{2\lambda}} \|x\| \xrightarrow[\epsilon \to 0]{} 0.$$

In other words, for any sequence $\epsilon_n, n \geq 1$, such that $\lim_{n\to\infty} \epsilon_n = 0$ the semigroups $\{T_n(t), t \geq 0\}, n \geq 1$, generated by $A_{\epsilon_n}, n \geq 1$, satisfy the assumptions of the Trotter–Kato Theorem. Also, we see that $R_\lambda x(\tau) = 0$ and that for any $x \in \mathbb{X}$ such that $x(0) = 0$,

$$\lim_{\lambda \to \infty} \lambda R_\lambda x = x. \tag{8.73}$$

Hence, $\mathbb{X}' = \{x \in \mathbb{X} | x(0) = 0\}$.

**8.4.6 Exercise**    Check that the semigroup related to Brownian motion described in 7.5.1 leaves $C[-\infty, \infty]$ invariant and is a strongly continuous semigroup in this space, and the resolvent of the restricted semigroup is still given by (7.41). Conclude that, for $x \in C[-\infty, \infty]$,

$$\lim_{\lambda \to \infty} \sqrt{\frac{\lambda}{2}} \int_{-\infty}^\infty e^{-\sqrt{2\lambda}|\tau - \sigma|} x(\sigma) \, d\sigma = x(\tau)$$

uniformly in $\tau$. Use this to prove (8.73). Moreover, show that the limit semigroup in Example 8.4.5 is the semigroup related to the minimal Brownian motion, restricted to $\mathbb{X}'$.    $\Box$

If condition (8.70) holds and $x \in \mathcal{D}(A)$, then for $\lambda > 0$ there exists a $y$ in $\mathbb{X}'$ such that $x = R_\lambda y$. Also, $x_n = R_{\lambda,n} y, n \geq 1$, belong to $\mathcal{D}(A_n)$ and we have $\lim_{n\to\infty} x_n = x$ and $\lim_{n\to\infty} A_n x_n = \lim_{n\to\infty} \lambda R_{\lambda,n} y - y = \lambda R_\lambda y - y = A R_\lambda y = Ax$. In Subsection 8.4.9 a version of the approximation theorem is presented where convergence of semigroups is characterized in terms of such convergence of their generators.

**8.4.7 Definition**    Given a sequence $A_n, n \geq 1$, of (in general, unbounded) operators in a Banach space $\mathbb{X}$, the domain of the **extended limit** $A_{ex}$ of this sequence is defined as the set of $x \in \mathbb{X}$ with the property that there exist $x_n \in \mathcal{D}(A_n)$ such that $\lim_{n\to\infty} x_n = x$ and the limit $\lim_{n \to \infty} A_n x_n$ exists. Although the latter limit is not uniquely determined by $x$ (see 8.4.11), we write $A_{ex} x = \lim_{n \to \infty} A_n x_n$. In other words, $A$ is a so-called multi-valued operator.

**8.4.8 Exercise** Let $A_n, n \geq 1$, be as in the above definition and let $c(\mathbb{X})$ be the space of all convergent sequences with values in $\mathbb{X}$. Let us define the operator $\mathcal{A}$ in $c(\mathbb{X})$ by $\mathcal{A}(x_n)_{n \geq 1} = (A_n x_n)_{n \geq 1}$ with domain

$$\mathcal{D}(\mathcal{A}) = \{(x_n)_{n \geq 1} \, | \, x_n \in \mathcal{D}(A_n), (A_n x_n)_{n \geq 1} \in c(\mathbb{X})\}.$$

Also let $L : c(\mathbb{X}) \to \mathbb{X}$ be given by $L(x_n)_{n \geq 1} = \lim_{n \to \infty} x_n$. Show that $x$ belongs to the domain of the extended limit of $A_n, n \geq 1$, iff there exists an $(x_n)_{n \geq 1} \in \mathcal{D}(\mathcal{A})$ such that $L(x_n)_{n \geq 1} = x$.

**8.4.9** *The Sova–Kurtz version of the Trotter–Kato Theorem* Let $\mathbb{X}$ be a Banach space. Suppose, as in 8.4.3, that $\{T_n(t), t \geq 0\}, n \geq 1$, is a sequence of strongly continuous semigroups with generators $A_n$, and that there exists an $M > 0$ such that $\|T_n(t)\| \leq M$. Also, suppose that for some $\lambda > 0$ the set of $y$ that can be expressed as $\lambda x - A_{ex} x$, where $A_{ex}$ is the extended limit of $A_n, n \geq 1$, is dense in $\mathbb{X}$. Then, the limit (8.70) exists for all $\lambda > 0$. Moreover, $\mathbb{X}' = cl(\mathcal{D}(A_{ex}))$ and the part $A_p$ of $A_{ex}$ in $\mathbb{X}$ is single-valued and is the infinitesimal generator of the semigroup defined by (8.71).

*Proof* By saying that $y$ may be expressed as $\lambda x - A_{ex} x$ we mean that there exists a sequence $x_n \in \mathcal{D}(A_n)$ such that $\lim_{n \to \infty} x_n = x$ and $\lim_{n \to \infty} A_n x_n$ exists and equals $\lambda x - y$. Clearly $R_{\lambda,n}(\lambda x_n - A_n x_n) = x_n$. Also $\|R_{\lambda,n} y - x_n\| = \|R_{\lambda,n}(y - \lambda x_n + A_n x_n)\| \leq M \lambda^{-1} \|y - \lambda x_n + A_n x_n\|$. Since $\lim_{n \to \infty}(\lambda x_n - A_n x_n) = y$, the sequence $R_{\lambda,n} y, n \geq 1$, converges and its limit equals $\lim_{n \to \infty} x_n = x$. This shows that the limit (8.70) exists for the $\lambda$ described in the assumption of our theorem and $x$ from a dense subspace of $\mathbb{X}$. Since $\|R_{\lambda,n}\| \leq M \lambda^{-1}$ this limit exists for all $x \in \mathbb{X}$.

The same argument applies now to show that

$$R_\lambda(\lambda x - A_{ex} x) = x \tag{8.74}$$

for all $x \in \mathcal{D}(A_{ex})$ and $\lambda > 0$. Hence, $\mathcal{D}(A_{ex}) \subset Range \, R_\lambda \subset \mathbb{X}'$. Let $A$ be the generator of the semigroup defined by (8.70). The remark made before 8.4.7 shows that $\mathcal{D}(A) \subset \mathcal{D}(A_{ex})$. Since the former set is dense in $\mathbb{X}'$, $cl(\mathcal{D}(A_{ex})) = \mathbb{X}'$.

Finally, if $x \in \mathcal{D}(A_p)$, i.e. if $x \in \mathcal{D}(A_{ex})$ and $A_{ex} x \in \mathbb{X}'$, then, by

(8.74), $x$ belongs to $\mathcal{D}(A)$ and since $R_\lambda(\lambda x - Ax) = x, x \in \mathcal{D}(A)$, we have $Ax = A_{\text{ex}}x, x \in \mathcal{D}(A_{\text{p}})$. On the other hand, if $x \in \mathcal{D}(A)$, then $x \in \mathcal{D}(A_{\text{ex}})$ and $A_{\text{ex}}x$ belongs to $\mathbb{X}'$ so that $\mathcal{D}(A) \subset \mathcal{D}(A_{\text{p}})$.     □

**8.4.10 Corollary**     Suppose that $\{T_n(t), t \geq 0\}, n \geq 1$, is a sequence of strongly continuous semigroups in $\mathbb{X}$ with generators $A_n, n \geq 1$, and that there exists an $M > 0$ such that $\|T_n(t)\| \leq M$. Let $A$ be a linear, in general unbounded operator such that $Range(\lambda - A)$ is dense in $\mathbb{X}$ for some $\lambda > 0$. If for every $x \in \mathcal{D}(A)$ there exists a sequence $x_n \in \mathcal{D}(A_n), n \geq 1$, such that $\lim_{n\to\infty} x_n = x$ and $\lim_{n\to\infty} A_n x_n = Ax$, then the limit (8.70) exists, the part $A_{\text{p}}$ of $A$ in $\mathbb{X}'$ is closable and its closure is the generator of the semigroup given by (8.71). In particular, if $A$ is the generator of a semigroup $\{S(t), t \geq 0\}$ in $\mathbb{X}$ (in which case $Range(\lambda - A) = \mathbb{X}, \lambda > 0$), $S(t)$ coincides with $T(t)$ given by (8.71).

**8.4.11 Example** *Telegraph equation with small parameter*     Let $0 < \epsilon < 1$ be given. The equation

$$\epsilon \frac{\partial^2 x(t,\tau)}{\partial t^2} + \frac{\partial x(t,\tau)}{\partial t} = \frac{1}{2}\frac{\partial^2 x(t,\tau)}{\partial \tau^2}, \quad x(0,\tau) = x(\tau), \frac{\partial x}{\partial t}(0,\tau) = y(\tau)$$
$$(8.75)$$

is called the **telegraph equation with small parameter**. Of course, this equation is obtained from (7.55) when we put $2a = 2v^2 = \epsilon^{-1}$.

It is reasonable to expect that, as $\epsilon \to 0$, solutions to (8.75) tend to the solutions of the diffusion equation

$$\frac{\partial x(t,\tau)}{\partial t} = \frac{1}{2}\frac{\partial^2 x(t,\tau)}{\partial \tau^2}, \quad x(0,\tau) = x(\tau).$$

To prove this conjecture, we consider the operators $A_\epsilon = A_{(2\epsilon)^{-1},(2\epsilon)^{-1/2}}$ where $A_{a,v}$ has been defined in 8.2.24. As proved there, $A_\epsilon$ is the generator of a strongly continuous semigroup, say $\{T_\epsilon(t), t \geq 0\}$, in $\mathbb{X} \times \mathbb{Y}$ where $\mathbb{X} = BUC_1(\mathbb{R})$ and $\mathbb{Y} = BUC(\mathbb{R})$. We have

$$T_\epsilon(t) = \begin{pmatrix} S_{00}(\epsilon,t) & S_{01}(\epsilon,t) \\ S_{10}(\epsilon,t) & S_{11}(\epsilon,t) \end{pmatrix}$$

where, by 8.2.26,

$$\|S_{00}(\epsilon,t)\| \leq 1, \qquad \|S_{01}(\epsilon,t)\| \leq \epsilon + \sqrt{2\epsilon} \leq 3\sqrt{\epsilon},$$
$$\|S_{10}(\epsilon,t)\| \leq \frac{1}{\sqrt{2\epsilon}} \leq \frac{1}{\sqrt{\epsilon}}, \quad \|S_{11}(\epsilon,t)\| \leq 1. \qquad (8.76)$$

We note that the Trotter–Kato Theorem cannot be applied to the semigroups $\{T_\epsilon(t), t \geq 0\}, 0 < \epsilon < 1$, since they are not equibounded. However, putting

$$V_\epsilon(t) = \begin{pmatrix} 1 & 0 \\ 0 & \sqrt{\epsilon} \end{pmatrix} T_\epsilon(t) \begin{pmatrix} 1 & 0 \\ 0 & \frac{1}{\sqrt{\epsilon}} \end{pmatrix} = \begin{pmatrix} S_{00}(\epsilon, t) & \frac{1}{\sqrt{\epsilon}} S_{01}(\epsilon, t) \\ \sqrt{\epsilon} S_{10}(\epsilon, t) & S_{11}(\epsilon, t) \end{pmatrix}$$

we obtain the family of equibounded semigroups $\{V_\epsilon(t), t \geq 0\}, 0 < \epsilon < 1$. Indeed, for $(x, y) \in \mathbb{X} \times \mathbb{Y}$,

$$\| S_{00}(\epsilon, t)x + \frac{1}{\sqrt{\epsilon}} S_{01}(\epsilon, t)y \| + \| \sqrt{\epsilon} S_{10}(\epsilon, t)x + S_{11}(\epsilon, t)y \|$$

$$\leq \|x\| + 3\|y\| + \|x\| + \|y\| \leq 4\|(x, y)\|,$$

so that $\|V_\epsilon(t)\| \leq 4$.

The domain of the generator, say $B_\epsilon$, of $\{V_\epsilon(t), t \geq 0\}$ is the same as the domain of $A_\epsilon$ $(= \mathcal{D}(\frac{\mathrm{d}^2}{\mathrm{d}\tau^2}) \times \mathbb{X})$ and

$$B_\epsilon \begin{pmatrix} x \\ y \end{pmatrix} = \begin{pmatrix} 1 & 0 \\ 0 & \sqrt{\epsilon} \end{pmatrix} A_\epsilon \begin{pmatrix} 1 & 0 \\ 0 & \frac{1}{\sqrt{\epsilon}} \end{pmatrix} \begin{pmatrix} x \\ y \end{pmatrix} = \begin{pmatrix} \frac{1}{\sqrt{\epsilon}} y \\ \frac{1}{\sqrt{\epsilon}} \frac{1}{2} x'' - \frac{1}{\epsilon} y \end{pmatrix}. \qquad (8.77)$$

We know that the operator $\frac{1}{2} \frac{\mathrm{d}^2}{\mathrm{d}\tau^2}$ with domain $\mathcal{D}(\frac{\mathrm{d}^2}{\mathrm{d}\tau^2})$ is the generator of the Brownian motion semigroup $\{T_B(t), t \geq 0\}$ in $\mathbb{Y}$. This semigroup leaves the space $\mathbb{X}$ invariant and $(T_B(t)x)' = T_B(t)x'$. Hence, the restricted semigroup $\{V(t), t \geq 0\}$ $(V(t) = T_B(t)_{|\mathbb{X}})$ is strongly continuous in $\mathbb{X}$, and its generator is $\frac{1}{2} \frac{\mathrm{d}^2}{\mathrm{d}\tau^2}$ with domain $\mathcal{D}_1(\frac{\mathrm{d}^2}{\mathrm{d}\tau^2})$ composed of three times differentiable functions with all three derivatives in $\mathbb{Y}$.

Let us take $x \in \mathcal{D}_1(\frac{\mathrm{d}^2}{\mathrm{d}\tau^2})$ and for every $\epsilon$ consider $(x, \sqrt{\epsilon} \frac{1}{2} x'' - \epsilon y)$ where $y \in \mathbb{X}$. Since $x''$ belongs to $\mathbb{X}$, $(x, \sqrt{\epsilon} \frac{1}{2} x'' - \epsilon y)$ belongs to $\mathcal{D}(B_\epsilon)$. Moreover, $B_\epsilon(x, \sqrt{\epsilon} \frac{1}{2} x'' - \epsilon y) = (\frac{1}{2} x'' - \sqrt{\epsilon} y, y) \to (\frac{1}{2} x'', y)$ and $(x, \sqrt{\epsilon} \frac{1}{2} x'' - \epsilon y) \to (x, 0)$, as $\epsilon \to 0$. This shows that, for any sequence $\epsilon_n, n \geq 1$, converging to zero, $\mathcal{D}_1(\frac{\mathrm{d}^2}{\mathrm{d}\tau^2}) \times \{0\}$ is contained in the domain of the extended limit $B_{ex}$ of $B_{\epsilon_n}$. Our calculation shows also that the set of vectors of the form $\lambda(x, y) - B_{ex}(x, y)$, where $(x, y) \in \mathcal{D}(B_{ex})$, contains vectors $(\lambda x - \frac{1}{2} x'', (\lambda - 1)y)$. Since the vectors of the form $\lambda x - \frac{1}{2} x''$ where $x \in \mathcal{D}_1(\frac{\mathrm{d}^2}{\mathrm{d}\tau^2})$ exhaust the whole of $\mathbb{X}$ and $\mathbb{X}$ is a dense subspace of $\mathbb{Y}$, the conditions of the Sova–Kurtz version of the Trotter–Kato Theorem are fulfilled with any $\lambda > 0, \lambda \neq 1$. (See Exercise 8.4.12.)

Moreover, if $(x_\epsilon, y_\epsilon)$ converges, as $\epsilon \to 0$, to $(x, y)$, in such a way that $B_\epsilon(x_\epsilon, y_\epsilon)$ converges, then, by (8.77), $\frac{1}{\sqrt{\epsilon}} y_\epsilon$ converges, and so $y_\epsilon$ converges to 0. This shows that $\mathcal{D}(B_{ex})$ is contained in $\mathbb{X} \times \{0\}$. Combining this with our previous findings we obtain that the subspace where the limit semigroup of $\{V_\epsilon(t), t \geq 0\}$ is defined equals $\mathbb{X} \times \{0\}$. By 8.4.9, we know that the part of $B_{ex}$ in $\mathbb{X} \times \{0\}$ is single-valued, and we have already seen that one of its possible values on $(x, 0)$ where $x \in \mathcal{D}_1(\frac{d^2}{d\tau^2})$ is $(\frac{1}{2}x'', 0)$. This means that the generator of the limit semigroup and the operator $A(x, 0) = (\frac{d^2}{d\tau^2}, 0)$ coincide on the set $\mathcal{D}_1(\frac{d^2}{d\tau^2}) \times \{0\}$, which is the domain of the latter operator. Since both operators are generators of semigroups, they must be equal (use 7.4.32, for example). Hence, we have

$$\lim_{\epsilon \to 0} V_\epsilon(t) \begin{pmatrix} x \\ 0 \end{pmatrix} = \lim_{\epsilon \to 0} \begin{pmatrix} S_{00}(\epsilon, t)x \\ \sqrt{\epsilon}S_{10}(\epsilon, t)x \end{pmatrix} = \begin{pmatrix} V(t)x \\ 0 \end{pmatrix}, \qquad x \in \mathbb{X}. \quad (8.78)$$

Therefore, in view of the inequality involving $\|S_{01}(\epsilon, t)\|$ contained in (8.76), $S_{00}(\epsilon, t)x + S_{01}(\epsilon, t)y$ converges to $V(t)x$ strongly in $\mathbb{X}$, for $x \in \mathbb{X}$ and $y \in \mathbb{Y}$. A density argument then applies to show that it converges to $V(t)x$ in $\mathbb{Y}$ for $x$ and $y$ in $\mathbb{Y}$, and this is what we have set out to prove.

**8.4.12 Exercise**    Modify the argument from the previous subsection to show that the conditions of 8.4.9 are fulfilled, as they must, with any $\lambda > 0$.

**8.4.13 Exercise**    Prove that the derivative of the solution to the telegraph equation with small parameter converges to that of the diffusion equation provided $x \in \mathcal{D}_1(\frac{d^2}{d\tau^2})$ and $y = \frac{1}{2}x''$.

**8.4.14 Exercise**    Prove convergence of semigroups from 8.4.5 using 8.4.9.

**8.4.15 Exercise**    Prove convergence of semigroups from 8.4.11 using the Trotter–Kato Theorem 8.4.3. To this end solve the system of equations

$$\lambda x - \frac{1}{\sqrt{\epsilon}}y = w,$$

$$\lambda y - \frac{1}{\sqrt{\epsilon}}\frac{1}{2}x'' + \frac{1}{\epsilon}y = z,$$

with given $w \in \mathbb{X}$ and $z \in \mathbb{Y}$ and unknown $x \in \mathcal{D}_1(\frac{d^2}{d\tau^2})$ and $y \in \mathbb{X}$ to show that

$$(\lambda - B_\epsilon)^{-1} = \begin{pmatrix} (\epsilon\lambda + 1)(\lambda^2\epsilon + \lambda - \frac{1}{2}\frac{d^2}{d\tau^2})^{-1} & \sqrt{\epsilon}(\lambda^2\epsilon + \lambda - \frac{1}{2}\frac{d^2}{d\tau^2})^{-1} \\ \sqrt{\epsilon}\frac{1}{2}\frac{d^2}{d\tau^2}(\lambda^2\epsilon + \lambda - \frac{1}{2}\frac{d^2}{d\tau^2})^{-1} & \epsilon\lambda(\lambda^2\epsilon + \lambda - \frac{1}{2}\frac{d^2}{d\tau^2})^{-1} \end{pmatrix},$$

where $(\lambda - \frac{1}{2}\frac{d^2}{d\tau^2})^{-1}$ is the resolvent of the operator $\frac{1}{2}\frac{d^2}{d\tau^2}$ in $\mathbb{X}$.

**8.4.16** *Proof of* 8.3.20  We have $R_{\lambda,r} \leq R_{\lambda,r'}$ for $r \leq r'$, and $\|R_{\lambda,r}\| \leq \lambda^{-1}$. Hence, by 8.3.17, there exists the strong limit $R_\lambda = \lim_{r \uparrow 1} R_{\lambda,r}$. Clearly, for any $N \in \mathbb{N}$, $\sum_{n=0}^{N} (\lambda - D)^{-1} r^n B_\lambda^n \leq R_{\lambda,r} \leq R_\lambda$. Hence, letting $r \uparrow 1$, we obtain $\sum_{n=0}^{N} (\lambda - D)^{-1} B_\lambda^n \leq R_\lambda$. Applying 8.3.17 again, the series $\sum_{n=0}^{\infty} (\lambda - D)^{-1} B_\lambda^n$ converges and we have $\sum_{n=0}^{\infty} (\lambda - D)^{-1} B_\lambda^n \leq R_\lambda$. On the other hand, $R_{\lambda,r} \leq \sum_{n=0}^{\infty} (\lambda - D)^{-1} B_\lambda^n$ and so, letting $r \uparrow 1$, $R_\lambda \leq \sum_{n=0}^{\infty} (\lambda - D)^{-1} B_\lambda^n$, proving that the two are equal. (Note that we *do not* claim that $R_\lambda = (\lambda - D)^{-1} \sum_{n=0}^{\infty} B_\lambda^n$; in fact, the series $\sum_{n=0}^{\infty} B_\lambda^n$ in general diverges.)

Next, we note that for $x \in \mathcal{D}(D)$, $\sum_{n=0}^{N+1} (\lambda - D)^{-1} B_\lambda^n (\lambda - D)x = x + \sum_{n=1}^{N+1} (\lambda - D)^{-1} B_\lambda^{n-1} Ox = x + \sum_{n=0}^{N} (\lambda - D)^{-1} B_\lambda^n Ox$. Letting $N \to \infty$, we obtain $R_\lambda(\lambda - D)x = x + R_\lambda Ox$, i.e. $R_\lambda(\lambda - D - O)x = x$. In particular, the range of $R_\lambda$ contains $\mathcal{D}(D)$ and so $cl(Range R_\lambda) = l^1$. Therefore, the semigroups $\{S_r(t), t \geq 0\}$ converge as $r \uparrow 1$ to a strongly continuous semigroup. The limit semigroup is composed of sub-Markov operators, the operators $S_r(t), t \geq, 0 \leq r < 1$, being sub-Markov. Finally, for $x \in \mathcal{D}(D), \lim_{r \uparrow 1} Dx + rOx = Dx + Ox$, proving that the extended limit of $D + rO$, which is the generator of the limit semigroup, is an extension of $D + O$.

**8.4.17** *Approximation by discrete-parameter semigroups*  Let us suppose that $T_n, n \geq 1$, are contractions in a Banach space $\mathbb{X}$ and $h_n, n \geq 1$, are positive numbers with $\lim_{n \to \infty} h_n = 0$. Then, the operators $A_n = h_n^{-1}(A_n - I)$ are generators of contraction semigroups $\{T_n(t), t \geq 0\}$ where $T_n(t) = e^{-h_n^{-1} t} e^{-h_n^{-1} t T_n}$. If the extended limit $A_{ex}$ of $A_n, n \geq 1$, has the property that for some $\lambda > 0$ the vectors of the form $\lambda x - A_{ex} x$ form a dense set in $\mathbb{X}$, then, by 7.4.48, 8.4.3 and 8.4.9, there exist the limits

$$U(t)x = \lim_{n \to \infty} \int_0^t T_n(u)x \, du, \qquad x \in \mathbb{X},$$

$$T(t)x = \lim_{n \to \infty} T_n(t)x, \qquad x \in \mathbb{X}.$$

We will show that $U(t)$ and $T(t)$ may also be approximated as follows:

$$U(t)x = \lim_{n \to \infty} \int_0^t T_n^{[u/h_n]} x \, du, \qquad x \in \mathbb{X}, \tag{8.79}$$

$$T(t)x = \lim_{n \to \infty} T_n^{[t/h_n]} x, \qquad x \in \mathbb{X}. \tag{8.80}$$

To this end, we note first that by 8.4.9 there exists the strong limit $R_\lambda = \lim_{n \to \infty} (\lambda - A_n)^{-1}$. Hence, by 7.4.50, there exists the limit on the right-hand side of (8.79). Equality must hold, because both sides are continuous and their Laplace transforms coincide. Moreover, from the proof of the Trotter–Kato Theorem 8.4.3 we know that $U(t)x \in \mathbb{X}'$ for all $t \geq 0$ and $x \in \mathbb{X}$; on the other hand, for $x \in \mathbb{X}'$, $U(t)x = \int_0^t T(s)x\,ds$, so that by the strong continuity of $\{T(t), t \geq 0\}$, $\lim_{t\to 0} \frac{1}{t} U(t)x = x$. This shows that it suffices to show (8.80) for $x$ of the form $x = U(s)y, y \in \mathbb{X}, s > 0$. In view of (8.79) and $\|T^{[t/h_n]}\| \leq 1$, this will be done once we show that $\lim_{n\to\infty} T_n^{[t/h_n]} \int_0^s T_n^{[u/h_n]} y\,du$ exists and equals $T(t)U(s)y$. We have:

$$T_n^{\left[\frac{t}{h_n}\right]} \int_0^s T_n^{\left[\frac{u}{h_n}\right]} y\,du = \int_0^s T_n^{\left[\frac{u+h_n\left[\frac{t}{h_n}\right]}{h_n}\right]} y\,du = \int_{h_n\left[\frac{t}{h_n}\right]}^{s+h_n\left[\frac{t}{h_n}\right]} T_n^{\left[\frac{u}{h_n}\right]} y\,du$$

$$= \int_0^{s+h_n\left[\frac{t}{h_n}\right]} T_n^{\left[\frac{u}{h_n}\right]} y\,du - \int_0^{h_n\left[\frac{t}{h_n}\right]} T_n^{\left[\frac{u}{h_n}\right]} y\,du$$

$$\xrightarrow[n\to\infty]{} U(t+s)y - U(t)y.$$

On the other hand, $U(t+s)y - U(t)y = \lim_{n\to\infty} \int_0^{t+s} T_n(u)y\,du - \lim_{n\to\infty} \int_0^t T_n(u)y\,du = \lim_{n\to\infty} T_n(t) \int_0^s T_n(u)y\,du = T(t)U(s)$, as desired.

**8.4.18 Corollary** *Central Limit Theorem again*    As in the proof of 5.5.2, we assume without loss of generality that $X_n, n \geq 1$, are independent, identically distributed random variables with mean zero and variance 1. Let $T_n = T_{\frac{1}{\sqrt{n}} X_n} = T_{\frac{1}{\sqrt{n}} X_1}$ be the related operators in $C[-\infty, \infty]$. By Lemma 5.5.1, $\lim_{n\to\infty} n(T_n - I)x = \frac{1}{2}x''$ for all twice differentiable functions $x \in C[-\infty, \infty]$ with $x'' \in C[-\infty, \infty]$. The operator $x \to \frac{1}{2}x''$ defined on the set of such functions is the generator of the semigroup $\{T(t), t \geq 0\}$ related to Brownian motion. Hence, by 8.4.10, $\lim_{n\to\infty} T_n^{[nt]}x = T(t)x$ for all $x \in C[-\infty, \infty]$. Taking $t = 1$ and noting that $T_n^n = T_{\frac{1}{\sqrt{n}} \sum_{i=1}^n X_i}$ we obtain the claim by 5.4.18.

**8.4.19 Example**    *A random walk approximating Brownian motion* Brownian motion is often approximated by the following random walk. Given a sequence of independent random variables $Y_i$ assuming values $+1$ and $-1$ with equal probability $\frac{1}{2}$, we define the simple random walk $W_k, k \geq 1$, by $W_0 = 0$ and $W_k = \sum_{i=1}^k Y_i$. Next, we define continuous-time processes $X_n(t), n \geq 1$, by $X_n(t) = \frac{1}{\sqrt{n}} W_{[nt]}, t \geq 0$. In other words, with $n$ increasing, we increase the number of steps of the random walk in a finite time, while decreasing their length; the steps are being taken at times $t = \frac{k}{n}$. We note that $E\,X_n(t) = 0$ and $D^2 X_n(t) = \frac{1}{n} \sum_{i=1}^{[nt]} Y_i^2 = \frac{[nt]}{n} \longrightarrow t$, as $n \to \infty$.

To see that $X_n(t), n \geq 1$, approximates a Brownian motion $w(t), t \geq 0$, we note that $T_{X_n(t)} = T_{\frac{1}{\sqrt{n}} Y_1}^{[nt]}$. By 5.5.1, $\lim_{n \to \infty} (T_{\frac{1}{\sqrt{n}} Y_1} x - x) = \frac{1}{2} x''$, for suitable class of functions $x$. As in 8.4.18, this implies $\lim_{n \to \infty} T_{X_n(t)} x(\tau) = E\,x(\tau + w(t)), \tau \in \mathbb{R}, x \in C[-\infty, \infty]$.

**8.4.20 Example** For an $n \geq 1$, we define a discrete-time Markov process $X_n(k), k \geq 0$, in $[0, 1]$ by requiring that given $X_n(k) = s$ we have $X_n(k+1) = Y/n$ where $Y$ is a binomial random variable with parameter $s$. We will show that as $n \to \infty$, $X_n([2nt]), n \geq 0$, approximates the continuous-time process in $[0, 1]$ with generator $\overline{A}$ introduced in 8.3.15 and 8.3.16. To this end it suffices to show that for a twice differentiable function $x \in C[0, 1]$ with $x'' \in C[0, 1]$, $\lim_{n \to \infty} 2n[A_n x - x] = \overline{A}x$ strongly in $C[0, 1]$, where $A_n x$ are Bernstein polynomials (see 2.3.29). Using the Taylor formula (5.21), $2n[A_n x - x](s) = 2nx'(s)E\,(Y/n - s) + nE\,(Y/n - s)^2 x''[s + \theta(Y/n - s)] = nE\,(Y/n - s)^2 x''[s + \theta(Y/n - s)]$. Also $\overline{A}x(s) = nx''(s)E\,(Y/n - s)^2$ since $E\,(Y - ns)^2 = D^2 Y = ns(1 - s)$.

For a given $\epsilon > 0$, a $\delta > 0$ may be chosen so that $|\tau - \sigma| < \delta$ implies $|x''(\sigma) - x''(\tau)| < \epsilon$. Hence, by Chebyshev's inequality, $\|2n[A_n x - x] - \overline{A}x\|$ does not exceed

$$\sup_{s \in [0,1]} \epsilon n E\,(Y/n - s)^2 1_{|Y/n - s| < \delta} + 2n\|x''\|\,E\,(Y/n - s)^2 1_{|Y/n - s| \geq \delta}$$

$$\leq \epsilon \sup_{s \in [0,1]} s(1 - s) + 2n^{-3}\delta^{-2}\|x''\| \sup_{s \in [0,1]} E\,(Y - ns)^4.$$

Since the first supremum equals $\frac{1}{4}$, we are left with proving that

$$\lim_{n \to \infty} n^{-3} \sup_{s \in [0,1]} E\,(Y - ns)^4 = 0.$$

This, however, follows by a straightforward calculation of $E\,(Y - ns)^4$; the details are left to the reader.

**8.4.21 Example** *The pure death process related to the $n$-coalescent of Kingman* Let us consider a population of $N$ individuals which evolves according to the following rules. The generations are discrete and non-overlapping, and each member of the $(n + 1)$st generation chooses his parent from the $n$th generation at random and independently from the other members, the probability of choosing any parent being equal to $N^{-1}$. We observe a sample of $n$ individuals from this population at time 0 and are interested in the number $X_N(k), k \geq 1$, of ancestors $k$ generations back; we assume that the process is well-defined for all $k \geq 0$, i.e. that the population has evolved in the manner described above for an infinitely long time. $X_N(k), k \geq 0$, is a discrete-time Markov chain with values in $\{1, ..., n\}$ and transition probabilities $p_{i,j} = p_{i,j}(N) = N^{-i} \begin{Bmatrix} i \\ j \end{Bmatrix} \binom{N}{j} j!$, where $\begin{Bmatrix} i \\ j \end{Bmatrix}$ is the Stirling number of the second kind – see 8.4.33. Indeed, $N^i$ is the number of all possible ways $i$ members may choose their parents, and the number of ways exactly $j$ parents may be chosen is the product of three numbers. The first of them is the number of ways the set of $i$ elements may be partitioned into $j$ subsets, i.e. the Stirling number of the second kind. The second is the number of ways $j$ parents may be chosen from the population of $N$ individuals – the binomial coefficient $\binom{N}{j}$, and the third is the number of possible assignments of $j$ parents to $j$ subsets.

The process $X_N(k), k \geq 0$, is a **pure death process** in that its paths are non-increasing sequences. We will show that $X_N([tN]), t \geq 0$, converges to a continuous-time (pure death) process with intensity matrix $Q = (q_{i,j})$, where

$$q_{i,i} = -\binom{i}{2}, i = 1, ..., n, \; q_{i,i-1} = \binom{i}{2}, i = 2, ...., n \qquad (8.81)$$

and $q_{i,j} = 0$ otherwise. To this end we note first that to prove that $N\left[(p_{i,j})_{1 \leq i,j \leq n} - I\right]$ converges to $Q$ it suffices to show that the corresponding entries of these matrices converge. Moreover,

$$p_{i,i} = \prod_{k=1}^{i-1} \left(1 - \frac{k}{N}\right) = 1 - \sum_{k=1}^{i-2} \frac{k}{N} + h_1$$

where $|h_1| \leq 2^{i-1} \sum_{l=2}^{i-1} \frac{(i-1)^l}{N^l}$, so that $\lim_{N \to \infty} N h_1 = 0$. Similarly,

$$p_{i,i-1} = \binom{i}{2} \frac{1}{N} \prod_{k=1}^{i-2} \left(1 - \frac{k}{N}\right) = \frac{1}{N} \binom{i}{2} + h_2 \qquad (8.82)$$

where $|h_2| \leq 2^{i-1} \binom{i}{2} \frac{1}{N} \sum_{l=1}^{i-2} \frac{(i-2)^l}{N^l}$ so that $\lim_{N \to \infty} N h_2 = 0$. This shows that $\lim_{N \to \infty} N[p_{i,i} - 1] = -\binom{i}{2} = -\lim_{N \to \infty} N p_{i,i-1}$. Moreover, since

$$\sum_{j=1}^{n} p_{i,j} = \sum_{j=1}^{i} p_{i,j} = 1, \qquad (8.83)$$

for $j \neq i, i-1$, $Np_{i,j} \leq N \sum_{l \neq i, i-1} p_{i,j} = N(1 - p_{i,i} - p_{i,i-1}) \leq N|h_1| + N|h_2| \longrightarrow 0$ as $N \to \infty$, as desired.

As formula (8.81) shows, if $N$ is large and the time is measured in units of $N$ generations, the distribution of the time (counted backwards) when there are $j$ ancestors of the given sample is approximately exponential with parameter $j(j-1)/2$. In particular, the expected time $T_{\text{MRCAS}}$ to the most recent common ancestor of a sample of $n$ individual is $ET_{\text{MRCAS}} \approx 2N \sum_{i=2}^{n} \frac{1}{i(i-1)} = 2N(1 - \frac{1}{n})$. In particular, the expected time $T_{\text{MRCAP}}$ to the most recent common ancestor of the whole population is $ET_{\text{MRCAP}} \approx 2N$.

**8.4.22 Example** *Kingman's n-coalescent*   A modification of the reasoning from the previous subsection allows tracing of the whole genealogy of a sample. To this end, for a sample of $n$ individuals we consider the Markov chain $\mathcal{R}_N(k), k \geq 0$, of equivalence relations in $\{1, ..., n\}$; the pair $(i,j)$ belongs to the equivalence relation $\mathcal{R}_N(k)$ iff the individuals $i$ and $j$ have a common ancestor $k$ generations ago. Each equivalence class corresponds to a member of a population that lived $k$ generations ago, yet the opposite statement is not true because some members of this generation may have not have descendants. $\mathcal{R}_N(0)$ is the main diagonal in the square $\{(i,j)|1 \leq i, j \leq n\}$ and by 8.4.21, $\mathcal{R}_N(k)$ eventually reaches the full equivalence relation, i.e. the whole square.

We follow Kingman [67], [68] to show that $\mathcal{R}_N([Nt]), t \geq 0$, converges, as $N \to \infty$, to the continuous-time Markov chain with intensity matrix $Q$ given by

$$q_{\mathcal{E},\mathcal{E}'} = \begin{cases} -\binom{|\mathcal{E}|}{2}, & \text{if } \mathcal{E} = \mathcal{E}', \\ 1, & \text{if } \mathcal{E} \prec \mathcal{E}', \\ 0, & \text{otherwise,} \end{cases} \quad . \quad (8.84)$$

where $|\mathcal{E}|$ denotes the number of equivalence classes in an equivalence relation $\mathcal{E}$ and we write $\mathcal{E} \prec \mathcal{E}'$ iff $\mathcal{E} \subset \mathcal{E}'$ and $\mathcal{E}'$ is formed by amalgamating (exactly) two equivalence classes of $\mathcal{E}$. The Markov chain with intensity matrix (8.84) is called the $n$-**coalescent of Kingman**.

To this end we note that $p_{\mathcal{E},\mathcal{E}'}$, the transition probability of the chain $\mathcal{R}_N$, is zero if $\mathcal{E} \not\subset \mathcal{E}'$. Also if $\mathcal{E} \subset \mathcal{E}'$ yet $\mathcal{E} \not\prec \mathcal{E}'$, then $|\mathcal{E}| - |\mathcal{E}'| \geq 2$, and $p_{\mathcal{E},\mathcal{E}'} \leq p_{|\mathcal{E}|,|\mathcal{E}'|}$ where $p_{i,j}$ is the transition probability of the pure death chain from the previous subsection. Hence, $\lim_{N \to \infty} Np_{\mathcal{E},\mathcal{E}'} = 0$. Moreover, $p_{\mathcal{E},\mathcal{E}} = p_{|\mathcal{E}|,|\mathcal{E}|}$, so that $\lim_{N \to \infty} N(p_{\mathcal{E},\mathcal{E}} - 1) = -\binom{|\mathcal{E}|}{2}$. Finally, if $\mathcal{E} \prec \mathcal{E}'$, $p_{\mathcal{E},\mathcal{E}'} = \binom{|\mathcal{E}|}{2}^{-1} p_{|\mathcal{E}|,|\mathcal{E}|-1}$ (we do know which two equivalence classes are to be amalgamated), so that $\lim_{N\to\infty} N p_{\mathcal{E},\mathcal{E}'} = 1$, as desired.

This result may be used to derive many classic formulae for sampling distributions in population genetics (see [39]) and in particular the famous Ewens sampling formula (see [69]).

**8.4.23** *A generalization of 8.4.17*   In applications it is often the case that semigroups, especially discrete parameter semigroups, approximating a semigroup do not act in the same space as the limit semigroup does. Typically, we have a sequence of Banach spaces, say $\mathbb{X}_n, n \geq 1$, approximating a Banach space $\mathbb{X}$ in the sense that there exist operators $P_n : \mathbb{X} \to \mathbb{X}_n$ such that $\lim_{n\to\infty} \|P_n x\|_n = \|x\|$, for all $x \in \mathbb{X}$ ($\|\cdot\|_n$ is the norm in $\mathbb{X}_n$). In such a case, we say that a sequence $x_n \in \mathbb{X}_n, n \geq 1$, converges to an $x \in \mathbb{X}$ if $\lim_{n\to\infty} \|x_n - P_n x\| = 0$. With such a convention, the Trotter–Kato Theorem remains true with the proof requiring only cosmetic changes. In Subsection 8.4.25 we present a typical example of such an approximation.

**8.4.24 Exercise**   Show that in the situation described above we have $\sup_{n\geq 1} \|P_n\| < \infty$.

**8.4.25 Example** *A random walk approximating the Ornstein–Uhlenbeck process*   Imagine an urn with $2n$ balls, say green and red. One ball is drawn at random; if it is red then a green ball is put into the urn and if it is green a red ball is put into the urn. This procedure is then repeated – this is the famous Ehrenfest model from statistical mechanics. The state of this process may be described by a singe number, for example by the difference between the number of red balls and $n$. In other words, we are dealing with a discrete-time Markov chain with values in the set $S_n = \{i \in \mathbb{Z} | |i| \leq n\}$ and transition probabilities $p_{i,i+1} = 1 - \frac{i+n}{2n}, -n \leq i \leq n - 1$, $p_{i,i-1} = \frac{i+n}{2n}, -n + 1 \leq i \leq n$.

If $(p_i)_{i\in S_n}$ is the initial distribution of the process, then at time $k$, the distribution is $U_n^k (p_i)_{i\in S_n}$ where $U_n (p_i)_{i\in S_n} = (q_i)_{i\in S_n}$ with $q_n = \frac{1}{2n} p_{n-1}, q_{-n} = \frac{1}{2n} p_{-n+1}$ and $q_{-n+i+1} = (1 - \frac{i}{2n}) p_{-n+i} + \frac{i+2}{2n} p_{-n+i+2}, 0 \leq i \leq n - 2$. The operator $U_n$ acts in the space of sequences $(p_i)_{i\in S_n}$ equipped with the norm $\| (p_i)_{i\in S_n} \| = \sum_{i\in S_n} |p_i|$. The related dual operator $T_n$ is defined in the space $\mathbb{X}_n$ of sequences $(\xi_i)_{i\in S_n}$ equipped with the norm $\| (\xi_i)_{i\in S_n} \|_n = \max_{i\in S_n} |\xi_i|$ as follows: $T_n (\xi_i)_{i\in S_n} = (\eta_i)_{i\in S_n}$ where $\eta_{-n+i} = \frac{i}{2n} \xi_{-n+i-1} + (1 - \frac{i}{2n}) \xi_{-n+i+1}, 1 \leq i \leq 2n - 1, \eta_{\mp n} = \xi_{\pm n \pm 1}$.

We will show that if the length of a single step of the process is taken to be $\frac{a}{\sqrt{n}}$ and the time scale is changed so that steps will be taken at times $\frac{k}{bn}$, then, as $n \to \infty$, these Markov chains approximate the Ornstein–Uhlenbeck process. We put $a^2 = \gamma^2 \alpha^{-1}$ and $b = \alpha$ where $\alpha$ and $\gamma$ are parameters of the Ornstein–Uhlenbeck process (see 8.1.14). In other words, we show that

$$\lim_{n\to\infty} T_n^{[bnt]} x = T(t)x, \tag{8.85}$$

where $\{T(t), t \geq 0\}$ is the Ornstein–Uhlenbeck semigroup in $C_0(\mathbb{R})$; we note that $\mathbb{X}_n, n \geq 1$, approximate the space $\mathbb{X} = C_0(\mathbb{R})$ in the sense of 8.4.23 if we let $P_n x = (\xi_i)_{i \in S_n}$ where $\xi_i = x(\frac{ai}{\sqrt{n}}), -n \leq i \leq n$.

To prove (8.85) we need the following result. Consider the operator $Ax(\tau) = -\alpha \tau x'(\tau) + \frac{\gamma^2}{2} x''(\tau)$ defined on $\mathcal{D}(A)$ composed of twice differentiable functions in $\mathbb{X}$ with both derivatives in $\mathbb{X}$ and such that $\tau \to Bx(\tau) = \tau x'(\tau)$ belongs to $\mathbb{X}$. We want to show that for $x \in \mathcal{D}(A)$, $\lim_{t \to 0+} \frac{1}{t}(T(t)x - x) = Ax$, i.e. that the infinitesimal generator of the Ornstein–Uhlenbeck semigroup is an extension of $A$. To this end, we recall that by 8.1.21, $T(t)x(\tau) = E\, x(e^{-at}\tau + w(\beta(t)))$ where $w(t), t \geq 0$, is a Brownian motion and $\beta(t) = \frac{\gamma^2}{2a}(1 - e^{-2at})$. Using the Taylor formula (5.21), we write $x(e^{-at}\tau + w(\beta(t)))$ as $x(\tau e^{-at}) + x'(\tau e^{-at}\tau)w(\beta(t)) + x''(\tau e^{-at} + \theta w(\beta(t)))w(\beta(t))/2$; note that $\theta$ is a function of $\tau, t$ and $\omega$. Using $E\, w(\beta(t)) = 0$ and $E\, w^2(\beta(t)) = \beta(t)$, and applying the Lagrange formula to $x(\tau e^{-at})$, we obtain $T(t)x(\tau) - x(\tau) = \tau(e^{-at} - 1)x'(\tau + \theta\tau(e^{-at} - 1)) + \frac{1}{2}\beta(t)E\, x''(\tau e^{-at} + \theta w(\beta(t)))$. Using the triangle inequality in a straightforward manner the task reduces to showing that

$$\lim_{t\to 0} \sup_{\tau \in \mathbb{R}} \left| \tau x'(\tau + \theta\tau(e^{-at} - 1)) - \tau x'(\tau) \right| = 0, \tag{8.86}$$

and

$$\lim_{t\to 0} \sup_{\tau \in \mathbb{R}} \left| E\, x''(\tau e^{-at} + \theta w(\beta(t))) - x''(\tau) \right| = 0, \tag{8.87}$$

uniformly in $\tau \in \mathbb{R}$. We will prove the first of these relations, leaving the proof of the other as an exercise.

Let $\epsilon > 0$ be given. We choose an $M > 0$ so that $\sup_{\tau \in \mathbb{R}} |\tau x'(\tau)| < \frac{9}{19}\epsilon$ for $|\tau| > \frac{9}{10}M$. Also, we choose a $\delta > 0$ so that $|x'(\tau) - x'(\sigma)| < \frac{\epsilon}{M}$ for $|\tau - \sigma| < \delta, |\tau|, |\sigma| \leq \frac{11}{10}M$. Finally, we take a $t$ small enough to have $1 - e^{-at} < \min(\frac{1}{10}, \frac{\delta}{M})$. Then, for $|\tau| > M$, $|\tau x'(\tau + \theta\tau(e^{-at} - 1))| < \left| \frac{\tau}{\tau + \theta\tau(e^{-at} - 1)} \right| \frac{9}{19}\epsilon \leq \frac{10}{19}\epsilon$, so that the absolute value in (8.86) is less than $\epsilon$. Similarly, for $|\tau| \leq M$ the absolute value in (8.86) does not exceed

$M|x'(\tau+\theta\tau(e^{-at}-1))-x'(\tau)| < \epsilon$ since $|\theta\tau(e^{-at}-1)| \le M(e^{-at}-1) <$ $\min(\delta, \frac{M}{10})$.

Using 8.1.21 we check that the Ornstein–Uhlenbeck semigroup leaves $\mathcal{D}(A)$ invariant. Since $\mathcal{D}(A)$ is dense in $\mathbb{X}$, $\mathcal{D}(A)$ is a core for the infinitesimal generator of $\{T(t), t \ge 0\}$. In particular, $Range(\lambda-A)$ is dense in $\mathbb{X}$. Therefore, (8.85) will be shown once we prove that $\lim_{n\to\infty} \|nb(T_n P_n x - P_n x) - P_n Ax\|_n = 0$ for $x \in \mathcal{D}(A)$.

By the Taylor formula $nb(T_n P_n x - P_n x)\left(a\frac{-n+i}{\sqrt{n}}\right)$ equals

$$\frac{ib}{2}\left[x\left(a\frac{-n+i-1}{\sqrt{n}}\right) - x\left(a\frac{-n+i}{\sqrt{n}}\right)\right]$$
$$+ b\left(\frac{2n-i}{2}\right)\left[x\left(a\frac{-n+i+1}{\sqrt{n}}\right) - x\left(a\frac{-n+i}{\sqrt{n}}\right)\right]$$
$$= -ab\frac{-n+i}{\sqrt{n}}x'\left(a\frac{-n+i}{\sqrt{n}}\right) + \frac{a^2bi}{4n}x''\left(a\frac{-n+i}{\sqrt{n}} - \theta_1\frac{a}{\sqrt{n}}\right)$$
$$+ \frac{a^2b(2n-i)}{4n}x''\left(a\frac{-n+i}{\sqrt{n}} + \theta_2\frac{a}{\sqrt{n}}\right), \qquad 1 \le i \le 2n-1,$$

and $nb(T_n P_n x - P_n x)(\mp a\sqrt{n})$ equals

$$\pm ab\sqrt{n}x'\left(\mp a\sqrt{n}\right) + \frac{a^2b}{2}x''\left(\mp a\sqrt{n} + \theta_3\frac{a}{\sqrt{n}}\right),$$

where $0 \le \theta_i \le 1, i = 1, 2, 3$. Since

$$P_n Ax\left(a\frac{-n+i}{\sqrt{n}}\right) = -\alpha a\frac{-n+i}{\sqrt{n}}x'\left(a\frac{-n+i}{\sqrt{n}}\right) + \frac{\gamma^2}{2}x''\left(a\frac{-n+i}{\sqrt{n}}\right),$$

$0 \le i \le 2n$, and $b = \alpha$ and $a^2b = \gamma^2$, we obtain

$$\|nb(T_n P_n x - P_n x) - P_n Ax\|_n \le \gamma^2 \sup_{\tau,\sigma,|\tau-\sigma|\le\frac{2a}{\sqrt{n}}} |x''(\tau) - x''(\sigma)|$$

which converges to 0 as $n \to \infty$ by uniform continuity of $x''$.

**8.4.26 Exercise**    Prove (8.87).

Suppose that we have two Feller processes $X_A$ and $X_B$ with generators $A$ and $B$ respectively. Sometimes, for example when $B$ is bounded, $A+B$ is well defined and generates a Feller semigroup. What does the process related to $A + B$ look like at time $t$? It is reasonable to expect that for large $n$ a good approximation is given by the following discrete-time process: we allow the process to evolve for time $t/n$ according to the transition probability of $X_A$ and then for the same time according to the distribution of $X_B$, and then repeat the whole circle $n$ times. Formally, we have the following theorem.

**8.4.27** *The Trotter product formula* Suppose that $A$ and $B$ and $C$ are generators of $c_0$ semigroups $\{S(t), t \geq 0\}$, $\{T(t), t \geq 0\}$ and $\{U(t), t \geq 0\}$ of contractions, respectively, in a Banach space $\mathbb{X}$. Suppose also that $\mathcal{D}$ is a core for $C$ and $\mathcal{D} \subset \mathcal{D}(A) \cap \mathcal{D}(B)$ and $Cx = Ax + Bx$ for $x \in \mathcal{D}$. Then,

$$U(t) = \lim_{n \to \infty} \left[ S\left(\frac{t}{n}\right) T\left(\frac{t}{n}\right) \right]^n, \qquad t \geq 0,$$

strongly.

*Proof* Since $\mathcal{D}$ is a core, by 8.4.9 it suffices to show that

$$\lim_{n \to \infty} \frac{n}{t}(S(t/n)T(t/n)x - x) = Ax + Bx$$

or, which is the same, $\lim_{n \to \infty} \left[ nt^{-1}(S(t/n)T(t/n)x - x) - S(t/n)Ax \right]$ $= Bx$. To this end we write $\|nt^{-1}[S(t/n)T(t/n)x - x] - S(t/n)Ax - Bx\| \leq \|S(t/n)\{nt^{-1}[T(t/n) - x] - Ax\}\| + \|nt^{-1}[S(t/n)x - x] - Bx\| \leq \|nt^{-1}[T(t/n) - x] - Ax\| + \|nt^{-1}[S(t/n) - x] - Ax\| \longrightarrow 0$, as $n \to \infty$. $\qquad \square$

**8.4.28 Exercise** Suppose that $\{T(t), t \geq 0\}$ and $\{S(t), t \geq 0\}$ are two contraction semigroups. Show that the limit $\lim_{n \to \infty}[S(t/n)T(t/n)]^n$ exists iff there exists the limit $\lim_{n \to \infty}[T(t/n)S(t/n)]^n$, and then both are equal.

**8.4.29 Exercise** Find an example showing that, in the notations of 8.4.27, even if the semigroups $\{S(t), t \geq 0\}$ and $\{T(t), t \geq 0\}$ commute and $\mathcal{D}(A) = \mathcal{D}(B)$, the domain $\mathcal{D}(C)$ may be strictly larger than $\mathcal{D}(A)$.

**8.4.30 Corollary** *The Feynman–Kac formula* Let $X_t, t \geq 0$, be a Lévy process, and let $\{T(t), t \geq 0\}$ be the related semigroup in $\mathbb{X} = C_0(\mathbb{R})$ or $\mathbb{X} = C[-\infty, \infty]$. Moreover, let $A$ be the infinitesimal generator of $\{T(t), t \geq 0\}$ and $B$ be the operator in $\mathbb{X}$ given by $Bx = bx$ where $b$ is a fixed member of $\mathbb{X}$. The semigroup $\{U(t), t \geq 0\}$ generated by $A + B - \beta I$ where $\beta = \|b\|$ is given by

$$U(t)x = \mathrm{e}^{-\beta t} E \, \mathrm{e}^{b(\tau + \int_0^t X_s \, \mathrm{d}s)} x(\tau + X_t). \tag{8.88}$$

*Proof* By 8.4.27, it suffices to show that $e^{-\beta t}[T(t/n)e^{(t/n)B}]^n x(\tau)$ converges pointwise in $\tau \in \mathbb{R}$ to the right-hand side in (8.88). We have

$$T(t)e^{tB}x(\tau) = \int_{\mathbb{R}} e^{tb(\tau+\sigma)}x(\tau + \sigma)\,\mu_t(\,d\sigma).$$

Hence, by induction,

$$[T(t)e^{tB}]^n x(\tau)$$
$$= \int_{\mathbb{R}}...\int_{\mathbb{R}} e^{t\sum_{i=1}^{n}b(\tau+\sum_{j=i}^{n}\sigma_j)}x(\tau + \sum_{i=1}^{n}\sigma_i)\,\mu_t(\,d\sigma_1)...\mu_t(\,d\sigma_n).$$

Since $\mu_{t/n}\otimes...\otimes\mu_{t/n}$ is a joint distribution of $X_{t/n}, X_{2t/n}-X_{t/n}, ..., X_t - X_{(n-1)t/n}$, we obtain

$$e^{-\beta t}[T(t/n)e^{(t/n)B}]^n x(\tau) = E\,e^{-\beta t}e^{(t/n)\sum_{k=1}^{n}b(\tau+X_{kt/n})}x(\tau + X_t).$$

This implies our result by the Lebesgue Dominated Convergence Theorem because, for almost all $\omega$, the map $t \to b(\tau + X_t(\omega))$ is Riemann integrable and $\lim_{n\to\infty}(t/n)\sum_{k=1}^{n}b(\tau+X_{kt/n}(\omega)) = \int_0^t b(\tau+X_s(\omega))\,ds$.

$\square$

**8.4.31 Corollary** *The characteristic function of the telegraph process* In 7.7.5 we have proved that the infinitesimal generator of the semigroup related to the process $\mathbf{g}_t, t \geq 0$, defined in (7.62) is the sum of two operators: $B$ and $C - aI$. The operator $B$ generates the semigroup of operators related to the convolution semigroup $\mu_t = \delta_{(vt,0)}, t \geq 0$, on the Kisyński group $\mathbb{G}$. The operator $C - aI$ generates the semigroup of operators related to the convolution semigroup $\mu_t^{\sharp} = e^{-at}\exp(at\delta_{(0,-1)}) = e^{-at}\sum_{n=0}^{\infty}\frac{(at)^n}{n!}\delta_{(0,-1)}^{*n} = e^{-at}\cosh(at)\delta_{(0,1)} + e^{-at}\sinh(at)\delta_{(0,-1)}, t \geq 0$, (convolution in $\mathbb{G}$.) Therefore, by 8.4.27, the distribution of $\mathbf{g}_t, t \geq 0$, is the limit of $(\mu_{t/n} * \mu_{t/n}^{\sharp})^{*n}$ as $n \to \infty$. Identifying a measure on $\mathbb{G}$ with a pair of measures on $\mathbb{R}$, by (1.16), we see that $(\mu_{t/n} * \mu_{t/n}^{\sharp})^{*n}$ is the pair of measures being the entries of the first column of the $n$th power of the matrix

$$A = A(t,n) = \begin{bmatrix} e^{-at/n}\cosh(at/n)\delta_{vt/n} & e^{-at/n}\sinh(at/n)\delta_{-vt/n} \\ e^{-at/n}\sinh(at/n)\delta_{vt/n} & e^{-at/n}\cosh(at/n)\delta_{-vt/n} \end{bmatrix}.$$

Since in calculating powers of this matrix we use convolution (in $\mathbb{R}$) as multiplication, it may be hard to find explicit formulae for $A^n$. The task, however, becomes much easier to achieve if we turn to characteristic functions and note that the characteristic function of the entry of $A^n$, a function of $\tau \in \mathbb{R}$, say, is the corresponding entry in the $n$th power of the matrix (with scalar entries)

$$A_{\mathrm{c}}(t, n, \tau) = \mathrm{e}^{-a\frac{t}{n}} \begin{bmatrix} \cosh \frac{at}{n} e^{\tau \frac{ivt}{n}} & \sinh \frac{at}{n} e^{-\tau \frac{ivt}{n}} \\ \sinh \frac{at}{n} e^{\tau \frac{ivt}{n}} & \cosh \frac{at}{n} e^{-\tau \frac{ivt}{n}} \end{bmatrix}$$

of characteristic functions of entries of $A$. In other words, our task reduces to that of finding $\lim_{n\to\infty} [A_{\mathrm{c}}(t, n, \tau)]^n$. In what follows we restrict ourselves to the case where $v = 1$; this will simplify our calculations and the general case may be easily recovered from this particular one.

To this end we will use some basic linear algebra. For any non-negative numbers $p$ and $q$ such that $p^2 - q^2 = 1$ and any complex $z = r + iu$ with $u, r \geq 0$ and $|z| = 1$, the matrix $M = \begin{pmatrix} zp & \overline{z}q \\ zq & \overline{z}p \end{pmatrix}$ has two complex eigenvalues $\lambda_j = pr + s_j$ with corresponding right eigenvectors $v_j = (-\overline{z}q, ipu - s_j)$ where $s_j$ are two (in general complex) roots of $p^2 r^2 - 1, j = 1, 2$. Since $p^2 r^2 - 1$ is real, $s_1 + s_2 = 0$. Therefore, $M$ equals

$$\begin{bmatrix} -\overline{z}q & -\overline{z}q \\ ipu - s_1 & ipu + s_1 \end{bmatrix} \begin{bmatrix} \lambda_1 & 0 \\ 0 & \lambda_2 \end{bmatrix} \frac{-1}{2\overline{z}qs_1} \begin{bmatrix} ipu + s_1 & \overline{z}q \\ s_1 - ipu & -\overline{z}q \end{bmatrix}. \tag{8.89}$$

The point in representing $M$ in such a form is that the $n$th power of (8.89) is easily computed to be

$$\begin{bmatrix} -\overline{z}q & -\overline{z}q \\ ipu - s_1 & ipu + s_1 \end{bmatrix} \begin{bmatrix} \lambda_1^n & 0 \\ 0 & \lambda_2^n \end{bmatrix} \frac{-1}{2\overline{z}qs_1} \begin{bmatrix} ipu + s_1 & \overline{z}q \\ s_1 - ipu & -\overline{z}q \end{bmatrix} \tag{8.90}$$

because in (8.89) the rightmost matrix is the inverse of the leftmost matrix. Of course, $\mathrm{e}^{\frac{at}{n}} A_{\mathrm{c}}(t, n, \tau)$ is of the form of $M$ with $z = \mathrm{e}^{\frac{it}{n}\tau}$, $p = \cosh \frac{at}{n}$ and $q = \sinh \frac{at}{n}$.

To find our limit we need to consider three cases: (a) $\tau^2 < a^2$, (b) $\tau^2 > a^2$, and (c) $\tau^2 = a^2$. Calculating the first two derivatives of $x \mapsto \cosh ax \cos \tau x - 1$ at $x = 0$ we see that for sufficiently large $n$, in case (a) $\cosh a\frac{t}{n} \cos a\frac{t}{n} > 1$ and in case (b), $\cosh a\frac{t}{n} \cos a\frac{t}{n} < 1$.

In both cases we represent $[A_{\mathrm{c}}(t, n, \tau)]^n$ as

$$\frac{1}{2}\mathrm{e}^{-at} \begin{bmatrix} 1 & 1 \\ \frac{-ipu+s_1}{\overline{z}q} & -\frac{ipu+s_1}{\overline{z}q} \end{bmatrix} \begin{bmatrix} \lambda_1^n & 0 \\ 0 & \lambda_2^n \end{bmatrix} \begin{bmatrix} 1 + ipus_1^{-1} & \overline{z}qs_1^{-1} \\ 1 - ipus_1^{-1} & -\overline{z}qs_1^{-1} \end{bmatrix}. \tag{8.91}$$

When $n \to \infty$, $\overline{z}$ converges to 1 and $puq^{-1}$ converges to $\tau a^{-1}$. In case (a), $p^2 r^2 - 1 = (pr + 1)(pr - 1) > 0$ for sufficiently large $n$ and we take $s_1 = \sqrt{p^2 r^2 - 1}$. Then $qs_1^{-1} = \sqrt{\frac{q^2}{(pr+1)(pr-1)}}$ has the same limit as $\sqrt{\frac{\sinh^2 ax}{2(\cosh ax \cos \tau x - 1)}}$ as $x \to 0$, and by de l'Hospital's rule this last limit equals $\frac{a}{\sqrt{a^2-\tau^2}}$. Hence $\frac{pu}{s_1} = \frac{pu}{q}\frac{q}{s_1}$ converges to $\frac{\tau}{\sqrt{a^2-\tau^2}}$. In case (b), $p^2 r^2 < 1$ for sufficiently large $n$ and we take $s_1 = \mathrm{i}\sqrt{1 - p^2 r^2}$. Then $\frac{q}{s_1}$ converges to $\frac{-\mathrm{i}a}{\sqrt{\tau^2-a^2}}$, and so $\frac{pu}{s_1}$ converges to $\frac{-\mathrm{i}\tau}{\sqrt{\tau^2-a^2}}$. Therefore, $[A_c(t, n, \tau)]^n$ converges to

$$\frac{1}{2}\mathrm{e}^{-at}\begin{bmatrix}\frac{1}{\frac{\sqrt{a^2-\tau^2}-\mathrm{i}\tau}{a}} & \frac{1}{-\frac{\sqrt{a^2-\tau^2}+\mathrm{i}\tau}{a}}\end{bmatrix}\begin{bmatrix}\alpha_1 & 0 \\ 0 & \alpha_2\end{bmatrix}\begin{bmatrix}1 + \frac{\mathrm{i}\tau}{\sqrt{a^2-\tau^2}} & \frac{a}{\sqrt{a^2-\tau^2}} \\ 1 - \frac{\mathrm{i}\tau}{\sqrt{a^2-\tau^2}} & -\frac{a}{\sqrt{a^2-\tau^2}}\end{bmatrix}$$

in the case (a), and to

$$\mathrm{e}^{-at}\frac{1}{2}\begin{bmatrix}\frac{1}{\mathrm{i}\frac{\sqrt{\tau^2-a^2}-\mathrm{i}\tau}{a}} & \frac{1}{-\mathrm{i}\frac{\sqrt{\tau^2-a^2}+\mathrm{i}\tau}{a}}\end{bmatrix}\begin{bmatrix}\alpha_1 & 0 \\ 0 & \alpha_2\end{bmatrix}\begin{bmatrix}1 + \frac{\tau}{\sqrt{\tau^2-a^2}} & \frac{-\mathrm{i}a}{\sqrt{\tau^2-a^2}} \\ 1 - \frac{\tau}{\sqrt{\tau^2-a^2}} & \frac{\mathrm{i}a}{\sqrt{\tau^2-a^2}}\end{bmatrix}$$

in case (b), where $\alpha_j = \lim_{n\to\infty} \lambda_j^n, j = 1, 2$. If $f(x)$ is a positive function such that $\lim_{x\to 0} f(x) = 0$ then $\lim_{x\to 0}[1 + f(x)]^{\frac{1}{x}} = \mathrm{e}^{\lim_{x\to 0}\frac{f(x)}{x}}$. Hence, in case (a), $\ln\alpha_1 = \lim_{x\to 0} t\sqrt{\frac{2(\cosh ax\cos ax - 1)}{x^2}}$ and by de l'Hospital's rule $\alpha_1 = \mathrm{e}^{\sqrt{a^2-\tau^2}t}$. Analogously, $\alpha_2 = \mathrm{e}^{-\sqrt{a^2-\tau^2}t}$. In case (b), $\alpha_1 = \mathrm{e}^{\mathrm{i}\sqrt{\tau^2-a^2}t}$ and $\alpha_2 = \mathrm{e}^{-\mathrm{i}\sqrt{\tau^2-a^2}t}$. Therefore,

$$\lim_{n\to\infty} A_c(t, n, \tau)^n = \mathrm{e}^{-at}\begin{bmatrix}\phi_1(t, \tau, a), & \phi_2(t, -\tau, a) \\ \phi_2(t, \tau, a), & \phi_1(t, -\tau, a)\end{bmatrix}$$

where, in case (a),

$$\phi_1(t, \tau, a) = \cosh\sqrt{a^2 - \tau^2}t + \mathrm{i}\frac{\tau}{\sqrt{a^2-\tau^2}}\sinh\sqrt{a^2 - \tau^2}t,$$

$$\phi_2(t, \tau, a) = \frac{a}{\sqrt{a^2-\tau^2}}\sinh\sqrt{a^2 - \tau^2}t,$$

and in case (b)

$$\phi_1(t, \tau, a) = \cos\sqrt{\tau^2 - a^2}t + \mathrm{i}\frac{\tau}{\sqrt{\tau^2-a^2}}\sin\sqrt{\tau^2 - a^2}t,$$

$$\phi_2(t, \tau, a) = \frac{a}{\sqrt{\tau^2-a^2}}\sin\sqrt{\tau^2 - a^2}t.$$

Case (c) may be treated analogously, or we may note that the functions $\phi_i, i = 1, 2$ must be continuous in $\tau \in \mathbb{R}$. This gives $\phi_1(t, \pm a, a) = 1 \pm \mathrm{i}at$ and $\phi_2(t, \pm a, a) = at$.

By (7.67), the characteristic function of $\xi(t) = \int_0^t (-1)^{N(s)}\,\mathrm{d}s$ equals $\phi_1 + \phi_2$. Hence it is given by (6.32) – compare [12], [96].

**8.4.32 Exercise**     Use 8.4.31, 6.6.18 and (8.65) to give a direct proof of convergence of solutions of the telegraph equation with small parameter to the solution of the diffusion equation that does not rely on the Trotter–Kato Theorem.

**8.4.33 Appendix** *Stirling numbers* (see [45], [101])

**1** *The definition* Let $(t)_n$ denote the polynomial

$$(t)_0 = 1, \quad (t)_1 = t, \quad (t)_n = t(t-1)\cdots(t-n+1), n \geq 2.$$

For each $n \in \mathbb{N}_0$, the polynomials $(t)_k, 0 \leq k \leq n$, are linearly independent, and so are the $t^k, 0 \leq k \leq n$. Hence, the Stirling numbers $s(n,k)$ of the first kind may be defined by the formula

$$(t)_n = \sum_{k=0}^{n} s(n,k)t^k; \quad \text{we put } s(n,k) = 0 \text{ if } k \notin \{0, ..., n\}. \tag{8.92}$$

Analogously, by definition, the Stirling numbers $S(n,k)$ of the second type are uniquely determined by

$$t^n = \sum_{k=0}^{n} S(n,k)(t)_k; \quad S(n,k) = 0 \text{ if } k \notin \{0, ..., n\}. \tag{8.93}$$

It is easy to see that $s(n,n) = S(n,n) = 1, s(n,0) = S(n,0) = 0$.

**2** *A recurrence relation* From now on we will focus on Stirling numbers of the second kind since they are of greater importance for us. Note that $(t)_{k+1} = (t)_k(t-k) = t(t)_k - k(t)_k$. Thus

$$tt^n = t\sum_{k=0}^{n} S(n,k)(t)_k = \sum_{k=0}^{n} S(n,k)(t)_{k+1} + \sum_{k=0}^{n} S(n,k)k(t)_k$$

$$= \sum_{k=1}^{n+1} S(n,k-1)(t)_k + \sum_{k=0}^{n} S(n,k)k(t)_k$$

$$= \sum_{k=1}^{n+1} [S(n,k-1) + kS(n,k)](t)_k$$

since $S(n,n+1) = 0$. Comparing this with (8.93) where $n$ was replaced by $n+1$ we get

$$S(n+1,k) = S(n,k-1) + kS(n,k), \quad k = 1, 2, ..., n+1. \tag{8.94}$$

Using this relation allows us to calculate the entries $S(n,k)$ of the matrix given in Table 8.1. To do that we have to take into account that $S(0,0) = 1$ and $S(n,0) = 0, n \geq 1$, i.e. that the first column (except for the first entry) is composed of zeros and that $S(n,n) = 1$, which gives the entries on the diagonal; (8.94) then allows us to fill consecutive columns (a bigger table can be found on page 258 of [45] or page 48 of [101]).

The main point, however, is that the recurrence relation (8.94) and "boundary conditions" $S(n,0) = 0, n \geq 1$, and $S(n,n) = 1$ determine all Stirling numbers of the second type.

| $n \backslash k$ | 0 | 1 | 2 | 3 | 4 | 5 | ... |
|---|---|---|---|---|---|---|---|
| 0 | 1 | 0 | 0 | 0 | 0 | 0 | ... |
| 1 | 0 | 1 | 0 | 0 | 0 | 0 | ... |
| 2 | 0 | 1 | 1 | 0 | 0 | 0 | ... |
| 3 | 0 | 1 | 3 | 1 | 0 | 0 | ... |
| 4 | 0 | 1 | 7 | 6 | 1 | 0 | ... |
| 5 | 0 | 1 | 15 | 25 | 10 | 1 | ... |
| 6 | 0 | 1 | 31 | 90 | 65 | 15 | ... |

Table 8.1

**3** *Relation to combinatorics* Let, as in 8.4.21, $\left\{ {n \atop k} \right\}$ denote the number of possible ways a set of $n$ elements may be partitioned into $k$ non-empty subsets. It is clear that $\left\{ {n \atop 0} \right\} = 0$ for $n \geq 1$, and $\left\{ {n \atop n} \right\} = 1$ and we could agree on $\left\{ {0 \atop 0} \right\} = 1$. Hence, to show that $S(n,k) = \left\{ {n \atop k} \right\}$ it suffices to show that

$$\left\{ {n+1 \atop k} \right\} = \left\{ {n \atop k-1} \right\} + k \left\{ {n \atop k} \right\}.$$

This can be achieved as follows. Think of a set with $n + 1$ elements as having $n$ ordinary elements and a special one. When we divide our set into $k$ subsets, this special element either forms a one-element set, and there are $\left\{ {n \atop k-1} \right\}$ partitions like that since then the remaining $n$ elements are partitioned into $k - 1$ subsets, or is a member of a subset with at least two elements. In the latter case the remaining elements are partitioned into $k$ subsets and the special element has been added to one of them – this last step may be done in $k$ ways.

**8.4.34 Exercise**     Relation (8.83) is clear because of the probabilistic interpretation. Prove it analytically.

# 9

# Appendixes

## 9.1 Bibliographical notes

**9.1.1 Notes to Chapter 1**    Rudiments of measure theory may be found in [103]. Classics in this field are [28] and [49]; see also [87]. A short but excellent account on convex functions may be found in [41], Chapter V, Section 8. A classical detailed treatment may be found in [85]. The proof of the Steinhaus Theorem is taken from [76].

**9.1.2 Notes to Chapter 2**    There are many excellent monographs devoted to Functional Analysis, including [2], [22], [32], [37], [54], [98], [112]. Missing proofs of the statements concerning locally compact spaces made in 2.3.25 may be found in [22] and [55].

**9.1.3 Notes to Chapter 3**    Among the best references on Hilbert spaces are [90] and [111]. The proof of Jensen's inequality is taken from [34]; different proofs may be found in [5] and [87]. Some exercises in 3.3 were taken from [20] and [34]. An excellent and well-written introductory book on martingales is [114]; the proof of the Central Limit Theorem is taken from this book. Theorems 3.6.5 and 3.6.7 are taken from [90]. A different proof of 3.6.7 may be found e.g. in [98].

**9.1.4 Notes to Chapter 4**    Formula (4.11) is taken from [59]. Our treatment of the Itô integral is largely based on [113]. For detailed information on matters discussed in 4.4.8 see e.g. [93], [64] and [38]. To be more specific: for integrals with respect to square integrable martingales see e.g. Proposition 3.4 p. 67, Corollary 5.4 p. 78, Proposition 6.1. p. 79, Corollary 5.4, and pp. 279–282 in [38], or Chapter 3 in [64] or Chapter 2 in [34]. See also [57], [61], [102] etc.

**9.1.5 Notes to Chapter 5** *From Bourbaki and Kuratowski to Bourbaki and Kuratowski* According to [37] p. 7, Lemma 5.1.10 "was discovered independently in 1923 by R. L. More and by K. Kuratowski, was rediscovered by Zorn in 1935 and then rediscovered yet again by Teichmüller a little later. The name *Zorn Lemma* was coined by Bourbaki, who was one of the first to make systematic use of the principle." The reader should be warned that Bourbaki is not a single person, but a group of (outstanding) French mathematicians, who publish all their work under one name. Körner [75] reports that (once upon a time) Mr. Bourbaki applied for membership in the AMS, but was replied that he should apply as an institutional member (and pay higher dues). He never wrote back. The proof of 5.2.6 is due to Banach. Theorem 5.2.16 is due to H. Steinhaus. As shown in [51], the assumption that $\mu$ is $\sigma$-finite may be relaxed if $\Omega$ is a locally compact topological space. Theorem 5.4.9 and its proof are due to Banach [3]; other proofs may be found in [22], [32], [37]. The proof of Prohorov's Theorem follows closely classic Ikeda and Watanabe [57]. The proof of Donsker's Theorem is a blending of arguments presented in [61] and [100], see also [5] and [107]. The proof of Tichonov's Theorem is taken from Kuratowski [77]. The absolute classic on convergence of probability measures is of course the first edition of Billingsley [6]. A very nice chapter on this subject may be found in Stroock [107], as well. Concerning Polish spaces, Stroock writes that this is "a name coined by Bourbaki in recognition of the contribution made to this subject by the Polish school in general and C. Kuratowski in particular".

**9.1.6 Notes to Chapter 6**    I. Gelfand was the first to notice and prove in the 1940s the importance of Banach algebras. Now the theory is flourishing with applications (see e.g. [27]). A different proof of 6.2.6, based on the Riesz Theorem, may be found in [22] p. 219. The idea of a character of a group is one of the basic notions of the rich and beautiful theory of abstract harmonic analysis [51]. The proof of the Factorization Theorem is due to P. Koosis [74]. For supplementary reading for this chapter see e.g. [22], [51], [63], [82], [117], [115]. In particular, in [51] a much more general version of the factorization theorem may be found.

**9.1.7 Notes to Chapter 7**    Example 7.3.6 is taken from [65]. There are a number of excellent books on semigroups of operators, some of them are listed in the bibliography. The absolute classics in this field are [54] and [112]. A thorough treatment of Lévy processes may be found in

[4]. For stochastic process with values in a (topological, locally compact) group see [55], see also [56]. Our proof of (7.47) is a simplified version of the argument leading to the theorem of Hunt, as presented in [55]. On the other hand, to arrive at (7.53) while avoiding technicalities involved in analyzing general Lie groups, I have used the argument of [41] and this part of reasoning apparently does not work in the general case of a Lie group. An explicit formula (in terms of the Hilbert transform) for the infinitesimal generator of the Cauchy semigroup can be given if we consider it in $L^p(\mathbb{R}), p > 1$ – see [89]. The probabilistic formula for the solution of the two dimensional Dirac equation is due to Ph. Blanchard *et al* [7], [8], [9], [10]. Group-theoretical aspects of the formula were discussed in [11].

**9.1.8 Notes to Chapter 8** The vast literature on Markov processes that covers various aspects of the theory includes [4], [20], [21], [29], [35], [34], [38], [41], [42], [46], [47], [48], [57], [58], [61], [62], [84], [88], [92], [93], [99], [100], [102], [105], [107], [109], [113], [114].

Theorem 8.2.1 is due to Hille, Yosida, Feller, Phillips and Miyadera, and generalizes the earlier result of Hille and Yosida where $\omega = 0$ and $M = 1$. The proof of this theorem as presented here differs from the original one, and there were many who contributed their ideas to simplification and clarification of the argument. The decisive steps, however, seem to be due to W. Arendt whose paper [1] inspired the whole literature on so-called integrated semigroups (where also 8.2.3 was established – with a different proof) and J. Kisyński, who noticed relations with the theory of representations of Banach algebras and introduced the algebraic version 8.2.16 of the theorem. In particular, thanks to W. Chojnacki's reference to Cohen's Factorization Theorem, he was the first to show (8.43). The whole research on non-densely defined operators was also greatly influenced by the paper by G. Da Prato and E. Sinestrari [25].

One of the most important cases of the Hille–Yosida theorem not discussed in this book is the Stone Theorem on generation of unitary groups. The famous Bochner Theorem characterizing Fourier transforms of bounded Borel measures on $\mathbb{R}^n, n \in \mathbb{N}$, may be proved to follow from the Stone Theorem [98], [112].

With the exception of the proof of 8.4.16, the second part of Section 8.3 follows [54] closely.

The equivalence of (a) and (b) in 7.4.48 was apparently first noticed by T. G. Kurtz [78], but at that time it seemed to be merely a side-remark (compare, however, [26] pp. 123–124 and probably hundreds of other places). Arendt's article [1] added a new dimension to this result; see [14]; see also the paper by Lizama [86] for convergence theorems for integrated semigroups. By the way, this part of Kurtz's article seems to be completely forgotten by many specialists of the theory of integrated semigroups.

Subsections 8.4.21 and 8.4.22 are of course based on Kingmans' original papers [67], [68]. Examples 8.4.1, 8.4.2, 8.4.5 and 8.4.11 are taken from [13] and [15]. A straightforward way of computing the characteristic function of the telegraph process may be found in [96].

## 9.2 Solutions and hints to exercises

*Hint to Exercise 1.2.4* By 1.2.3, it suffices to show that $\mathcal{B}(\mathbb{R}) = \sigma(\mathcal{G})$ where $\mathcal{G}$ is the class of intervals $(-\infty, t]$, $t \in \mathbb{R}$. To this end, prove first that intervals $(s, t)$, $s, t \in \mathbb{R}$, belong to $\sigma(\mathcal{G})$. Then show that every open set in $\mathbb{R}$ is a countable union of such intervals, and deduce that $\sigma(\mathcal{G})$ contains all Borel sets. For the example we may take $(\Omega, \mathcal{F}) = (\mathbb{R}, \mathcal{B}(\mathbb{R}))$ and $f(\tau) = \tau$.

*Exercise 1.2.6* The family $\mathcal{F}$ defined by the right-hand side of (1.3) is a $\sigma$-algebra, and contains open sets in $S'$. Hence, $\mathcal{F} \supset \mathcal{B}(S')$. Moreover, the family $\mathcal{G}$ of subsets $A$ of $S$ such that $A \cap S'$ is Borel in $S'$ is a $\sigma$-algebra and contains open sets in $S$. Therefore $\mathcal{G} \supset \mathcal{B}(S)$. This implies $\mathcal{F} \subset \mathcal{B}(S')$ and completes the proof.

*Exercise 1.2.10* Let $\mathcal{G}$ be the class of sets of the form $A \cup B$ where $A \in \mathcal{F}$ and $B \in \mathcal{F}_0$. Of course $\mathcal{G} \subset \mathcal{F}_\mu$. We see that $\Omega$ belongs to $\mathcal{G}$ and so do countable unions of elements of $\mathcal{G}$. Also, if $A \in \mathcal{F}$ and $B \in \mathcal{F}_0$ and $C$ is as in the definition of $\mathcal{F}_0$, then $(A \cup C)^{\complement} \in \mathcal{F}$, and $C \setminus (A \cup B) \in \mathcal{F}_0$. Since $C^{\complement} \subset B^{\complement}$,

$$
\begin{aligned}
(A \cup B)^{\complement} &= [(A \cup B)^{\complement} \cap C] \cup [A^{\complement} \cap B^{\complement} \cap C^{\complement}] \\
&= [C \setminus (A \cup B)] \cup [A^{\complement} \cap C^{\complement}] = [C \setminus (A \cup B)] \cup (A \cup C)^{\complement}
\end{aligned}
$$

proving that $(A \cup B)^{\complement} \in \mathcal{F}_0$. Thus, $\mathcal{G}$ is a $\sigma$-algebra and we must have $\mathcal{F}_\mu = \mathcal{G}$. Suppose that $A, A' \in \mathcal{F}$, $B, B' \in \mathcal{F}_0$ and $C, C'$ are chosen as in the definition of $\mathcal{F}_0$. If $A \cup B = A' \cup B'$, then the symmetric difference of $A$ and $A'$ is a subset of $B \cup B' \subset C \cup C'$, and so $\mu(A)$ equals $\mu(A')$. Therefore, we may define $\mu(A \cup B)$ as $\mu(A)$, $A \in \mathcal{F}, B \in \mathcal{F}_0$. The rest is clear.

*Exercise 1.2.12* For almost all $\omega \in \Omega$ and $h \neq 0$, we have $\frac{1}{h}[x(\tau+h,\omega) - x(\tau,\omega)] = x'(\tau + \theta h, \omega)$ where $\theta$ depends on $\tau, h$ and $\omega$. The absolute value of this expression is no greater than $y(\omega)$. The claim thus follows from the Lebesgue Dominated Convergence Theorem.

*Exercise 1.2.31* If $l$ is the left-hand side of the relation we are to prove, then, changing to polar coordinates,

$$l^2 = \int_{\mathbb{R}} \int_{\mathbb{R}} e^{-\frac{s^2+t^2}{2}} \, ds \, dt = \int_0^{\pi} \int_0^{\infty} r e^{-\frac{r^2}{2}} \, dr \, d\theta$$

$$= [-e^{\frac{r^2}{2}}]_0^{\infty} \, 2\pi = 2\pi.$$

*Exercise 1.2.36*

$$E\,X = \int_{X\geq\epsilon} X \, d\mathbb{P} + \int_{X<\epsilon} X \, d\mathbb{P} \geq \epsilon \int_{X\geq\epsilon} d\mathbb{P} = \epsilon\mathbb{P}\{X \geq \epsilon\}.$$

*Exercise 1.2.37* By the Fubini Theorem, since $\mathbb{P}\{X > s, Y > t\} = \int_{\Omega} 1_{\{X>t,Y>t\}} \, d\mathbb{P}$, the left-hand side in (1.17) equals

$$\int_{\Omega} \int_{[0,X(\omega))\times[0,Y(\omega))} \alpha s^{\alpha-1}\beta t^{\beta-1} \, leb_2(d(s,t)) \, P(d\omega).$$

Using the Fubini Theorem again, the inner (double) integral equals

$$\int_0^{X(\omega)} \alpha s^{\alpha-1} \, ds \int_0^{Y(\omega)} \beta t^{\beta-1} \, dt = X^{\alpha}(\omega)Y^{\beta}(\omega),$$

and we are done. (1.18) is now straightforward, and so is (1.20) if we have (1.19). To prove (1.19) note that $\mathbb{P}\{Y1_{\{X>s\}} > t\} = \mathbb{P}\{X > s, Y > t\}$.

*Hint to Exercise 1.3.4* The right-hand limit of $v_+(t)$ exists and is no less than $v_+(t)$, for this function is non-decreasing. May $v_+(t+)$ be strictly bigger than $v_+(t)$? The following argument shows that this is impossible. If $v_+(t+) > v_+(t)$ then there exists a $\delta > 0$ such that for any sufficiently small $\epsilon > 0$, $var[y,t,t_0] > \delta$, where $t_0 = t + \epsilon$. We may assume that this $\epsilon > 0$ is so small that $|y(t) - y(s)| < \delta' = \delta/2$ for $s \in [t,t_0]$. This implies that we may find a $t < t_1 < t_0$ and a partition of $[t_1,t_0]$ such that the appropriate sum of absolute values of differences between values of $y$ at partitioning points is greater than $\delta'$. The interval $[t,t_1]$, however, enjoys the same properties as $[t,t_0]$. Therefore we may find a $t < t_2 < t_1$, and a partition of $[t_2,t_1]$, (and consequently a partition of $[t_2,t_0]$) such that the appropriate sum is bigger than $2\delta'$.

Since this is supposed to be a hint and not a complete solution, we shall say no more.

*Exercise 1.3.8* Write

$$S(\mathcal{T},\Xi,x,y) = x(\xi_{k-1})y(b) - x(\xi_0)y(a) - \sum_{i=1}^{k-1}[x(\xi_i) - x(\xi_{i-1})]y(t_i),$$

to see that

$$|S(\mathcal{T}_n,\Xi_n,x,y) - S(\mathcal{T}_n,\Xi_n,x,y_r) - x(\xi_{0,n})[y(a^+) - y(a)]|$$

is less than

$$var[a,b,y] \cdot \sup_{|\xi-\eta|\leq 2\Delta(\mathcal{T}_n)} |x(\xi) - x(\eta)|,$$

which tends to zero, by the uniform continuity of $x$. Here $\xi_{0,n}$ is the first element of $\Xi_n$ and $\lim_{n\to\infty} x(\xi_{0,n}) = x(a)$.

*Exercise 1.4.4* It is enough to note that $\sigma(f(X)) \subset \sigma(X)$ and $\sigma(g(Y)) \subset \sigma(Y)$.

*Exercise 1.4.5* The "if" part is trivial, the "only if" part follows from 1.2.7.

*Exercise 1.4.10* Note that $Z$ is an exponential random variable with parameter $\lambda$ iff $\mathbb{P}[Z > s] = e^{-\lambda s}, s \geq 0$. Now, $\mathbb{P}[Y > s] = \mathbb{P}[X_1 > s, X_2 > s] = \mathbb{P}[X_1 > s]\mathbb{P}[X_2 > s] = e^{-(\lambda_1+\lambda_2)s}$.

*Hint to Exercise 1.4.11*

$$\mathbb{P}[X \leq Y] = \lambda\mu \int_0^\infty \int_t^\infty e^{-\mu s}e^{-\lambda t}\, ds\, dt.$$

*Hint to Exercise 1.4.18* Find a recurrence for $Z_n$.

*Exercise 1.5.2* $u_2$ belongs to $[u_1, u_3]$ iff there exists an $0 \leq \alpha \leq 1$ such that $u_2 = \alpha u_3 + (1 - \alpha)u_1$. This $\alpha$ equals $\frac{u_2-u_1}{u_3-u_1}$. See 2.1.26.

*Exercise 1.5.3* We calculate $\tilde{\phi}(\alpha u + \beta v) = \phi(a+b-\alpha u - \beta v) = \phi(\alpha(a+b-u) + \beta(a+b-v)) \leq \alpha\phi(a+b-u) + \beta\phi(a+b-v) = \alpha\tilde{\phi}(u) + \beta\tilde{\phi}(v)$, where $\beta = 1 - \alpha$. The claim concerning $\overline{\phi}$ is proved similarly.

*Exercise 1.6.1* Since $\mu(A \cap B) = \mu(A \cap B \cap C) + \mu(A \cap B \cap C^{\complement})$, and

$\mu(C \cap B) = \mu(C \cap B \cap A) + \mu(C \cap B \cap A^{\complement})$, then

$$|\mu(A \cap B) - \mu(C \cap B)| = |\mu(A \cap B \cap C^{\complement}) - \mu(B \cap C \cap A^{\complement})|$$
$$\leq \mu(A \cap B \cap C^{\complement}) + \mu(B \cap C \cap A^{\complement})$$
$$\leq \mu(A \cap C^{\complement}) + \mu(C \cap A^{\complement}) = \mu(A \div C).$$

*Exercise 1.6.3* Suppose that $\delta = 0$. Then there exist sequences $a_n \in A$ and $b_n \in B$ such that $\lim_{n \to \infty} d(a_n, b_n) = 0$. Since $A$ is compact we may choose a converging subsequence $a_{n_k}$; $\lim_{k \to \infty} a_{n_k} = a \in A$. However, this implies that $b_{n_k}$ also converges to $a$, and since $B$ is closed, $a \in B$. This is a contradiction.

An appropriate example may be constructed in $\mathbb{R}$, and the trick is to choose $A$ and $B$ unbounded. For example $A = \bigcup_{n \in \mathbb{N}} [2n, 2n + 1]$, $B = \{2n + 1 + \frac{1}{n}, n \geq 2\}$.

*Exercise 1.6.6* For $s > 0$, $x(ks) = x(\sum_{i=1}^{k} s) = \sum_{i=1}^{k} x(s) = kx(s)$. Taking $s = \frac{t}{k}$, and an arbitrary $t > 0$, $x(\frac{t}{k}) = \frac{1}{k} x(k \frac{t}{k}) = \frac{1}{k} x(t)$. Combining these two relations, we get the claim.

*Exercise 2.1.2* Suppose that there are two vectors $\Theta_1$ and $\Theta_2$ satisfying (a2). Then $\Theta_1 = \Theta_1 + \Theta_2 = \Theta_2 + \Theta_1 = \Theta_2$.

*Exercise 2.1.3* Suppose $x + x'' = \Theta$. Then $x'' = x'' + \Theta = x'' + (x + x') = (x'' + x) + x' = (x + x'') + x' = \Theta + x' = x' + \Theta = x'$.

*Exercise 2.1.4* We have $0x = (0+0)x = 0x + 0x$. Thus, $\Theta = 0x + (0x)' = (0x + 0x) + (0x)' = 0x + (0x + (0x)') = 0x + \Theta = 0x$.

*Exercise 2.1.5* $\Theta = 0x = [1 + (-1)]x = 1x + (-1)x$. Thus, by 2.1.3, $(-1)x = x'$.

*Exercise 2.1.14* (a) follows directly from the definition of a linear map. (b) If $x$ and $y$ belong to $Ker\, L$, then $L(\alpha x + \beta y) = \alpha Lx + \beta Ly = 0$. (c) is proved similarly.

*Exercise 2.1.17* If a class is not equal to $\mathbb{Y}$, it contains an element, say $x$, that does not belong to $\mathbb{Y}$. An element $y$ belongs to this class iff $x \sim y$, i.e. $x - y \in \mathbb{Y}$, so that the class equals $x + \mathbb{Y}$.

*Hint to Exercise 2.1.20* Pick $p \in S$ and show that the set of functions that vanish at this point is an algebraic subspace of $\mathbb{R}^S$. Proceed as in 2.1.19 to show that this subspace is algebraically isomorphic to $\mathbb{R}^S / \mathbb{Y}$.

*Hint to Exercise 2.1.21* Show that $I[x] = Lx$ does not depend on the choice of $x$ from the class $[x]$ and that $I$ is the desired algebraic isomorphism.

*Exercise 2.1.23* The inverse image of $\emptyset$ is $\emptyset$ and the inverse image of $\Omega$ is $\Omega$, but the range of $f$ equals $\{\frac{1}{2}\} \notin \mathcal{F}'$.

*Exercise 2.1.28* Observe (and then check) that both $\mathbb{Y}_1$ and $\mathbb{Y}_2$ are convex, and that $span\,\mathbb{Y}_1 = span\,\mathbb{Y}_2 = \mathbb{R}^2$.

*Hint to Exercise 2.1.29* Show that the set of $z$ of the form (2.1) is convex, and note that any convex set that contains $y_i$ must contain elements of such a form. As for the counterexample, think of the letter "x" as a subset of $\mathbb{R}^2$, and take $\mathbb{Y}_i$ to be the left-hand part of the letter and $\mathbb{Y}_2$ to be the right-hand part of the letter. The convex hull of the "x" is a square and the set of $z$ of the form (2.1) is clepsydra-shaped.

*Exercise 2.1.32* The inclusion $span\,\mathbb{Z}_n \subset span\,\mathbb{Y}_n$ is easy. Moreover, $span\,\mathbb{Z}_n \subset \mathbb{Z}_{n+1}$. We will show that $span\,\mathbb{Y}_n \subset span\,\mathbb{Z}_n$ by induction. For $n = 1$ this follows from

$$z_0 = y_{0,1} + y_{1,1}, \quad z_1 = y_{0,1} - y_{1,1}.$$

Assume that $\mathbb{Y}_n \subset \mathbb{Z}_n$, so that $y_{l,n} \in span\,\mathbb{Z}_n \subset span\,\mathbb{Z}_{n+1}$, for all $0 \le l < 2^n$. Since

$$y_{l,n} = y_{2l,n+1} + y_{2l+1,n+1}, \quad z_{2^n+l} = y_{2l,n+1} - y_{2l+1,n+1},$$

or

$$y_{2l,n+1} = \frac{1}{2}(y_{l,n} + z_{2^n+l}), \quad y_{2l+1,n+1} = \frac{1}{2}(y_{l,n} - z_{2^n+l}),$$

$y_{k,n+1} \in span\,\mathbb{Z}_{n+1}$, for $0 \le k < 2^{n+1}$, as desired.

*Exercise 2.2.2* By (n4), $\|x\| - \|y\| = \|x \pm y \mp y\| - \|y\| \le \|x \pm y\|$, which gives that claim, except for the absolute value sign on the left-hand side. We complete the proof by changing the roles of $x$ and $y$.

*Exercise 2.2.7* By 2.2.2,

$$\big|\|x_n\| - \|x\|\big| \le \|x_n - x\|.$$

*Hint to Exercise 2.2.14* The "only if" part is immediate. To show the other part, assuming that $x_n$ is a Cauchy sequence, find a subsequence

$x_{n_k}$, such $\|x_{n_{k+1}} - x_{n_k}\| < \frac{1}{2^k}$. Then the series $\sum_{k=1}^{\infty}(x_{n_{k+1}} - x_{n_k}) + x_{n_1}$ converges. Prove that $x_n$ converges to the sum of this series.

*Hint to Exercise 2.2.17* Proceed as in 2.2.16: the uniform limit of a sequence of bounded continuous functions is bounded and continuous (use 2.2.9). Analogously, the limit (even pointwise) of a sequence of measurable functions is measurable.

*Hint to Exercise 2.2.42* Take $\Omega = \mathbb{R}_*^+$ with Lebesgue measure and $x(t) = \frac{1}{t}$.

*Hint to Exercise 2.3.4* Use

$$\|S(\mathcal{T}, \Xi, Ax.) - S(\mathcal{T}', \Xi', Ax.)\| \leq \|A\| \, \|S(\mathcal{T}, \Xi, x.) - S(\mathcal{T}', \Xi', x.)\|.$$

Cf. 7.3.4.

*Hint to Exercise 2.3.12* Proof by induction. For the induction step write $P_{n+1} - R_{n+1}$ as $(A_{n+1}P_n - A_{n+1}R_n) + (A_{n+1}R_n - B_{n+1}R_n)$ where $P_n = A_n A_{n-1}...A_1$ and $R_n = B_n B_{n-1}...B_1$.

*Hint to Exercise 2.3.13* Prove that $S_n(t) = \sum_{i=0}^{n} \frac{t^i A^i}{i!}$ is a Cauchy sequence in $\mathcal{L}(\mathbb{X})$. To this end, use Exercise 2.3.11 to see that $\|A^n\| \leq \|A\|^n$.

*Exercise 2.3.14*
$e^{tA} e^{tB}$ equals:

$$\sum_{k=0}^{\infty} \sum_{n=0}^{\infty} \frac{(tA)^n}{n!} \frac{(tB)^k}{k!} = \sum_{i=0}^{\infty} \sum_{j=0}^{i} \frac{(tA)^j}{j!} \frac{(tB)^{i-j}}{(i-j)!} \quad (i = n + k)$$

$$= \sum_{i=0}^{\infty} \frac{1}{i!} \sum_{j=0}^{i} \binom{i}{j} (tA)^j (tB)^{i-j} = \sum_{i=0}^{\infty} \frac{t^i(A+B)^i}{i!} = e^{t(A+B)}. \quad (9.1)$$

*Hint to Exercise 2.3.16*
Note that our exponent equals $e^c \sum_{n=0}^{\infty} \frac{(C)^n}{n!}$, where $C = aL + bR$. Thus it is enough to show that $\|e^C\| = e^{a+b}$. Furthermore, $\|e^C\| \leq e^{\|C\|} = e^{a+b}$, for obviously $\|C\| = a + b$. Also, we have

$$C^k(\xi_n)_{n \geq 1} = \left( \sum_{i=0}^{k} \binom{k}{i} a^i b^{k-i} \xi_{n-k+2i} \right)_{n \geq 1} \quad (9.2)$$

where $\xi_0 = 0, \xi_{-i} = -\xi_i$, for $i \geq 1$. In particular, if $(\xi_n)_{n \geq 1} = (\delta_{n,m})_{n \geq 1}$

for some $m \geq 1$, then for $k < m$, $\|C^k(\xi_n)_{n\geq 1}\|_{l^1}$ equals

$$\sum_{n=m-k}^{m+k} \sum_{i=0}^{k} \binom{k}{i} a^i b^{k-i} \delta_{n-k+2i,m} = \sum_{n=0}^{2k} \sum_{j=0}^{k} \binom{k}{j} a^{k-j} b^j \delta_{n+m-2j,m}$$

$$= \sum_{n=0}^{k} \sum_{j=0}^{k} \binom{k}{j} a^{k-j} b^j x_{2n+m-2j} = \sum_{i=0}^{k} \binom{k}{i} a^{k-i} b^i = (a+b)^k.$$

Thus, since the operator $C$ is positive, we get for all $m \geq 2$,

$$\|e^C\| \geq \|e^C(\delta_{n,m})_{n\geq 1}\|_{l^1} \geq \left\|\sum_{k=0}^{m-1} \frac{C^k}{k!}(\delta_{n,m})_{n\geq 1}\right\| = \sum_{k=0}^{m-1} \frac{(a+b)^k}{k!},$$

and, consequently, $\|e^C\| \geq e^{a+b}$.

*Hint to Exercise 2.3.19*  The formula $E\,Xg(X) = \lambda E\,g(X+1)$ holds iff $p_n = \mathbb{P}[X = n]$ satisfies $(n+1)p_n = \lambda p_n$. The formula $E\,g(X) = qE\,g(X+1)$ holds iff $p_{n+1} = qp_n$.

*Exercise 2.3.40*  This problem is a particular case of 2.3.43.

*Exercise 2.3.43*  The integral in the definition is finite almost everywhere by Fubini's Theorem. Moreover, the image of $x$ lies in $L^1(\mathbb{R}, \mathcal{M}, leb)$, for changing the order of integration we obtain:

$$\int_{\mathbb{R}} |Kx(\tau)|\,d\tau \leq \int_{\mathbb{R}} \int_{\mathbb{R}} k(\tau,\sigma)|x(\sigma)|\,d\sigma\,d\tau = \int_{\mathbb{R}} |x(\sigma)|\,d\sigma = \|x\|. \quad (9.3)$$

The reader should check that $K$ maps classes into classes. Moreover, for non-negative $x$, omitting absolute values signs in (9.3) we obtain a sequence of equalities.

*Hint to Exercise 2.3.44*  Taking $x = 1_A$ where $A \in \mathcal{F}'$, we see that $\mu'(A) = \int_{\Omega'} 1_A\,d\mu'$ must be equal to $\int_{\Omega} 1_A \circ f\,d\mu = \int_{\Omega} 1_{f^{-1}(A)}\,d\mu = \mu(f^{-1}(A))$. Thus the only measure that will satisfy the required property is the transport of $\mu$ via $f$: $\mu' = \mu_f$. It remains to check that $\mu_f$ is indeed the proper choice.

*Exercise 3.1.9*  Subtract both sides of (3.1) with $t = 1$ and $t = -1$, respectively.

*Exercise 3.1.19*  If $P$ is a projection on a subspace $\mathbb{H}_1$, then

$$(Px, y) = (Px, y - Py + Py) = (Px, y - Py) + (Px, Py) = (Px, Py)$$

since $y - Py$ is perpendicular to $\mathbb{H}_1$.

*Hint to Exercise 3.1.16* Take $x = Pz$ and $y = z - Pz$ where $P$ is the projection on $\mathbb{H}_1$. If $x + y = x' + y'$, then $x - x' = y' - y$ is perpendicular to itself.

*Hint to Exercise 3.1.24* Use 3.1.23. Note that $P_3^2 = P_3$ iff $(a)$ holds; in such case $P_3$ is self-adjoint. Analogously, $P_4$ is self-adjoint iff $(x, P_1 P_2 y) = (x, P_2 P_1 y)$ for all $x$ and $y$ in $\mathbb{H}$, i.e. iff $(a)$ holds; in such a case $P_4^2 = P_4$.

Moreover, the range of $P_3$ is contained in $\mathbb{H}_1 + \mathbb{H}_2$ for $P_3 = P_1 + P_2(I - P_1)$, and $\mathbb{H}_1 + \mathbb{H}_2$ is a subset of the range because a direct calculation shows that if $x = P_1 y_1 + P_2 y_2$ for some $y_1$ and $y_2$ in $\mathbb{H}$ then $P_3 x = x$, which implies that $x$ belongs to the range of $P_3$.

Finally, the range of $P_4 = P_1 P_2 = P_2 P_1$ is contained in $\mathbb{H}_1$ and in $\mathbb{H}_2$ and it must be equal the intersection of these two subspaces since for $x$ in the intersection we have $x = P_4 x$.

*Hint to Exercise 3.2.2* For the converse show that $x(t) = \ln \mathbb{P}(T > t), t \geq 0$ satisfies the Cauchy equation.

*Exercise 3.3.4* The sum $\sum_{i=1}^{n} b_i 1_{B_i}$ is $\mathcal{G}$ measurable, and we check that

$$\int_{B_j} \sum_{i=1}^{n} b_i 1_{B_i} = c_j \int_{B_j} d\mathbb{P} = \int_{B_j} X \, d\mathbb{P}.$$

*Exercise 3.3.6* The square of the norm of $\phi$ in $L^2(\Omega, \mathcal{F}, \mathbb{P})$ is

$$\int_{\Omega} \phi^2 \, d\mathbb{P} = \sum_{i=1}^{\infty} \int_{B_i} \phi^2 \, d\mathbb{P} = \sum_{i=1}^{\infty} b_i^2 \mathbb{P}(B_i)$$

so that $\phi \in L^2(\Omega, \mathcal{F}, \mathbb{P})$ iff the last series converges. As in (3.7) we show that

$$\int_{\Omega} (\phi - 1_A)^2 \, d\mathbb{P} = \sum_{i=1}^{\infty} \left[ b_i^2 \mathbb{P}(B_i) - 2 b_i \mathbb{P}(B_i \cap A) + \mathbb{P}(A) \right],$$

so that we have to choose $b_i = \frac{\mathbb{P}(A \cap B_i)}{\mathbb{P}(B_i)} = \mathbb{P}(A|B_i), i = 1, 2, \dots$ Since

$$\sum_{i=1}^{\infty} \left( \frac{\mathbb{P}(A \cap B_i)}{\mathbb{P}(B_i)} \right)^2 \mathbb{P}(B_i) = \sum_{i=1}^{\infty} \frac{\mathbb{P}^2(A \cap B_i)}{\mathbb{P}(B_i)} \leq \sum_{i=1}^{\infty} \frac{\mathbb{P}(A \cap B_i) \mathbb{P}(B_i)}{\mathbb{P}(B_i)}$$

$$= \sum_{i=1}^{\infty} \mathbb{P}(A \cap B_i) = P(A),$$

$\phi$ with such coefficients belongs to $L^2(\Omega, \mathcal{F}, \mathbb{P})$ (in particular the minimal distance is finite). Also, for such a $\phi$,

$$\int_{B_i} \phi \, d\mathbb{P} = \int_{B_i} b_i \, d\mathbb{P} = b_i \mathbb{P}(B_i) = \mathbb{P}(A \cap B_i) = \int_{B_i} 1_A \, d\mathbb{P}.$$

*Exercise 3.3.7* For any $t \in \mathbb{R}$,

$$t^2 \mathbb{E}(Y^2|\mathcal{G}) - 2t\mathbb{E}(XY|\mathcal{G}) + \mathbb{E}(X^2|\mathcal{G}) = \mathbb{E}((tY - X)^2|\mathcal{G}) \qquad (9.4)$$

except for the set $A_t$ of probability zero. But just because of that we should not claim that this holds for all $t \in \mathbb{R}$, except on a set of probability zero. We need to be more careful and proceed as follows.

For any rational $t \in \mathbb{Q}$, (9.4) holds except on a set $A_t$ of probability zero. Moreover, $\mathbb{E}((tY - X)^2|\mathcal{G}) \geq 0$, except on a set $B_t$ of probability zero. Thus

$$t^2 \mathbb{E}(Y^2|\mathcal{G}) - 2t\mathbb{E}(XY|\mathcal{G}) + \mathbb{E}(X^2|\mathcal{G}) \geq 0 \qquad (9.5)$$

except on $A_t \cup B_t$. Furthermore, $A = \bigcup_{t \in \mathbb{Q}} (A_t \cup B_t)$ *is* a set of probability zero. For all $\omega \in \Omega \setminus A$, (9.5) holds for all $t \in \mathbb{Q}$, and by continuity of the right-hand side, for all $t \in \mathbb{R}$. Therefore, except for a set of probability zero, the discriminant of the right-hand side is non-positive, and our claim follows.

*Exercise 3.3.9* Using 3.3.1 (e), we obtain that $\mathbb{E}(X|\mathcal{G}) \geq \mathbb{E}(Y|\mathcal{G})$ whenever $X \geq Y$. Hence $\mathbb{E}(1_{X \geq a}|\mathcal{G}) \leq \mathbb{E}\left(\frac{X}{a}1_{X \geq a}|\mathcal{G}\right) \leq \mathbb{E}\left(\frac{X}{a}|\mathcal{G}\right)$.

*Exercise 3.3.10* We have $E \, \mathrm{VAR}(X|\mathcal{G}) = E\left[\mathbb{E}(X^2|\mathcal{G}) - \mathbb{E}(X|\mathcal{G})^2\right] = E \, X^2 - E\left[\mathbb{E}(X|\mathcal{G})\right]^2$ and $D^2\left[\mathbb{E}(X|\mathcal{G})\right] = E\left[\mathbb{E}(X|\mathcal{G})\right]^2 - \left[E \, \mathbb{E}(X|\mathcal{G})\right]^2 = E\left[\mathbb{E}(X|\mathcal{G})\right]^2 - (E \, X)^2$. Adding up we get the claim.

*Exercise 3.3.16* $\mathbb{E}(Z_1|Z_2) = E \, X_1 1_\Omega + \mathbb{E}(Y|Z_2) = E \, X_1 1_\Omega + \mathbb{E}(Z_2 - X_2|Z_2) = \mathbb{E}(Z_2|Z_2) = Z_2$.

*Exercise 3.3.19* We have $E \, XY = E\left[\mathbb{E}(XY|\sigma(Y))\right] = E\left(Y\mathbb{E}(X|\sigma(Y))\right)$ $= E\left(Y E \, X\right) = E \, X \cdot E \, Y$.

*Exercise 3.5.2* Define $A_{n+1} = \sum_{i=1}^{n} \left[\mathbb{E}(X_{i+1}|\mathcal{F}_i) - X_i\right]$.

*Exercise 3.5.7* Set $\|X\| = \|X_k^n\|_{L^2}, k, n \geq 1$. We have

$$E \, Z_{n+1}^2 = E \sum_{k=0}^{\infty} \left(\sum_{i=1}^{k} X_i^{n+1}\right)^2 1_{Z_n = k} = \sum_{k=1}^{\infty} \mathbb{P}\{Z_n = k\} E \, (\sum_{i=1}^{k} X_i^{n+1})^2$$

$$= \sum_{k=1}^{\infty} \mathbb{P}\{Z_n = k\} \left[ k\|X\|^2 + k(k-1)m^2 \right]$$

$$= \|X\|^2 E\, Z_n + m^2 E\, (Z_n^2 - Z_n)$$

$$= \left(\|X\|^2 - m^2\right) E\, Z_n + m^2 D^2(Z_n) + m^{2(n+1)},$$

whence (3.25) follows.

The case $m = 1$ is clear, since $D^2(Z_1) = \sigma^2$. In the other case we note that $\frac{\sigma^2}{m(1-m)} m^n$ is a particular solution of (3.25) and that the general form of the solution to the homogeneous recurrence associated with (3.25) is $\alpha m^{2n}$; the constant $\alpha$ is chosen so that $D^2(Z_1) = \sigma^2$.

*Hint to Exercise 3.6.6* Apply 3.6.5 to $A_n' = -A_n$.

*Exercise 3.7.9* The condition of the first definition implies the two conditions of the second definitions because (i) for all $n, k \geq 1$ we have $E\,|X_n| \leq E\,|X_n|1_{|X_n|\geq k} + k\mu(\Omega)$ which proves that $X_n, n \geq 1$, is bounded, and (ii) in view of the fact that $E\,|X_n|1_A = E\,|X_n|1_{A\cap|X_n|\geq k} + E\,|X_n|1_{A\cap|X_n|<k} \leq E\,|X_n|1_{|X_n|\geq k} + k\mu(A)$ we may make $E\,|X_n|1_A$ less than an arbitrary $\epsilon > 0$ by choosing a large $k$ first, and then taking, say, $\delta < \frac{\epsilon}{2k}$ (provided $\mu(A) < \delta$). On the other hand, if our sequence is bounded in $L^1$ then $E\,|X_n|1_{|X_n|\geq k}$ is bounded by the same constant, say $M$, and so the measure of each of the sets $A_{m,k} = \{|X_m| \geq k\}$ is less than $\frac{M}{k}$. Hence, given $\epsilon > 0$ and $\delta = \delta(\epsilon)$ spoken of in the second definition we may take $k > \frac{M}{\delta}$ to make $\sup_{n\geq 1} E\,|X_n|1_{|X_n|\geq k} \leq \sup_{n\geq 1} \sup_{m\geq 1} E\,|X_n|1_{A_{m,k}}$ less than $\epsilon$.

*Hint to Exercise 3.7.10* Note that random variables $\phi_k(X)$, where $\phi_k$ was defined in the last paragraph of 3.7.3, converge to $X$ in $L^1$.

*Hint to Exercise 3.7.13* For part (c), given $A \in \mathcal{F}_\tau$, consider the proces $X_n = 1_{A\cap\{\tau=n\}}, n \geq 1$, and show that it is adapted.

*Exercise 4.2.3* For any real $\alpha_k, 1 \leq k \leq n$,

$$\left\|\sum_{k=1}^{n} \alpha_k x_k - x\right\|^2 = \left(\sum_{k=1}^{n} \alpha_k x_k - x, \sum_{k=1}^{n} \alpha_k x_k - x\right) \tag{9.6}$$

$$= \|x\|^2 - 2\sum_{k=1}^{n} \alpha_k(x_k, x) + \sum_{k=1}^{n} \alpha_k^2$$

$$= \|x\|^2 - \sum_{k=1}^{n} (x_k, x)^2 + \sum_{k=1}^{n} [\alpha_k - (x, x_k)]^2.$$

The first two terms above do not depend on $\alpha_k$, and thus the minimum is attained exactly when the last term equals 0, i.e. iff $\alpha_k = (x_k, x)$.

*Exercise 4.2.5* For any $n \geq 1$, $\sum_{k=1}^{n}(x_n, x)^2$ is the square of the norm of the projection of $x$ onto $span\{x_k, 1 \leq k \leq n\}$. Thus it is less than $\|x\|^2$ by 3.1.13 (it also follows directly from (9.6) above if we take $\alpha_k = (x, x_k)$). Since $n$ is arbitrary, (4.3) follows.

*Exercise 4.2.15* Note that $\|\sum_{i=k}^{l} a_i z_i\|^2 = \sum_{i=k}^{l} a_i^2$ so that the sequence $\sum_{i=1}^{n} a_i z_i$ is Cauchy.

*Hint to Exercise 4.3.12* It suffices to show that

$$\mathbb{E}\left(e^{a[w(t+h)-w(t)]^2} | \mathcal{F}_t\right) = E\, e^{a[w(t+h)-w(t)]^2} = e^{\frac{a^2 h}{2}},$$

or that $E\, e^X = e^{\frac{\sigma^2}{2}}$, provided $X \sim N(0, \sigma^2)$, which can be checked directly. (Recall that $M_X(t) = E\, e^{tX}$ is a so-called moment generating function of a random variable $X$; if $X \sim N(0, \sigma^2)$ then $M_X(t) = e^{\frac{\sigma^2 t^2}{2}}$.)

*Exercise 5.1.11* Use the Kuratowski–Zorn Lemma.

*Hint to Exercise 5.2.5* The isomorphism is given in 7.4.24. A functional on $l_r^1$ may be represented as $Fx = F(\xi_n)_{n \geq 1} = \sum_{n=1}^{\infty} \alpha_n r^n \xi_n$ where $(\alpha_n)_{n \geq 1}$ belongs to $l^\infty$. To prove this, use 5.2.3 and the argument from 5.2.10.

*Hint to Exercise 5.2.12* By 2.2.38, $C_0(\mathbb{G})$ is isometrically isomorphic to $C_0(\mathbb{R}) \times C_0(\mathbb{R})$.

*Hint to Exercise 5.1.17* Check that finite combinations $y = \sum_{i=1}^{n} \xi_i e_i$ of $e_i := (\delta_{i,n})_{n \geq 1}, i \geq 1$, belong to $\mathbb{Y}$.

*Hint to Exercise 5.4.3* To prove weak convergence use 5.2.16. To show that the sequence does not converge strongly note that if it converges it must converge to zero and that $\|y_n\| = 1$.

*Hint to Exercise 5.5.4* Argue as in the proof of the Markov inequality (see 1.2.36).

*Hint to Exercise 5.5.5* Use the estimate:

$$E\,|X_k - \mu_k|^{2+\alpha} 1_{\{|X_k - \mu_k| > \delta s_n\}} \geq s_n^\alpha \delta^\alpha E\,|X_k - \mu_k|^2 1_{\{|X_k - \mu_k| > \delta s_n\}}.$$

*Hint to Exercise 5.7.20* For any $\epsilon > 0$ there exists an $n \geq 1$ and

$x_1, ..., x_n \in A$ such that $\min_{i=1,...,n} \|x - x_i\| < \frac{\epsilon}{3}$ for all $x \in A$. Moreover, there is a $\delta$ such that $d(p, p') < \delta$ implies $|x_i(p) - x_i(p')| \leq \frac{\epsilon}{3}$.

*Hint to Exercise 6.2.8* Use the argument from the proof of Alaoglu's Theorem to show that $F(xy) = Fx \, Fy$ for all $x$ and $y$ in $\mathbb{A}$ and $F$ in the closure of $\mathcal{M} \cup \{0\}$. Note that we may not exclude the zero functional from being in the closure of $\mathcal{M}$. If $\mathcal{A}$ has a unit $u$, though, we show that $F(u) = 1$ for all $F$ in the closure of $\mathcal{M}$, so that the zero functional does not belong to $cl \, \mathcal{M}$.

*Hint to Exercise 6.3.5* Note first that it is enough to determine $\alpha_n, n \geq 2$. Now, choose $n = 2k + 1$ and $m = 2k - 1$ in (6.14) to obtain

$$\alpha_{2k+1} = 2\alpha_1\alpha_{2k} - \alpha_{2k-1}.$$

Similarly, choose $n = 2k$ and $n = 2k + 2$, to obtain $\alpha_{2k+2} = 2\alpha_1\alpha_{2k+1} - \alpha_{2k}$. This proves that values of $\alpha_n$ and $\alpha_{n+1}$ determine $\alpha_{n+2}$, so the proof is completed by the induction argument.

*Hint to Exercise 6.4.6* (b) Use (a) and (6.24) with $r = \frac{1}{2}$, and $\alpha$ replaced by $4\alpha pq$. Note that $(2n - 1)!!2^n n! = (2n)!$ so that $\binom{n-1+\frac{1}{2}}{n} 4^n = \binom{2n}{n}$. (f) Use Abel's theorem to show that $\sum_{n=1}^{\infty} r_n = \lim_{\alpha \to 1} g(\alpha)$, and note that $1 - 4pq = (p + q)^2 - 4pq$. Details may be found in [40] or [48].

*Exercise 6.4.7* For $\tau = 0$, this result follows directly from 1.2.31 by the substitution $s = \sqrt{\lambda t}$.
  For $\tau > 0$,

$$\frac{\partial g(\tau, t)}{\partial \tau} = -\frac{\tau}{t} g(\tau, t),$$

and

$$\frac{\partial g(\tau, t)}{\partial t} = (-\frac{1}{t} + \frac{\tau^2}{t^2}) g(\tau, t) = \frac{1}{2} \frac{\partial^2 g(\tau, t)}{\partial \tau^2}.$$

Let $(a, b) \subset \mathbb{R}$ be a *finite* interval. For $\tau$ in this interval,

$$\left| \frac{\partial g(\tau, t)}{\partial \tau} \right| \leq \frac{c}{t} g(d, t), \qquad \left| \frac{\partial^2 g(\tau, t)}{\partial \tau^2} \right| \leq 2(\frac{1}{t} + \frac{c^2}{t^2}) g(d, t)$$

where $c = |a| \vee |b|$ and $d = |a| \wedge |b|$. Taking $\mu$ in Exercise 1.2.12 to be the measure on $\mathbb{R}^+$ with density $e^{-\lambda t}$ we obtain

$$\frac{\partial}{\partial \tau} G(\lambda, \tau) = \int_{\mathbb{R}^+} \frac{\partial}{\partial \tau} g(\lambda, \tau) \, \mu(d\omega), \qquad \tau > 0.$$

Similarly,

$$\frac{\partial^2}{\partial \tau^2} G(\lambda, \tau) = \int_{\mathbb{R}^+} \frac{\partial^2}{\partial \tau^2} g(\lambda, \tau) \, \mu(d\omega) = 2 \int_0^\infty e^{-\lambda t} \frac{\partial}{\partial t} g(\tau, t) \, dt.$$

Integrating by parts, this equals $2\lambda G(\tau, t)$. The general solution to the equation $\frac{d^2}{d\tau^2} G(\lambda, \tau) = 2\lambda G(\lambda, \tau)$ in $\mathbb{R}^+_*$ is given by

$$G(\lambda, \tau) = C_1 e^{-\sqrt{2\lambda}\tau} + C_2 e^{\sqrt{2\lambda}\tau}.$$

(This is an ODE with $\lambda$ treated as a parameter.) Since $\lim_{\tau \to \infty} G(\lambda, \tau) = 0$, $C_2 = 0$, and since $\lim_{\tau \to 0} G(\lambda, \tau) = \frac{1}{\sqrt{2\lambda}}$, $C_1 = \frac{1}{\sqrt{2\lambda}}$.

A different, more direct computation of this integral may be found e.g. in [109].

*Exercise 6.4.8* By the Lebesgue Dominated Convergence Theorem, and the estimate $|e^{ihs} - 1| \le |sh|$ (draw a picture!) the difference quotient $\frac{\phi(\alpha+h)-\phi(\alpha)}{h} = \int_{-\infty}^\infty e^{i\alpha s} \frac{e^{ihs}-1}{h} e^{-\frac{s^2}{2}} \, ds$, converges to $i \int_{-\infty}^\infty s e^{i\alpha s} e^{-\frac{s^2}{2}} \, ds$. Integrating by parts, this equals i times $[-e^{i\alpha s} e^{-\frac{s^2}{2}}]_{s-\infty}^\infty + i\alpha\phi(\alpha)$. Since the first term above equals zero, (6.25) is proved. A general solution to the equation $\frac{d}{d\alpha}\phi(\alpha) = -\alpha\phi(\alpha)$ is $\phi(\alpha) = Ce^{-\frac{\alpha^2}{2}}$. Since we must have $\phi(0) = 1$, $C = 1$, and we obtain $\phi(\alpha) = e^{-\frac{\alpha^2}{2}}$.

*Exercise 6.5.4* Use the relations

$$x \vee y = \frac{1}{2}|x - y| + \frac{1}{2}(x + y), \quad x \wedge y = -[(-x) \vee (-y)],$$

or

$$x \wedge y = -\frac{1}{2}|x - y| + \frac{1}{2}(x + y), \quad x \vee y = -[(-x) \wedge (-y)],$$

or their combination.

*Hint to Exercise 6.5.5* Show first that the set of real parts of the elements of $\mathbb{A}$ is equal to the algebra of real continuous functions on $S$.

*Exercise 6.5.7* Since $\|x(X, Y) - y(X, Y)\|_{L^1(\Omega)} \le \|x - y\|_{C(\overline{\mathbb{R}^2})}$, by linearity it suffices to show the formula for $x(\tau, \sigma) = y_1(\tau)y_2(\sigma)$ where $y_i \in C(\overline{\mathbb{R}}), i = 1, 2$. We have $\mathbb{E}(y_1(X)y_2(Y)|\mathcal{F}) = (Ey_1(X))y_2(Y)$, $Y$ being $\mathcal{F}$ measurable and $X$ being independent of $\mathcal{F}$. On the other hand, since $y_2(Y)$ does not depend on $\tau$,

$$\int_{\mathbb{R}} y_1(s)y_2(Y) \, \mathbb{P}_X(d\tau) = y_2(Y) \int_{\mathbb{R}} y_1(s) \, \mathbb{P}_X(d\tau) = y_2(Y)Ey_1(X),$$

as desired.

*Hint to Exercise 6.6.6* Show the the following calculation is correct:

$$\int_{-\pi}^{\pi} \hat{x}(t) e^{itm} \, dt = \sum_{n=-\infty}^{\infty} \xi_n \int_{-\pi}^{\pi} e^{itn} e^{itm} \, dt = 2\pi \xi_{-m}, \quad m \in \mathbb{Z}.$$

*Exercise 6.6.14* The pgf of the binomial distribution equals $(p\alpha + q)^n = [1 + (1 - \alpha)p]^n$. In the limit, as $p \to 0$ and $n \to \infty$ in such a way that $np \to \lambda$, we obtain $e^{-\lambda(1-\alpha)}$ as desired.

*Exercise 7.1.1* Note that $x_n$ is a Cauchy sequence, for if $m \geq n$ then $x_m \in B_m \subset B_n$, so that $\|x_n - x_m\| \leq r_n$. Let $x = \lim_{n \to \infty} x_n$. For any $n \in N$, $x_m \in B_n$, as long as $m \geq n$ and $B_n$ is closed. Thus, $x \in B_n$, as desired.

*Exercise 7.1.9* By 7.1.3, $A_n$ are equibounded. Hence, there exists an $M$ such that

$$\|A_n x_n - Ax\| \leq \|A_n x_n - A_n x\| + \|A_n x - Ax\|$$
$$\leq M\|x_n - x\| + \|A_n x - Ax\|$$

which implies our claims.

*Exercise 7.1.7* If the supremum of norms is not finite, then $\mathbb{T}$ must contain an infinite number of elements, and we may choose $t_n \in \mathbb{T}$ such that $\|A_{t_n}\| \geq n$. This is a contradiction by the Banach–Steinhaus Theorem.

*Hint to Exercise 7.3.4* The key is the fact that if the sequences $x_n, n \geq 1$, and $Cx_n$ converge, then so do $Bx_n$ and $Ax_n$.

*Exercise 7.4.6* By 7.1.6, there exists a $\delta > 0$ and $M \geq 1$ such that for $0 \leq t \leq \delta$, $\|T_t\| \leq M$ ($M$ is bigger than 1 since $\|T_0\| = 1$). Thus, for any $x \in \mathbb{X}$, and $0 \leq t \leq \delta$,

$$\|T_t S_t x - x\| \leq \|T_t S_t x - T_t x\| + \|T_t x - x\| \leq M\|S_t x - x\| + \|T_t x - x\|.$$

The claim follows by taking the limit as $t \to 0$.

*Hint to Exercise 7.4.10* Use 5.4.12 and a result analogous to 5.4.18, where $\mathbb{R}$ is replaced by the unit circle.

*Exercise 7.4.13* Note that $S_t = e^{-\omega t} T_t$ is a semigroup, and that $\|S_t\| \leq M$. By the semigroup property, for any $t > 0$ and $n \geq 1$,

$$\|S_t\| = \|S_{\frac{t}{n}}^n\| \leq \|S_{\frac{t}{n}}\|^n \leq M^n.$$

Since $0 \leq M < 1$, $\|S_t\| = 0$, i.e. $S_t = 0$, and $T_t = 0$.

*Hint to Exercise 7.4.25* The characteristic values of $A$ are $0, -1$ and $-6$ with eigenvectors $[1, 1, 1]$, $[0, 2, 3]$ and $[0, 1, -1]$ respectively. Furthermore,

$$\begin{bmatrix} 1 & 0 & 0 \\ 1 & 2 & 1 \\ 1 & 3 & -1 \end{bmatrix}^{-1} = \frac{1}{5} \begin{bmatrix} 5 & 0 & 0 \\ -2 & 1 & 1 \\ -1 & 3 & -2 \end{bmatrix}.$$

*Exercise 7.4.23* The semigroup property of $\{S_t, t \geq 0\}$ and its strong continuity are trivial. Moreover, $x$ belongs to $\mathcal{D}(B)$ iff the limit

$$\lim_{t \to 0+} \frac{S_t x - x}{t} = \lim_{t \to 0+} \frac{T_t x - x}{t}$$

exists. If it exists, however, it belongs to $\mathbb{X}_1$, for $\mathbb{X}_1$ is closed.

*Exercise 7.4.27* Let $e = (1, \cdots, 1)^{\mathrm{T}}$ be the column-vector with all entries equal to 1. The condition $\sum_{j \in \mathbb{I}} q_{i,j} = 0, i \in \mathbb{I}$, is equivalent to $Qe = 0$. Analogously, $\sum_{j \in \mathbb{I}} p_{i,j} = 1, i \in \mathbb{I}$, iff $P(t)e = e$. The key to the proof is the fact that $Q = \frac{\mathrm{d}}{\mathrm{d}t} P(t)|_{t=0}$. If $P(t), t \geq 0$, are stochastic matrices then $q_{i,j} = \frac{\mathrm{d}}{\mathrm{d}t} p_{i,j}(t)|_{t=0} = \lim_{t \to 0+} \frac{p_{i,j}(t)}{t} \geq 0$, $i \neq j$, and $q_{i,i} = \lim_{t \to 0+} \frac{p_{i,i}(t) - 1}{t} \leq 0$. Moreover, $Qe = \frac{\mathrm{d}}{\mathrm{d}t} [P(t)e]|_{t=0} = [\frac{\mathrm{d}}{\mathrm{d}t} e]|_{t=0} = 0$. Conversely, if $Q$ is an intensity matrix, then all the entries of $\exp tQ$ are non-negative because $rI + Q$ where $r = -\min_{i \in \mathbb{I}} q_{i,i}$ has non-negative entries and we have $e^{Qt} = e^{-rt} e^{(rI+Q)t}$, matrices $-rI$ and $Q$ commuting. Moreover, $e^{Qt} e = e$ since $e^{Q0} e = e$ and $\frac{\mathrm{d}}{\mathrm{d}t} e^{Qt} e = e^{Qt} Qe = 0$.

*Hint to Exercise 7.4.37* Use the Taylor formula (5.21). For example, for $x \in \mathcal{D}(\frac{\mathrm{d}^2}{\mathrm{d}\tau^2})$, $C(t)x(\tau) - x(\tau) = \frac{1}{4} v^2 t^2 [x''(\tau + \theta_1 vt) + x''(\tau - \theta_2 vt)]$ and $(C(t)x)'(\tau) - x'(\tau) = \frac{1}{2} tv[x''(\tau + \theta_3 vt) - x''(\tau + \theta_4 vt)]$, where $0 \leq \theta_i \leq 1, i = 1, ..., 4$. This implies $\lim_{t \to 0} \frac{1}{t}(C(t)x - x) = 0$ for $x \in \mathcal{D}(\frac{\mathrm{d}^2}{\mathrm{d}\tau^2})$ strongly in $\mathbb{X}$.

*Hint to Exercise 7.4.47* Show that without loss of generality we may assume that the semigroup is bounded and consider elements of the form $H(\phi)x$ where $\phi$ is $C^\infty$ with bounded support.

*Hint to Exercise 7.4.48* Both sets $\{e_\lambda, \lambda > 0\}$ and $\{1_{[0,t)}, t > 0\}$ are linearly dense in $L^1(\mathbb{R}^+)$.

*Exercise 7.5.3* See (7.41) and the paragraph following it.

*Hint to Exercise 7.5.4* Repeat the analysis given for the Brownian motion. The semigroup is given by $T_t x(\tau) = E\, x(\tau + at + w(t))$. Prove that the generator equals $\frac{1}{2}x'' + ax'$, and has the domain $\mathcal{D}(\frac{d^2}{d\tau^2})$.

*Hint to Exercise 7.5.10* $\lambda \int_0^\infty e^{-\lambda t} e^{-at} \frac{a^k t^k}{k!}\, dt = \lambda \frac{a^k}{(\lambda+a)^{k+1}}$.

*Exercise 7.7.2* $y(t,\tau) = 1$ solves (7.55)–(7.56) with $y(\tau) = 1$.

*Hint to Exercise 7.7.6* If $x_k$, $k = 1, -1$ are differentiable and their derivatives belong to $BUC(\mathbb{R})$ then $y(\tau, k) = kvx'(\tau, k)$ belongs to $BUC(\mathbb{G})$. Moreover,

$$S_t x(\tau, k) - x(\tau, k) = x(vtk + \tau, k) - x(\tau, k) = vtkx'(\tau + \theta vtk, k)$$

where $\theta = \theta(v, t, k, x, \tau)$ belongs to the interval $[0, 1]$. Thus

$$\frac{1}{t}\|S_t x - x - ty\| \le \sup_{\tau \in \mathbb{R}, k = 1, -1} |x'(\tau + \theta tkv, k) - x'(\tau, k)|$$

$$\le \sup_{|\tau - \sigma| \le tv} v|x'(\sigma, k) - x'(\tau, k)|$$

which tends to zero, as $t \to 0$, by uniform continuity of $\tau \to x'(\tau, k)$, $k = 1, -1$. The rest is proven as in 7.4.16.

*Hint to Exercise 7.7.8* Let $y(\tau) = \frac{1}{v}\int_0^\tau z(\sigma)\, d\sigma$, and write $z(\tau + v\sigma)$ as $\frac{\partial}{\partial \sigma} y(\tau + v\sigma)$, then integrate $\int_{-t}^t I_0\left(a\sqrt{t^2 - \sigma^2}\right) \frac{\partial}{\partial \sigma} y(\tau + v\sigma)\, d\sigma$ by parts to obtain

$$y(\tau + vt) - y(\tau - vt) + a\int_{-t}^t I_1\left(a\sqrt{t^2 - \sigma^2}\right) \frac{\sigma}{\sqrt{t^2 - \sigma^2}} y(\tau + v\sigma)\, d\sigma.$$

Next note that the function $x_\tau(\sigma) := y(\tau + v\sigma) - y(\tau - v\sigma)$ is odd, and so is $\sigma \mapsto I_1\left(a\sqrt{t^2 - \sigma^2}\right)\frac{\sigma}{\sqrt{t^2-\sigma^2}}$ while $\sigma \mapsto I_1\left(a\sqrt{t^2 - \sigma^2}\right)\frac{t}{\sqrt{t^2-\sigma^2}}$ and $\sigma \mapsto I_0\left(a\sqrt{t^2 - \sigma^2}\right)$ are even. Hence,

$$y(t, \tau) = e^{-at}x_\tau(t) + e^{-at}\frac{a}{2}\int_{-t}^t I_0\left(a\sqrt{t^2 - \sigma^2}\right) x_\tau(\sigma)\, d\sigma$$

$$+ e^{-at}\frac{a}{2}\int_{-t}^t I_1\left(a\sqrt{t^2 - \sigma^2}\right)\frac{t + \sigma}{\sqrt{t^2 - \sigma^2}}x_\tau(\sigma)\, d\sigma.$$

Also, using (7.61),

$$e^{-at} + \frac{a}{2} e^{-at}\int_{-t}^t I_0\left(a\sqrt{t^2 - \sigma^2}\right) d\sigma$$

$$+ \frac{a}{2} e^{-at}\int_{-t}^t \frac{t + \sigma}{\sqrt{t^2 - \sigma^2}} I_1\left(a\sqrt{t^2 - \sigma^2}\right) d\sigma = 1.$$

*Hint to Exercise 7.7.9* (b) Let $y(\tau) = \frac{1}{v} \int_0^\tau z(\sigma) \, d\sigma$. Note that $y(\tau + v\xi(t)) - y(\tau - v\xi(t)) = \int_{-\xi(t)}^{\xi(t)} z(\tau + v\sigma) \, d\sigma$. Use linearity of the telegraph equation.

*Exercise 8.1.2* As shown in 2.3.17, operators $T_\mu$ commute with translations. Let $S_t$ denote translation: $S_t x(\tau) = x(\tau + t)$. Then

$$(U_t S_s x)(\tau) = E \, x(|s + \tau + w(t)|),$$

while

$$S_s U_t x)(\tau) = E \, x(s + |\tau + w(t)|).$$

These two are not equal for all $x \in BUC(\mathbb{R}^+)$: take e.g.

$$x_s(\tau) = \begin{cases} \tau - s, & \tau < s, \\ 0, & \tau \geq 0. \end{cases}$$

*Exercise 8.1.7* Easy.

*Exercise 8.1.8* In 1.2.20 it was shown that there exist continuous functions $f_h$ converging pointwise, as $h \to 0$, to $1_{(a,b]}$ where $a < b$. By the Lebesgue Dominated Convergence Theorem, $f_h(X(s))$ converges in $L^1(\Omega)$ to $1_{X(s)\in(a,b]}$. Hence $\mathbb{E}(1_{X(s)\in(a,b]}|\mathcal{F}_t) = \mathbb{E}(1_{X(s)\in(a,b]}|X(t))$, i.e., for any $A \in \mathcal{F}_t$, $\int_A 1_{X(s)\in B} \, d\mathbb{P} = \int_A \mathbb{E}(1_{X(s)\in B}|X(t)) \, d\mathbb{P}$ for $B$ of the form $B = (a,b]$. However, both sides in this equality are finite measures in $B$. By the $\pi$–$\lambda$ theorem, the equality holds for all $B$, proving (8.10).

*Hint to Exercise 8.1.10* Use 6.5.7 to conclude that $\mathbb{E}(f(w_r(s))|\mathcal{F}_t) = \int_{\mathbb{R}} f(|\tau + \sigma + w_r(t)|) \, \mathbb{P}_{w(s)-w(t)}(d\sigma)$.

*Exercise 8.1.12* The only non-trivial part is the fact that, for any Borel set $B$ and a measure $\mu$, the function $f_B(\tau) := \mu(B - \tau)$ is measurable. If $B = (-\infty, \sigma]$ for some $\sigma \in \mathbb{R}$, then $f_B(\tau) = \mu(-\infty, \sigma - \tau]$. Hence, $F_B$ is bounded, left-continuous and non-increasing, and in particular measurable. Using linearity of the integral, we check that $f_B$ is measurable for $B$ of the form $(a,b], a < b$. Such sets form a $\pi$-system. Moreover, the class $\mathcal{H}$ of Borel sets such that $f_B$ is measurable is a $\lambda$-system. Indeed, $\mathbb{R} \in \mathcal{H}$ and if $A, B \in \mathcal{H}$, and $A \subset B$ then $A - \tau \subset B - \tau$ and the relation $(B \setminus A) - \tau = (B - \tau) \setminus (A - \tau)$ (see 2.1.18) results in $f_{B\setminus A}(\tau) = f_B(\tau) - f_A(\tau)$, proving conditions (a) and (b) in 1.2.7. To prove (c) it is enough, by the already proved part (b), that for any *disjoint* $A_i \in \mathcal{H}$ we have $\bigcup_{i=1}^\infty A_i \in \mathcal{H}$ (for if $A_i$ are not disjoint then we may consider $A_i^\sharp = A_i - \bigcup_{j=1}^{i-1}$ instead). For such $A_i$, however, $f_{\bigcup_{i=1}^\infty A_i} = \sum_{i=1}^\infty f_{A_i}$. By the $\pi$–$\lambda$ theorem, $\mathcal{B}(\mathbb{R}) \subset \mathcal{H}$, i.e. $f_B$ is measurable for all Borel sets $B$.

*Exercise 8.2.13* By (8.30) and the Binomial Theorem we have

$$\left\| \left( \frac{1}{\lambda + \mu} + \left( \frac{\lambda}{\lambda + \mu} \right)^2 R_{\frac{\lambda\mu}{\lambda+\mu}} \right)^n \right\| \leq \left( \frac{1}{\lambda + \mu} + \left\| \left( \frac{\lambda}{\lambda + \mu} \right)^2 R_{\frac{\lambda\mu}{\lambda+\mu}} \right\| \right)^n$$

$$\leq M \sum_{i=0}^n \left( \frac{1}{\lambda + \mu} \right)^{n-i} \left( \frac{\lambda}{\lambda + \mu} \right)^{2i} \frac{(\lambda + \mu)^i}{(\lambda\mu)^i}$$

$$\leq \frac{M}{\mu^n} \sum_{i=0}^n \left( \frac{\mu}{\lambda + \mu} \right)^{n-i} \left( \frac{\lambda}{\lambda + \mu} \right)^i = \frac{M}{\mu^n}.$$

*Hint to Exercise 8.2.15* On both sides of this equality, we have continuous functions. Calculate their Laplace transforms and use the Hilbert equation (7.33) to show that they are equal.

*Hint to Exercise 8.3.16* The operator $x \to x''$ with natural domain in $C[\epsilon, 1 - \epsilon], 0 < \epsilon < \frac{1}{2}$ is closed.

*Hint to Exercise 8.3.17* Show coordinate-wise convergence first; then use Scheffé's Theorem.

*Hint to Exercise 8.4.6* Note that $R_\lambda x(\tau) = \frac{1}{\sqrt{2\lambda}} \int_{-\infty}^\infty e^{-\sqrt{2\lambda}|\tau - \sigma|} x^*(\sigma) \, d\sigma$ where $x^*(\tau) = (\text{sgn } \tau) \, x(|\tau|)$.

*Hint to Exercise 8.4.24* Use the Uniform Boundedness Principle on the operators $A_n, n \geq 1$ :

$$A_n x = (P_1 x, ..., P_n x, 0, 0, ...)$$

mapping $\mathbb{X}$ into the Banach space of bounded sequences $(x_n)_{n \geq 1}$ with $x_n \in \mathbb{X}_n, n \geq 1$.

*Exercise 8.4.34* Put $t = N, n = i$ and $k = j$ in 8.93.

## 9.3 Some commonly used notations

$[\tau]$ or $\llcorner \tau \lrcorner$ – the largest integer not exceeding $\tau$,
$\lceil \tau \rceil$ – the smallest integer no smaller than $\tau$,
$\tau^+ = \max(0, \tau), \tau^- = (-\tau)^+$,
$A, B, C$ either a set or an operator,

$A^{\complement}$ – the complement of a set $A$,

$BC(S)$ – the space of bounded continuous functions on $S$,

$BM(S)$ – the space of bounded measurable functions on $S$,

$\mathbb{BM}(S)$ – the space of bounded Borel charges (signed measures) on a topological space $S$,

$BUC(\mathbb{G})$ – the space of bounded uniformly continuous functions on a locally compact group $\mathbb{G}$ – see 2.3.25,

$C(S)$ – the space of continuous functions on $S$,

$C([-\infty, \infty])$ – the space of continuous functions on $\mathbb{R}$ with limits at both $\infty$ and $-\infty$,

$C(\overline{\mathbb{R}})$ – the subspace of $C([-\infty, \infty])$ of functions with the same limit at $\infty$ and $-\infty$,

$\mathcal{D}(A)$ – the domain of an operator $A$,

$e_\lambda - e_\lambda(\tau) = e^{-\lambda \tau}$, see 2.2.49,

$\mathcal{F}, \mathcal{G}$, etc. – $\sigma$-algebras,

$m, \lambda, \mu, \nu$ – measures, but

$\lambda, \mu, \nu$ – are also often used to denote a positive (or non-negative) number, especially in the context of the Laplace transform, while $m$ may denote a mean or a median of a random variable,

$\mathbb{G}$ – a (semi-)group,

$\mathcal{M}$ – Lebesgue measurable sets (see 1.2.1),

$\mathbb{N}$ – the set of natural numbers,

$\mathbb{N}_0 - \mathbb{N} \cup \{0\}$,

$\Omega, (\Omega, \mathcal{F}, \mathbb{P})$ – a probability space,

$p, q$ – points of $S$,

$\mathbb{Q}$ – the set of rational numbers,

$\mathbb{R}$ – the set of reals,

$\mathbb{R}^+$ – the set of non-negative reals,

$\mathbb{R}_*^+ - \mathbb{R}^+ \setminus \{0\}$,

$S$ – a set, or a space, probably a topological or metric space,

sgn – the signum function, see (1.46),

$\tau, \sigma$ – usually a real number, sometimes a Markov time,

$\mathbb{X}, \mathbb{Y}, \mathbb{Z}$ etc. – a linear space, a Banach space or a subset, perhaps a subspace of such a space, but

$\mathbb{Z}$ – is sometimes used to denote the set of integers, see e.g. 6.1.1,

$x, y, z$ – elements of a Banach space, or of a linear space.

# References

[1]  Arendt, W., 1987, "Vector-valued Laplace transforms and Cauchy problem", *Israel J. Math.* **59**, 327–352.

[2]  Balakrishnan, A. V., 1981, *Applied Functional Analysis*, Springer.

[3]  Banach, S., 1987, *Theory of Linear Operations*, translation of 1932 original, North-Holland.

[4]  Bertoin, J., 1996, *Lévy Processes*, Cambridge University Press.

[5]  Billingsley, P., 1979, *Probability and Measure*, Wiley.

[6]  Billingsley, P., 1999, *Convergence of Probability Measures*, Wiley.

[7]  Blanchard, Ph., Combe, Ph., Sirugue, M., Sirugue-Collin, M., "Path integral representation of the Dirac equation in presence of an external electromagnetic field", pp. 396–413 in *Path integrals from meV to MeV (Bielefeld, 1985)*, Bielefeld Encount. Phys. Math., VII, World Scientific.

[8]  ———— 1987, "Jump processes related to the two–dimensional Dirac equation", pp. 1–13 in *Stochastic processes–mathematics and physics, II, (Bielefeld, 1985)*, Lecture Notes in Mathematics 1250, Springer.

[9]  ———— 1986, "Stochastic jump processes associated with Dirac equation", pp. 87–104 in *Stochastic processes in classical and quantum system, (Ascona, 1985)*, Lecture Notes in Physics 262, Springer.

[10]  ———— 1988, "Jump processes. An introduction and some applications to quantum physics", pp. 47–104 in *Functional integration with emphasis on the Feynman integral*, Rend. Circ. Math. Palermo (2) Suppl. No. 17.

[11]  Bobrowski, A., 1992, "Some remarks on the two-dimensional Dirac equation", *Semigroup Forum* **45**, 77–91.

[12]  ———— 1993, "Computing the distribution of the Poisson–Kac process", *Annales UMCS* **XLVII, 1 Sec. A**, 1-17.

[13]  ———— 1994, "Degenerate convergence of semigroups", *Semigroup Forum* **49**, 303–327.

[14]  ———— 1994, "Integrated semigroups and the Trotter–Kato Theorem", *Bull. Polish Acad. Sci.* **41 No. 4**, 297–303.

[15]  ———— 1996, "Generalized Telegraph Equation and the Sova–Kurtz version of the Trotter–Kato Theorem", *Ann. Polon. Math.* **LXIV.1**, 37–45.

[16]  ———— 2004, "Quasi-stationary distributions of a pair of Markov chains related to time evolution of a DNA locus", *Adv. Appl. Prob.* **36**, 57–77.

[17]  Bobrowski, A., Kimmel, M., Chakraborty, R., Arino, O., 2001, "A

semigroup representation and asymptotic behavior of the Fisher–Wright–Moran coalescent", Chapter 8 in *Handbook of Statistics 19: Stochastic Processes: Theory and Methods*, C. R. Rao and D. N. Shanbhag, eds., Elsevier Science.

[18] Bobrowski, A., Wang, N., Kimmel, M., Chakraborty, R., 2002, "Nonhomogeneous Infinite Sites Model under demographic change: mathematical description and asymptotic behavior of pairwise distributions", *Mathematical Biosciences* **175**, 83–115.

[19] Breiman, L., 1986, *Probability*, Addison-Wesley.

[20] Brzeźniak, Z., Zastawniak, T., 1999, *Basic Stochastic Processes*, Springer.

[21] Chung, K. L., 1980, *Lectures from Markov processes to Brownian motion*, Springer.

[22] Conway, J. B., 1990, *A Course in Functional Analysis*, Springer.

[23] Courant, R., 1962, *Partial Differential Equations*, Interscience Publishers.

[24] Daley, D. J., Vere-Jones, D., 2003, *An Introduction to the Theory of Point Processes*, second edition of *Point Processes* published in 1988, Springer.

[25] Da Prato, G., Sinestrari, E., 1987, "Differential operators with non-dense domain", *Ann. Squola Norm. Sup. Pisa* **14(2)**, 285–344.

[26] Davies, E. B., 1980, *One-Parameter Semigroups*, Academic Press.

[27] Douglas, R. G., 1998, *Banach Algebra Techniques in Operator Theory*, second edition, Springer.

[28] Doob, J. L., 1944, *Measure Theory*, Springer.

[29] ——— 1953, *Stochastic processes*, Wiley.

[30] ——— 1984, *Classical Potential Theory and its Probabilistic Counterpart*, Springer.

[31] Dudley, R. M., 1976, *Probabilities and Metrics*, Aarhus Universitet.

[32] Dunford, N., Schwartz, T., 1958, *Linear Operators*, Interscience Publishers.

[33] Durrett, R., 1996, *Probability: Theory and Examples*, Wadsworth.

[34] ——— 1996, *Stochastic Calculus*, CRC Press.

[35] Dynkin, E. B., 1965, *Markov Processes*, Springer.

[36] Edgar, G. A., 1998, *Itegral, Probability and Fractal Geometry*, Springer.

[37] Edwards, R. E., 1995, *Functional Analysis. Theory and Applications*, Dover Publications.

[38] Ethier, S. N., Kurtz, T. G., 1986 *Markov Processes. Characterization and Convergence*, Wiley.

[39] Ewens, W. J., 2004, *Mathematical Population Genetics. I. Theoretical Introduction*, second edition, Springer.

[40] Feller, W., 1950, *An Introduction to Probability Theory and Its Applications*, Vol I., (third edition, 1970), Wiley.

[41] ——— 1966, *An Introduction to Probability Theory and Its Applications*, Vol II., (second edition, 1971), Wiley.

[42] Freidlin, M., 1985, *Functional Integration and Partial Differential Equations*, Princeton.

[43] Goldstein, J.A., 1985, *Semigroups of Linear Operators and Applications*, Oxford University Press.

[44] Goldstein, S., 1951, "On diffusion by discontinuous movements and on the telegraph equation", *Quart. J. Mech. App. Math.* **4**, 129–156.

[45] Graham, R. L., Knuth, D. E., Patashnik, O., 1994, *Concrete Mathematics*, second edition, Addison-Wesley.

[46] Gihman, I. I., Skorohod, A. V., 1969, *Introduction to the theory of random processes*, W. B. Saunders.

[47] ———, 1979, *The theory of stochastic processes*, I–III, Springer.

[48] Grimmet, G. R., Stirzaker, D. R., 2001, *Probability and Random Processes*, third edition, Oxford.

[49] Halmos, P. R., 1974, *Measure theory*, Springer.

[50] Hamermesh, M., 1962, *Group Theory and Its Application to Physical Problems*, Addison-Wesley.

[51] Hewitt, E., Ross, K. A., 1963, *Abstract Harmonic Analysis I*, (second edition, 1979), Springer.

[52] ———, 1970, *Abstract Harmonic Analysis II*, Springer.

[53] Hilbert, D., Courant R., 1952, *Methods of Mathematical Physics*, Vol. 1, Interscience Publ.

[54] Hille, E., Phillips, R. S., 1957, *Functional Analysis and Semigroups*, Providence.

[55] Heyer, H., 1977, *Probability Measures on Locally Compact Groups*, Springer.

[56] Högnäs, G., Mukhreya, A., 1995, *Probability Measures on Semigroups*, Plenum Press.

[57] Ikeda, N., Watanabe, S., 1981, *Stochastic Differential Equations and Diffusion Processes*, North-Holland.

[58] Ito, K., McKean, H. P., Jr, 1996, *Diffusion Processes and their Sample Paths*, reprint of the 1974 edition, Classics in Mathematics Series, Springer.

[59] Jakubowski, J., Sztencel, R., 2001, *An Introduction to the Theory of Probability*, (in Polish), Script Press.

[60] Kac, M., 1956, *Some Stochastic Problems in Physics and Mechanics*, Magnolia Petrolum Co. Colloq. Lect. 2.

[61] Kallenberg, O., 1997, *Foundations of Modern Probability*, Springer.

[62] Kalyanpur, G., 1980, *Stochastic Filtering Theory*, Springer.

[63] Kawata, T., 1972, *Fourier Analysis in Probability Theory*, Academic Press.

[64] Karatzas, I., Shreve S. E., 1991, *Brownian Motion and Stochastic Calculus*, Springer.

[65] Kato, T., 1995, *Perturbation Theory for Linear Operators*, reprint of the 1980 edition, Classics in Mathematics Series, Springer.

[66] Kingman, J. F. C., 1993, *Poisson Processes*, Oxford.

[67] ———, 1982, "On the Genealogy of Large Populations", *J. Appl. Prob.* **19A**, 27–43.

[68] ———, 1982, "The coalescent", *Stoch. Proc. Applns.* **13**, 235–248.

[69] ———, 1982, "Exchangeability and the evolution of large populations", pp. 97–112 in *Exchangeability in Probability and Statistics*, Koch, G., Spizzichino, F., eds., North-Holland.

[70] Kisyński, J., 1970, "On second order Cauchy's equation in a Banach space", *Bull. Polish Acad Sci.* **18 No. 7**, 371–374.

[71] ———, 1974, "On M. Kac's probabilistic formula for the solution of the telegraphist's equation", *Ann. Polon. Math.* **29**, 259–272.

[72] ———, 2000, "On Cohen's proof of the Factorization Theorem", *Ann. Polon. Math.* **75.2**, 177–192.

[73] Kolmogorov, A. N., 1931, "Über die analytishen Methoden in Wahrsheinlichkeitsrechnung" *Math. Ann.* **104**, 415–258.

[74] Koosis, P., 1964, "Sur un théorème de Paul Cohen", *C. R. Acad. Sci. Paris* **259**, 1380–182.

[75] Körner, T. W., 1988, *Fourier Analysis*, Cambridge University Press.

[76] Kuczma, M., 1985, *An Introduction to the Theory of Functional Equations and Inequalities. Cauchy's Equation and Jensen's Inequality*, Polish Scientific Press and Silesian University.

[77] Kuratowski, C., 1972, *Introduction to Set Theory and Topology*, Elsevier.

[78] Kurtz, T. G., 1970, "A general theorem on the convergence of operator semigroups", *Trans. Amer. Math. Soc.* **148**, 23–32.

[79] Kwapień, S., Woyczyński, W. A., 1992, *Random Series and Stochastic Integrals: Single and Multiple*, Birkhäuser.

[80] Lasota, A., Mackey, M. C., 1994, *Chaos, Fractals, and Noise. Stochastic Aspects of Dynamics*, Springer.

[81] Lasota, A., Yorke J. A., 1994, "Lower bound technique for Markov operators and Iterated Function Systems", *Random Computational Dynamics* **2(1)**, 41–47.

[82] Larsen, R., 1973, *Banach Algebras. An Introduction*, Marcel Dekker.

[83] Lebedev, N. N., 1972, *Special Functions and Their Applications*, Dover.

[84] Liggett, T. M., 1985, *Interacting Particle Systems*, Springer.

[85] Hardy, G. H., Littlewood, J. E., Polya, G., 1934, *Inequalities*, Cambridge University Press.

[86] Lizama, C., 1994, "On the convergence and approximation of integrated semigroups", *J. Math. Anal. Appl.* **181**, 89–103.

[87] Malliavin, P., 1995, *Integration and probability*, Springer.

[88] Mandl, P., 1968, *Analytical Treatment of One-Dimensional Markov Processes*, Springer.

[89] McBride, A. C., 1987, *Semigroups of Linear Operators: an Introduction*, Pitman Research Notes in Mathematics Series, Longman.

[90] Mlak, W., 1990, *Hilbert Spaces and Operator Theory*, Kluwer.

[91] Nagel, R., (ed.), 1986, *One-Parameter Semigroups of Positive Operators*, Springer.

[92] Norris, J. R., 1997, *Markov Chains*, Cambridge.

[93] Øksendal, B., 1998, *Stochastic Differential Equations*, fifth edition, Springer.

[94] Pazy, A., 1983, *Semigroups of Linear Operators and Applications to Partial Differential Equations*, Springer.

[95] Petrovsky, I. G., 1954, *Lectures on Partial Differential Equations*, Interscience Publ.

[96] Pinsky, M., 1991, *Lectures on Random Evolutions*, World Scientific.

[97] ——— 1998, *Partial Differential Equations and Boundary-Value Problems with Applications*, third edition, McGraw-Hill.

[98] Reed, M., Simon, B., 1980, *Methods of Modern Mathematical Physics I: Functional Analysis*, revised and enlarged edition, Academic Press.

[99] Rao, M. M., 1979, *Stochastic Processes and Integration*, Sijthoff & Noordhoff.

[100] Revuz, D., Yor, M., 1999, *Continuous Martingales and Brownian Motion*, third edition, Springer.

[101] Riordan, J., 1958, *An Introduction to Combinatorial Analysis*, Wiley.

[102] Rogers, L. C. G., Williams, D., 2000, *Diffusions, Markov Processes and Martingales, Vol. 1, Foundations*, Cambridge University Press.

[103] Rudin, W., 1974, *Real and Complex Analysis*, McGraw-Hill.
[104]     1991, *Functional Analysis*, McGraw-Hill.
[105] Schuss, Z., 1980, *Theory and Applications of Stochastic Differential Equations*, Wiley.
[106] Shiryaev, A. N., 1989, *Probability*, second edition, Springer.
[107] Stroock, W., 1993, *Probability Theory. An Analytic View*, Cambridge University Press.
[108] Świerniak, A., Polański, A., Kimmel, M., "Control problems arising in chemotherapy under evolving drug resistance", *Preprints of the 13th Worlds Congress of IFAC*, Vol. B, pp. 411–416.
[109] Taylor, H. M., Karlin, S., 1981, *Second Course in Stochastic Processes*, Academic Press.
[110] Tiuryn, J., Rudnicki, R., Wójtowicz, D., 2004, "A case study of genome evolution: from continuous to discrete time model", pp. 1-26 in *Proceedings of Mathematical Foundations of Computer Sciences 2004*, Fiala, J., Koubek, V., Kratochwil, J., eds., Lecture Notes in Computer Sciences 3153, Springer.
[111] Young, N., 1988, *An Introduction to Hilbert Space*, Springer.
[112] Yosida, K., 1965, *Functional Analysis*, Springer.
[113] Wentzel, A. D., 1981, *A Course in the Theory of Stochastic Processes*, McGraw-Hill.
[114] Williams, A. D., 1991, *Probability with Martingales*, Cambridge University Press.
[115] Żelazko, W., 1973, *Banach Algebras*, Elsevier.
[116] Zolotarev, W. M., 1986, *A General Theory of Summing Independent Random Variables*, (in Russian), Nauka.
[117] Zygmund, A., 1959, *Trigonometric Series*, Cambridge.

# Index

Printed in the United States
By Bookmasters